David E.

2010

# VACCINE ADJUVANTS AND DELIVERY SYSTEMS

BICENTENNIAL
BICENTENNIAL

1807
⊛WILEY
2007

BICENTENNIAL
BICENTENNIAL

## THE WILEY BICENTENNIAL–KNOWLEDGE FOR GENERATIONS

*E*ach generation has its unique needs and aspirations. When Charles Wiley first opened his small printing shop in lower Manhattan in 1807, it was a generation of boundless potential searching for an identity. And we were there, helping to define a new American literary tradition. Over half a century later, in the midst of the Second Industrial Revolution, it was a generation focused on building the future. Once again, we were there, supplying the critical scientific, technical, and engineering knowledge that helped frame the world. Throughout the 20th Century, and into the new millennium, nations began to reach out beyond their own borders and a new international community was born. Wiley was there, expanding its operations around the world to enable a global exchange of ideas, opinions, and know-how.

For 200 years, Wiley has been an integral part of each generation's journey, enabling the flow of information and understanding necessary to meet their needs and fulfill their aspirations. Today, bold new technologies are changing the way we live and learn. Wiley will be there, providing you the must-have knowledge you need to imagine new worlds, new possibilities, and new opportunities.

Generations come and go, but you can always count on Wiley to provide you the knowledge you need, when and where you need it!

WILLIAM J. PESCE
PRESIDENT AND CHIEF EXECUTIVE OFFICER

PETER BOOTH WILEY
CHAIRMAN OF THE BOARD

# VACCINE ADJUVANTS AND DELIVERY SYSTEMS

Edited by

**MANMOHAN SINGH**, M. Pharm., Ph.D.
Novartis Vaccines
Emeryville, California

**WILEY-
INTERSCIENCE**

A JOHN WILEY & SONS, INC., PUBLICATION

For general information on our other products and services or for technical support, please
contact our Customer Care Department within the United States at (800) 762-2974, outside the
United States at (317) 572-3993 or fax (317) 572-4002.

Wiley also publishes its books in a variety of electronic formats. Some content that appears in
print may not be available in electronic formats. For more information about Wiley products,
visit our web site at www.wiley.com.

Wiley Bicentennial Logo: Richard J. Pacifico

*Library of Congress Cataloging-in-Publication Data:*

Vaccine adjuvants and delivery systems / [edited by] Manmohan Singh.
     p.  ;  cm.
   Includes bibliographical references and index.
   ISBN 978-0-471-73907-4 (cloth)
   1. Vaccines.  2. Immunological adjuvants.  I. Manmohan Singh, 1964 Nov. 8 –
[DNLM:  1. Vaccines – administration & dosage.  2. Adjuvants,
Immunologic – administration & dosage.  3. Drug Delivery Systems – methods.
QW 805 V115 2007]
   QR189.V232  2007
   615′.372 – dc22

                                                              2006100243

Printed in the United States of America

10  9  8  7  6  5  4  3  2  1

# CONTENTS

# CONTRIBUTORS

**Alexander K. Andrianov,** Apogee Technology, Inc., Norwood, Massachusetts

**Jory R. Baldridge,** GlaxoSmithKline-Biologicals, Hamilton, Montana

**James Chesko,** Novartis Vaccines, Emeryville, California

**Debbie Currie,** Laboratory of Experimental Immunology, National Cancer Institute, Frederick, Maryland

**Erik D'Hondt,** Rupelmondestraat 16, B-9150 Bazel, Belgium

**Nejat K. Egilmez,** School of Medicine and Biomedical Sciences, University at Buffalo/SUNY, Buffalo, New York

**Stanley L. Hem,** Department of Industrial and Physical Pharmacy, Purdue University, West Lafayette, Indiana

**Harm HogenEsch,** Department of Comparative Pathobiology, Purdue University, West Lafayette, Indiana

**Kefei Hu,** Department of Medical Sciences, Section of Virology, University Hospital, Uppsala University, and Iscovent AB, Uppsala, Sweden

**David A. Johnson,** GlaxoSmithKline-Biologicals, Hamilton, Montana

**Catherine V. Kaplanski,** Discovery and Biologics Safety Assessment, Merck Research Laboratories, West Point, Pennsylvania

**Jina Kazzaz,** Novartis Vaccines, Emeryville, California

**Dennis M. Klinman,** Laboratory of Experimental Immunology, National Cancer Institute, Frederick, Maryland

**Lakshmi Krishnan,** Institute for Biological Sciences, National Research Council of Canada, Ottawa, Ontario, Canada

**Lawrence B. Lachman,** Department of Experimental Therapeutics, University of Texas M.D. Anderson Cancer Center, Houston, Texas

**Jose A. Lebron,** Discovery and Biologics Safety Assessment, Merck Research Laboratories, West Point, Pennsylvania

**Brian J. Ledwith,** Discovery and Biologics Safety Assessment, Merck Research Laboratories, West Point, Pennsylvania

**Erik B. Lindblad,** Brenntag Biosector, Frederikssund, Denmark

**Karin Lövgren,** Isconova AB, Dag Hammarskjölds väg 54A, Uppsala, Sweden

**Nils Lycke,** Department of Microbiology and Immunology, Institute of Biomedicine, University of Göteborg, Göteborg, Sweden

**Padma Malyala,** Novartis Vaccines, Emeryville, California

**Amit Misra,** Pharmaceutics Division, Central Drug Research Institute, Lucknow, India

**Bror Morein,** Department of Medical Sciences, Section of Virology, University Hospital, Uppsala University, and Iscovent AB, Uppsala, Sweden

**Derek T. O'Hagan,** Novartis Vaccines, Emeryville, California

**Gary Ott,** Dynavax Technologies Corporation, Berkeley, California

**Bulent Ozpolat,** Department of Experimental Therapeutics, University of Texas M.D. Anderson Cancer Center, Houston, Texas

**Subash Sad,** Institute for Biological Sciences, National Research Council of Canada, Ottawa, Ontario, Canada

**Hidekazu Shirota,** Laboratory of Experimental Immunology, National Cancer Institute, Frederick, Maryland

**Manmohan Singh,** Novartis Vaccines, Emeryville, California

**G. Dennis Sprott,** Institute for Biological Sciences, National Research Council of Canada, Ottawa, Ontario, Canada

**Mildred Ugozolli,** Novartis Vaccines, Emeryville, California

**Jeffrey B. Ulmer,** Novartis Vaccines, Emeryville, California

**Michael Vajdy,** Novartis Vaccines, Emeryville, California

**Nicholas M. Valiante,** Novartis Vaccines, Emeryville, California

**Gary Van Nest,** Dynavax Technologies Corporation, Berkeley, California

**Jayanthi J. Wolf,** Discovery and Biologics Safety Assessment, Merck Research Laboratories, West Point, Pennsylvania

**Fengfeng Xu,** Novartis Vaccines, Emeryville, California

# PREFACE

Prevention of infectious diseases, allergies, malignancies, fertility, and immune disorders using vaccination technologies has been explored extensively in the past decade. Also, the discovery of new antigens through the host genome, which are predominantly recombinant proteins, will require the use of potent immunopotentiators and suitable delivery systems to engender strong responses.

Alum remains the most common adjuvant used in the vaccine market globally. Apart from its safety profile, its use had expanded due to the lack of availability of a suitable alternative. In the last few years, the awareness of how some vaccine adjuvants work has led to a dramatic increase of focus in this area. Whether through activation of innate immune responses or delivery to the targeted site, these novel adjuvant formulations can now be better characterized and optimized for their function. Formulations can now be designed to induce both cellular and humoral responses. Local responses using the nasal and oral routes can now be generated using selective mucosal adjuvants. Evaluation of synergistic effects and repeated use are also being explored. However, these new technologies will have to demonstrate a safety profile that is acceptable for mass immunization and prophylactic use.

This book highlights some of these newly emerging vaccine technologies, some of which will be part of licensed products in the near future. The book evaluates in depth all factors that govern induction of an optimal immune response. Chapters on adjuvant history, antigen presentation, mechanism of action, and the safety profile build a sound base for addressing specific vaccine formulation issues. Detailed descriptions of all leading vaccine formulations and technologies, together with their limitations, should help both researchers and students to enhance their understanding of these technologies. Some of these formulations are purely delivery systems; others comprise immune potentiators with or without delivery systems. The book also has chapters on clinical and nonclinical safety evaluation of vaccine formulations which should serve as prerequisites in moving vaccine research from preclinical to clinical testing. Overall, the book highlights most recent advances in the field of adjuvant and vaccine research.

MANMOHAN SINGH

# 1

# DEVELOPMENT OF VACCINE ADJUVANTS: A HISTORICAL PERSPECTIVE

GARY OTT AND GARY VAN NEST

## 1.1 INTRODUCTION

Since the earliest attempts to raise significant immune responses against nonliving agents, investigators have tried to identify useful additives that can be combined with antigens to enhance immune responses. Such immune-enhancing additives are known as *adjuvants*. Virtually all adjuvant systems developed to date have focused on one of two mechanisms: specific immune activation or the delivery–depot effect. Although many adjuvant systems have been developed and tested in preclinical models, few have actually proved useful for human vaccines. The primary limitations for the use of new adjuvant systems with human vaccines revolve around safety issues. Whereas the toxicity of adjuvants has been reduced systematically through research and development efforts over the last 80 years, the safety barriers presented by regulatory and liability issues have continued to increase. Adjuvants to be used with prophylactic vaccines in normal, healthy populations need to have virtually pristine safety profiles. The fact that most vaccines today are given to infants or children heightens the safety concerns of vaccine adjuvants.

In this chapter we review the history of vaccine adjuvant development from the beginning studies of the early twentieth century through to the present day. We recognize four periods of adjuvant development: (1) the initial

*Vaccine Adjuvants and Delivery Systems,* Edited by Manmohan Singh
Copyright © 2007 John Wiley & Sons, Inc.

development of adjuvants for toxoid vaccines from the 1920s to the 1940s, (2) the broadened use of oils and aluminum adjuvants from the 1940s to the 1970s, (3) the development of synthetic adjuvants and second-generation delivery–depot systems from the 1970s to the 1990s, and (4) the development of rational receptor-associated adjuvants that active the innate immune system from the 1990s until the present day. We provide perspectives in the areas of work in preclinical systems, clinical evaluation and the use of adjuvants, and the interplay between immunology and adjuvant development in each of these periods.

## 1.2   INITIAL DEVELOPMENT OF ADJUVANTS FOR TOXOID VACCINES: 1920s–1940s

Some of the earliest studies leading to the development of adjuvants for active vaccines involved live [1] or killed bacterial vaccines in which the antigen and immune-stimulating agents were both provided by the bacteria [2,3]. Protection against diphtheria by passive transfer of horse antidiphtheria antiseria was a Nobel Prize–winning advance by von Behring [4]. The concept of an active subunit vaccine was first demonstrated in 1907 by Smith, who demonstrated that administration of toxin/antitoxin in immunoprecipitating ratios could provide protection, and von Behring used this approach in people with some success in the period 1910–1920 [4]. Addition of oil or lanolin with killed salmonella is the first documented study with a delivery–depot substance used with a killed bacterial vaccine [5]. Adjuvant research began in earnest with the development of diphtheria subunit toxoid [6] vaccines due to the weak immunogenicity observed with these vaccines [7–9]. As noted by Freund: "Interest in promoting antibody formation by addition of unrelated substances to antigens has never been lacking" [10]. Substances such as agar, tapioca, lecithin starch oil, saponin, salts of calcium and magnesium, killed *Salmonella typhi*, and even bread crumbs were tested [6,11,12].

The most significant vaccine adjuvants to be developed are the aluminum salt adjuvants: generically, but not correctly, referred to as *alums*. The first alum-adjuvanted vaccine was formulated by coprecipitation of diphtheria toxoid dissolved in carbonate buffer (pH 8.0) with aluminum (a purification trick), resulting in a coprecipitate of aluminum hydroxide and diphtheria toxoid [13,14]. The alum adjuvant was developed on the basis of faster and higher antitoxoid antibody responses in guinea pigs. The results of human trials with diphtheria toxoid precipitated with alum were published as early as 1934 [15]. Coprecipitated alum–toxoid nearly eradicated diphtheria in Canada in the 1920s and 1930s. Successful trials with tetanus toxoid were completed in the same time frame [16]. However, some early alum formulations showed poor reproducibility, and results of failed clinical trials were also published by Volk [17]. The alternative approach of adsorbing antigen to the surface of "naked" alum particles was demonstrated as early as 1931 [18] and later came into common use. Only occasional and moderate toxicities were reported with these early alum–toxoid vaccines. The levels of toxicity seen were deemed acceptable given the dramatic decreases in diphtheria and tetanus disease resulting from use of the vaccines.

While the low-toxicity depot approach with alum went forward in clinical applications, efforts were made to generate more potent vaccines using several approaches. One such approach was the use of toxin–antitoxin mixtures [19]. Another approach involved work with tuberculosis (TB) vaccines which demonstrated that the inflammation induced by TB could enhance immune responses to other antigens. As early as 1924, Lewis noted that intraperitoneal injection of live TB a few days before immunization with a variety of antigens dramatically increased antibody responses to those antigens [20]. Presentation of antigen at inflammatory TB foci resulted in elevated antibody titers [21]. These observations pushed forward the immunostimulatory adjuvant approach, which in the 1930s meant the generation of inflammation.

The next advance in adjuvant development involved the combination of killed tubercle with oils. Initial combinations of killed tubercle with paraffin oil produced sensitization to TB but no increased protection from disease [22,23]. Freund demonstrated similar increased antibody responses using live TB with oils. Freund made two jumps in the technology in the 1930s with the substitution of killed TB for live TB and the use of a water-in-oil emulsion [24], inspired by repository formulation techniques being used at the time [10]. The water-in-oil emulsion was formed by the mixture of one volume of 10% Arlacel A (mannide monooleate) and 90% mineral oil with one volume of antigen solution. This system became the standard for adjuvant activity when Freund demonstrated that the emulsion without killed TB was almost as potent as the emulsion with killed TB when used as an adjuvant with diphtheria toxoid and far exceeded the potency of an alum–toxoid formulation [10]. These emulsions went on to become the standard potent adjuvant systems used in preclinical settings and became known as *complete Freund's adjuvant* (CFA, with killed TB) and *incomplete Freund's adjuvant* (IFA, without TB). The emulsion adjuvant was shown to have activity with a variety of antigens, including those from Japanese encephalitis and influenza virus being developed in the same period [25,26]. Water-in-oil emulsion without TB was tested in early human trials with influenza vaccine and demonstrated faster and higher antibody responses than those of vaccine alone [27].

By the mid-1940s, two major adjuvant systems had emerged: the low-reactogenic, modestly effective, and difficult-to-reproduce alum systems, and the new, more potent water-in-oil emulsion systems. It was postulated that alum worked by means of a slow-release depot system [14]. Freund attributed the activity of the water-in-oil emulsion in some part to extended antigen presentation [10]. In this era, adjuvant discovery scientists appeared to be closely involved with immunologists of the day, with adjuvant mechanisms contributing to immunological theory.

## 1.3   BROADENED USE OF ALUMINUM AND OIL ADJUVANTS: 1940s–1970s

Although the definition of any real scientific boundary in adjuvant development in the mid-1940s is somewhat artificial, the period before that time was largely characterized by initial formulation of alum and Freund's systems, while the period from the 1940s through the 1970s can be characterized by extensive efforts to develop these systems for safe, reproducible use in human vaccines and realization of the limitations of their use.

One of the first alum precipitate vaccines [diphtheria, tetanus, pertussis (DTP)] was licensed in 1948 [28], just as the first report of diphtheria vaccine adsorbed to aluminum phosphate was published showing a more controllable composition [29]. A number of studies on the use of alum with pertussis vaccines reported varying success [30–33]. Variability of the potency of the alum-adjuvanted pertussis vaccines seemed to be a common problem. It was demonstrated that alum provided increased antibody titers [34], but another study showed that alum provided no advantage in protection [35]. Whereas the results with alum-adjuvanted pertussis were variable, results with alum-adjvuanted DPT vaccines were more consistent and favorable. It has been suggested that the pertussis component of these combination vaccines actually served as an adjuvant for the diphtheria and tetanus components [31,32]. Alum has continued as the nearly universal adjuvant for DPT vaccines.

Several limitations of alum were becoming clear with continued human use. Alum was observed not to be useful in boosting immunizations with diphtheria and tetanus antigens [36] or influenza hemagglutinin (HA) [37]. Granulomas were often observed at the injection site [38–40]. Occasional erythema was observed [31,32] as well as increases in IgE [41,42]. By the early 1980s, aluminum adjuvants were a major part of human vaccines, but the limited potency, lack of biodegradability, and IgE responses left room for other approaches.

Development of the oil adjuvants continued in the same time period. Freund was demonstrating the wide range of potency of water-in-oil emulsions in the 1940s and 1950s. Use of the original oil formulations containing the commonly available mineral oil Drakeol and the surfactant Arlacel A continued due to the conclusion that nonmetabolizable oil was required for full activity [43]. The use of Freund's adjuvant proceeded in several directions. A number of basic studies (utilizing both IFA and CFA) defined the range of antigens that were made highly immunogenic by presentation with the adjuvant and addressed the mechanism of action [44]. Freund and co-workers demonstrated their usefulness with additional viral antigens, such as rabies and polio [45], as well as sensitization to small molecules (e.g., picryl chloride) [46] and self-antigens [44]. Production of allergic aspermatogenesis, allergic encephalomyelitis, neuritis, and uveitis were described. These studies contributed to fears that the use of potent adjuvants could lead to accidental generation of autoimmunity.

Mechanistic work by McKinney and Davenport [47] demonstrated that the mode of action of mineral oil adjuvants was complex. They concluded that the mechanism involved an initial antibody stimulus resulting from antigen dispersal; the slow release of antigen, which maintains antibody levels; and the inflammatory response, which promotes better utilization of antigen. Studies, including irradiation at periods after vaccination in the presence of adjuvant, excision and reimplanting of granulomas, implantation of virus-saturated cotton plugs, and daily injection experiments, indicated that an early response (<16 days) is critical for the generation of antibody titer. This early response has been linked with both attraction of certain cells to the inoculation site [48] and development of inflammation at the injection site [44]. The long-term maintenance of antibody was correlated with presence of the adjuvant depot for a period of months after injection [47]. These basic precepts of adjuvant activity remained through the 1960s and 1970s.

Whereas the more toxic CFA adjuvant was not deemed appropriate for human use nor was it required for antibody generation, the mycobacterial component had been shown to be necessary for cellular and tubercular sensitization [49]. Attempts to fractionate the active material from killed cells showed that a wax fraction, not the protein fraction, was responsible for the generation of tubercular hypersensitivity [50–52]. This fraction was subsequently shown to be composed of mycolic acid, polysaccharides, and amino acids [53]. This marked the beginning studies of immune agonists that moved beyond the consideration of adjuvant function as an antigen reservoir and granulomatous source of inflammation.

Large-scale testing of IFA for human vaccines was made practical when Salk et al. [48,54] produced highly purified mineral oil and Arlacel A surfactant for use in studies on influenza and polio vaccines. Very large scale evaluation of the adjuvant was done both in the public sector [55] and with the U.S. military in influenza vaccine trials [56]. These studies, as well as studies by Salk on polio [57,58], validated the potency of the adjuvant for enhancement of antibody titers in human subjects. Although failure of efficacy was reported for adenovirus [59], the potency of Freund's adjuvant became established as the "gold standard" for most vaccines. The adjuvant was applied to allergy therapy as well [60], but the special hazards of incompletely controlled exposure of allergic persons to allergen in the presence of adjuvant was unacceptable.

Issues with toxicology made acceptance of IFA controversial in the 1950s and 1960s. Intense inflammation and formation of granulomatous lesions at the injection site were documented [61], but perhaps more alarming was the finding that the emulsion was not entirely retained at the injection site [44,47] and that the poorly metabolized mineral oils might be a risk as carcinogens. The subject of the risk associated with vaccination using the adjuvant and the acceptability of the local reactions was reviewed very extensively by Hilleman [62], who noted that only 109 reactions were reported from 23,917 doses of adjuvanted poliovirus vaccine [63] and commented: "The remaining and most questionable aspects in relation to decision making rest largely on speculative grounds extrapolated from effects which have been observed in animals in connection with experiments designed for other purposes."

Licensure for IFA did not occur, and a tone of extreme caution with respect to adjuvants extended through the 1970s. While attempts to formulate water-in-oil adjuvants with metabolizable oils had been made [43,64], Hilleman and his Merck collaborators [65–67] introduced an efficient peanut oil–based adjuvant using purified Arlacel A and aluminum stearate as stabilizers. Adjuvant 65, as this formulation was named, was reported to be of similar potency to Freund's in both animal and human vaccination with influenza virus [68–70], although in a British influenza trial it was also reported to be significantly less potent [71]. Despite extensive review of safety over 10 years, data showing induction of tumors in mice by Arlacel A [72] kept this system from achieving licensure. The approach of water-in-oil emulsions for adjuvant purposes was set back severely, but would reappear in the 1990s with the Seppic-produced systems.

Work in the aftermath of the water-in-oil adjuvant experience was marked by extreme caution. A seminal review by Edelman [73] cautioned that adjuvants should not risk induction of autoimmunity or allergy, produce no terratogenic effects, and have a very low incidence of adverse events. Chemical composition should be well defined, demonstrated to be carcinogen-free and biodegradable, and the type of immunity induced should be specific for the vaccine and not generally activating.

The next generation of adjuvants was composed of two classes of agents: small molecules often derived from bacterial fractions shown previously to be stimulatory with water-in-oil emulsion adjuvants, and particulate vehicles of dimensions similar to either bacteria or viruses where the agonists are naturally found. Both of these approaches were encouraged by concurrent advances in immunololgy [74] which indicated that much more complex interactions with a number of cell types, including Langerhans cells, macrophages, and dendritic cells, might be important.

A significant part of the small molecule agonist library was derived from bacterial extracts from *Mycobacterium tuberculosis*, *M. avium*, and saphrophytic strains of mycobacteria. White et al. [75] screened fractions from a variety of bacterial sources that showed activity in Wax D, phosphatide, Wax C, and cord factor fractions of mycobacterial strains, which increased antibody titers. Additional activity was found in DNA and RNA digests [76]. A large body of work was devoted to a peptide-containing fraction isolated from Wax D and characterized in the 1940s [50,52]. The composition of the active fraction appeared to be analogous to that of the water-soluble cell wall peptidoglycan [77], and the major part of the activity was ultimately isolated from the cell wall by lysozyme digestion [78,79]. Although a broad range of bacteria exhibited varying adjuvant activity, suggesting a variety of possible variants [76], structural work on a few key strains, including *M. bovis*, *Nocardia rubra*, and *Listeria monocytogenes*, was accumulated and the minimal active subunit of the cell wall was defined as *N*-acetylmuramyl-L-alanyl-D-isoglutamine (MDP) [80]. It was noted by both major groups characterizing MDP activity that in vivo activity required administration in water-in-oil emulsions; saline solutions were inactive. In addition, mycobacterial MDP was shown to be pyrogenic [81]. Attempts to optimize activity led to chemical synthesis of novel MDP derivatives [82] for both vaccine application and induction of nonspecific resistance [83]. Structure–function studies were undertaken [84] to reduce toxicity, optimally activate an as yet undiscovered receptor, and create compatibility with a variety of delivery systems to be discussed later.

While additional work with mycobacterial fractions such as the cord factor first described by Bloch [85] and identified as trehalose dimycolate [86] continued to find application in experimental adjuvants [87], activities of agonists from other sources were also being characterized. Antibody-enhancing adjuvant activity of both poly A:U [88] and polyribo I:C was demonstrated with rabies vaccine [89], along with reports of interferon induction in primates by polylysine/carboxymethylcellulose–stabilized poly I:C [90].

The ability of gram-negative bacilli to enhance antibody titer had also long been established [91], and the adjuvant and endotoxic properties of the purified agent endotoxin were characterized [92]. The adjuvant and endotoxic activities were shown to be separable by both acylation [93] and desterification [94] of the liposaccharide mixtures, and detailed structural work was under way in the 1970s [95].

Finally, saponins, first noted as having adjuvant activity by Ramon [8], were rediscovered and found to be useful in foot-and-mouth vaccines [96]. The modern era of saponin use began with the discovery that extracts from Quillaja saponaria are the most adjuvant active of the saponins [97] and that partial purification of the extracts produced the fraction Quil A, which, although still reactogenic, was markedly better than the crude Quil saponin [98].

In addition to the development of small molecule agonists, several new delivery vehicle approaches that targeted phagocytic cells and did not produce granulomas were demonstrated. Liposomes adopted as carriers for a variety of molecules [99,100] were shown to be adjuvants as carriers of antigen and adjuvant agonists [101,102]. An alternative approach to targeting phagocytic cells, use of very slowly biodegradable methacrylate polymer nanospheres, was introduced by Kreuter and co-workers [103,104]. They demonstrated induction of antibody-mediated protection against influenza in mice with antigen either incorporated into the particles or bound to the particle surface [105].

By 1980 the adjuvant field was beginning to recover from the very serious setbacks incurred when water-in-oil emulsions did not achieve licensure with influenza vaccine. Although the dominant correlate for protection by vaccine remained neutralizing titer, many adjuvant approaches moved away from granuloma-inducing depots to targeting of phagocytic cells, emphasizing both chemotaxis and uptake by macrophages. The move toward micro-/nanoparticle delivered agonists and antigen association had begun.

## 1.4   RATIONAL RECEPTOR-DRIVEN ADJUVANTS THAT ACTIVATE THE INNATE IMMUNE SYSTEM: 1990s–PRESENT

### 1.4.1   Major Breakthroughs of the Era

The early 1980s brought two major changes to the vaccine and adjuvant world. First, the first recombinant DNA–generated vaccine made against hepatitis B [106] was successfully demonstrated and ultimately achieved commercial licensure. This signaled the beginning of recombinant production of a spectrum of recombinant subunit antigens, many of which, like diphtheria and tetanus toxoids, would prove to be active only with adjuvant. Second, the discovery of the HIV virus responsible for AIDS [107,108] and definition of the gp120/140 and gag antigens from the virus set in motion an unprecedented wave of investigation into adjuvants suitable for protection against AIDS with subunit vaccines.

Caution with respect to toxicity continued to be the major factor in moving materials into clinical trials [109]. Potent adjuvants of low toxicity were developed and achievement of commercial licensure for the MF59-adjuvanted flu vaccine Fluad [110] in the European Economic Community and the IRIV(immunopotentiating reconstituted influenza virosome)-based hepatitis A vaccine [111] in Switzerland finally brought acceptance of post-alum adjuvants.

The fields of adjuvant development and immunology became tightly intertwined as the professional antigen-presenting cells were characterized [112–114], the cytokine profiles responsible for generating Th2 versus Th1 immunity were demonstrated [115], the requirements for MHC (major histocompatability complex) class I versus class II presentation were defined [116], and the relationship between the innate immune system and many of the known adjuvant-active molecules was demonstrated with the characterization of Toll-like receptors (TLRs) responsible for signaling innate immune activity [117].

### 1.4.2   Historical Progression

*Molecular Adjuvants*   Two mycobacterial components with a history of adjuvant activity but marginal toxicity profiles received further attention. Trehalose dimycolate (TDM) was investigated further by Masihi et al. [118], and a greater effort was made with muramyl peptides, where less toxic or pyrogenic derivatives were synthesized. The water-soluble Murabutide had no toxicity problems in humans but was not convincingly active in clinical trials with tetanus toxoid [119]. A second water-soluble candidate, threonyl MDP, was nonpyrogenic, did not induce uveitis, and was potent in animal studies [120]. Additional derivatives, including MDP-lys and the lipophilic muramyltripeptide phosphatidylethanolamine (MTP-PE), were tested in human clinical trials for either vaccine adjuvant or chemotherapeutic activities [121,122].

Development of lipid A–related adjuvants was a key activity in this period. While it was shown that there was significant heterogeneity in lipid A components from a variety of gram-negative bacteria [123], significant progress was made in development of adjuvant based on lipid A from *Salmonella minnesota*. Purification and structure were determined by the Ribi group [124], who demonstrated that toxicity could be attenuated dramatically by hydrolysis of the 1-phosphate [125] and 3-hydroxytetradecanoyl groups [126] generating 3D-monophosphoral lipid A (MPL). Data on the safety of MPL in humans was generated quite early in tumor therapy application [127], and the biological activities were shown to include stimulation of synthesis of a number of cytokines, including γ-interferon [128,129]. Both 3D-MPL and later the aminoalkylglucosamine phosphates (e.g., RC529) have further application in humans when combined with particulate delivery systems to be discussed later.

Among the most promising adjuvant actives to be discovered in the post-1980 period are the immunostimulatory DNA sequences comprising an unmethylated CpG. Antitumor activity first demonstrated in bacterial DNA [130] was shown to result from unique palindromic sequences containing unmethylated CG sequences [131] with selected flanking sequences [132] [immunostimulatory sequences (ISSs)]. The activity of the ISS DNA was characterized by induction of interferons, activation of natural killer (NK) cells, production of Th1-biased antibody response [133], and direct activation of B cells [132], murine macrophages [134], and both murine [135] and human plasmacytoid dendritic cells [136]. Plasmid DNA sequences containing certain CpG motifs have been shown to be active as adjuvants for a number of antigen-expressing DNA vaccines [137] as well as protein antigens [133,138]. Use of synthetic phosphorothioate oligonucleotide ISSs [139] as vaccine adjuvants has been investigated for the three classes of immunostimulatory sequences identified [140–142] as well as for other applications. The demonstration of TLR9 as an ISS receptor [143] is allowing studies on TLR distribution and signaling to aid in rational development of ISS-based adjuvants. A spectrum of vaccines utilizing soluble ISSs have been evaluated in preclinical models using protein [144], peptide [145], polysaccharide conjugate [146], and viruslike particles [147]. Vaccines have been administered by mucosal [148,149] as well as intramuscular routes. ISS conjugates of fusion peptides have been employed to generate cytotoxic lymphocytes [150]. Conjugation of ISS to the ragweed protein allergen Amb a 1 has been shown to both increase immunogenicity and decrease allergenicity [151]. ISS oligonucleotides have shown an excellent toxicity profile [152], have been applied to hepatitis B vaccine in human clinical trials both with hepatitis B surface antigen (HBsAg) alone [153,154] and with HBsAg–alum [155]. Additional clinical trials have been performed with soluble ISS in combination with influenza vaccine [155] and the Amb a 1–ISS conjugate (AIC) [156]. Additional work on combination of ISS with delivery systems is discussed later. The use of RNA adjuvant molecules has been difficult despite the advent of stabilized RNA derivatives [157]. However, the activity of the imidazoquinoline derivatives, which also stimulate

RNA receptors TLR7 and TLR8 [158], have shown preclinical potential as Th1-directing adjuvants for herpes simplex vaccines [159–161]. One such product, Imiquimod, has been licensed for topical treatment of herpes simplex [162], but use of imidazoquinoline derivatives as adjuvants remains at preclinical stages. As for the ISS system, antigen conjugates of R848 (another imidazoquinoline derivative) are reported to offer greater activity than that of the soluble mixtures [163]. A number of other TLR7 and TLR8 agonists are under development.

Several nonparticulate adjuvants that have not been identified as TLR agonists have been characterized. Further fractionation of Quil A saponin isolated by Dalsgaard [98] revealed at least 24 peaks [164]. Analysis of adjuvant activity and toxicity revealed the much less toxic compound triterpene glycoside QS-21, which was shown to be an active adjuvant in mice [165], producing Th1 antibody isotypes [166] and CD8 cytotoxic T lympocytes in mice [167]. QS-21 has been used clinically for both cancer immunotherapy [168,169] and prophylactic vaccination against HIV [170] and the malaria peptide SPF66 [171]. However, injection-site pain was a notable problem, making the system unacceptable except in extreme circumstances. Particulate saponin constructs are discussed below.

The mucosal adjuvants cholera enterotoxin (CT) and *Escherichia coli* heat-labile enterotoxin (LT) are potent mucosal adjuvants and have about 80% sequence homology [172]. Their activity has been linked to ADP–ribosyltransferase activity [173]. While CT has been the standard for mucosal adjuvant activity, the toxicity of the A subunit has discouraged clinical use [174]. LT also has toxicity associated with its A subunit, but mutants with significantly lower toxicity have been generated [175,176]. The mutants LTK63 and LTR72 have been shown to have potent activity in the generation of mucosal antibody in preclinical models against a variety of antigens when administered by oral, nasal [177], or transdermal routes [178] and appear ready for clinical testing [177].

The molecular adjuvants thus far described give rise to chemokine and cytokine synthesis. The basic paradigm first described by Mossmann et al. [115] is that two basic types of immune response can be generated. The Th2 response is characterized by a cytokine profile dominated by interleukin-4 and interleukin-5 activates principally B cells. The Th1 response is characterized by γ-interferon, granulocyte monocyte colony stimulating factor, and interleukin-12 activates macrophages and cytotoxic T cells [179]. The direct approach of using cytokines as vaccine adjuvants first concentrated on three cytokines, interleukin-1, interleukin-2, and γ-interferon [180], followed by successful application of interleukin-12 to leishmania vaccine [181] and use of granulocyte monocyte colony-stimulating factor [182] with both peptide and protein tumor antigens. The most extensive efforts on infectious disease vaccines were made with interleukin-2, a T-cell-activating agent that was used alone with rabies vaccines, where it increased protection 25- to 50-fold [183,184] and in combination with several vehicles to be discussed later. The natural activities of cytokines as short-range very low concentration signals between cells are quite different from those of an injection agent in a bolus at high concentrations. The toxicity of interleukin-2 [185], interleukin-12 [186], and γ-interferon [187], and often a need for complex dosing regimens, has led to restricting the use of cytokines to tumor vaccines [188] and to exploration of DNA vaccines in which antigen and cytokine are coexpressed, such as a herpes simplex virus (HSV)-2gD/interleukin-12 system [189] or an HIV env/interleukin-12 vaccinia system [190].

***Particulate Adjuvants*** The primary delivery systems before 1980 were characterized in large part as depot systems. The aluminum salt adjuvants, until recently the only licensed adjuvants for human use [191], continue to be regarded as safe [192] and are in common use with tetanus, diphtheria, pertussis, and poliomyelitis vaccines as well as more recent use with hepatitis B (HBV), hepatitis A (HAV), and anthrax vaccines [193]. Although these systems are not workable for a number of proteins and peptides [37,194], considerable progress has been made in understanding binding parameters [195]. The limitations of alum are a driving force for research into new adjuvant systems. The characterization of aluminum salt adjuvants as Th2-directing systems that stimulate IgE production and very poor cellular immunity is well documented [196,197]. Thus, alum alone is inappropriate for use against a variety of diseases that require Th1/cellular immune responses. The combination of alum with molecular adjuvants (discussed later) may overcome some of these problems. Alum suffers additionally from some reactogenicity at the injection site, giving rise to swelling and cutaneous nodules [109,198]. Although these effects are tolerable, adjuvants that disperse more quickly or do not give rise to inflammation were desired.

Alternative particulate adjuvants giving an extended presence of antigen have been described. Replacement of aluminum salts with calcium phosphate has long been described [199]. Efforts with calcium adjuvants have continued [200], and work with calcium phosphate nanoparticles has had some preclinical success [201]. Use of stearyl tyrosine has been described for a variety of antigens now in use with aluminum adjuvants, including tetanus toxoid [200], diphtheria toxoid [202], and recombinant hepatitis B [203]. Although residence time is shorter than for aluminum salt systems, the benefits have not yet given rise to clinical trials. Use of tyrosine as an adjuvant in allergy vaccines has a long clinical history and a good record of safety [204].

A number of groups have invested effort in controlled release of antigen by polymeric particles aimed at single-dose vaccines. The first demonstration of a single immunization system [205] with nondegradable ethylvinyl acetate showed six-month antibody maintenance against bovine serum albumin (BSA). While several classes of biodegradable polymers, including polyanhydrides and polyorthoesters, have been described for medical applications [206], more recent efforts have used the well-characterized biodegradable poly(lactide-co-glycolide) polymeric microparticle systems [207,208]. This approach advanced to use of very active sub-10-μm particles taken up by antigen-presenting cells combined with 30- to 100-μm particles giving long-term release of antigen [208] and pulsed release of antigen using mixtures of particles of varying molecular weight and lactide/glycolide ratio [209].

Although the manufacturing hurdles and protein stability issues have been solved for some systems and controlled-release formulations have been licensed [210], they have not been clinically tested for vaccines.

Interesting pre-1980s formulations in development remain the water-in-oil emulsions originated by Freund and Hosmer [24]. Major factors in their potency included a long-term depot effect of a mineral oil bolus, which often resulted in cutaneous nodules along with longer-term immunity. Additionally, these emulsions attracted a variety of immune cells, resulting in a long-term reactive center. Efforts to continue with the water-in-oil emulsion (a particularly effective approach for peptides) have been made by Seppic. The Montanide ISA adjuvants (utilizing mineral oil and mannide monooleate, which emulsifies water with a low energy input) [211] have a substantial record in veterinary applications. More recently, the ISA 720 formulation using vegetable oil has been tested in clinical trials with HIV peptide [212] and a malaria–HBV core antigen [213].

Oil-in-water emulsions judged to be ineffective when using mineral oil [10] were reexamined when Ribi and co-workers found antitumor activity with trehalose dimycolate surfaces on drakeol oil-in-water emulsions [214]. The somewhat-toxic trehalose dimycolate surface was replaced with pluronic polyol block polymer surfactants, and a correlation was established between the hydrophile–lipophile balance (HLB) and activity [215]. Advances in synthetic techniques allowed production of higher-molecular-weight block copolymers. The copolymer CRL 1005 showed good adjuvant activity when mixed directly with inactivated whole virus flu vaccine in mouse studies [216]. Reynolds took a different step from mineral oil–based oil-in-water systems, showing that phospholipid-stabilized lipid emulsions (relatives of nutritional emulsions) had adjuvant activity with viral antigens [217]. Significant progress was made when several groups applied low-HLB surfactants with squalane/squalene oil-in-water emulsions. Ribi and co-workers [118] used squalane–water emulsions with trehalose dimycolate surfaces for veterinary applications. The Syntax adjuvant formulation (SAF) [196] used the potent block copolymer L-121 to generate a squalane-in-water formulation. The SAF M formulation developed for manufacturing was shown to be effective in several primate systems [218–220]. Use of the nontoxic low-HLB spreading agent Span 85 [221] and Tween 80 as stabilizers for a squalene–water emulsion produced the adjuvant MF59 [222,223]. This formulation stimulated neutralizing antibody (but not convincing protection) successfully with recombinant HSV surface antigens in phase III clinical trials [224]. Phase III and IV trials with the commercially available MF59/influenza vaccine Fluad [225,226] showed this vaccine to be particularly effective in the elderly. MF59 also appears to have very good adjuvant activity with H5 influenza vaccines [227].

Polymeric nanoparticles with either encapsulated antigen or protein-binding surfaces were used by Kreuter et al. in the 1970s [103]. Much later work [228] has shown that these easily prepared and well-tolerated poly(methylmethacrylate) nanoparticles are a superior adjuvant to a large array of particulates when used with HIV-2 split virus. A related set of approaches using poly(lactide-co-glycolide) microparticles (<10μm) employing either encapsulation of antigen [207,208] or utilization of surface-charged microspheres [229–231] have shown significant promise in preclinical models. This approach has also been applied to delivery of DNA vaccines, with encouraging preclinical results [229].

As noted previously, liposomes adopted as carriers for a variety of molecules [99,100] were shown to be effective as carriers of antigen and adjuvant compounds [101,102]. An extensive amount of work was completed in the 1980s and 1990s in a quest to optimize the adjuvant effects of liposomes. This work has been well reviewed by several authors [232–234]. Efforts to optimize adjuvant efficacy have included comparisons of multilamelllar versus unilamellar systems, variation in size and fluidity of lipids, incorporation of antigen by encapsulation versus surface interaction, and alteration of surface with PEG [poly(ethylene glycol)]-ylated lipids. Preclinical testing has been done with at least 20 antigens. After extensive testing the general conclusion is that liposomal delivery of subunit protein–peptide antigen alone is not a powerful method for enhancement of immunogenicity [235]. Use of liposomes for delivery of DNA vaccines both by encapsulation and by interaction with cationic lipid components has been well studied [236–238]. Development of more effective and less toxic cationic lipids has allowed testing of lipid-adjuvanted DNA vaccines in primate studies [239,240]. However, human trials of DNA vaccines have proceeded with naked DNA and appear to need viral boosts for best [241] effects. It is important to note that liposomal formulations have shown promise in the generation of cellular immunity, particularly cytotoxic lymphocytes, which are suspected to be critical in the protection of a variety of infectious diseases and particularly in cancer therapy. Introduction of pH-sensitive liposomes capable of introducing ovalbumin into the MHC class I pathway in mice and generating cytotoxic T lymphocytes (CTLs) [242,243] marked the beginning of series of CTL-generating formulations using liposome delivery.

The earliest liposome-related vaccines to be licensed, the IRIV and IRIV/hepatitis A vaccines [232], are based on 150-nm unilamellar vesicles created by reconstitution of detergent-extracted influenza surface glycoproteins and phospholipids with egg yolk phosphotidylcholine and phosphatidylethanolamine (PE). For the HepA vaccine (and a number of others in preclinical studies, including tetanus toxoid, poliovirus VP2 peptide, and HBsAg peptide), the PE moiety is cross-linked covalently to the antigen. Cellular entry of the vaccine particle and endosomal fusion are thought to use the neuraminidase (NA)- and HA-mediated entry systems evolved by the influenza virus [244]. The system has intriguing possibilities, but complex composition and formulation issues may limit its application. A second lipid-based particle with extensive application is the ISCOM/ISCOMATRIX system, based on Quillaja saponin, cholesterol, and phospholipids. Dissolution of these components in the presence of the detergent Mega 10 followed by removal of the detergent by dialysis results in a 30- to 40-nm cagelike structure [245] which will incorporate amphiphilic antigens to produce the original immune-stimulating complex (ISCOM) structure [246]. This structure is an active producer of antibody but is set apart from most of the adjuvant systems developed thus far in that it effectively delivers antigen to the MHC class I pathway [247], although not by the TAP (peptide transporter) pathway. ISCOMs do not interact with any of the known Toll receptors. ISCOMs are effective CTL-generating systems in primate systems, including humans [248]. ISCOMs have been shown to be effective against a broad spectrum of viruses [249] in veterinary applications, with the first commercial use in an equine flu vaccine [250], where continued success has been reported with intranasal boosting [251].

***Delivery System/Molecular Adjuvant Combinations***   A major tactic from the first days of adjuvant development has been to use delivery systems to protect, deliver, and extend the therapeutic lifetime of molecular adjuvants (the most famous being complete Freund's adjuvant). Use of alternative water-in-oil emulsion systems have been examined with a birth control vaccine based on peptide antigen and a saline/or MDP–squalene emulsion showing clinical promise [252]. A similar approach for veterinary vaccines was taken with Titer Max, where saline was emulsified into squalene with the block copolymer CRL8941 [253]. For the period since 1980, most of the combination adjuvants have been designed with chemically defined Toll agonists delivered with sub-10-μm particulates. Although criticisms of aluminum salt adjuvants are often made, combinations of Toll agonists with aluminum adjuvants continue to be evaluated. Alving began work on vaccines for HIV and malaria using a combination of alum, liposomes, and lipid A [254]. Use of alum–MPL with HSV gD2 showed significantly better performance than that of alum alone in both mice [255] and humans [256], where it remains a possible candidate for future application. An MPL–alum combination has also been tested with hepatitis B vaccine in humans [257,258] and was recently licensed in Europe for use in dialysis patients. A combination of CpG1826 with aluminum salt was shown to be equipotent to IFA with a malarial peptide in mice [259] and may represent an interesting new direction.

A number of groups have used oil-in-water emulsions as carriers or coadjuvants for molecular adjuvants. Extensive work has been done with an SAF squalane/L121 block copolymer system in conjunction with threonyl-MDP pioneered by Allison and Byars [260]. Preclinical efficacy was demonstrated with a spectrum of antigens [261], including HIV in chimpanzees [220]. The squalene–water emulsion MF59 was tested with MTP–PE and HIV vaccines in a phase I clinical trial [262] and with influenza vaccine [263] where reactogenicity in the presence of MTP–PE was unacceptable. Use of catatonically modified MF59 with CpG oligonucleotides has shown preclinical promise without unusual reactogenicity [264]. When tested in cancer patients, use of oil-in-water emulsions with cell wall skeleton (CWS)–MPL in squalane (Ribi Detox), showed no systemic toxicity but some reactogenicity at the injection site and a few granulomas [265]. Later approaches have omitted the CWS component and used either the monophosphoryl lipid A or later, monophosphoryl lipid A derivative RC529 with the proprietary oil-in-water emulsion SE with HIV peptides in primates [266] or anthrax protective antigen with a squalene–water emulsion in primates showing a strong immune response [267]. The MPL–emulsion approach appears to be a good candidate for next-generation human vaccines [268].

The use of sub-10-μm polymeric particles as a delivery system for adjuvants has been tested using both encapsulation and surface binding of adjuvants. Cationic poly(lactide-co-glycoside) microspheres shown previously to be effective carriers for DNA vaccines [229] have also been shown to enhance the activity of surface-bound CpG-containing oligonucleotides [269]. The alternative approach of encapsulation of molecular adjuvants approximates well-developed drug delivery techniques. This has been shown to potentiate MPL and derivatives [270], and the approach could be used in combination with surface-bound antigen systems [271].

Liposomes are modestly potent delivery systems for antibody induction and have shown potential for CTL generation [242] and as carriers for molecular antigens. The adjuvant delivered may be either encapsulated or surface bound, with results depending on that distribution. Use of liposomes for lipid A and its derivatives has shown potential with malaria vaccine in phase I trials [272] and with malarial peptides in a liposome–alum formulation [273].

Incorporation of cytokines onto liposomes was first demonstrated [274] showing that IL-2 could be incorporated into dehydration and rehydration vesicles, increasing the activity of IL-2. The approach was used in preclinical studies for HSV [275] but was not pursued further. Incorporation of CpG oligonucleotides into cationic liposomal delivery formulations was shown to enhance CpG activity [276]. It has been noted that liposomal delivery of CpG to the endosomal TLR9 receptor may have complex effects [277]. There is indication that CpG liposome formulations can enable CTL generation with HIV antigens [278] or cancer antigens [279]. Liposomal formulations utilizing encapsulated membrane-traversing systems such as listeriolysin O, not usually considered as a molecular adjuvant, can generate anti-ovalbumin CTL [280].

## 1.5  CONCLUSIONS

Vaccine adjuvant research and development has been an ongoing activity for more that a century. The need for methods to enhance vaccine immunogenicity has been recognized from the days of the very first testing of nonliving vaccines. The development of successful vaccine adjuvants has been a constant balancing act between safety and immunogenicity, delivery and immunostimulation, and simplicity and complexity. The fact that after over 100 years so few adjuvants have been approved for human vaccines attests to the difficulty of this research and development activity. We appear to be at the beginning of a new era in which a variety of new adjuvants are being approved or are about to be approved for human vaccines. In this chapter we have described the steps involved in the process over the last century that have led to these new vaccine adjuvants.

## REFERENCES

1. Rabinowitsch, L. Zur Frage des Vorkommens von Tuberkelbazillen in Marktbutter. *Hygiene* 1897, **26**, 90.
2. Kitasato, S. The bacillus of bubonic plague. *Lancet* 1894, **2**, 428–430.
3. Pfeiffer, R. & Kolle, W. Frage der Schutzimpfung des Menschen gegen Typhus abdominalis. *Dtsch Med Wochenschr* 1896, **22**, 735–737.
4. Holmes, W. *Diphtheria: History in Bacillary and Rickettsial Infections*, Macmillan, New York, 1940, pp. 291–305.
5. Pinoy, L.M. Les vaccins en émulsion dans les corps gras ou "lipo-vaccins." *C R Soc Biol* 1916, **79**, 201–203.
6. Ramon, G. Procédures pour acroitre la production des antitoxines. *Ann Inst Pasteur Paris* 1926, **40**, 1–10.
7. Glenny, A.T. & Hopkins, B.E. Diphtheria toxoid as an immunising agent. *Br J Exp Pathol* 1923, **4**, 283–288.
8. Ramon, G. Sur l'augmentation anormale de l'antitoxin chez le chevaux producteurs de serum antidiphtherique. *Bull Soc Centr Med Vet* 1925, **101**, 227–234.
9. Ramon, G. & Zoeller, C. De la valeur antigenique de l'anatoxinetetanique chez l'homme. *Ann Inst Pasteur Paris* 1926, **247**, 245–247.
10. Freund, J. Some Aspects of Active Immunizatiion. *Annu Rev Microbiol* 1947, **1**, 291–308.
11. Landsteiner, K. *J Exp Med* 1923, **38**, 127.
12. Jacobs, J. *J Exp Med* 1934, **59**, 479–490.
13. Glenny, A.T., Waddington, H. & Wallace, V. The antigenic value of toxoid precipitated by potassium alum. *J Pathol Bact* 1926, **29**, 38–45.
14. Glenny, A.T. Alum toxoid precipitates as antigens. *J Pathol Bact* 1931, **34**, 118.
15. White, J. Diphtheria toxoid: comparative immunizing value with and without alum as indicated by Schick test. *J Am Med Assoc* 1934, **102**, 915.
16. Jones, F. Studies on Tetanus Toxoid I: the antitoxic titer of human subjects following immunization with tetanus toxoid and tetanus alum precipitated toxoid. *J Immunol* 1936, **30**, 115.
17. Volk, V. Diphtheria immunization with fluid toxoid and alum precipated toxiod. *Am J Public Health* 1942, **32**, 690.
18. Maschmann, E., Kuster, E. & Fischer, W. Uber die fähigheitdes Tonerde–Praparates B Diphtherie–toxinzu adsorbieren. *Ber Dtsch Chem Ges* 1931, **64**, 2174–2178.
19. Swift, H.F. The synergic action of staphylotoxin and beef lens extract in rabbit. *J Exp Med* 1936, **63**, 703–724.

20. Lewis, P. Allergic irritability: the formation of anti-sheephemolytic amboceptor in the normal and tuberculous guinea pig. *J Exp Med* 1924, **40**, 503.

21. Dienes, L. *J Immunol* 1927, **14**, 61.

22. Coulaud, E. *Rev Tuberc* 1934, **2**, 850.

23. Saenz, A. *C R Soc Biol* 1937, **125**, 495.

24. Freund, J. & Hosmer, E.P. Sensitization and antibody formation after injection of tubercle bacilli and paraffin oil. *Proc Soc Exp Biol Med* 1937, **37**, 509–513.

25. Stokes, J., Chenoweth, A., Waltz, A. et al. Results of immunization by means of virus of active influenza. *J Clin Invest* 1937, **16**, 237–243.

26. Warren, J. *Proc Soc Exp Biol Med* 1946, **61**, 109–113.

27. Henle, W. Effect of adjuvants on vaccination of human beings against influenza. *Proc Soc Exp Biol NY* 1945, **59**, 179–181.

28. Decker, M.D., Edwards, K.M. & Bogaerts, H.H. Combination vaccines. In *Vaccines*, Plotkin, W.A., Ed., Elsevier, Philadelphia, PA, 2004, pp. 825–861.

29. van Ramshost, J.D. The adsorption of diphtheria toxoid on aluminum phosphate. *Recl Trav Chim Pays Bas* 1949, **68**, 169–180.

30. Pittman, M. Variability of the potency of pertussis vaccine in relation to the number of bacteria. *J Pediatr* 1954, **45**(1), 57–69.

31. Aprile, M.A. & Wardlaw, A.C. Aluminium compounds as adjuvants for vaccines and toxoids in man: a review. *Can J Public Health* 1966, **57**(8), 343–360.

32. Wardlaw, A.C. & Jakus, C.M. The inactivation of pertussis protective antigen, histamine sensitizing factor, and lipolysaccharide by sodium metaperiodate. *Can J Microbiol* 1966, **12**(6), 1105–1114.

33. Cameron, J. Problems associated with the control testing of pertussis vaccine. *Adv Appl Microbiol* 1976, **20**, 57–80.

34. Preston, N.W., Mackay, R.I., Bomford, F.N., Crofts, J.E. & Burland, W.L. Pertussis agglutinins in vaccinated children: better response with adjuvant. *J Hyg* 1974, **73**, 119–125.

35. Butler, N.R. et al. Response of infants to pertussis vaccine at one week and to poliomyelitis, diphtheria, and tetanus vaccine at six months. *Lancet* 1962, **2**, 112–114.

36. Ipsen, J. Immunization of adults with diphtheria and tetanus. *N Engl J Med* 1954, **251**, 459–466.

37. Davenport, F.M., Hennessy, A.V. & Askin, F.B. Lack of adjuvant effect of A1PO4 on purified influenza virus hemagglutinins in man. *J Immunol* 1968, **100**(5), 1139–1140.

38. Beck, W. [Aluminum hydroxide granuloma after vaccination against tetanus]. *Medizinische* 1954, **11**, 363–365.

39. White, R.G., Coons, A.H. & Connolly, J.M. Studies on antibody production: III. The alum granuloma. *J Exp Med* 1955, **102**(1), 73–82.

40. Erdohazi, M. & Newman, R.L. Aluminium hydroxide granuloma. *Br Med J* 1971, **3**(775), 621–623.

41. Nagel, J. et al. IgE synthesis in man: I. Development of specific IgE antibodies after immunization with tetanus–diphtheria (Td) toxoids. *J Immunol* 1977, **118**(1), 334–341.

42. Nagel, J.E. et al. IgE synthesis in man: II. Comparison of tetanus and diphtheria IgE antibody in allergic and nonallergic children. *J Allergy Clin Immunol* 1979, **63**(5), 308–314.

43. Freund, J. The effect of paraffin oil and mycobacteria on antibody formation and sensitization: a review. *Am J Clin Pathol* 1951, **21**(7), 645–656.

44. Freund, J. The mode of action of immunologic adjuvants. *Bibl Tuberc* 1956(10), 130–148.

45. Lipton, M.M. & Freund, J. The formation of complement fixing and neutralizing antibodies after the injection of inactivated rabies virus with adjuvants. *J Immunol* 1950, **64**(4), 297–303.

46. Freund, J. Potentiating effect of *Nocardia asteroides* on sensitization to picryl chloride and on production of isoallergic encephalomyelitis. *Soc Exp Biol Med* 1948, **68**, 373.

47. McKinney, R.W. & Davenport, F.M. Studies on the mechanism of action of emulsified vaccines. *J Immunol* 1961, **86**, 91–100.

48. Salk, J.E., Bailey, M.L. & Laurent, A.M. The use of adjuvants in studies on influenza immunization: II. Increased antibody formation in human subjects inoculated with influenza virus vaccine in a water-in-oil emulsion. *Am J Hyg* 1952, **55**(3), 439–456.

49. Freund, J. *Proc Exp Biol Med* 1942, **49**, 548–553.

50. Raffel, S. The components of the tubercle bacillus responsible for the delayed type of "infectious" allergy. *J Infect Dis* 1948, **82**, 267.

51. Raffel, S.A., Dukes, C.D. & Huang J.S. The role of wax of the tubercle bacteria in establishing delayed type hypersensitivity II hypersensitivity to a protein antigen, egg albumin. *J Exp Med* 1949, **90**, 53.

52. Choucroun, N. Tubercle bacillus antigens: biological properties of two substances isolated from paraffin oil extract of dead tubercle bacteria. *Am Rev Tuberc* 1947, **56**, 203.

53. Lederer, E. Chimie et biochimie des mycobacteries. Presented at 2'eme Congrès International de Biochemie, Paris, 1952.

54. Salk, J.E. Principles of immunization as applied to poliomyelitis and influenza. *Am J Public Health* 1953, **43**(11), 1384–1398.

55. Hennessy, A.V. & Davenport, F.M. Relative merits of aqueous and adjuvant influenza vaccines when used in a two-dose schedule. *Public Health Rep* 1961, **76**, 411–419.

56. Davenport, F.M., Hennessy, A.V. & Bell, J.A. Immunologic advantages of emulsified influenza virus vaccines. *Mil Med* 1962, **127**, 95–100.

57. Salk, J.E. et al. The use of adjuvants to facilitate studies on the immunologic classification of poliomyelitis viruses. *Am J Hyg* 1951, **54**(2), 157–173.

58. Salk, J.E. Studies in human subjects on active immunization against poliomyelitis: I. A preliminary report of experiments in progress. *J Am Med Assoc* 1953, **151**(13), 1081–1098.

59. Miller, L.F. et al. Efficacy of adjuvant and aqueous adenovirus vaccines in prevention of naval recruit respiratory disease. *Am J Public Health Nations Health* 1965, **55**, 47–59.

60. Loveless, M.H. Repository immunization in pollen allergy. *J Immunol* 1957, **79**(1), 68–79.

61. Steiner, J.W., Langer, B. & Schatz, D.L. The local and systemic effects of Freund's adjuvant and its fractions. *Arch Pathol* 1960, **70**, 424–434.

62. Hilleman, M.R. Critical appraisal of emulsified oil adjuvants applied to viral vaccines. *Prog Med Virol* 1966, **8**, 131–182.

63. Cutler, J.C., Lesesne, L. & Vaughn, I. Use of poliomyelitis virus vaccine in light mineral oil adjuvant in a community immunization program and report of reactions encountered. *J Allergy* 1962, **33**, 193–209.

64. Woodhour, A.F., Jensen, K.E. & Warren, J. Development and application of new parenteral adjuvants. *J Immunol* 1961, **86**, 681–689.

65. Hilleman, M.R. A forward look at viral vaccines: with special reference to a new immunologic adjuvant. *Am Rev Respir Dis* 1964, **90**, 683–706.

66. Woodhour, A.F. et al. New metabolizable immunologic adjuvant for human use: I. Development and animal immune response. *Proc Soc Exp Biol Med* 1964, **116**, 516–523.

67. Stokes, J., Jr. et al. New metabolizable immunologic adjuvant for human use: 3. Efficacy and toxicity studies in man. *N Engl J Med* 1964, **271**, 479–487.

68. Weibel, R.E. et al. New metabolizable immunologic adjuvant for human use: 5. Evaluation of highly purified influenza-virus vaccine in adjuvant 65. *N Engl J Med* 1967, **276**(2), 78–84.

69. Weibel, R.E. et al. Ten-year follow-up study for safety of adjuvant 65 influenza vaccine in man. *Proc Soc Exp Biol Med* 1973, **143**(4), 1053–1056.

70. Hilleman, M.R. et al. The clinical application of adjuvant 65. *Ann Allergy* 1972, **30**(3), 152–158.

71. Stuart-Harris, C.H. Adjuvant influenza vaccines. *Bull World Health Organ* 1969, **41**(3), 617–621.

72. Murray, R., Cohen, P. & Hardegree, M.C. Mineral oil adjuvants: biological and chemical studies. *Ann Allergy* 1972, **30**(3), 146–151.

73. Edelman, R. Vaccine adjuvants. *Rev Infect Dis* 1980, **2**(3), 370–383.

74. Waksman, B.H. Adjuvants and immune regulation by lymphoid cells. In *Springer Seminars in Immunopathology*, Springer-Verlag, New York, 1979, pp. 5–33.

75. White, R.G. et al. The influence of components of *M. tuberculosis* and other mycobacteria upon antibody production to ovalbumin. *Immunology* 1958, **1**(1), 54–66.

76. White, R.G. The adjuvant effect of microbial products on the immune response. *Annu Rev Microbiol* 1976, **30**, 579–600.

77. White, R.G. et al. Correlation of Adjuvant Activity and Chemical Structure of Wax D Fractions of Mycobacteria. *Immunology* 1964, **17**, 158–171.

78. Azuma, I. et al. Adjuvanticity of mycobacterial cell walls. *Jpn J Microbiol* 1971, **15**(2), 193–197.

79. Adam, A. et al. Preparation and biological properties of water-soluble adjuvant fractions from delipidated cells of *Mycobacterium smegmatis* and *Nocardia opaca*. *Infect Immun* 1973, **7**(6), 855–861.

80. Ellouz, F. et al. Minimal structural requirements for adjuvant activity of bacterial peptidoglycan derivatives. *Biochem Biophys Res Commun* 1974, **59**(4), 1317–1325.

81. Riveau, G. et al. Central pyrogenic activity of muramyl dipeptide. *J Exp Med* 1980, **152**(4), 869–877.

82. Azuma, I. et al. Adjuvant activity of synthetic 6-*O*-"mycoloyl"-*N*-acetylmuramyl-L-alanyl-D-isoglutamine and related compounds. *Infect Immun* 1978, **20**(3), 600–607.

83. Chedid, L. et al. Enhancement of nonspecific immunity to *Klebsiella pneumoniae* infection by a synthetic immunoadjuvant (*N*-acetylmuramyl-L-alanyl-D-isoglutamine) and several analogs. *Proc Natl Acad Sci USA* 1977, **74**(5), 2089–2093.

84. Chedid, L. & Lederer, E. Past, present and future of the synthetic immunoadjuvant MDP and its analogs. *Biochem Pharmacol* 1978, **27**(18), 2183–2186.

85. Bloch, H. A component of tubercle bacilli concerned with their virulence. *Bull NY Acad Med* 1950, **26**(7), 506–507.

86. Noll, H. et al. The chemical structure of the cord factor of *Mycobacterium tuberculosis*. *Biochim Biophys Acta* 1956, **20**(2), 299–309.

87. Vosika, G.J. et al. Intralesional immunotherapy of malignant melanoma with *Mycobacterium smegmatis* cell wall skeleton combined with trehalose dimycolate (P3). *Cancer* 1979, **44**(2), 495–503.

88. Branche, R. & Renoux, G. Stimulation of rabies vaccine in mice by low doses of polyadenylic:polyuridylic complex. *Infect Immun* 1972, **6**(3), 324–325.

89. Fenje, P. & Postic, B. Prophylaxis of experimental rabies with the polyriboinosinic–polyribocytidylic acid complex. *J Infect Dis* 1971, **123**(4), 426–428.

90. Levy, H.B. et al. A modified polyriboinosinic–polyribocytidylic acid complex that induces interferon in primates. *J Infect Dis* 1975, **132**(4), 434–439.

91. Greenberg, L.F. Increased efficiency of diphtheria toxoidwhen combined with pertussis vaccine. *Can J Public Health* 1947, **38**, 279–286.

92. Landy, M. & Johnson, A.G. Studies on the O antigen of *Salmonella typhosa*. IV. Endotoxic properties of the purified antigen. *Proc Soc Exp Biol Med* 1955, **90**(1), 57–62.

93. Freedman, H.H. & Sultzer, B.M. Dissociation of the biological properties of bacterial endotoxin by chemical modification of the molecule. *J Exp Med* 1962, **116**, 929–942.

94. Noll, H. & Braude, A.I. Preparation and biological properties of a chemically modified *Escherichia coli* endotoxin of high immunogenic potency and low toxicity. *J Clin Invest* 1961, **40**, 1935–1951.

95. Galanos, C., Luederitz, O., Rietschel, T. & Westphal, O. Newer aspects of the chemistry and biology of bacterial lipopolysaccharides with special reference to their lipid A component. In *International Review of Biochemistry*, Goodwin, T., Ed., University Park Press, Baltimore, MD, 1977, pp. 239–335.

96. Espinet, R.G. Nouveau vaccin antiaphteux a complex glucoviral. *Gac Vet* 1951, **13**, 268.

97. Dalsgaard, K. Thinlayer chromatographic fingerprinting of commercially available saponins. *Dan Tidsskr Farm* 1970, **44**(8), 327–331.

98. Dalsgaard, K. Saponin adjuvants: 3. Isolation of a substance from *Quillaja saponaria* Molina with adjuvant activity in food-and-mouth disease vaccines. *Arch Gesamte Virusforsch* 1974, **44**(3), 243–254.

99. Gregoriadis, G. & Ryman, B.E. Liposomes as carriers of enzymes or drugs: a new approach to the treatment of storage diseases. *Biochem J* 1971, **124**(5), 58P.

100. Papahadjopoulos, D. et al. Incorporation of lipid vesicles by mammalian cells provides a potential method for modifying cell behaviour. *Nature* 1974, **252**(5479), 163–166.

101. Allison, A.C. & Gregoriadis, G. Liposomes as immunological adjuvants. *Recent Results Cancer Res* 1976(56), 58–64.

102. Allison, A.G. & Gregoriadis, G. Liposomes as immunological adjuvants. *Nature* 1974, **252**(5480), 252.

103. Kreuter, J. & Speiser, P.P. In vitro studies of poly(methyl methacrylate) adjuvants. *J Pharm Sci* 1976, **65**(11), 1624–1627.

104. Kreuter, J. & Speiser, P.P. New adjuvants on a polymethylmethacrylate base. *Infect Immun* 1976A, **13**(1), 204–210.

105. Kreuter, J. & Liehl, E. Protection induced by inactivated influenza virus vaccines with polymethylmethacrylate adjuvants. *Med Microbiol Immunol (Berl)* 1978, **165**(2), 111–117.

106. Valenzuela, P. et al. Synthesis and assembly of hepatitis B virus surface antigen particles in yeast. *Nature* 1982, **298**(5872), 347–350.

107. Gelmann, E.P. et al. Proviral DNA of a retrovirus, human T-cell leukemia virus, in two patients with AIDS. *Science* 1983, **220**(4599), 862–865.

108. Barre-Sinoussi, F. et al. Isolation of a T-lymphotropic retrovirus from a patient at risk for acquired immune deficiency syndrome (AIDS). *Science* 1983, **220**(4599), 868–871.

109. Gupta, R.K. et al. Adjuvants: a balance between toxicity and adjuvanticity. *Vaccine* 1993, **11**(3), 293–306.

110. Podda, A. & Del Giudice, G. MF59-adjuvanted vaccines: increased immunogenicity with an optimal safety profile. *Expert Rev Vaccines* 2003, **2**(2), 197–203.

111. Wegmann, A. et al. [Immunogenicity and stability of an aluminum-free liposomal hepatitis A vaccine (Epaxal Berna)]. *Schweiz Med Wochenschr* 1994, **124**(45), 2053–2056.

112. Steinman, R.M., Witmer-Pack, M. & Inaba, K. Dendritic cells: antigen presentation, accessory function and clinical relevance. *Adv Exp Med Biol* 1993, **329**, 1–9.

113. Banchereau, J. et al. Will the making of plasmacytoid dendritic cells in vitro help unravel their mysteries? *J Exp Med* 2000, **192**(12), F39–F44.

114. Caux, C. et al. Characterization of human CD34+ derived dendritic/Langerhans cells (D-Lc). *Adv Exp Med Biol* 1995, **378**, 1–5.

115. Mosmann, T.R. et al. Two types of murine helper T cell clone: I. Definition according to profiles of lymphokine activities and secreted proteins. *J Immunol* 1986, **136**(7), 2348–2357.

116. Morrison, L.A. et al. Differences in antigen presentation to MHC class I– and class II–restricted influenza virus-specific cytolytic T lymphocyte clones. *J Exp Med* 1986, **163**(4), 903–921.

117. Medzhitov, R. & Janeway, C., Jr. The Toll receptor family and microbial recognition. *Trends Microbiol* 2000, **8**(10), 452–456.

118. Masihi, K.N. et al. Immunobiological activities of nontoxic lipid A: enhancement of nonspecific resistance in combination with trehalose dimycolate against viral infection and adjuvant effects. *Int J Immunopharmacol* 1986, **8**(3), 339–345.

119. Telzak, E. et al. Clinical evaluation of the immunoadjuvant murabutide, a derivative of MDP, administered with a tetanus toxoid vaccine. *J Infect Dis* 1986, **153**(3), 628–633.

120. Waters, R.V., Terrell, T.G. & Jones, G.H. Uveitis induction in the rabbit by muramyl dipeptides. *Infect Immun* 1986, **51**(3), 816–825.

121. Tsubura, E. [Muramyl dipeptide derivative and its clinical application]. *Kekkaku* 1989, **64**(11), 731–739.

122. Kahn, J.O. et al. Clinical and immunologic responses to human immunodeficiency virus (HIV) type 1SF2 gp120 subunit vaccine combined with MF59 adjuvant with or without muramyl tripeptide dipalmitoyl phosphatidylethanolamine in non-HIV-infected human volunteers. *J Infect Dis* 1994, **170**(5), 1288–1291.

123. Mayer, H., Krauss, J.H., Yokota, A., Weckesser, J. & Natural, J. Natural variants in lipid A. In *Advances in Experimental Medicine and Biology*, Friedman, T.K.H., Nakano, M. & Nowotny, A., Eds., Plenum Press, Boca Raton, FL, 1989, pp. 45–70.

124. Qureshi, N., Takayama, K. & Ribi, E. Purification and structural determination of nontoxic lipid A obtained from the lipopolysaccharide of *Salmonella typhimurium*. *J Biol Chem* 1982, **257**(19), 11808–11815.

125. Qureshi, N. et al. Monophosphoryl lipid A obtained from lipopolysaccharides of *Salmonella minnesota* R595: purification of the dimethyl derivative by high performance liquid chromatography and complete structural determination. *J Biol Chem* 1985, **260**(9), 5271–5278.

126. Myers, K., Truchot, A.T. & Ward, J. A critical determinant of lipid A endotoxic activity. In *Cellular and Molecular Aspects of Endotoxin Reactions*, Nowotny, J.S.A. & Ziegler, E.J., Eds., Elsevier, Amsterdam, 1990, pp. 145–156.

127. Vosika, G., Giddings, C. & Gray, G.R. Phase I study of intravenous mycobacterial cell wall skeleton and trehalose dimycolate attached to oil droplets. *J Biol Response Mod* 1984, **3**(6), 620–626.

128. Henricson, B.E., Benjamin, W.R. & Vogel, S.N. Differential cytokine induction by doses of lipopolysaccharide and monophosphoryl lipid A that result in equivalent early endotoxin tolerance. *Infect Immun* 1990, **58**(8), 2429–2437.

129. Odean, M.J. et al. Involvement of gamma interferon in antibody enhancement by adjuvants. *Infect Immun* 1990, **58**(2), 427–432.

130. Tokunaga, T. et al. Antitumor activity of deoxyribonucleic acid fraction from *Mycobacterium bovis* BCG: I. Isolation, physicochemical characterization, and antitumor activity. *J Natl Cancer Inst* 1984, **72**(4), 955–962.

131. Yamamoto, S. et al. Unique palindromic sequences in synthetic oligonucleotides are required to induce IFN [correction of INF] and augment IFN-mediated [correction of INF] natural killer activity. *J Immunol* 1992, **148**(12), 4072–4076.

132. Krieg, A.M. et al. CpG motifs in bacterial DNA trigger direct B-cell activation. *Nature* 1995, **374**(6522), 546–549.

133. Roman, M. et al. Immunostimulatory DNA sequences function as T helper-1-promoting adjuvants. *Nat Med* 1997, **3**(8), 849–854.

134. Stacey, K.J., Sweet, M.J. & Hume, D.A. Macrophages ingest and are activated by bacterial DNA. *J Immunol* 1996, **157**(5), 2116–2122.

135. Sparwasser, T. et al. Bacterial DNA and immunostimulatory CpG oligonucleotides trigger maturation and activation of murine dendritic cells. *Eur J Immunol* 1998, **28**(6), 2045–2054.

136. Krug, A. et al. Identification of CpG oligonucleotide sequences with high induction of IFN-alpha/beta in plasmacytoid dendritic cells. *Eur J Immunol* 2001, **31**(7), 2154–2163.

137. Sato, Y. et al. Immunostimulatory DNA sequences necessary for effective intradermal gene immunization. *Science* 1996, **273**(5273), 352–354.

138. Klinman, D.M., Barnhart, K.M. & Conover, J. CpG motifs as immune adjuvants. *Vaccine* 1999, **17**(1), 19–25.

139. Zhao, Q., Yu, D. & Agrawal, S. Site of chemical modifications in CpG containing phosphorothioate oligodeoxynucleotide modulates its immunostimulatory activity. *Bioorg Med Chem Lett* 1999, **9**(24), 3453–3458.

140. Krieg, A.M. CpG motifs in bacterial DNA and their immune effects. *Annu Rev Immunol* 2002, **20**, 709–760.

141. Marshall, J.D. et al. Identification of a novel CpG DNA class and motif that optimally stimulate B cell and plasmacytoid dendritic cell functions. *J Leukoc Biol* 2003, **73**(6), 781–792.

142. Vollmer, J. et al. Characterization of three CpG oligodeoxynucleotide classes with distinct immunostimulatory activities. *Eur J Immunol* 2004, **34**(1), 251–262.

143. Bauer, S. et al. Human TLR9 confers responsiveness to bacterial DNA via species-specific CpG motif recognition. *Proc Natl Acad Sci USA* 2001, **98**(16), 9237–9342.

144. Lipford, G.B. et al. CpG-containing synthetic oligonucleotides promote B and cytotoxic T cell responses to protein antigen: a new class of vaccine adjuvants. *Eur J Immunol* 1997, **27**(9), 2340–2344.

145. Oxenius, A. et al. CpG-containing oligonucleotides are efficient adjuvants for induction of protective antiviral immune responses with T-cell peptide vaccines. *J Virol* 1999, **73**(5), 4120–4126.

146. Chu, R.S. et al. CpG oligodeoxynucleotides act as adjuvants for pneumococcal polysaccharide–protein conjugate vaccines and enhance antipolysaccharide immunoglobulin G2a (IgG2a) and IgG3 antibodies. *Infect Immun* 2000, **68**(3), 1450–1456.

147. Storni, T. et al. Nonmethylated CG motifs packaged into virus-like particles induce protective cytotoxic T cell responses in the absence of systemic side effects. *J Immunol* 2004, **172**(3), 1777–1785.

148. Magone, M.T. et al. Systemic or mucosal administration of immunostimulatory DNA inhibits early and late phases of murine allergic conjunctivitis. *Eur J Immunol* 2000, **30**(7), 1841–1850.

149. McCluskie, M.J. et al. CpG DNA is an effective oral adjuvant to protein antigens in mice. *Vaccine* 2000, **19**(7–8), 950–957.

150. Daftarian, P. et al. Novel conjugates of epitope fusion peptides with CpG–ODN display enhanced immunogenicity and HIV recognition. *Vaccine* 2005, **23**(26), 3453–3468.

151. Tighe, H. et al. Conjugation of immunostimulatory DNA to the short ragweed allergen Amb a 1 enhances its immunogenicity and reduces its allergenicity. *J Allergy Clin Immunol* 2000, **106**(1 Pt 1), 124–134.

152. Weeratna, R.D. et al. CpG DNA induces stronger immune responses with less toxicity than other adjuvants. *Vaccine* 2000, **18**(17), 1755–1762.

153. Halperin, S.A. et al. A phase I study of the safety and immunogenicity of recombinant hepatitis B surface antigen co-administered with an immunostimulatory phosphorothioate oligonucleotide adjuvant. *Vaccine* 2003, **21**(19–20), 2461–2467.

154. Halperin, S.A. et al. Comparison of the safety and immunogenicity of hepatitis B virus surface antigen co-administered with an immunostimulatory phosphorothioate oligonucleotide and a licensed hepatitis B vaccine in healthy young adults. *Vaccine* 2006, **24**(1), 20–26.

155. Cooper, C.L. et al. Safety and immunogenicity of CPG 7909 injection as an adjuvant to Fluarix influenza vaccine. *Vaccine* 2004, **22**(23–24), 3136–3143.

156. Simons, F.E. et al. Selective immune redirection in humans with ragweed allergy by injecting Amb a 1 linked to immunostimulatory DNA. *J Allergy Clin Immunol* 2004, **113**(6), 1144–1151.

157. Scheel, B. et al. Immunostimulating capacities of stabilized RNA molecules. *Eur J Immunol* 2004, **34**(2), 537–547.

158. Prins, R.M. et al. The TLR-7 agonist, imiquimod, enhances dendritic cell survival and promotes tumor antigen-specific T cell priming: relation to central nervous system antitumor immunity. *J Immunol* 2006, **176**(1), 157–164.

159. Bernstein, D.I. et al. Effect of imiquimod as an adjuvant for immunotherapy of genital HSV in guinea-pigs. *Vaccine* 1995, **13**(1), 72–76.

160. Bernstein, D.I. Miller, R.L. & Harrison, C.J. Adjuvant effects of imiquimod on a herpes simplex virus type 2 glycoprotein vaccine in guinea pigs. *J Infect Dis* 1993, **167**(3), 731–735.

161. Wu, J.J., Huang, D.B. & Tyring, S.K. Resiquimod: a new immune response modifier with potential as a vaccine adjuvant for Th1 immune responses. *Antiviral Res* 2004, **64**(2), 79–83.

162. Brummitt, C.F. Imiquimod 5% cream for the treatment of recurrent, acyclovir-resistant genital herpes. *Clin Infect Dis* 2006, **42**(4), 575.

163. Wille-Reece, U. et al. Immunization with HIV-1 gag protein conjugated to a TLR7/8 agonist results in the generation of HIV-1 gag-specific Th1 and CD8+ T cell responses. *J Immunol* 2005, **174**(12), 7676–7683.

164. Kersten, G.F. et al. Incorporation of the major outer membrane protein of *Neisseria gonorrhoeae* in saponin–lipid complexes (ISCOMs): chemical analysis, some structural features, and comparison of their immunogenicity with three other antigen delivery systems. *Infect Immun* 1988, **56**(2), 432–438.

165. Kensil, C.R. et al. Separation and characterization of saponins with adjuvant activity from *Quillaja saponaria* Molina cortex. *J Immunol* 1991, **146**(2), 431–437.

166. Kensil, C., Newman, M.J., Coughlin, R.T., Soltysik, S., Bedore, D., Rechia, J., Wu, J. & Marciani, D. The use of Stimulon adjuvant to boost vaccine response. *Vaccine Res* 1993, **2**, 273–281.

167. Shirai, M. et al. Helper-cytotoxic T lymphocyte (CTL) determinant linkage required for priming of anti-HIV CD8+ CTL in vivo with peptide vaccine constructs. *J Immunol* 1994, **152**(2), 549–556.

168. Livingston, P. et al. Tumor cell reactivity mediated by IgM antibodies in sera from melanoma patients vaccinated with GM2 ganglioside covalently linked to KLH is increased by IgG antibodies. *Cancer Immunol Immunother* 1997, **43**(6), 324–330.

169. Livingston, P.O. Augmenting the immunogenicity of carbohydrate tumor antigens. *Semin Cancer Biol* 1995, **6**(6), 357–366.

170. Evans, T.G. et al. QS-21 promotes an adjuvant effect allowing for reduced antigen dose during HIV-1 envelope subunit immunization in humans. *Vaccine* 2001, **19**(15–16), 2080–2091.

171. Kashala, O. et al. Safety, tolerability and immunogenicity of new formulations of the *Plasmodium falciparum* malaria peptide vaccine SPf66 combined with the immunological adjuvant QS-21. *Vaccine* 2002, **20**(17–18), 2263–2277.

172. Spicer, E.K. et al. Sequence homologies between A subunits of *Escherichia coli* and *Vibrio cholerae* enterotoxins. *Proc Natl Acad Sci U S A* 1981, **78**(1), 50–54.

173. Lycke, N., Tsuji, T. & Holmgren, J. The adjuvant effect of *Vibrio cholerae* and *Escherichia coli* heat-labile enterotoxins is linked to their ADP–ribosyltransferase activity. *Eur J Immunol* 1992, **22**(9), 2277–2281.

174. Johnson, A.G. Molecular adjuvants and immunomodulators: new approaches to immunization. *Clin Microbiol Rev* 1994, **7**(3), 277–289.

175. Pizza, M. et al. A genetically detoxified derivative of heat-labile *Escherichia coli* enterotoxin induces neutralizing antibodies against the A subunit. *J Exp Med* 1994, **180**(6), 2147–2153.

176. Chong, C., Friberg, M. & Clements, J.D. LT(R192G), a non-toxic mutant of the heat-labile enterotoxin of *Escherichia coli*, elicits enhanced humoral and cellular immune responses associated with protection against lethal oral challenge with *Salmonella* spp. *Vaccine* 1998, **16**(7), 732–740.

177. Pizza, M. et al. Mucosal vaccines: nontoxic derivatives of LT and CT as mucosal adjuvants. *Vaccine* 2001, **19**(17–19), 2534–2541.

178. Tierney, R. et al. Transcutaneous immunization with tetanus toxoid and mutants of *Escherichia coli* heat-labile enterotoxin as adjuvants elicits strong protective antibody responses. *J Infect Dis* 2003, **188**(5), 753–758.

179. Giedlin, M., Ed. Cytokines as vaccine adjuvants: the use of interleukin-2. In *Methods in Molecular Medicine*, Walker, J., Ed., Vol. 42, Humana Press, Totowa, NJ, 2000, pp. 283–297.

180. Heath, A.W. & Playfair, J.H. Cytokines as immunological adjuvants. *Vaccine* 1992, **10**(7), 427–434.

181. Afonso, L.C. et al. The adjuvant effect of interleukin-12 in a vaccine against *Leishmania major*. *Science* 1994, **263**(5144), 235–237.

182. Disis, M.L. et al. Granulocyte-macrophage colony-stimulating factor: an effective adjuvant for protein and peptide-based vaccines. *Blood* 1996, **88**(1), 202–210.

183. Nunberg, J. Interleukin-2 as an adjuvant to vaccination. In *Technological Advances in Vaccine Development*, Lasky, L., Ed., Alan R. Liss, New York, 1988, pp. 4240–4243.

184. Schijns, V.E. et al. Modulation of antiviral immune responses by exogenous cytokines: effects of tumour necrosis factor-alpha, interleukin-1 alpha, interleukin-2 and interferon-gamma on the immunogenicity of an inactivated rabies vaccine. *J Gen Virol* 1994, **75**(Pt 1), 55–63.

185. Ueno, M. Lymphokine-activated killer cells induced in vivo in mice showing IL-2 toxicity have cytoplasmic granules containing perforin and its hemolytic activity. *Immunopharmacology* 1998, **39**(2), 75–82.

186. Sacco, S. et al. Protective effect of a single interleukin-12 (IL-12) predose against the toxicity of subsequent chronic IL-12 in mice: role of cytokines and glucocorticoids. *Blood* 1997, **90**(11), 4473–4479.

187. Sriskandan, K. et al. A toxicity study of recombinant interferon-gamma given by intravenous infusion to patients with advanced cancer. *Cancer Chemother Pharmacol* 1986, **18**(1), 63–68.

188. Portielje, J.E. et al. IL-12: a promising adjuvant for cancer vaccination. *Cancer Immunol Immunother* 2003, **52**(3), 133–144.

189. Sin, J.I. et al. IL-12 gene as a DNA vaccine adjuvant in a herpes mouse model: IL-12 enhances Th1-type CD4+ T cell-mediated protective immunity against herpes simplex virus-2 challenge. *J Immunol* 1999, **162**(5), 2912–2921.

190. Gherardi, M.M., Ramirez, J.C. & Esteban, M. Interleukin-12 (IL-12) enhancement of the cellular immune response against human immunodeficiency virus type 1 env antigen in a DNA prime/vaccinia virus boost vaccine regimen is time and dose dependent: suppressive effects of IL-12 boost are mediated by nitric oxide. *J Virol* 2000, **74**(14), 6278–6286.

191. Gupta, R.K. et al. Adjuvant properties of aluminum and calcium compounds. *Pharm Biotechnol* 1995, **6**, 229–248.

192. Goldenthal, K., Cavagnaro, J.L., Alving, C. & Vogel, F.R. Safety evaluation of vaccine adjuvants. *NCDVG Work. Groups AIDS Res Hum Retroviruses* 1993, **9**(Suppl 1), S47–S51.

193. Lindblad, E.B. Aluminium adjuvants: in retrospect and prospect. *Vaccine* 2004, **22**(27–28), 3658–3668.

194. Francis, M.J. et al. Immune response to uncoupled peptides of foot-and-mouth disease virus. *Immunology* 1987, **61**(1), 1–6.

195. Hem, S.L. & White, J.L. Characterization of aluminum hydroxide for use as an adjuvant in parenteral vaccines. *J Parenter Sci Technol* 1984, **38**(1), 2–10.

196. Kenney, J.S. et al. Influence of adjuvants on the quantity, affinity, isotype and epitope specificity of murine antibodies. *J Immunol Methods* 1989, **121**(2), 157–166.

197. Hamaoka, T. et al. Hapten-specific IgE antibody responses in mice: I. Secondary IgE responses in irradiated recipients of syngeneic primed spleen cells. *J Exp Med* 1973, **138**(1), 306–311.

198. Frost, L. et al. Persistent subcutaneous nodules in children hyposensitized with aluminium-containing allergen extracts. *Allergy* 1985, **40**(5), 368–372.

199. Relyveld, E.H. et al. [Vaccination with calcium phosphate adsorbed antigens]. *Prog Immunobiol Stand* 1970, **4**, 540–547.

200. Gupta, R.K. & Siber, G.R. Comparison of adjuvant activities of aluminium phosphate, calcium phosphate and stearyl tyrosine for tetanus toxoid. *Biologicals* 1994, **22**(1), 53–63.

201. He, Q. et al. Calcium phosphate nanoparticle adjuvant. *Clin Diagn Lab Immunol* 2000, **7**(6), 899–903.

202. Penney, C.L. et al. Further studies on the adjuvanticity of stearyl tyrosine and ester analogues. *Vaccine* 1993, **11**(11), 1129–1134.

203. Nixon-George, A. et al. The adjuvant effect of stearyl tyrosine on a recombinant subunit hepatitis B surface antigen. *J Immunol* 1990, **144**(12), 4798–4802.

204. Baldrick, P., Richardson, D. & Wheeler, A.W. Review of L-tyrosine confirming its safe human use as an adjuvant. *J Appl Toxicol* 2002, **22**(5), 333–344.

205. Preis, I. & Langer, R.S. A single-step immunization by sustained antigen release. *J Immunol Methods* 1979, **28**(1–2), 193–197.

206. Heller, J. Polymers for controlled parenteral delivery of peptides and proteins. *Adv Drug Deliv Res* 1993, **10**, 163–204.

207. O'Hagan, D.T. et al. Biodegradable microparticles as controlled release antigen delivery systems. *Immunology* 1991, **73**(2), 239–242.

208. Eldridge, J.H. et al. Biodegradable microspheres as a vaccine delivery system. *Mol Immunol* 1991, **28**(3), 287–294.

209. Cleland, L. Single-immunization vacciners Using PLGA. In *Vaccine Design: The Subunit and Adjuvant Approach*, Powell, M.J., Ed., Plenum Press, New York, 1995, pp. 439–462.

210. Jones, A.J. et al. Recombinant human growth hormone poly(lactic-co-glycolic acid) microsphere formulation development. *Adv Drug Deliv Res* 1997, **28**(1), 71–84.

211. Johnston, B.A., Eisen, H. & Fry, D. An evaluation of several adjuvant emulsion regimens for the production of polyclonal antisera in rabbits. *Lab Anim Sci* 1991, **41**(1), 15–21.

212. Toledo, H. et al. A phase I clinical trial of a multi-epitope polypeptide TAB9 combined with Montanide ISA 720 adjuvant in non-HIV-1 infected human volunteers. *Vaccine* 2001, **19**(30), 4328–4336.

213. Oliveira, G.A. et al. Safety and enhanced immunogenicity of a hepatitis B core particle *Plasmodium falciparum* malaria vaccine formulated in adjuvant Montanide ISA 720 in a phase I trial. *Infect Immun* 2005, **73**(6), 3587–3597.

214. McLaughlin, C.A. et al. Regression of tumors in guinea pigs after treatment with synthetic muramyl dipeptides and trehalose dimycolate. *Science* 1980, **208**(4442), 415–416.

215. Hunter, R., Strickland, F. & Kezdy, F. The adjuvant activity of nonionic block polymer surfactants: I. The role of hydrophile–lipophile balance. *J Immunol* 1981, **127**(3), 1244–1250.

216. Katz, J.M. et al. A nonionic block co-polymer adjuvant (CRL1005) enhances the immunogenicity and protective efficacy of inactivated influenza vaccine in young and aged mice. *Vaccine* 2000, **18**(21), 2177–2187.

217. Reynolds, J.A. et al. Adjuvant activity of a novel metabolizable lipid emulsion with inactivated viral vaccines. *Infect Immun* 1980, **28**(3), 937–943.

218. Morgan, A.J. et al. Validation of a first-generation Epstein–Barr virus vaccine preparation suitable for human use. *J Med Virol* 1989, **29**(1), 74–78.

219. Byars, N.E., Nakano, G., Welch, M. & Allison, A.C. Use of syntex adjuvant formulations enhance immune responses to viral antigens. In *Vaccines*, New York, Plenum Press, 1991, pp. 33–42.

## 28    DEVELOPMENT OF VACCINE ADJUVANTS: A HISTORICAL PERSPECTIVE

220. Girard, M. et al. Immunization of chimpanzees confers protection against challenge with human immunodeficiency virus. *Proc Natl Acad Sci USA* 1991, **88**(2), 542–546.

221. Woodard, L.F. & Jasman, R.L. Stable oil-in-water emulsions: preparation and use as vaccine vehicles for lipophilic adjuvants. *Vaccine* 1985, **3**(2), 137–144.

222. Ott, G. et al. MF59. Design and evaluation of a safe and potent adjuvant for human vaccines. *Pharm Biotechnol* 1995, **6**, 277–296.

223. Ott, G., Barchfeld, G.L. & Van Nest, G. Enhancement of humoral response against human influenza vaccine with the simple submicron oil/water emulsion adjuvant MF59. *Vaccine* 1995, **13**(16), 1557–1562.

224. Langenberg, A.G. et al. A recombinant glycoprotein vaccine for herpes simplex virus type 2: safety and immunogenicity [corrected]. *Ann Intern Med* 1995, **122**(12), 889–898.

225. Pregliasco, F. et al. Immunogenicity and safety of three commercial influenza vaccines in institutionalized elderly. *Aging (Milano)* 2001, **13**(1), 38–43.

226. Podda, A. The adjuvanted influenza vaccines with novel adjuvants: experience with the MF59-adjuvanted vaccine. *Vaccine* 2001, **19**(17–19), 2673–2680.

227. Stephenson, I. et al. Detection of anti-H5 responses in human sera by HI using horse erythrocytes following MF59-adjuvanted influenza A/Duck/Singapore/97 vaccine. *Virus Res* 2004, **103**(1–2), 91–95.

228. Stieneker, F. et al. Comparison of 24 different adjuvants for inactivated HIV-2 split whole virus as antigen in mice: induction of titres of binding antibodies and toxicity of the formulations. *Vaccine* 1995, **13**(1), 45–53.

229. Singh, M. et al. Cationic microparticles: a potent delivery system for DNA vaccines. *Proc Natl Acad Sci USA* 2000, **97**(2), 811–816.

230. Otten, G. et al. Induction of broad and potent anti-human immunodeficiency virus immune responses in rhesus macaques by priming with a DNA vaccine and boosting with protein-adsorbed polylactide coglycolide microparticles. *J Virol* 2003, **77**(10), 6087–6092.

231. O'Hagan, D.T., Jeffery, H. & Davis, S.S. Long-term antibody responses in mice following subcutaneous immunization with ovalbumin entrapped in biodegradable microparticles. *Vaccine* 1993, **11**(9), 965–969.

232. Gluck, R. Liposomal presentation of antigens for human vaccines. In *Vaccine Design: The Subunit and Adjuvant Approach*, Powell, M.J., Ed., Plenum Press, New York, 1995, pp. 325–345.

233. Pietrobon, B. Liposome design and vaccine development. In *Vaccine Design: The Subunit and Adjuvant Approach*, Powell, M.J., Ed., Plenum Press, New York, 1995, pp. 347–361.

234. Gregoriadis, G., McCormack, B., Obrenovic, M., Perrie, Y. & Saffie, R. In *Vaccine Adjuvants: Preparation Methods and Research Protocols*, O'Hagan, D.T., Ed., Humana Press, Totowa, NJ, 2000, pp. 137–150.

235. Singh, M. & O'Hagan, D. Advances in vaccine adjuvants. *Nat Biotechnol* 1999, **17**(11), 1075–1081.

236. Gregoriadis, G., Saffie, R. & de Souza, J.B. Liposome-mediated DNA vaccination. *FEBS Lett* 1997, **402**(2–3), 107–110.

237. Gregoriadis, G., Saffie, R. & Hart, S.L. High yield incorporation of plasmid DNA within liposomes: effect on DNA integrity and transfection efficiency. *J Drug Target* 1996, **3**(6), 469–475.

238. Perrie, Y., Frederik, P.M. & Gregoriadis, G. Liposome-mediated DNA vaccination: the effect of vesicle composition. *Vaccine* 2001, **19**(23–24), 3301–3310.

239. Gramzinski, R.A. et al. Immune response to a hepatitis B DNA vaccine in *Aotus* monkeys: a comparison of vaccine formulation, route, and method of administration. *Mol Med* 1998, **4**(2), 109–118.

240. Locher, C.P. et al. Human immunodeficiency virus type 2 DNA vaccine provides partial protection from acute baboon infection. *Vaccine* 2004, **22**(17–18), 2261–2272.

241. Cui, Z. DNA vaccine. *Adv Genet* 2005, **54**, 257–289.

242. Nair, S. et al. Soluble proteins delivered to dendritic cells via pH-sensitive liposomes induce primary cytotoxic T lymphocyte responses in vitro. *J Exp Med* 1992, **175**(2), 609–612.

243. Huang, L. et al. Liposomal delivery of soluble protein antigens for class I MHC-mediated antigen presentation. *Res Immunol* 1992, **143**(2), 192–196.

244. Gluck, R. Immunopotentiating reconstituted influenza virosomes (IRIVs). In *Vaccine Adjuvants: Preparation Methods and Research Protocols*, O'Hagan, D.T. Ed., Humana Press, Totowa, NJ, 2000, pp. 151–178.

245. Lovgren, K. & Morein, B. The requirement of lipids for the formation of immunostimulating complexes (ISCOMs). *Biotechnol Appl Biochem* 1988, **10**(2), 161–172.

246. Morein, B. et al. ISCOM, a novel structure for antigenic presentation of membrane proteins from enveloped viruses. *Nature* 1984, **308**(5958), 457–460.

247. van Binnendijk, R.S. et al. The predominance of CD8+ T cells after infection with measles virus suggests a role for CD8+ class I MHC-restricted cytotoxic T lymphocytes (CTL) in recovery from measles: clonal analyses of human CD8+ class I MHC-restricted CTL. *J Immunol* 1990, **144**(6), 2394–2399.

248. Rimmelzwaan, G.F. et al. A randomized, double blind study in young healthy adults comparing cell mediated and humoral immune responses induced by influenza ISCOM vaccines and conventional vaccines. *Vaccine* 2000, **19**(9–10), 1180–1187.

249. Rimmelzwann, G.F. ADME, A novel generation of viral vaccines based on the ISCOM matrix. In *Vaccine Design: The Subunit and Adjuvant Approach*, Powell, M.J., Ed., Plenum Press, New York, 1995.

250. Sundquist, B., Lovgren, K. & Morein, B. Influenza virus ISCOMs: antibody response in animals. *Vaccine* 1988, **6**(1), 49–53.

251. Crouch, C.F. et al. The use of a systemic prime/mucosal boost strategy with an equine influenza ISCOM vaccine to induce protective immunity in horses. *Vet Immunol Immunopathol* 2005, **108**(3–4), 345–355.

252. Jones, W.R. et al. Phase I clinical trial of a World Health Organisation birth control vaccine. *Lancet* 1988, **1**(8598), 1295–1298.

253. Bennett, B. et al. A comparison of commercially available adjuvants for use in research. *J Immunol Methods* 1992, **153**(1–2), 31–40.

254. Alving, C.R. Lipopolysaccharide, lipid A, and liposomes containing lipid A as immunologic adjuvants. *Immunobiology* 1993, **187**(3–5), 430–446.

255. DeWilde, M. Preclinical and clinical experience with MPL and QS21 adjuvanted recombinant subunit vaccines. In *Novel Vaccine Strategies*, IBC Conferences, Bethesda, MD, 1994.

256. Leroux-Roels, G., Moreaux, E., Verhasselt, B., Biernaux, S., Brulein, B., Francotte, M., Pala, P., Slaoui, M. & Vandepapeliere, P. Immunogenicity and reactogenicity of a recombinant HSV-2 glycoprotein D vaccine with or without monophosphoryl lipid a in HSV seronegative and seropositive subjects. Presented at the 33rd Interscience Congress on Antimicrobial Agents and Chemotherapy, 1993.

257. Thoelen, S. et al. Safety and immunogenicity of a hepatitis B vaccine formulated with a novel adjuvant system. *Vaccine* 1998, **16**(7), 708–714.

258. Tong, N.K. et al. Immunogenicity and safety of an adjuvanted hepatitis B vaccine in prehemodialysis and hemodialysis patients. *Kidney Int* 2005, **68**(5), 2298–2303.

259. Cunha, M.G., Rodrigues, M.M. & Soares, I.S. Comparison of the immunogenic properties of recombinant proteins representing the *Plasmodium vivax* vaccine candidate MSP1(19) expressed in distinct bacterial vectors. *Vaccine* 2001, **20**(3–4), 385–396.

260. Allison, A.C. & Byars, N.E. An adjuvant formulation that selectively elicits the formation of antibodies of protective isotypes and of cell-mediated immunity. *J Immunol Methods* 1986, **95**(2), 157–168.

261. Lidgate, D. Development of an emulsion-based muramyl dipeptide adjuvant formulation for vaccines. In *Vaccine Design: The Subunit and Adjuvant Approach*, Powell, M.J., Ed., Plenum Press, New York, 1995, pp. 313–324.

262. Keefer, M.C. et al. Safety and immunogenicity of Env 2–3, a human immunodeficiency virus type 1 candidate vaccine, in combination with a novel adjuvant, MTP-PE/MF59. NIAID AIDS Vaccine Evaluation Group. *AIDS Res Hum Retroviruses* 1996, **12**(8), 683–693.

263. Keitel, W. et al. Pilot evaluation of influenza virus vaccine (IVV) combined with adjuvant. *Vaccine* 1993, **11**(9), 909–913.

264. Ott, G. et al. A cationic sub-micron emulsion (MF59/DOTAP) is an effective delivery system for DNA vaccines. *J Control Release* 2002, **79**(1–3), 1–5.

265. Mitchell, M.S. et al. Active specific immunotherapy for melanoma: phase I trial of allogeneic lysates and a novel adjuvant. *Cancer Res* 1988, **48**(20), 5883–5893.

266. Egan, M.A. et al. A comparative evaluation of nasal and parenteral vaccine adjuvants to elicit systemic and mucosal HIV-1 peptide-specific humoral immune responses in *Cynomolgus* macaques. *Vaccine* 2004, **22**(27–28), 3774–3788.

267. Ivins, B.E. et al. Comparative efficacy of experimental anthrax vaccine candidates against inhalation anthrax in rhesus macaques. *Vaccine* 1998, **16**(11–12), 1141–1148.

268. Baldridge, J.R. & Crane, R.T. Monophosphoryl lipid A (MPL) formulations for the next generation of vaccines. *Methods* 1999, **19**(1), 103–107.

269. Singh, M. et al. Cationic microparticles are an effective delivery system for immune stimulatory cpG DNA. *Pharm Res* 2001, **18**(10), 1476–1479.

270. Kazzaz, J. et al. Encapsulation of the immune potentiators MPL and RC529 in PLG microparticles enhances their potency. *J Control Release* 2006, **110**(3), 566–573.

271. Singh, M. et al. Polylactide-co-glycolide microparticles with surface adsorbed antigens as vaccine delivery systems. *Curr Drug Deliv* 2006, **3**(1), 115–120.

272. Alving, C.R. et al. Novel adjuvant strategies for experimental malaria and AIDS vaccines. *Ann N Y Acad Sci* 1993, **690**, 265–275.

273. Heppner, D.G. et al. Safety, immunogenicity, and efficacy of *Plasmodium falciparum* repeatless circumsporozoite protein vaccine encapsulated in liposomes. *J Infect Dis* 1996, **174**(2), 361–366.

274. Tan, L.G. Effect of interleukin-2 on the immunoadjuvant action of liposomes. *Biochem Soc Trans* 1989, **17**, 693–694.

275. Ho, R.J., Burke, R.L. & Merigan, T.C. Liposome-formulated interleukin-2 as an adjuvant of recombinant HSV glycoprotein gD for the treatment of recurrent genital HSV-2 in guinea-pigs. *Vaccine* 1992, **10**(4), 209–213.

276. Gursel, I. et al. Sterically stabilized cationic liposomes improve the uptake and immunostimulatory activity of CpG oligonucleotides. *J Immunol* 2001, **167**(6), 3324–3328.

277. Wilson, A., Pitt, B. & Li, S. Complex roles of CpG in liposomal delivery of DNA and oligonucleotides. *Biosci Rep* 2002, **22**(2), 309–322.

278. Rao, M. et al. Immunostimulatory CpG motifs induce CTL responses to HIV type I oligomeric gp140 envelope protein. *Immunol Cell Biol* 2004, **82**(5), 523–530.

279. Jerome, V. et al. Cytotoxic T lymphocytes responding to low dose TRP2 antigen are induced against B16 melanoma by liposome-encapsulated TRP2 peptide and CpG DNA adjuvant. *J Immunother* 2006, **29**(3), 294–305.

280. Mandal, M. & Lee, K.D. Listeriolysin *O*-liposome-mediated cytosolic delivery of macromolecule antigen in vivo: enhancement of antigen-specific cytotoxic T lymphocyte frequency, activity, and tumor protection. *Biochim Biophys Acta* 2002, **1563**(1–2), 7–17.

# 2

# ANTIGEN PROCESSING AND PRESENTATION

Subash Sad

## 2.1  INTRODUCTION

Development of vaccines against pathogens provides the safest and most cost-effective method of ensuring long-term protection against pathogens [1]. However, development of vaccines against infectious diseases such as AIDS, tuberculosis, malaria, and herpes has been challenging [2–4]. Pathogens responsible for these diseases induce chronic infection and involve multiple immune evasive strategies [4,5]. Development of vaccines against such pathogens will involve a thorough understanding of the key virulence strategies of the pathogens [6]. While the ultimate goal of vaccination is the development of potent T-cell memory response to the pathogen, it appears that the programming for the development of T-cell memory occurs rapidly within the first week of vaccination or infection [1]. Antigen presentation by infected cells occurs rapidly after infection or vaccination and is over within the first few days [7]. The magnitude of the acquired immune response is governed in large part by how efficiently antigen is processed and presented within the first few days. Although various cell types can present antigens to T cells, dendritic cells (DCs) have the unique ability to induce antigen presentation that translates to the induction of potent T-cell memory [8]. Thus, antigen processing and presentation forms the foundation of acquired immunity, and efficient vaccine design relies extensively on devising strategies to ensure potent antigen presentation.

Integration of two distinct arms of the immune system, the innate and adaptive, is critical for ensuring rapid curtailment of the pathogen and subsequent generation of long-lasting protection [1]. Recently, the role of antigen-nonspecific recognition of pathogens by cells of the innate immune system through pattern-recognition receptors (PRRs) has come to light [9,10]. It is clear that such innate recognition of pathogens by the host not only curtails pathogens rapidly until maturation of adaptive immunity but critically sets an inflammatory milieu for mobilization and development of appropriate adaptive immunity [11]. Thus, although vaccines ultimately aim to evoke long-lived adaptive immunity, optimal acquired immunity cannot be achieved in the absence of appropriate PRR interactions [12]. Among the adaptive immune effectors, B and T cells have specialized functions in combating infections. Traditionally, extracellular pathogens can be eliminated by strong antibody responses with help from CD4$^+$ T cells. In the presence of appropriate cytokines, CD4$^+$ T cells further differentiate into Th1 and Th2 cells, aiding cell-mediated and antibody responses, respectively [13]. On the other hand, combating intracellular pathogens requires strong cell-mediated immune response, particularly from CD8$^+$ T cells [2]. CD8$^+$ T cells mediate cytotoxicity toward infected cells and tumors and produce various cytokines, including IFN-$\gamma$ and tumor necrosis factor (TNF), which aid long-term immunity [2,14–17]. Furthermore, CD8$^+$ T cells have been shown to produce granulysin, which is a component of the granules of cytotoxic T lymphocytes (CTLs) with strong antimicrobial activity [18]. It is the latter type of response that has remained a challenge for safer subunit vaccines.

Vaccine formulations comprise the following major components: specific antigenic determinants, effective adjuvanting, and optimal delivery strategy. In the context of vaccines against the major global infectious diseases, such as tuberculosis (TB), AIDS, and malaria [19], several conditions must be fulfilled for a vaccine formulation to elicit optimal T-cell memory:

1. Vaccine formulations should activate DCs sufficiently to induce adaptive immunity, but not to cause overt toxicity.
2. Antigen should be trafficked within the antigen-presenting cells (APCs) for presentation by both MHC (major histocompatibility complex) classes I and II to stimulate CD8$^+$ and CD4$^+$ T cells.
3. The delivery system should accommodate several antigens for widespread effectiveness.
4. Ideally, vaccines should evoke long-lasting memory responses after a single injection.

In the past, adjuvants were chosen empirically for their ability to evoke inflammation without disease induction (such as killed particles of bacteria) or to provide an antigenic depot at the site of injection (alum). Only two adjuvants, alum and MF59, are currently approved for universal use in humans [20]. Both these adjuvants are poor at evoking cell-mediated immunity.

## 2.2 RECOGNITION AND UPTAKE

It is now apparent that DCs, the most potent antigen-presenting cells use a dual "key" to activate T cells [8]. Particulate and soluble antigens are efficiently internalized by phagocytosis and macropinocytosis, respectively [21]. Both processes are actin dependent, require membrane ruffling, and result in the formation of large vacuoles. Phagocytosis is generally receptor mediated, whereas macropinocytosis is a cytoskeleton-dependent type of fluid-phase endocytosis [8]. In macrophages and epithelial cells, macropinocytosis is transiently induced by growth factors. In immature DCs, in contrast, macropinocytosis is constitutive. Macropinocytosis represents a critical antigen-uptake pathway, allowing DCs to sample large amounts of surrounding fluid rapidly and nonspecifically. Phagocytosis, in contrast, is initiated by the engagement of specific receptors, triggering a cascade of signal transduction, which is required for actin polymerization and effective engulfment. Receptor-mediated endocytosis allows the uptake of macromolecules through specialized regions of the plasma membrane termed *coated pits*. This process is initiated by a signal in the cytoplasmic tail of the endocytic receptor, which is recognized by a family of adaptors responsible for the recruitment of clathrin lattices [22]. A large number of endocytic receptors are expressed selectively by subpopulations of immature DCs. These include the Fc portion of immunoglobulin (FcR), CD91, scavenger receptors (SRs), macrophage mannose receptor (MMR), DEC205, CD23, low-affinity IgE receptor, langerin, and dendritic cell–specific ICAM3-grabbing nonintegrin (DC-SIGN) [23]. These receptors facilitate uptake of a wide variety of pathogens or macromolecules in different anatomical compartments by dendritic cells. Immature DCs also internalize apoptotic and necrotic cells and present antigens to T cells [24].

## 2.3 DENDRITIC CELL ACTIVATION

Pathogens express pathogen-associated molecular patterns (PAMPs) which are recognized by specific receptors (PRRs) on dendritic cells, which results in appropriate signaling that potentiates T-cell priming and memory generation [10,25]. The Toll-like receptor (TLR) family constitutes a classical PRR that critically directs innate immune system response to the PAMPs [25]. Activation of cells through TLRs elicits a variety of inflammatory cytokines and chemokines, depending on the cell type and specific TLR being stimulated [10]. TLR knock-out mice exhibit a dramatic susceptibility to infection and compromised generation of acquired immune responses [9,26–30]. Over the past few years, various TLR ligands have been identified [31] and offer an opportunity to guide memory development.

Dendritic cells respond to two types of signals: direct recognition of pathogens through specific pathogen-recognition receptors (PPRs), and indirect sensing of infection through inflammatory cytokines. In response to these signals, dendritic cells are activated to enter an integrated development program called *maturation*, which transforms dendritic cells to efficient T-cell stimulators.

Five types of surface receptors have been reported to trigger dendritic cell maturation [8]: (1) Toll-like receptors, (2) cytokine receptors, (3) TNF-receptor family molecules, (4) FcR, and (5) sensors for cell death.

1. Dendritic cells mature in response to various pathogenic compounds, including several bacterial wall compounds [such as lipopolysaccharide (LPS)], unmethylated CpG motifs, and double-stranded RNA. Most of these molecules are recognized by the large family of surface receptors called Toll-like receptors. In different cells of the immune system, different PPRs are specifically recognized by one or by a combination of TLRs [25]. For example, TLR4 determines responses of gram-negative bacteria through binding to LPS; TLR2 is involved in responses to different gram-positive cell wall compounds and to bacterial lipoproteins; TLR5 recognizes flagellin from both gram-positive and gram-negative bacteria; and TLR9 binds to unmethylated CpG motifs.

2. Dendritic cells sense danger and infections indirectly through inflammatory mediators such as TNFα, IL-1, and PGE-2, whose secretion is triggered by pathogens.

3. CD4 T-cells induce dendritic cell maturation. Triggering of CD40 by CD40L on T cells induces dendritic cell maturation, but CD40-independent mechanisms also exist.

4. Dendritic cells can also be activated through receptors for immunoglobulins. Indeed, engagement of most FcR by immune complexes or specific antibodies induces DC maturation.

5. Cell death can be sensed as a danger signal by dendritic cells [8]. Cell injury releases adjuvant compounds that enhance T-cell responses [32]. Necrotic but not apoptotic cell death induces mouse and human DC maturation [8].

## 2.4   ANTIGEN PRESENTATION AND CROSS-PRESENTATION

CD8[+] and CD4[+] T cells express clonally distributed receptors that recognize fragments of antigens (peptides) associated with MHC class I and II molecules, respectively [33]. Antigen degradation and peptide loading onto MHC molecules occurs intracellularly in APCs. The intracellular pathways for peptide loading on MHC class I and II molecules have been analyzed in detail [21,22]. Strict compartmentalization of MHC class I and II biogenesis results in the loading of exogenous and endogenous antigens on MHC class II molecules in the endocytic pathway and the selective loading of endogenous but not exogenous antigens on MHC class I molecules in the endoplasmic reticulum

(ER) [34]. This model accounts at the effector level for the selective killing by MHC class I–restricted CD8[+] T cells of virus-infected cells (expressing endogenous viral antigens), but not of neighboring cells that have internalized inactive virus. This model, however, is also in conflict with other reports, where it was shown that the priming of CD8[+] T-cell responses in vivo can occur after presentation of exogenous antigens by MHC class I molecules, a phenomenon referred to as *cross-priming* [35]. Recent studies in dendritic cells reconcile these two series of studies by showing that in addition to MHC class II, internalized antigens may also be presented by MHC class I molecules, a phenomenon referred to as *cross-presentation* [35,36].

## 2.5 MHC CLASS I–RESTRICTED ANTIGEN PRESENTATION

### 2.5.1 Endogenous Pathway

Most peptides to be loaded on MHC class I molecules are generated by proteasome degradation of newly synthesized ubiquitinated proteins [22]. The resulting peptides are transferred to the ER by specialized peptide transporters (TAP) and loaded on new MHC class I molecules under the control of a loading complex composed of several ER resident chaperons (including tapasin, calnexin, and calreticulin) [37]. Once associated with peptides, MHC class I molecules are transferred rapidly through the Golgi apparatus to the plasma membrane. Like other cells, dendritic cells present self- or virus-derived endogenous antigens. It is interesting that in contrast to MHC class II molecules, MHC class I molecules are still efficiently synthesized and transported to the cell surface in mature dendritic cells, highlighting the functional differences in the regulation of antigen presentation to CD4[+] and CD8[+] T cells. In addition, dendritic cells constitutively express low levels of the immunoproteasome, which becomes the main type of proteasome in mature dendritic cells; LMP2, LMP7, and PA28 are all induced during dendritic cell maturation [22]. This change in proteasome composition may up-regulate the efficiency of presentation of some epitopes while decreasing the presentation of others. Targeting of proteins for proteasomal degradation requires their ubiquitination.

### 2.5.2 Cross-Presentation Pathway

Exogenous antigens internalized by various pathways, however, may also be presented by MHC class I molecules [35,36,38]. Macropinocytosis allows receptor-independent cross-presentation of soluble antigens by dendritic cells. Physiologically, phagocytosis is probably a major route for antigen uptake and cross-presentation. It has been shown that linking antigens to latex beads, thus forcing internalization by phagocytosis, strongly increased the efficiency of cross-presentation [21]. Similarly, entrapment of antigens into lipid vesicles results in potent antigen presentation and T-cell priming [39]. Phagocytosis of bacteria results in efficient cross-presentation in dendritic cells [40]. Phagocytosis of apoptotic cells also results in efficient cross-presentation of viral and tumor antigens [24]. Whether apoptosis per se is required for cross-presentation is a matter of debate. Death by necrosis or inhibitors of apoptosis resulted in inefficient cross-presentation. Exosomes, small membrane vesicles secreted by tumor cells, or infected phagocytes contain antigens that are taken up by dendritic cells to induce cross-presentation [41].

Initially, two main intracellular pathways for cross-presentation were reported, resulting in either endocytic or ER MHC class I peptide loading [21,34]. Loading in endocytic compartments is in general insensitive to inhibitors of protein neosynthesis (Brefeldin A) and to inhibitors of the proteasome (lactacystine). In addition, loading is independent of TAP transporters and sensitive to inhibitors of lysosomal function (e.g., ammonium chloride, chloroquine). Conversely, ER peptide loading is blocked by proteasome inhibitors and requires the expression of TAP. This pathway requires transport of internalized antigens from the endocytic compartments to the cytosol [21,34].

More recently it was reported that the membrane of ER fuses with nascent phagosomes, suggesting that this peripheral compartment (phagosomes) in macrophages and dendritic cells may serve as an organelle optimized for MHC class I–restricted cross-presentation of exogenous antigens [42]. The process allows intersection of the endosomal system with the ER, the classical site of MHC class I peptide loading, and may reconcile the seemingly conflicting evidence indicating that both of these sites are crucial in cross-presentation.

## 2.6 MHC CLASS II–RESTRICTED ANTIGEN PRESENTATION

Peptide loading on MHC class II molecules occurs through a different pathway [43]. Soon after synthesis in the ER, three α/β MHC class II dimmers associate to a trimer of invariant (Ii) chains [44]. These nonamers exit the ER and pass through the Golgi apparatus before being transported to the endocytic pathway under the influence of transport signals present in the cytoplasmic region of the Ii chain. Once in endosomes and lysosomes, the nonameric complexes meet an acidic, protease-rich environment, where the Ii chain is degraded by several proteolytic enzymes of the cathtepsin family. MHC class II dimmers become competent to bind antigenic peptides under the control of two nonpolymorphic MHC class II molecules, HLA-DM/H-2M and HLA-DO/H2-O (in human and mouse) [43]. Once loaded with peptides, Ii chain-free MHC class II/peptide complexes reach the plasma membrane. Antigen degradation in the endocytic pathway and the generation of antigenic peptides require several proteases, including cathepsins and asparaginyl endopeptidases [44]. Thus, MHC class II molecules bind mainly to peptides derived from antigens present in the endocytic pathway (internalized or membrane proteins).

## 2.7   ANTIGEN PROCESSING AND ADJUVANTS

Antigens incorporated onto traditional adjuvants, while gaining access to the MHC class II pathway for CD4$^+$ T cell activation, fail to gain access for MHC class I processing [34]. Thus, such vaccines fail to stimulate CD8$^+$ T cells. Recombinant attenuated intracellular viruses and bacteria that can enter the cytosol act as antigen carriers for induction of CD8$^+$ T cells [45,46]. Although this strategy has been successful to some extent, many risks are associated with the use of live vaccines. Another strategy of targeting antigens to the cytosol is to tag them with bacterial toxins that interact with specific receptors on cell surfaces and translocate into the cytosol. In this context, diptheria toxin, *Shigella*-like toxin, and heat-labile enterotoxin have been shown to evoke CD8$^+$ T-cell immunity [47]. However, the utility for vaccination remains questionable, as they do not target specific APCs, have toxicity, and are often ineffective on repeated use. Particulate antigen delivery systems that provide an antigenic depot are phagocytosed and may be modulated to deliver antigen to the cytosol. For example, pH-sensitive liposomes deliver their cargo into the cytosol by fusion with phagolysomal membrane at acidic pH ([48]). More promising approaches include viruslike particles [49] and archaeosomes [50] that follow a phagosome–cytosol route and deliver antigen efficiently to the classical processing machinery [51].

## 2.8   INFLUENCE OF THE TYPE OF ANTIGEN-PRESENTING CELL

In addition to the differences in antigen-trafficking pathways for CD4$^+$ and CD8$^+$ T-cell activation, the types of APCs that process and present antigens can have a profound influence in the development of T-cell response and memory. Compared to dendritic cells, macrophages and B cells are very poor in antigen-presenting ability [8,52,53]. The unique ability of dendritic cells to induce potent antigen presentation is due partially to their expression of high levels of co-stimulatory molecules CD40 and B7, which provide strong T-cell priming ability [8]. Even within dendritic cells, there is considerable heterogeneity [54], and recently, dendritic cells expressing CD8$\alpha$ have been shown to be superior to CD8$\alpha^-$ DCs in antigen processing and presenting ability [55,56]. However, dendritic cells cannot induce potent T-cell responses [57,58] unless they are activated by inflammatory stimuli (such as type I IFN), which occurs when dendritic cells respond to LPS-like molecules on pathogens, suggesting that perception of "danger" by the innate immune system is important for the generation of a strong T-cell response [12]. In this context, the role of Toll-like receptors on APCs in "policing" danger signals for appropriate activation has been highlighted [59]. Further, co-stimulatory interactions between CD28-CD80, CD40-CD40L, 4-1BB-4-1BBL, and OX40-OX40L provide additional signals that increase the expansion of T-cell populations [60–64].

## 2.9 T-CELL PRIMING

T-cell response can be categorized into three phases. The first phase involves T-cell priming, which is followed by an apoptotic phase where the majority of the primed effectors die. T cells that survive the death phase enter the memory phase, in which the number of memory T cells stabilizes and these cells are maintained for prolonged periods of time. In the primary response to antigen, naive CD4$^+$ and CD8$^+$ T cells undergo a massive burst of expansion, which in turn governs the extent of memory generated [65]. Against some intracellular bacteria (e.g., *Listeria monocytogenes*), the burst of expansion appears to be of a much higher magnitude for CD8$^+$ T cells compared to CD4$^+$ T cells [66]. As the expression of MHC class I molecules is nearly ubiquitous in comparison to MHC class II molecules that are selectively expressed on fewer cell types, CD8$^+$ T cells would have a greater opportunity to encounter antigen than would CD4$^+$ T cells. However, it is not clear whether this is true for all infections, as antigen presentation is induced mainly by professional APCs that express high levels of MHC class II molecules. CD4$^+$ and CD8$^+$ T cells also exhibit differences in the requirements of co-stimulatory molecules [67], suggesting that differential thresholds might exist for CD4$^+$ versus CD8$^+$ T-cell activation [1]. The extent of proliferation of T cells also appears to be governed by the amount of antigen available in vivo. Infection with higher doses of bacteria or viruses results in a stronger, protective, T-cell memory [68,69]. The influence of the antigen levels on memory commitment was shown clearly in elegant experiments with recombinant vaccinia virus engineered to express low or high levels of ovalbumin (OVA) wherein the response increased progressively with increasing antigen dose until a saturation in epitope levels was achieved [70]. Besides the amount of antigen, immunodominance of the peptides also plays an important role in development of CD8$^+$ T-cell memory. Immunodominance appears to be governed mainly by the affinity, stability, and dissociation of peptide from MHC molecules [71,72]. For example, LLO$_{91-99}$, which is the most immunodominant CTL epitope of *Listeria monocytogenes*, is the least abundant peptide in infected APCs.

The duration of antigen presentation required for inducing T-cell memory is still a subject of active studies. Recent investigations indicate that naive T cells soon after receiving appropriate antigen and costimulatory signals commit to clonal expansion and differentiation into effector cells. Using either antibiotics to terminate pathogen growth, or transferring activated T cells into recipient mice, it has been shown that memory T-cell commitment occurs within, but not after, the first 24 hours of infection [7,69,73,74]. Thus, the initial antigenic encounter appears to initiate an instructive program wherein CD8$^+$ T cells undergo a minimum of seven to 10 divisions without further antigenic stimulation, and once initiated, the antigen-specific T cells continue to differentiate and divide, even in the absence of further antigenic stimulation [7]. The continued persistence of antigen may result in a greater effector T-cell expansion [75–77]. However, when the duration of antigen presentation becomes too long, such as in various chronic viral infection models, effector populations fail to continue to expand and may undergo clonal exhaustion or deletion, indicating that antigen-driven proliferation of $CD8^+$ T cells does not occur indefinitely [78–80]. Regardless of whether the pathogen persists briefly and is cleared in a few days, or persists for a week, the peak of the $CD8^+$ T-cell response, particularly during viral infections, occurs around day 7, indicating that the timing of the peak is determined early in the encounter with the pathogen [76]. After activation by APCs, naive T cells will undergo seven to 10 divisions, which could take approximately five days assuming an average doubling time of 12 hours. After the second cell division, $CD8^+$ T cells appear to acquire effector function and contribute to control of infection [81]. For $CD4^+$ T cells, it was shown recently that whereas the initial antigenic encounter can induce T-cell expansion for up to seven divisions, repeated antigenic stimulation is required for generation of effector activity and further expansion [82]. The number of cell divisions required for a T cell to acquire effector activity has been controversial [83,84], as T cells appear to acquire effector activity even without entering the S phase [84]. Thus, whereas cell cycling may be necessary for enhancing the frequency of specific cells, it may not be essential for effector activity.

An important determinant during T-cell priming relates to the immune status of the host. The induction of a highly inflammatory environment (high numbers of activated dendritic cells, macrophages, NK (natural killer) cells, $\gamma\delta$ T cells) by the chronic pathogen *Mycobacterium bovis* (BCG) compromises T-cell priming against a subsequent pathogen, *Listeria monocytogenes* (LM) [85]. This impairment in T-cell priming was due mainly to the cessation in the initial expansion of the replicating immunogen, *L. monocytogenes*. Preexisting inflammation resulted in a selective impairment in T-cell priming against the replicating immunogen as $CD8^+$ T-cell response to OVA administered as an inert antigen (OVA archaeosomes [86]) was enhanced by BCG preimmunization, whereas priming toward OVA administered as a live immunogen (LM-OVA) was impaired [85]. Thus, depending on the nature of the immunogen, the presence of prior inflammatory responses may either impede or boost vaccine efficacy. This may provide some scientific understanding for the long-held clinical practice of avoiding vaccinations during fever associated with ongoing infections. Similarly, the presence of antibodies or T cells that cross-react with a given vaccine may limit the vaccine efficacy by reducing the growth of live vaccine in vivo, and hence result in poor T-cell and antibody response [87].

Another factor that can influence T-cell priming and subsequent memory generation relates to the cytokine microenvironment, which may influence the phenotype and function of memory T cells generated. As with the cytokine-secreting subsets of $CD4^+$ T cells (Th1 and Th2) [13], $CD8^+$ T cells can also differentiate into distinct type 1 and type 2 cytokine-secreting subsets [88,89]. These cytokine-secreting subsets of $CD8^+$ T cells can exhibit quantitative differences in their effector function and protective efficacies [90,91]. The cytokines (IL-12 and IL-4) that induce the differentiation of CD4$^+$ T cells (Th1 and Th2) also induce the differentiation of CD8$^+$ cells (Tc1 and Tc2).

## 2.10 GENERATION OF MEMORY

It has been debated whether memory T cells represent a separate lineage or whether they are the descendents of effector cells, and a key question that remains is how memory T cells avoid the apoptotic program, which eliminates more than 90% of the effectors generated. Although it could be argued that memory T cells represent a population of escapees that evade apoptotic instruction, the remarkably constant proportion of memory T cells generated (10% of peak effector numbers) irrespective of the dose and type of pathogen used suggests that memory commitment may not be random.

The duration of antigen exposure may be an important factor in memory generation since continued antigenic stimulation might switch on the death pathways. Conversely, T cells that are primed toward the end of an immune response might experience a brief exposure to antigen, which could still drive them to differentiate into effectors without activating the apoptotic pathways. It is possible that memory T cells are generated by incomplete differentiation of a subset of cells that avoids death (overstimulation) perhaps by ignoring or avoiding peptide-carrying APCs through some unknown mechanism, or by arriving late in the immune response. The finding that continued viral persistence can induce T-cell exhaustion and/or deletion is consistent with this model [78,79].

The development of immune response and subsequent memory can vary with each pathogen or vaccine, depending on such factors as the relative duration of antigen presentation, the amount of antigen expressed, and the degree of inflammation caused. This has allowed further elucidation of the generation of subsets of memory T cells. It has been reported that killed bacterial vaccines induce substantial numbers of central phenotype memory T cells without going through an effector stage [92]. Similarly, in another study, CD8$^+$ T cells cultured in IL-15 developed into central phenotype memory cells, whereas cells cultured in high-dose IL-2 developed into effector phenotype cells [93]. These studies suggest that the lineage development of memory may vary with the immunization protocol and that effector differentiation may not be a prerequisite for generation of memory CD8$^+$ T cells in all models. A reverse linear differentiation model has also been suggested wherein the central memory cells arise first and give rise to effector cells [94,95]. The notion that memory T cells can be maintained only as effector or central memory subsets may be too simplistic [96]. Indeed, in HIV, cytomegalovirus (CMV), Epstein–Barr virus (EBV), and hepatitis C virus (HCV) infection models, memory CD8$^+$ T cells exhibited a spectrum of phenotypes typical of uncommitted precursors to highly polarized terminally differentiated effectors [95,97], and memory

CD8$^+$ T cells against different viruses tend to accumulate at different points along the differentiation pathway.


## 2.11  PHENOTYPES OF MEMORY

Various cell surface markers have been used to identify memory T cells; however, the complexity of expression patterns of different proteins varies with the degree of activation, type of infection, and the differential isoform splicing of various molecules [98]. Of all the markers used for discriminating naive versus memory T cells, CD44 is by far the most reliable marker that is expressed at high levels in memory T cells of mice [99,100]. Naive T cells also express high levels of CD62L and lose the expression of this molecule upon activation by antigen–APCs [101]. However, toward the end of the primary immune response, as the antigen levels dwindle, some of these activated T cells appear to revert back to expressing high levels of CD62L [81,102], causing ambiguity in the phenotypic characterization of memory T cells. According to this model [81,102], various subsets of CD8$^+$ T cells can be discerned: CD62L$^{hi}$CD44$^{low}$ (naive), CD62L$^{hi}$CD44$^{hi}$ (resting memory), and CD62L$^{low}$CD44$^{hi}$ (effector memory). In addition to these markers, Ly-6C and CD122 have also been shown to be expressed selectively on memory CD8$^+$ T cells [103,104].

A well-defined phenotypic and functional model of the subsets of human memory CD4$^+$ and CD8$^+$ T cells has been proposed based on expression of CD62L and CC-chemokine receptor 7 (CCR7) [94]. CD62L is involved in attachment and rolling of T-cells on high endothelial venules [105], whereas CCR7 binds the chemokines CCL19 and CCL21 that are presented on the endothelial cells in the lymph nodes, which causes the initiation of extravasation [106]. As a result, CD62L$^{hi}$CCR7$^+$ and CD62L$^{low}$CCR7$^-$ T cells would be expected to have distinct migratory properties in vivo. Several studies have shown that CD62L$^{hi}$CCR7$^+$ T cells migrate efficiently to peripheral lymph nodes, whereas T cells lacking these molecules were found selectively in the nonlymphoid compartment [107,108]. Functionally, CD62L$^{hi}$CCR7$^+$ cells produced only IL-2 after activation in vitro, whereas CD62L$^{low}$CCR7$^-$ T cells contained intracellular perforin and produced high levels of various effector cytokines [94]. Based on these functional differences, tissue-homing *effector memory T cells*, capable of immediate effector function, were separated from the lymph-node homing *central memory T cells*, which were available in secondary lymphoid organs to generate a second wave of T-cell effectors. When restimulated in vitro, CD62L$^{hi}$CCR7$^+$ memory T cells became CD62L$^{low}$CCR7$^-$, suggesting that central memory cells can give rise to effector memory T cells [94]. Several recent reports have confirmed the presence of distinct effector versus resting memory T-cell subsets and shown that effector memory T cells extravasate selectively in the nonlymphoid compartment [109–111]. However, a recent report evaluating the subsets of memory CD8$^+$ T cells in mice questions

the segregation of effector versus resting memory CD8$^+$ T cells based on CCR7
expression, as both CCR7$^+$ and CCR7$^-$ subsets of CD8$^+$ T cells were equally
cytolytic and produced similar levels of IFN-$\gamma$ [112].

## 2.12   T-CELL MEMORY AND PROTECTIVE EFFICACY

An important step in vaccination relates to devising methods that ensure
optimal generation and maintenance of memory T cells. As effector memory
cells possess rapid effector function (cytotoxicity and cytokine production) and
the ability to extravasate rapidly to peripheral sites [94,110]. It was previously
believed that vaccinations ought to induce mainly effector memory cells [113–
115]. However, effector memory phase during most infections and vaccinations
appears to be very short-lived, which makes sense biologically, since the persistence of such effectors can result in immunopathology, particularly if there is
cross-reactivity with self-antigens. Interestingly, it was recently reported that
central phenotype memory and effector memory subsets of CD8$^+$ T cells in
mice do not differ in immediate effector function [112]. The strict segregation
of CD8$^+$ T cells into either central or effector cells can be misleading considering that during CD8$^+$ T-cell differentiation a continuum of cells from central to
effector phenotypes may be generated [96]. Not all cells within the effector or
central memory subset exhibit a uniform cell size or expression of various
molecules. An important factor in evaluating the protective efficacy of T cells
lies in the diseases targeted and the models used. During lymphocytic choriomeningitis virus (LCMV) infection of mice, it was recently reported that the
elevated susceptibility of effector memory CD8$^+$ T cells to apoptosis offsets
their usefulness, and it is indeed the central memory but not effector memory
CD8$^+$ T cells that mediate protective efficacy, due to their increased propensity
to proliferate and to convert into effector memory cells upon restimulation
[116]. This interpretation does not fit well with a previous report where high
numbers of central memory T cells were induced in mice by heat-killed *L.
monocytogenes*, but such memory cells were not protective [92]. In an influenza
virus infection model it was reported that only effector memory CD8$^+$ T cells
entered the lungs with rapid kinetics and in high cell numbers, which resulted
in stronger protection against a lethal viral challenge [117]. On the other hand,
central memory CD8$^+$ T cells, which mediated a slightly reduced protection,
were first seen to increase in cell numbers in the draining lymph nodes, where
they converted into effector memory cells before entering the lungs. Similarly,
in a murine respiratory syncitial viral model, CD8$^+$ T cells that mediate high
levels of cytolytic activity do not persist, resulting in a decrease in resistance to
reinfection to very low levels within two months following immunization [118].
Regardless of the controversy regarding the differential effector function and
protective efficacy of central memory versus effector memory CD8$^+$ T cells, it
is conceivable that in diseases where the pathogen disseminates rapidly, central
memory T cells may not get sufficient time to be effective. The extravasation,

expansion, and further differentiation of central memory T cells into highly functional effectors takes time, which could potentially be too late against some pathogens.

Another important factor relates to the degree of reduction in the pathogen burden that is sufficient for inducing host survival. In some situations a 10- to 100-fold reduction in bacterial load (e.g., *L. monocytogenes*) by specific memory T cells is sufficient to induce host survival, as the reduced bacterial burden can be cleared further by other immune mechanisms. However, against another pathogen, HIV, a 10- to 100-fold reduction in HIV burden does not translate to any protection. Similarly, a 10- to 100-fold reduction in tumor load by specific memory T cells is insufficient for host survival and translates to only a modest delay in death. The differential protection between central versus effector memory T cells may not be apparent in the conventional models that involve pathogen challenge at systemic sites, as such infections are often associated with danger signals [12] that induce strong inflammation sufficient to reactivate central memory T cells rapidly. This is more relevant when one considers protection against tumors since tumor cells, as self, do not induce the activation or danger signals that a pathogen will readily induce; hence, the maintenance of antigen-specific T cells in a signal 2–independent status seems essential for effective tumor control [69].

Although it may be difficult to determine accurately the immediate protective capability of central versus effector memory CD8+ T cells, it is reasonable to suggest that vaccines should induce significant proportions of both cell types. Manipulative strategies such as multiple vaccinations, prime–boost with various adjuvants, and controlled antigen release may induce balanced proportions of central versus effector memory T-cell pools, offering the host the advantages of both subsets and minimizing the limitations.

## REFERENCES

1. Kaech, S.M., Wherry, E.J. & Ahmed, R. Effector and memory T-cell differentiation: implications for vaccine development. *Nat Rev Immunol* 2002, **2**(4), 251–262.
2. Zinkernagel, R.M. Immunology taught by viruses. *Science* 1996, **271**(5246), 173–178.
3. Kaufmann, S.H. Is the development of a new tuberculosis vaccine possible? *Nat Med* 2000, **6**(9), 955–960.
4. McMichael, A. & Hanke, T. The quest for an AIDS vaccine: Is the CD8+ T-cell approach feasible? *Nat Rev Immunol* 2002, **2**(4), 283–291.
5. Flynn, J.L. & Chan, J. Immune evasion by *Mycobacterium* tuberculosis: living with the enemy. *Curr Opin Immunol* 2003, **15**(4), 450–455.
6. Cossart, P. & Sansonetti, P.J. Bacterial invasion: the paradigms of enteroinvasive pathogens. *Science* 2004, **304**(5668), 242–248.

7. Kaech, S.M. & Ahmed, R. Memory CD8+ T cell differentiation: initial antigen encounter triggers a developmental program in naive cells. *Nat Immunol* 2001, **2**(5), 415–422.

8. Banchereau, J. & Steinman, R.M. Dendritic cells and the control of immunity. *Nature* 1998, **392**(6673), 245–252.

9. Hemmi, H., Takeuchi, O., Kawai, T., Kaisho, T., Sato, S., Sanjo, H. et al. A Toll-like receptor recognizes bacterial DNA. *Nature* 2000, **408**(6813), 740–745.

10. O'Neill, L.A. Toll-like receptor signal transduction and the tailoring of innate immunity: A role for Mal? *Trends Immunol* 2002, **23**(6), 296–300.

11. Ahmed, R. & Gray, D. Immunological memory and protective immunity: understanding their relation. *Science* 1996, **272**(5258), 54–60.

12. Matzinger, P. An innate sense of danger. *Semin Immunol* 1998, **10**(5), 399–415.

13. Mosmann, T.R. & Coffman, R.L. TH1 and TH2 cells: different patterns of lymphokine secretion lead to different functional properties. *Annu Rev Immunol* 1989, **7**, 145–173.

14. Schuler, T., Qin, Z., Ibe, S., Noben-Trauth, N. & Blankenstein, T. T helper cell type 1–associated and cytotoxic T lymphocyte–mediated tumor immunity is impaired in interleukin 4–deficient mice. *J Exp Med* 1999, **189**(5), 803–810.

15. White, D.W. & Harty, J.T. Perforin-deficient CD8+ T cells provide immunity to *Listeria monocytogenes* by a mechanism that is independent of CD95 and IFN-gamma but requires TNF-alpha. *J Immunol* 1998, **160**(2), 898–905.

16. Bohm, W., Thoma, S., Leithauser, F., Moller, P., Schirmbeck, R. & Reimann, J. T cell-mediated, IFN-gamma-facilitated rejection of murine B16 melanomas. *J Immunol* 1998, **161**(2), 897–908.

17. Smyth, M.J., Thia, K.Y., Street, S.E., MacGregor, D., Godfrey, D.I. & Trapani, J.A. Perforin-mediated cytotoxicity is critical for surveillance of spontaneous lymphoma. *J Exp Med* 2000, **192**(5), 755–760.

18. Stenger, S., Hanson, D.A., Teitelbaum, R., Dewan, P., Niazi, K.R., Froelich, C.J. et al. An antimicrobial activity of cytolytic T cells mediated by granulysin. *Science* 1998, **282**(5386), 121–125.

19. Lalvani, A. & Hill, A.V. Cytotoxic T-lymphocytes against malaria and tuberculosis: from natural immunity to vaccine design. *Clin Sci (Lond)* 1998, **95**(5), 531–538.

20. O'Hagan, D.T. & Valiante, N.M. Recent advances in the discovery and delivery of vaccine adjuvants. *Nat Rev Drug Discov* 2003, **2**(9), 727–735.

21. Rock, K.L. A new foreign policy: MHC class I molecules monitor the outside world. *Immunol Today* 1996, **17**(3), 131–137.

22. Pamer, E. & Cresswell, P. Mechanisms of MHC class I–restricted antigen processing. *Annu Rev Immunol* 1998, **16**, 323–358.

23. Geijtenbeek, T.B., Van Vliet, S.J., Koppel, E.A., Sanchez-Hernandez, M., Vandenbroucke-Grauls, C.M., Appelmelk, B. et al. Mycobacteria target DC-SIGN to suppress dendritic cell function. *J Exp Med* 2003, **197**(1), 7–17.

24. Albert, M.L., Sauter, B. & Bhardwaj, N. Dendritic cells acquire antigen from apoptotic cells and induce class I–restricted CTLs. *Nature* 1998, **392**(6671), 86–89.

25. Akira, S., Takeda, K. & Kaisho, T. Toll-like receptors: critical proteins linking innate and acquired immunity. *Nat Immunol* 2001, **2**(8), 675–680.

26. Takeuchi, O., Hoshino, K., Kawai, T., Sanjo, H., Takada, H., Ogawa, T. et al. Differential roles of TLR2 and TLR4 in recognition of gram-negative and gram-positive bacterial cell wall components. *Immunity* 1999, **11**(4), 443–451.

27. Yamamoto, M., Sato, S., Hemmi, H., Sanjo, H., Uematsu, S., Kaisho, T. et al. Essential role for TIRAP in activation of the signalling cascade shared by TLR2 and TLR4. *Nature* 2002, **420**(6913), 324–329.

28. Alexopoulou, L., Thomas, V., Schnare, M., Lobet, Y., Anguita, J., Schoen, R.T. et al. Hyporesponsiveness to vaccination with *Borrelia burgdorferi* OspA in humans and in TLR1- and TLR2-deficient mice. *Nat Med* 2002, **8**(8), 878–884.

29. Horng, T., Barton, G.M., Flavell, R.A. & Medzhitov, R. The adaptor molecule TIRAP provides signalling specificity for Toll-like receptors. *Nature* 2002, **420**(6913), 329–333.

30. Adachi, O., Kawai, T., Takeda, K., Matsumoto, M., Tsutsui, H., Sakagami, M. et al. Targeted disruption of the MyD88 gene results in loss of IL-1 – and IL-18– mediated function. *Immunity* 1998, **9**(1), 143–150.

31. Kaisho, T. & Akira, S. Toll-like receptors as adjuvant receptors. *Biochim Biophys Acta* 2002, **1589**(1), 1–13.

32. Shi, Y., Zheng, W. & Rock, K.L. Cell injury releases endogenous adjuvants that stimulate cytotoxic T cell responses. *Proc Natl Acad Sci USA* 2000, **97**(26), 14590–14595.

33. Zinkernagel, R.M. & Doherty, P.C. The discovery of MHC restriction. *Immunol Today* 1997, **18**(1), 14–17.

34. Watts, C. & Powis, S. Pathways of antigen processing and presentation. *Rev Immunogenet* 1999, **1**(1), 60–74.

35. Moore, M.W., Carbone, F.R. & Bevan, M.J. Introduction of soluble protein into the class I pathway of antigen processing and presentation. *Cell* 1988, **54**(6), 777–785.

36. Yewdell, J.W., Bennink, J.R. & Hosaka, Y. Cells process exogenous proteins for recognition by cytotoxic T lymphocytes. *Science* 1988, **239**(4840), 637–640.

37. Bachmann, M.F., Oxenius, A., Pircher, H., Hengartner, H., Ashton-Richardt, P.A., Tonegawa, S. et al. TAP1-independent loading of class I molecules by exogenous viral proteins. *Eur J Immunol* 1995, **25**(6), 1739–1743.

38. Kovacsovics, B.M., Clark, K., Benacerraf, B. & Rock, K.L. Efficient major histocompatibility complex class I presentation of exogenous antigen upon phagocytosis by macrophages. *Proc Natl Acad Sci USA* 1993, **90**(11), 4942–4946.

39. Krishnan, L., Dicaire, C.J., Patel, G.B. & Sprott, G.D. Archaeosome vaccine adjuvants induce strong humoral, cell-mediated, and memory responses: comparison to conventional liposomes and alum. *Infect Immun* 2000, **68**(1), 54–63.

40. Pfeifer, J.D., Wick, M.J., Roberts, R.L., Findlay, K., Normark, S.J. & Harding, C.V. Phagocytic processing of bacterial antigens for class I MHC presentation to T cells. *Nature* 1993, **361**(6410), 359–362.

41. Schaible, U.E., Winau, F., Sieling, P.A., Fischer, K., Collins, H.L., Hagens, K. et al. Apoptosis facilitates antigen presentation to T lymphocytes through MHC-I and CD1 in tuberculosis. *Nat Med* 2003, **9**(8), 1039–1046.

42. Houde, M., Bertholet, S., Gagnon, E., Brunet, S., Goyette, G., Laplante, A. et al. Phagosomes are competent organelles for antigen cross-presentation. *Nature* 2003, **425**(6956), 402–406.

43. Guermonprez, P., Valladeau, J., Zitvogel, L., Thery, C. & Amigorena, S. Antigen presentation and T cell stimulation by dendritic cells. *Annu Rev Immunol* 2002, **20**, 621–667.

44. Bryant, P.W., Lennon-Dumenil, A.M., Fiebiger, E., Lagaudriere-Gesbert, C. & Ploegh, H.L. Proteolysis and antigen presentation by MHC class II molecules. *Adv Immunol* 2002, **80**, 71–114.

45. Medina, E. & Guzman, C.A. Use of live bacterial vaccine vectors for antigen delivery: potential and limitations. *Vaccine* 2001, **19**(13–14), 1573–1580.

46. Plotkin, S.A. Cytomegalovirus vaccine. *Am Heart J* 1999, **138**(5 Pt 2), S484–487.

47. Moron, G., Dadaglio, G. & Leclerc, C. New tools for antigen delivery to the MHC class I pathway. *Trends Immunol* 2004, **25**(2), 92–97.

48. Collins, D.S., Findlay, K. & Harding, C.V. Processing of exogenous liposome-encapsulated antigens in vivo generates class I MHC–restricted T cell responses. *J Immunol* 1992, **148**(11), 3336–3341.

49. Moron, V.G., Rueda, P., Sedlik, C. & Leclerc, C. In vivo, dendritic cells can cross-present virus-like particles using an endosome-to-cytosol pathway. *J Immunol* 2003, **171**(5), 2242–2250.

50. Sprott, G.D., Patel, G.B. & Krishnan, L. Archaeobacterial ether lipid liposomes as vaccine adjuvants. *Methods Enzymol* 2003, **373**, 155–172.

51. Krishnan, L., Sad, S., Patel, G.B. & Sprott, G.D. Archaeosomes induce long-term CD8+ cytotoxic T cell response to entrapped soluble protein by the exogenous cytosolic pathway, in the absence of CD4+ T cell help. *J Immunol* 2000, **165**(9), 5177–5185.

52. Inaba, K., Inaba, M., Romani, N., Aya, H., Deguchi, M., Ikehara, S. et al. Generation of large numbers of dendritic cells from mouse bone marrow cultures supplemented with granulocyte/macrophage colony-stimulating factor. *J Exp Med* 1992, **176**(6), 1693–1702.

53. Croft, M., Duncan, D.D. & Swain, S.L. Response of naive antigen-specific CD4+ T cells in vitro: characteristics and antigen-presenting cell requirements. *J Exp Med* 1992, **176**(5), 1431–1437.

54. Henri, S., Vremec, D., Kamath, A., Waithman, J., Williams, S., Benoist, C. et al. The dendritic cell populations of mouse lymph nodes. *J Immunol* 2001, **167**(2), 741–748.

55. den Haan, J.M., Lehar, S.M. & Bevan, M.J. CD8(+) but not CD8(−) dendritic cells cross-prime cytotoxic T cells in vivo. *J Exp Med* 2000, **192**(12), 1685–1696.

56. Maldonado-Lopez, R., De Smedt, T., Michel, P., Godfroid, J., Pajak, B., Heirman, C. et al. CD8alpha+ and CD8alpha- subclasses of dendritic cells direct the development of distinct T helper cells in vivo. *J Exp Med* 1999, **189**(3), 587–592.

57. Dhodapkar, M.V., Steinman, R.M., Krasovsky, J., Munz, C. & Bhardwaj, N. Antigen-specific inhibition of effector T cell function in humans after injection of immature dendritic cells. *J Exp Med* 2001, **193**(2), 233–238.

58. Hawiger, D., Inaba, K., Dorsett, Y., Guo, M., Mahnke, K., Rivera, M. et al. Dendritic cells induce peripheral T cell unresponsiveness under steady state conditions in vivo. *J Exp Med* 2001, **194**(6), 769–779.

59. Aderem, A. & Ulevitch, R.J. Toll-like receptors in the induction of the innate immune response. *Nature* 2000, **406**(6797), 782–787.

60. Maxwell, J.R., Campbell, J.D., Kim, C.H. & Vella, A.T. CD40 activation boosts T cell immunity in vivo by enhancing T cell clonal expansion and delaying peripheral T cell deletion. *J Immunol* 1999, **162**(4), 2024–2034.

61. Takahashi, C., Mittler, R.S. & Vella, A.T. Cutting edge: 4-1BB is a bona fide CD8 T cell survival signal. *J Immunol* 1999, **162**(9), 5037–5040.

62. Borrow, P., Tishon, A., Lee, S., Xu, J., Grewal, I.S., Oldstone, M.B. et al. CD40L-deficient mice show deficits in antiviral immunity and have an impaired memory CD8+ CTL response. *J Exp Med* 1996, **183**(5), 2129–2142.

63. Shahinian, A., Pfeffer, K., Lee, K.P., Kundig, T.M., Kishihara, K., Wakeham, A. et al. Differential T cell costimulatory requirements in CD28-deficient mice. *Science* 1993, **261**(5121), 609–612.

64. Kopf, M., Ruedl, C., Schmitz, N., Gallimore, A., Lefrang, K., Ecabert, B. et al. OX40-deficient mice are defective in Th cell proliferation but are competent in generating B cell and CTL responses after virus infection. *Immunity* 1999, **11**(6), 699–708.

65. Hou, S., Hyland, L., Ryan, K.W., Portner, A., Doherty, P.C., Foulds, K.E. et al. Virus-specific CD8+ T-cell memory determined by clonal burst size. *Nature* 1994, **369**(6482), 652–654.

66. Foulds, K.E., Zenewicz, L.A., Shedlock, D.J., Jiang, J., Troy, A.E. & Shen, H. Cutting edge: CD4 and CD8 T cells are intrinsically different in their proliferative responses. *J Immunol* 2002, **168**(4), 1528–1532.

67. Szabo, S.J., Sullivan, B.M., Stemmann, C., Satoskar, A.R., Sleckman, B.P., Glimcher, L.H. et al. Distinct effects of T-bet in Th1 lineage commitment and IFN-gamma production in CD4 and CD8 T cells. *Science* 2002, **295**(5553), 338–342.

68. Ochsenbein, A.F., Karrer, U., Klenerman, P., Althage, A., Ciurea, A., Shen, H. et al. A comparison of T cell memory against the same antigen induced by virus versus intracellular bacteria. *Proc Natl Acad Sci USA* 1999, **96**(16), 9293–9298.

69. Dudani, R., Chapdelaine, Y., Faassen H.H., Smith, D.K., Shen, H., Krishnan, L. et al. Multiple mechanisms compensate to enhance tumor-protective CD8(+) T cell response in the long-term despite poor CD8(+) T cell priming initially: comparison between an acute versus a chronic intracellular bacterium expressing a model antigen. *J Immunol* 2002, **168**(11), 5737–5745.

70. Wherry, E.J., Puorro, K.A., Porgador, A. & Eisenlohr, L.C. The induction of virus-specific CTL as a function of increasing epitope expression: responses rise steadily until excessively high levels of epitope are attained. *J Immunol* 1999, **163**(7), 3735–3745.

71. DiPaolo, R.J. & Unanue, E.R. Cutting edge: chemical dominance does not relate to immunodominance: studies of the CD4+ T cell response to a model antigen. *J Immunol* 2002, **169**(1), 1–4.

72. Kerksiek, K.M. & Pamer, E.G. T cell responses to bacterial infection. *Curr Opin Immunol* 1999, **11**(4), 400–405.

73. Mercado, R., Vijh, S., Allen, S.E., Kerksiek, K., Pilip, I.M. & Pamer, E.G. Early programming of T cell populations responding to bacterial infection. *J Immunol* 2000, **165**(12), 6833–6839.

74. van Stipdonk, M.J., Lemmens, E.E. & Schoenberger, S.P. Naive CTLs require a single brief period of antigenic stimulation for clonal expansion and differentiation. *Nat Immunol* 2001, **2**(5), 423–429.

75. Busch, D.H., Kerksiek, K.M. & Pamer, E.G. Differing roles of inflammation and antigen in T cell proliferation and memory generation. *J Immunol* 2000, **164**(8), 4063–4070.

76. Badovinac, V.P., Porter, B.B. & Harty, J.T. Programmed contraction of CD8(+) T cells after infection. *Nat Immunol* 2002, **3**(7), 619–626.

77. Gray, D. & Matzinger, P. T cell memory is short-lived in the absence of antigen. *J Exp Med* 1991, **174**(5), 969–974.

78. Zajac, A.J., Blattman, J.N., Murali-Krishna, K., Sourdive, D.J., Suresh, M., Altman, J.D. et al. Viral immune evasion due to persistence of activated T cells without effector function. *J Exp Med* 1998, **188**(12), 2205–2213.

79. Moskophidis, D., Lechner, F., Pircher, H. & Zinkernagel, R.M. Virus persistence in acutely infected immunocompetent mice by exhaustion of antiviral cytotoxic effector T cells. *Nature* 1993, **362**(6422), 758–761.

80. Mohri, H., Perelson, A.S., Tung, K., Ribeiro, R.M., Ramratnam, B., Markowitz, M. et al. Increased turnover of T lymphocytes in HIV-1 infection and its reduction by antiretroviral therapy. *J Exp Med* 2001, **194**(9), 1277–1287.

81. Oehen, S. & Brduscha-Riem, K. Differentiation of naive CTL to effector and memory CTL: correlation of effector function with phenotype and cell division. *J Immunol* 1998, **161**(10), 5338–5346.

82. Bajenoff, M., Wurtz, O. & Guerder, S. Repeated antigen exposure is necessary for the differentiation, but not the initial proliferation, of naive CD4(+) T cells. *J Immunol* 2002, **168**(4), 1723–1729.

83. Bird, J.J., Brown, D.R., Mullen, A.C., Moskowitz, N.H., Mahowald, M.A., Sider, J.R. et al. Helper T cell differentiation is controlled by the cell cycle. *Immunity* 1998, **9**(2), 229–237.

84. Ben-Sasson, S.Z., Gerstel, R., Hu-Li, J. & Paul, W.E. Cell division is not a "clock" measuring acquisition of competence to produce IFN-gamma or IL-4. *J Immunol* 2001, **166**(1), 112–120.

85. Dudani, R., Chapdelaine, Y., van Faassen, H., Smith, D.K., Shen, H., Krishnan, L. et al. Preexisting inflammation due to *Mycobacterium bovis* BCG infection differentially modulates T-cell priming against a replicating or nonreplicating immunogen. *Infect Immun* 2002, **70**(4), 1957–1964.

86. Krishnan, L., Sad, S., Patel, G.B. & Sprott, G.D. The potent adjuvant activity of archaeosomes correlates to the recruitment and activation of macrophages and dendritic cells in vivo. *J Immunol* 2001, **166**(3), 1885–1893.

87. Kundig, T.M., Kalberer, C.P., Hengartner, H. & Zinkernagel, R.M. Vaccination with two different vaccinia recombinant viruses: long-term inhibition of secondary vaccination. *Vaccine* 1993, **11**(11), 1154–1158.

88. Sad, S., Marcotte, R. & Mosmann, T.R. Cytokine-induced differentiation of precursor mouse CD8+ T cells into cytotoxic CD8+ T cells secreting Th1 or Th2 cytokines. *Immunity* 1995, **2**(3), 271–279.

89. Mosmann, T.R. & Sad, S. The expanding universe of T-cell subsets: Th1, Th2 and more. *Immunol Today* 1996, **17**(3), 138–146.

90. Carter, L.L. & Dutton, R.W. Relative perforin- and Fas-mediated lysis in T1 and T2 CD8 effector populations. *J Immunol* 1995, **155**(3), 1028–1031.

91. Dobrzanski, M.J., Reome, J.B. & Dutton, R.W. Therapeutic effects of tumor-reactive type 1 and type 2 CD8+ T cell subpopulations in established pulmonary metastases. *J Immunol* 1999, **162**(11), 6671–6680.

92. Lauvau, G., Vijh, S., Kong, P., Horng, T., Kerksiek, K., Serbina, N. et al. Priming of memory but not effector CD8 T cells by a killed bacterial vaccine. *Science* 2001, **294**(5547), 1735–1739.

93. Manjunath, N., Shankar, P., Wan, J., Weninger, W., Crowley, M.A., Hieshima, K. et al. Effector differentiation is not prerequisite for generation of memory cytotoxic T lymphocytes. *J Clin Invest* 2001, **108**(6), 871–878.

94. Sallusto, F., Lenig, D., Forster, R., Lipp, M. & Lanzavecchia, A. Two subsets of memory T lymphocytes with distinct homing potentials and effector functions. *Nature* 1999, **401**(6754), 708–712.

95. Champagne, P., Ogg, G.S., King, A.S., Knabenhans, C., Ellefsen, K., Nobile, M. et al. Skewed maturation of memory HIV-specific CD8 T lymphocytes. *Nature* 2001, **410**(6824), 106–111.

96. Roman, E., Miller, E., Harmsen, A., Wiley, J., Von Andrian, U.H., Huston, G. et al. CD4 effector T cell subsets in the response to influenza: heterogeneity, migration, and function. *J Exp Med* 2002, **196**(7), 957–968.

97. Appay, V., Dunbar, P.R., Callan, M., Klenerman, P., Gillespie, G.M., Papagno, L. et al. Memory CD8+ T cells vary in differentiation phenotype in different persistent virus infections. *Nat Med* 2002, **8**(4), 379–385.

98. Dutton, R.W., Bradley, L.M. & Swain, S.L. T cell memory. *Annu Rev Immunol* 1998, **16**, 201–223.

99. Budd, R.C., Cerottini, J.C., Horvath, C., Bron, C., Pedrazzini, T., Howe, R.C. et al. Distinction of virgin and memory T lymphocytes: stable acquisition of the Pgp-1 glycoprotein concomitant with antigenic stimulation. *J Immunol* 1987, **138**(10), 3120–3129.

100. Pihlgren, M., Arpin, C., Walzer, T., Tomkowiak, M., Thomas, A., Marvel, J. et al. Memory CD44(int) CD8 T cells show increased proliferative responses and IFN-gamma production following antigenic challenge in vitro. *Int Immunol* 1999, **11**(5), 699–706.

101. Bradley, L.M., Duncan, D.D., Tonkonogy, S. & Swain, S.L. Characterization of antigen-specific CD4+ effector T cells in vivo: immunization results in a transient population of MEL-14-, CD45RB-helper cells that secretes interleukin 2 (IL-2), IL-3, IL-4, and interferon gamma. *J Exp Med* 1991, **174**(3), 547–559.

102. Usherwood, E.J., Hogan, R.J., Crowther, G., Surman, S.L., Hogg, T.L., Altman, J.D. et al. Functionally heterogeneous CD8(+) T-cell memory is induced by Sendai virus infection of mice. *J Virol* 1999, **73**(9), 7278–7286.

103. Walunas, T.L., Bruce, D.S., Dustin, L., Loh, D.Y. & Bluestone, J.A. Ly-6C is a marker of memory CD8+ T cells. *J Immunol* 1995, **155**(4), 1873–1883.

104. Walzer, T., Arpin, C., Beloeil, L. & Marvel, J. Differential in vivo persistence of two subsets of memory phenotype CD8 T cells defined by CD44 and CD122 expression levels. *J Immunol* 2002, **168**(6), 2704–2711.

105. Arbones, M.L., Ord, D.C., Ley, K., Ratech, H., Maynard-Curry, C., Otten, G. et al. Lymphocyte homing and leukocyte rolling and migration are impaired in L-selectin-deficient mice. *Immunity* 1994, **1**(4), 247–260.

106. Campbell, J.J., Bowman, E.P., Murphy, K., Youngman, K.R., Siani, M.A., Thompson, D.A. et al. 6-C-kine (SLC), a lymphocyte adhesion-triggering chemokine expressed by high endothelium, is an agonist for the MIP-3beta receptor CCR7. *J Cell Biol* 1998, **141**(4), 1053–1059.

107. Weninger, W., Crowley, M.A., Manjunath, N. & von Andrian, U.H. Migratory properties of naive, effector, and memory CD8(+) T cells. *J Exp Med* 2001, **194**(7), 953–966.

108. Iezzi, G., Scheidegger, D. & Lanzavecchia, A. Migration and function of antigen-primed nonpolarized T lymphocytes in vivo. *J Exp Med* 2001, **193**(8), 987–993.

109. Marshall, D.R., Turner, S.J., Belz, G.T., Wingo, S., Andreansky, S., Sangster, M.Y. et al. Measuring the diaspora for virus-specific CD8+ T cells. *Proc Natl Acad Sci USA* 2001, **98**(11), 6313–6318.

110. Masopust, D., Vezys, V., Marzo, A.L. & Lefrancois, L. Preferential localization of effector memory cells in nonlymphoid tissue. *Science* 2001, **291**(5512), 2413–2417.

111. Reinhardt, R.L., Khoruts, A., Merica, R., Zell, T. & Jenkins, M.K. Visualizing the generation of memory CD4 T cells in the whole body. *Nature* 2001, **410**(6824), 101–105.

112. Unsoeld, H., Krautwald, S., Voehringer, D., Kunzendorf, U. & Pircher, H. Cutting edge: CCR7+ and CCR7– memory T cells do not differ in immediate effector cell function. *J Immunol* 2002, **169**(2), 638–641.

113. Kundig, T.M., Bachmann, M.F., Oehen, S., Hoffmann, U.W., Simard, J.J., Kalberer, C.P. et al. On the role of antigen in maintaining cytotoxic T-cell memory. *Proc Natl Acad Sci USA* 1996, **93**(18), 9716–9723.

114. Oehen, S., Waldner, H., Kundig, T.M., Hengartner, H. & Zinkernagel, R.M. Anti-virally protective cytotoxic T cell memory to lymphocytic choriomeningitis virus is governed by persisting antigen. *J Exp Med* 1992, **176**(5), 1273–1281.

115. Cerwenka, A., Morgan, T.M. & Dutton, R.W. Naive, effector, and memory CD8 T cells in protection against pulmonary influenza virus infection: homing properties rather than initial frequencies are crucial. *J Immunol* 1999, **163**(10), 5535–5543.

116. Wherry, E.J., Teichgraber, V., Becker, T.C., Masopust, D., Kaech, S.M., Antia, R. et al. Lineage relationship and protective immunity of memory CD8 T cell subsets. *Nat Immunol* 2003, **4**(3), 225–234.

117. Cerwenka, A., Morgan, T.M., Harmsen, A.G. & Dutton, R.W. Migration kinetics and final destination of type 1 and type 2 CD8 effector cells predict protection against pulmonary virus infection. *J Exp Med* 1999, **189**(2), 423–434.

118. Kulkarni, A.B., Connors, M., Firestone, C.Y., Morse, H.C., 3rd & Murphy, B.R. The cytolytic activity of pulmonary CD8+ lymphocytes, induced by infection with a vaccinia virus recombinant expressing the M2 protein of respiratory syncytial virus (RSV), correlates with resistance to RSV infection in mice. *J Virol* 1993, **67**(2), 1044–1049.

# 3

# MECHANISMS OF ADJUVANT ACTION

Nils Lycke

## 3.1 INTRODUCTION

Over the last decade our knowledge of various substances that exert adjuvant effects has grown immensely. By necessity this means that only a selection of adjuvants, typifying certain mechanisms, will be discussed in any greater detail here. However, to get an overview and a better understanding of the mechanisms involved and the types of strategies used to elicit these effects, a few revealing examples are presented. This is also usually more informative than reporting from a long, comprehensive list of substances with adjuvant activity.

Because of an almost absolute requirement for adjuvants in most commercial vaccines, there is growing interest in adjuvant development. Despite this, we must conclude that little information is available as to what adjuvants, in fact, do in vivo. Most of our mechanistic knowledge stems from in vitro studies using freshly isolated cells, cell lines, or ex vivo–generated cells, such as bone marrow–derived dendritic cells (DCS) [1]. Few reports have successfully described the effects of adjuvants on in vivo events, and scarcely any documentation on more complex interactions leading to priming of T cells or T/B-cell interactions exists. However, with the introduction of biphoton microscopy and other such techniques, allowing for real-time assessments, cellular movement and function due to adjuvant administration will soon be past of a rapidly growing literature [2,3]. This is, indeed, a much warranted development for future vaccine design and formulation.

*Vaccine Adjuvants and Delivery Systems*, Edited by Manmohan Singh
Copyright © 2007 John Wiley & Sons, Inc.

The word *adjuvare* is Latin and means "to help," which is exactly what adjuvants do. Most vaccines based on nonliving material lack the ability to stimulate a significant immune response. Therefore, substances or formulations that augment immunogenicity of a particular antigen are obligatory for any vaccine based on nonliving vectors. Although not needed for most attenuated live vaccines, adjuvants may still be required to improve efficacy or modulate immune responses. A growing literature presents microbial vectors that have been engineered to express cytokines or immunomodulators together with the vaccine antigens [4]. Historically, adjuvants are mostly of microbial origin and induce various degrees of inflammation. In fact, the adjuvant efficacy could in many ways be linked directly to the level of local inflammation. Thus, in this regard adjuvants can also be responsible for tissue destruction, pain, and distress, which are key components to consider in any vaccine development. Hence, the selection of adjuvant to use is often as critical as which antigen or combination of antigens to include in a vaccine. All types of immune responses can be augmented; antibody formation as well as cell-mediated immunity, including cytotoxic T-cell activity. Adjuvants may use many different mechanisms to exert an augmenting effect on immune responses: from establishing an antigen depot in the tissue, to direct or indirect immunomodulation and antigen-targeting effects. For example, nonliving adjuvants can be formulations of lipid or gel (alum) to create a depot effect of the vaccine following injection. Also, nonliving adjuvants can be both delivery systems, such as liposomes and poly(lactide-co-glycolide) (PLG) microspheres, or modulators, such as muramyl dipeptide (MDP) and monophosphoryl lipid A (MPL) [5]. Here we focus mostly on the immunomodulating and targeting effects of adjuvants, as antigen delivery and depot effects are discussed in more detail in other chapters.

## 3.2   INDUCTION OF IMMUNITY

Today, we have very precise information about necessary signals for the stimulation of adaptive immune responses: This involves the stepwise stimulation of naive T cells through recognition by the T-cell receptor of MHC (major histocompatibility complex) class I or II together with peptide, to the more complicated second signals that are provided by cytokines or co-stimulatory molecules on the membranes of antigen-presenting cells (APCs) [6]. The co-stimulatory molecules CD40, CD80, CD86, and OX40L especially are thought to contribute to an adjuvant function. The linking of innate responses to the induction of adaptive immunity is the key to successful adjuvant construction. It is the choice of adjuvant, which can affect differently innate immunity, that directly influences and shapes the adaptive immune response. The activation of naive CD4$^+$ T cells can lead to the development of either Th1 or Th2 cells, characterized by distinct cytokine production. Whereas Th1 cells are critical for cell-mediated immunity and produce IFN-$\gamma$, TNF$\alpha$, and IL-2, the Th2 phenotype secretes IL-4, IL-5, IL-6, IL-10, IL-13, and IL-15, which positively influences and forms the magnitude, quality, and isotype of the antibody response [7]. The decision to develop Th1 or Th2 dominance is made early in the course of an immune response. This stage also determines the usefulness of the adaptive response in relation to the triggering event. In accordance, helminth infections or extracellular bacterial infections are best eliminated by antibody production and Th2 skewing, whereas infections with intracellular parasites, bacteria, or viruses require cell-mediated immunity and Th1 domination [8]. Furthermore, in the course of an immune response, regulatory elements are usually induced. One such important factor is the differentiation of regulatory T cells (Tregs), which have a dampening effect on the effector phase of an immune response. For example, Tregs could reduce IFN-γ production, thereby reducing tissue destruction driven by Th1 cells [7]. These regulatory T cells express the transcription factor, Foxp-3-encoding gene, and produce transforming growth factor β (TGFβ) or IL-10, and contrary to natural thymus-derived Tregs, they are CD25-negative [9]. Modulation of the function of Tregs may therefore prove to be a promising avenue in the search for future vaccine adjuvants. But, at present, few adjuvant strategies have included modulation of Tregs as a means to achieve protective immunity [10,11]. Taken together, the identification of ways to modulate Th1, Th2, or Tregs will most certainly be key to the successful development of novel vaccine adjuvants.

## 3.3  OVERVIEW OF ADJUVANT MECHANISMS

In principle, adjuvants may exploit three types of modulating effects on the innate immune response which will affect the adaptive immune response and promote improved immunogenicity that can eventually convey protection against infection (Figure 3.1). What were discussed previously as danger signals are recognized today as defined molecules with distinct chemical and genetic properties [12]. Behind the activation of innate immunity by danger signals, we find a series of reactions strictly controlled by specific receptors, called *pattern-recognition receptors* (PRRs). These receptors bind microbial products, such as endotoxin or other microbial membrane products, or bacterial or viral DNA and RNA, respectively. One family of such receptors is that of the Toll-like receptors (TLRs), which bind lipopolysaccharide (LPS), flagellin, HSP60, CpG DNA, dsRNA, or peptidoglycans, all with unique and distinct receptors [13,14]. The family of TLRs is growing, and currently encompasses 10 members in humans and 11 in mice [15]. The nucleotide-binding oligomerization domain (NOD) proteins are another example of receptors that can sense microbial products [16,17]. However, some controversy remains as to whether NOD1 and NOD2 are to be considered to be true PRRs [18]. The TLRs are distinctly located in membranes or intracellular compartments, whereas NODs are intracellular PRRs. Both can be found in many types of cells, among them DCs, macrophages, and B cells, which are also the most important cells for

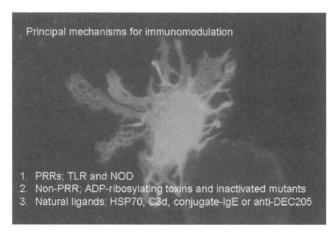

**Figure 3.1** Summary of the major pathways for immunomodulation and adjuvant function. A dendritic cell labeled with FITC-conjugated anti-CD11c antibodies. (*See insert for color representation of figure.*)

antigen presentation to naive T cells and stimulation of an adaptive immune response [19]. In the case of DCs, binding to PRRs will cause migration to regional lymph nodes and the maturation of the DCs into effective APCs, expressing the co-stimulatory molecules required for optimal T-cell priming [20]. Thus, most adjuvants are ligands for PRRs and binding results in signaling that eventuate in activation of the transcription factor NFκB. The NFκB translocates to the nucleus, where it may stimulate expression of an entire range of genes, many which drive inflammation [21]. However, we also know of adjuvants that do not appear to involve PRRs and subsequently should be expected to be less dependent on the NFκB pathway for their function.

To this second family of adjuvants, which probably affect innate immunity through receptors other than the PRRs, we associate the bacterial enterotoxins or derivatives thereof, typified by cholera toxin (CT) and *Escherichia coli* heat-labile (LT) toxin [22]. These molecules are known to be $AB_5$ complexes and carry an A1 subunit that is an ADP-ribosylating enzyme and five B subunits that bind distinct ganglioside receptors present on most nucleated mammalian cells. Because gangliosides reside in the cell membrane of all nucleated cells, the binding is very promiscuous and hence the enzyme can affect virtually all cells in the human body. As discussed later in the chapter, the holotoxins stimulate adenylate cyclase, leading to an increase in intracellular cAMP. This is what makes them highly toxic but is also a property that contributes to their excellent adjuvant function in vivo [23]. Furthermore, other bacterial toxins, such as *Bacillus anthracis* edema toxin, which also induces intracellular cAMP in target cells, has recently been found to act as an adjuvant in mice, strongly enhancing priming of specific CD4$^+$ T cells as well as antibody production after nasal administration [24]. Pertussis toxin (PT) is another ADP-ribosylating enzyme that induces cytosolic cAMP, but through a different G-protein (G$_i$) than CT and LT (G$_{s\alpha}$), and which is also known to act as an adjuvant [25–27]. Whereas PT is associated primarily with an augmenting effect on Th1 immunity, variable effects on the skewing to Th1 or Th2 differentiation has been reported for CTs, and, in particular, for LT [27–29].

A third group of immunomodulators use natural components of the immune system, such as complement factors, C3d, or antibodies, preferably IgE antibodies, or antibodies for targeting to innate cells, exemplified by anti-DEC205 antibody conjugates to target antigen to DCs. The latter strategy was shown to promote enhanced T-cell immunity by exploiting the endocytic DEC205 receptor on DCs in draining lymph nodes following subcutaneous administration [30]. Conjugation of antigen to immunoglobulins of certain isotypes may also enhance immune responses. Conjugates based on IgE antibodies especially have been found to have an adjuvant effect. The mechanism for this augmenting effect is dependent on CD23, the low-affinity receptor for IgE on B cells [31]. Moreover, bacterial products such as the 70-kDa microbial heat shock protein (mHSP70) uses a natural non-TLR pathway for activation of DCs and have a profound augmenting effect on the immune response. The mHSP70 interacts with the CD40 receptor on DCs and monocytes to produce pro-inflammatory cytokines and chemokines [32]. The mHSP70 induced maturation of DCs and hence replaced the natural ligand, CD40L, expressed by activated T cells. Of note, other heat shock proteins, such as HSP60, which acts through TLR2 and TLR4, also exert pro-inflammatory and immunoenhancing properties and may act preferentially on B cells [33]. Finally, the conjugation of antigen to complement factors such as C3d is a well-established strategy to achieve an augmenting effect on humoral responses. However, recent findings point to a complex dependence on complement receptors where C3d may enhance or inhibit antigen-specific humoral immune responses through both CD21-dependent and CD21-independent mechanisms, depending on the concentration and nature of the antigen–C3d complexes [34].

## 3.4 TARGETING AS A MEANS TO EXERT ADJUVANT FUNCTION

Because of a more detailed understanding of how to trigger innate immunity, through specific receptor recognition, adjuvant research is focused increasingly on vaccine targeting of the innate immune system. Targeting is also a way to restrict side effects, to reduce the dose of antigen, and to limit the risk of adverse reactions to vaccination. An adjuvant strategy that involves the specific delivery of an activation signal through a PRR in a given set of cells provides a unique opportunity to tailor the next generation of vaccine adjuvants with predicted action. Moreover, since adjuvants may also be live vectors, such as *Salmonella* and *Adenovirus*, which by gene recombination techniques have been modified to express relevant proteins, targeting PRRs could also be a property of such vectors [35]. For example, the live vector adjuvant can function both as an antigen delivery system and by immunomodulation (e.g., via TLR binding). However, live vectors are often unstable, and few vaccines based on this technology have reached the state of clinical practice. Attempts to target nonliving vaccines to TLR for internalization into APCs have also been successful. A synthetic vaccine against *Listeria monocytogenes* that consisted of a TLR2 ligand and an antigen-specific T-cell epitope gave augmented antibody and CTL immunity [36].

## 3.5   ADJUVANT FUNCTION AND TOXICITY

A fundamental aspect on any adjuvant development is toxicity. Whereas aluminum (Al)-based mineral salts are the most frequently used adjuvants in commercial human vaccines, they are relatively ineffective compared to many other experimentally used adjuvants. Although the antigen-depot effect of Al is associated with relatively mild tissue inflammation at the site of injection, there is increasing concern about allergic reactions to Al adjuvants [37]. Therefore, adjuvants that can replace Al in vaccines are much warranted. Also, a known limitation with Al adjuvant–based vaccines is the poor cytotoxic T lymphocyte (CTL)-promoting ability and skewing toward Th2, resulting in a better effect on humoral immunity but less strong stimulation of cell-mediated immunity [38]. Targeting may allow for enhanced elicitation of specific signaling through PRRs, for example, because of a higher local adjuvant concentration or better focused administration of the adjuvant. This strategy also appears promising as a measure to reduce toxicity and improve safety while retaining or augmenting strong cell-mediated and humoral immunity in future vaccines.

## 3.6   SPECIFIC MECHANISMS VIA THE TOLL-LIKE RECEPTORS

### 3.6.1   Lipopolysaccharide and Monophosphoryl Lipid A

Lipopolysaccharide (LPS) is the prototypic activator of innate immunity via TLRs. Its effects on DCs in vitro, for example, have been amply documented, and even detailed analyses of in vivo effects have been reported [39]. LPS is one of few adjuvants that have also been used for screening of gene expression in host APCs subsequent to exposure. Among several groups, Ricciardi-Castagnoli and co-workers have published extensively on the transcriptomes of DCs treated with LPS using Affymetrics technology [19]. These studies have revealed that more than 2000 genes appear to be regulated by LPS. A significant number of the genes being up-regulated are encoding pro-inflammatory cytokines and chemokines, explaining the strong augmenting effect of LPS on

Th1-dominated immunity [40]. The effect of LPS is strictly TLR4 dependent, and the signal transduction engages either of two pathways; the MyD88 or TRIF/TRAM pathways [41]. Subsequent to TLR signaling is activation of MAPkinase and eventually NFκB, which in turn regulates gene expression [42]. Not only are pro-inflammatory cytokines and chemokines up-regulated following exposure to LPS, but genes encoding co-stimulatory molecules are induced.

Whereas whole LPS is not a candidate for a vaccine adjuvant, monophosphoryl lipid A, a modified less toxic derivative of lipid A, has been tested extensively in human trials [43]. However, the effect is still dependent on the TLR4 pathway, and only in combination with other adjuvants does this material appear to be efficacious and clinically attractive for vaccine use.

### 3.6.2 Bacterial DNA

CpG oligodeoxynucleotide-(ODN) is a good example of an adjuvant with a receptor-restricted mechanism because its specificity is directly dependent on the binding and function of TLR9 in the cell [44]. The CpG motifs are common in bacterial DNA but are underrepresented and methylated in vertebrate DNA. The presence of bacterial DNA is sensed by TLR9 in many different types of cells, including APCs. However, the distribution of TLR9 can vary within, as well as between, different cell types. For example, in humans only B cells and plasmacytoid DCs express TLR9, whereas in mice, other cells also express this PRR [45]. This could lead to different effects and potency of a particular CpG motif in different species or between different types of cells within one species. Also, distinct species differences exist as to a given motif of nucleotides, GACGTT, being more potent in mice than in humans. As TLR9 signaling is one of the most potent activators of type I interferon (IFN), CpG is frequently used as a selective inducer of IFN-α, which has potent antiviral and pro-inflammatory Th1-stimulating effects . In fact, CpG ODN appears to be the strongest Th1 inducer of all adjuvants tested in mice [46]. Although not induced exclusively by CpG, IL12 has been found to be a major factor responsible for the Th1 dominance induced by TLR9 signaling [47]. Of note, though, as a consequence of the different TLR9 patterns and sensitivities, extrapolation of findings in mice may not apply directly to humans. Thus, the TLR9–adjuvant pathway is an example where caution must be entertained when promising results in mice are to be translated into vaccine development for human use.

The relative ineffective uptake of CpG ODN by APCs has been found to be helped by a formulation that focuses the CpG to a particle or carrier protein. Thus, the poly(lactide-co-glycide) (PLG)–CpG ODN combination was found to be an effective adjuvant [48]. Furthermore, the cholera toxin B-subunit (CTB)-CpG conjugate is an original and innovative combination that renders the CpG especially potent as a mucosal adjuvant. Importantly, it also provides a means to reduce cross-species variability in CpG ODN function. Both cell-mediated and humoral immune responses were strongly enhanced following parenteral as well as mucosal immunizations with CTB-CpG adjuvant, and the responses were Th1-dominated and dependent on TLR9/MyD88 expression [49].

## 3.7  NOD RECEPTORS

*Muramyl dipeptide* (MDP) is the minimum effective component of complete Freund's adjuvant derived from Mycobacteriae [50]. However, for experimental purposes it is an effective and reliable adjuvant, which is recognized through intracellular NOD2 [16]. This leads to activation of caspases and proinflammatory and type I cytokines and chemokines, which results in tissue inflammation [51]. The latter is, however, the limiting property of MDP, since it is generally viewed to be too toxic for human use. A new generation of synthetic modified MDPs have recently been tested with variable effects on different cytokines [51]. A recent development is to explore the potential adjuvant benefits of simultaneous activation of NOD and TLRs [52,53].

## 3.8  PARTICULATE ANTIGEN DELIVERY: MICROPARTICLES/ VIRUSLIKE PARTICLES/LIPOSOMES AND IMMUNOSTIMULATING COMPLEXES

Mechanistically, the adjuvants cited above are perhaps a heterogeneous group, but for the sake of simplicity we discuss all particulate adjuvants together. It is believed that the foremost effect of these particles is the effective delivery of antigen for enhanced uptake by APCs, and DCs in particular. Particulate structures based on amphiphilic molecules such as phospholipids and detergents are a large group of adjuvants to which liposomes as well as immunostimulating complexes (ISCOMs) belong [54]. The liposomes are submicron particles made up of bilayers of phospholipids, which allow for the incorporation of hydrophilic antigens (Figure 3.2). The efficacy of antigen loading varies greatly between various particles, but in general, members of the liposome group demonstrate better versatility with regard to size and physicochemical characteristics than do ISCOMs. The latter contain the saponin mixture Quil A as an integrated adjuvant together with phospholipids and cholesterol in a cagelike structure roughly 40 nm in size [55]. Incorporation of antigen into the ISCOMs has proven most effective, although not a prerequisite for an adjuvant effect, since mixing antigen with empty ISCOMs (ISCOM-Matrix) has consistently been found to give quite similar stimulation of immune responses against many different antigens [56]. Whereas liposomes have a modest-to-poor immunoenhancing effect, ISCOMs are known to be effective inducers of augmented humoral and cell-mediated immune responses [57]. In particular, the effective stimulation of cytotoxic T lymphocyte (CTL) responses is a hallmark of ISCOMs [58].

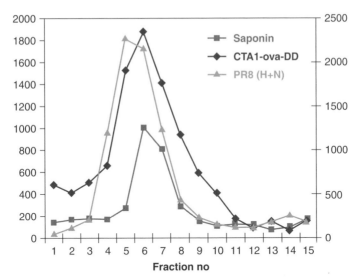

**Figure 3.2** ISCOMs as they can be studied in electron microscopy. Superimposed is the construction of ISCOMs containing two proteins: CTA1-DD adjuvant and PR8 influenza virus proteins. ISCOMs containing CTA1-DD and PR8 antigens were subjected to sucrose density gradient analysis and the fractions were analyzed for CTA1-DD/PR8 and saponin content. The peaks of incorporation of CTA1-DD, PR8 proteins, and saponin are superimposed, illustrating the successful formation of ISCOMS. The *y*-axis values represent arbitrary OD units at A450nm for immunodetection of HRP-labeled CTA1-DD or PR8 (left) and saponin at A214nm (right) using a spectrophotometer. (*See insert for color representation of figure.*)

The particulate adjuvants are most often taken up by classical APCs, such as DCs and macrophages, whereas B cells are not the prime target for these delivery systems [57]. The outcome of immunization with ISCOMs is a wide range of immune responses with augmented antigen-specific antibody production, delayed-type hypersensitivity (DTH), and Th1 immunity, which are all dependent on the early production of IL-12 [59].

A recent description of stable cationic liposomes, based on dimethyldioctadecylammonium (DDA)-cationic lipids, appears to be especially promising, as when used in the context of a tuberculosis (TB) vaccine, it induced strong specific Th1 immunity and IFN-γ production [60].

### 3.9 COMBINATION ADJUVANTS

To achieve a stronger effect or a more potent skewing of immune responses, adjuvants may be combined. This is also a way to reduce side effects and bring down the amount of adjuvant and antigen in a vaccine. Currently, much effort is devoted to combining CpG with alum, and promising results have recently been reported in humans [61]. In experimental models, other combinations of CpG (e.g., with QS21, MDP, or MPL) have proven very effective [62]. Admixing CpG with detoxified LTR72 adjuvant gave stronger Th1 responses than were seen with LTR72 alone when given intranasally [63]. Since alum is the classical and most used vaccine adjuvant in commercial vaccines, the rational for combining alum with other immunomodulators is strong. Coadministration with MDP or Quil A together with alum have been reported to provide good enhancing effects on both humoral and cell-mediated responses [64]. Transcutaneous immunizations using the combination of CT with CpG ODN was recently proven more effective than using either adjuvant alone for stimulating antichlamydial immunity in the genital tract [65]. The synergistic effect resulted in better protection against a live challenge infection in immunized mice compared to mice immunized with CpG or CT adjuvant alone.

Chitosan is a deacetylation derivative of the polysaccharide chitin, which is sampled from shells of crustaceans. The function of the chitosan is believed to be associated with its ability to open up tight junctions of polarized epithelial cells. However, used successfully as a mucosal adjuvant, the particulate form or antigen-trapping properties of chitosan may favor uptake of antigen over the mucosal barrier. In any case, colloidal carrier vectors consisting of a combination of chitosan with MDP, CTB, LTK63, and CTA1-DD have proven very effective in augmenting mucosal and systemic immune responses after nasal immunizations [66]. These results clearly argue in favor of combination adjuvants, which could drastically reduce toxicity and improve vaccine efficacy.

## 3.10  MUCOSAL VACCINE ADJUVANTS

Mucosal immunization was introduced in the 1950s with the Sabin live attenuated oral polio vaccine. The success of this vaccine, attributable not only to its needle-free character, but foremost to its ability to stimulate protective immunity, has not been followed by the series of efficient mucosal vaccines one might have anticipated. There are several explanations as to why this was not the case, the most important being the requirements for strong clinically acceptable vaccine adjuvants for mucosal administration. It is well documented that protein antigens given at mucosal sites, in most cases, are poor immunogens. Thus, nonliving vaccines given at mucosal sites are likely to be ineffective and can even result in mucosal tolerance, which is the naturally occurring immunological phenomenon that prevents harmful inflammatory responses to nondangerous proteins such as food components [67]. Mucosal immunization with antigen coadministered with a mucosally active adjuvant such as CT, however, commonly induces both systemic and mucosal immunity [68–70]. Other active adjuvant formulations at mucosal sites are the synthetic CpG ODN, MPL, and ISCOMs [71–74].

Our understanding of the requirements that render a mucosal vaccine adjuvant effective is at present incomplete. Both pro- and anti-inflammatory pathways have been reported to be activated following the use of these adjuvants, but in general, it is believed that effective mucosal adjuvants drive inflammatory responses. For the classical adjuvants such as Freund's complete adjuvant (FCA), there appears to be a good correlation between the dose, the degree of local inflammation, and the adjuvanticity it carries [75]. As aforementioned, in the interest of safety and clinical acceptance, it is important, however, to develop adjuvants with as little local reactogenicity as possible but with intact immunoenhancing ability. The study of mutant CT and LT or derivatives thereof have clearly pointed in that direction, showing convincingly that cell-targeted adjuvants enable the development of vaccines with limited local irritation while retaining enhanced immunogenicity [76].

Dendritic cells are considered the key APCs for priming of naive T cells [77,78]. Whether this is also the case at mucosal membranes awaits to be proven. The difficulty in targeting DCs in vivo has limited our knowledge about the priming events that determine whether antigen stimulation will result in a tolerogenic or an immunogenic outcome [79]. Immature DCs residing in tissues are known to take up antigen and, if maturation occurs, migrate to regional lymph nodes or the spleen [80, 81]. In the secondary lymphoid tissues the DC immigrants, expressing strong co-stimulation, may be inherently stimulatory. But whether resident or poorly activated immigrants are tolerogenic is currently a much debated issue. In particular, we lack in vivo information about DCs at specific anatomical sites such as the marginal zone of the spleen, the lamina propria of the mucosal membranes, or the conduit system in the peripheral lymph nodes and spleen.

## 3.11 CHOLERA TOXIN: THE PROTOTYPE ADJUVANT

Cholera toxin (CT) is perhaps the best studied and most effective experimental adjuvant known today. There is a vast literature on its structure and function [82]. It is the key element responsible for the diarrhea in infections caused by *Vibrio cholerae* bacteria, and it has been used extensively in the study of various cell biological topics involving the identification of cell membrane structures and lipid rafts [83,84]. As mentioned before, it is an $AB_5$ complex with the ADP-ribosyltransferase active A1 linked to a pentamer of B subunits via the A2 fragment. The B subunit of CT is responsible for binding to the GM1 receptor, a glycosphingolipid found ubiquitously on membranes of most mammalian cells. Several studies have shown that ADP-ribosyltransferase activity is required for optimal adjuvanticity, but that binding to cells of the immune system is mediated via the B subunit [85–87]. The mechanism of adjuvanticity of the ADP-ribosylating toxins has been the subject of considerable debate. Most investigators agree, though, that both elements, binding and enzymatic activity, can contribute to the immunomodulation. There is ample evidence in the literature to support this notion, and studies with the closely related LT and derivatives thereof, have documented this point [88]. However, the ADP-ribosyltransferase activity appears to be key to an optimal immu-noenhancing effect, as shown elegantly by Rappuoli and co-workers [89]. Studies with polarized epithelial cells have demonstrated that CT, following internalization in the target cell, is subject to retrograde transport into Golgi and endoplasmic reticulum, prior to its appearance as an A1 fragment in the cytosol, acting on the heterotrimeric GTP-binding protein (Gsα) in the cell membrane [86]. Whether this observation is also relevant for cells of the innate immune system, such as DCs, macrophages, and B lymphocytes, is assumed, but has never been investigated carefully [90]. The enzymatic activity in the cytosol is dependent on ADP-ribosylating factors (ARFs). Whether these factors are important for adjuvanticity is, however, poorly understood [91,92]. To this end, mutant holotoxins were found to host binding sites for ARFs, but no attempt to mutate these sites has yet been performed in the context of adjuvanticity [93]. Activation of the Gsα leads to an increase in adenylate cyclase and subsequent increase in intracellular cAMP. The participation of the latter elements in the adjuvant effect has never been addressed directly in experimental models, although studies by Tsai et al. would suggest that the LT (R192G) mutant, which is a strong adjuvant, cannot be unfolded and thus would not be transported to the membrane [94]. The significance of this assumption is, however, difficult to know.

The adjuvant effect is thought to involve the modulation of APCs, but it is poorly understood which APCs are functionally targeted by CT in vivo. All nucleated cells, including all professional APCs, can bind the toxin via the GM1-ganglioside receptor present in the cell membrane. Previous reports have documented both a pro- and an anti-inflammatory effect of CT. From several studies, including our own work, CT exposure of APCs has an augmenting effect on IL-1 and IL-6 production, whereas in other studies a down-regulating effect on IL-12 and TNFα, and a promoting effect on IL-10 production have been reported. Taken together, this would indicate an anti-inflammatory effect. In fact, investigators have used the CT adjuvant to generate Th1 cells, but most reports have shown a bias for Th2 cells, and recently also, IL-10 producing regulatory Tr1 cells [70,95]. In this context it is interesting to note that PT has been reported to reduce Treg activity, while CT was found to specifically stimulate Treg differentiation [70,96]. Although data reported by different groups on CT's effects on cytokine production, especially IL-1 and IL-12, are thus not consistent, it is clear that CT is a strong adjuvant and may induce both pro- and anti-inflammatory effects in vivo.

One recent study documented involvement of the NFκB pathway in the interaction between CT and the targeted APCs, suggesting effects on gene transcription associated with inflammatory responses. Whether or not this activation was a prerequisite for an adjuvant effect, was not discussed [87]. In preliminary experiments we failed to observe an impaired adjuvant effect in TLR4- and MyD88-deficient mice, suggesting that CT exerts adjuvant functions that are TLR independent, although signal transduction pathways other, than MyD88 and TRIF/TRAM cannot be excluded at present [41].

Moreover, CT, but not CTB, up-regulates the CXCR4 and CCR7 chemo-kine receptors, which may have fundamental importance for the adjuvant function, as these receptors favor the migration of immunocytes to the lymph nodes and spleen and promote attraction between DCs and naive T cells [97,98]. How this dual pro- and anti-inflammatory ability of CT affects its adjuvant function in vivo still appears unresolved. Interestingly, the CTB, which is devoid of the CTA1 enzyme, is a well-documented carrier for the induction of mucosal Ag-specific tolerance [99]. Therefore, CT and CTB represent a unique pair of modulating proteins that differ with regard to enzymatic activity but share the GM1-receptor binding ability, allowing them to be exploited for mechanistic studies aiming at unraveling the regulatory elements that control mucosal tolerance and active IgA immune responses.

## 3.12 SEARCHING FOR IN VIVO CORRELATES OF ADJUVANT ACTION

In a series of experiments we have used ovalbumin (OVA)-conjugated CT to study the deposition of the adjuvant in vivo following injections [100]. We found that after intravenous injections, most of the OVA accumulated in the marginal zone (MZ) of the spleen (Figure 3.3). No OVA could be detected in

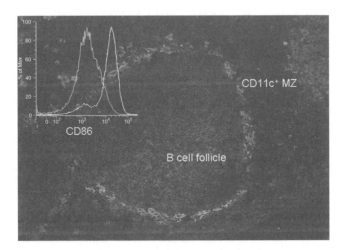

**Figure 3.3** To detect the deposition of CT in the spleen of mice following intravenous injection we conjugated CT to OVA and traced the OVA by specific labeling with an anti-OVA FITC (green) conjugated antibody. The CT-OVA localizes distinctly in the marginal zone of the spleen after 2 hours following intravenous injections. The B-cell follicle is labeled with TexasRed-conjugated anti-IgM antibody (red). Superimposed is the expression of CD86 (red) on CD11c⁺ cells that have been isolated from the spleen 24 hours after injection compared to cells from naive (blue) mice. (*See insert for color representation of figure.*)

this zone when unconjugated OVA was used, even when very high doses were injected. We could conclude that the deposition of CT-OVA was facilitated by the GM1-ganglioside receptor binding, because a nonbinding mutant toxin conjugate failed to deposit in the MZ. Antigen concentrations in the targeted cells were very high and appeared to form a depot of OVA in the MZ for up to 48 hours after injection. After 24 hours the concentration of OVA was reduced in the MZ, which coincided with the appearance of OVA-containing DCs in the T-cell zone. These DCs were high expressors of CD86, but not CD80, and appeared to provide enhanced APC function, as peptide-specific T cells greatly expanded in response to the CT-OVA conjugate. By contrast, CTB-OVA conjugates failed to promote T-cell expansion despite the fact that they delivered antigen to the MZ. As we failed to detect maturation of DCs after CTB-OVA conjugate treatment, we concluded that CT, but not CTB, instructs targeted DCs to mature and migrate into the T-cell zone. Thus, targeting of ADP-ribosylating enzymes to DCs in vivo appears to be a powerful mechanism to restrict unwanted reactions and, concomitantly, to greatly augment the priming of a specific immune response. As mentioned earlier, CTB conjugates are known to be highly tolerogenic when given orally [101]. Based on these and other observations, we have hypothesized that ADP ribosylation is the gate-keeper in determining whether APC function will result in active immunity or in tolerance, as these response patterns appear to be mutually exclusive [68].

## 3.13  TOXINS ARE TOO TOXIC FOR CLINICAL USE

A major limitation of holotoxins is their promiscuous binding to the GM1-ganglioside receptors present on all nucleated cells, including epithelial cells and nerve cells, which render these toxin adjuvants, or derivatives thereof, unattractive for clinical use [102,103]. Indeed, a commercial intranasal flu vaccine with LT as the adjuvant revealed increased incidence of cases with Bell's palsy in vaccinated subjects and led to withdrawal of the vaccine from the market [103–105]. An alternative strategy has proven that mutants of CT or LT, which host no to very little enzymatic activity, can act as mucosal adjuvants [106]. The single-amino acid mutations most frequently used, LTK63 and CTE112, are located in the A1 subunit and although dramatically less toxic, have been found to retain substantial adjuvant function [61,107]. However, these mutants also carry a risk of being accumulated in the central nervous tissues following intranasal immunizations. Since experimental data indicate that the adjuvant dose of the holotoxins is related to the toxic dose, one can anticipate that it will meet with difficulty to separate adjuvanticity from toxicity in a human vaccine based on mutants of the holotoxins.

## 3.14  NONTOXIC CTA1-DD ADJUVANT

To circumvent the toxicity problem discussed above, we have generated a novel immuno-modulating molecule by combining the enzymatic activity of

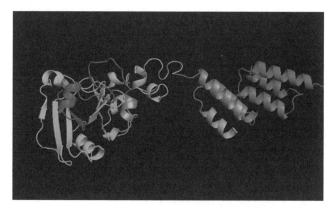

**Figure 3.4** Computer modeling of CTA1-DD adjuvant. To the left is the CTA1 and to the right is the DD-dimer. (*See insert for color representation of figure.*)

CTA1 with B-cell targeting. This was achieved by constructing a gene-fusion protein combining CTA1 with a synthetic dimer of the D fragment of *Staphylococcus aureus* protein A [108,109] (Figure 3.4). The resulting CTA1-DD adjuvant is devoid of GM1-ganglioside binding and is, instead, redirected to the APC population. The CTA1-DD was found to be completely nontoxic [108,110]. Mice and monkeys have been given doses of more than 200 μg of CTA1-DD without apparent side effects or signs of reactogenicity, while similar doses of CT are known to be lethal. It is noteworthy that humans can get overt diarrhea from doses as low as 2.5 to 10 μg of CT [111]. Thus, CTA1-DD appears to be a safe and nontoxic mucosal vaccine adjuvant, although it carries the same amount of ADP-ribosylating ability as CT holotoxin. Furthermore, it was shown to be as potent as the CT holotoxin in augmenting immune responses after systemic as well as mucosal immunizations [108,112,113]. The adjuvant effect of CTA1-DD was comparable to that of intact CT and shown to be dependent on an intact CTA1 enzymatic activity as well as on the Ig-binding ability of the DD dimer [112].

In view of the findings of direct binding and accumulation of CT or LT to cells of the nervous system following intranasal administration, we investigated the behavior of CTA1-DD [104,114]. We found that the CTA1-DD adjuvant did not bind or accumulate in the nervous tissues after intranasal administration [76]. In extended studies in mice using CTA1-DD as an intranasal vaccine adjuvant, we found no signs of local inflammation in the nasal mucosa or cellular deposition of inflammatory cells in the lamina propria or the organized nasal associated lymphoid tissues (NALTs). This finding corroborated our previous observation of no local edema when injected into the foot pads of mice, whereas CT elicited a substantial edema in injected foot pads [110]. Thus, we have developed a highly effective adjuvant that does not cause local inflammation, is safe, and does not accumulate in the nervous tissues. The CTA1-DD adjuvant is currently being tested in the context of several candidate vaccines as part of ongoing international collaborations. In particular, promising results have been obtained with candidate mucosal vaccines against *Influenza virus*, *Helicobacter pylori*, *Chlamydia trachomatis*, and rota virus [115,116]. Antiviral immunity against influenza virus was improved significantly in mice immunized intranasally with an M2e-based vaccine with the CTA1-DD adjuvant admixed. The adjuvant effect was clearly demonstrable as enhanced survival and reduced morbidity to a challenge infection in immunized mice [117].

## 3.15   GENE EXPRESSION PROFILING

A mechanistically important asset to the CTA1–adjuvant system is the possibility to understand the molecular mechanisms underlying adjuvanticity. For the first time we have the models and prior information to unravel which intracellular regulatory elements are involved in the adjuvant process. Since adjuvanticity is dependent on the ADP-ribosylating ability of CT and CTA1-DD, we might be able to establish which intracellular proteins are targets for the enzymatic activity. Our recent work has generated a list of target genes that are regulated by CT and the ADP-ribosylating enzyme. We have used Affymetrix technology to investigate global gene expression in B cells and DCs following exposure to CT, CTB, CTA1-DD, and the enzymatically inactive mutant CTA1R7K-DD. Our rational was to study gene expression patterns after exposure to various adjuvants to gain insights into what may be the difference at the gene transcriptional level between CT/CTA1-DD and the nonfunctioning enzymatically inactive CTA1R7K-DD mutant or CTB. These experiments have revealed between 40 and 65 genes of interest that might be associated with an adjuvant function. Our work plan for the future involves dissecting the role of these genes in the immunomodulation and adjuvant function. For example, by taking a reductionistic approach to the Affymetrix data, we have confirmed that the co-stimulatory molecules CD80, CD86, and CD83 are among the most interesting genes that are up-regulated by the ADP-ribosyltransferase active molecules. It is now believed that co-stimulation is one of the key elements in the adjuvant function of these adjuvants [68,70,118].

## 3.16   COMBINATION ADJUVANTS WITH CTA1-DD

The CTA1-DD/ISCOM vector is a rationally designed mucosal adjuvant that greatly potentiates humoral and cellular immune responses [119]. It was developed to incorporate the distinctive properties of either adjuvant alone in a combination that exerted additive enhancing effects on mucosal immune responses. We recently demonstrated that CTA1-DD and an unrelated antigen could be incorporated together into ISCOMs, resulting in greatly augmented immunogenicity of the antigen. To demonstrate its relevance for protection against infectious diseases, we tested the vector incorporating PR8 antigen from the influenza virus [120]. Following intranasal immunization, we found that the immunogenicity of the PR8 proteins was augmented significantly by a mechanism that was enzyme dependent, as the presence of the enzymatically inactive CTA1R7K-DD mutant largely failed to enhance the response above the enhancement seen with ISCOMs alone. The combined vector was a highly effective enhancer of a broad range of immune responses, including specific serum antibodies and a balanced Th1 and Th2 CD4$^+$ T-cell priming as well as a strong mucosal IgA response. Unlike unmodified ISCOMs, antigen incorporated into the combined vector could be presented by B cells in vitro and in vivo as well as by DCs. The combined vector unexpectedly accumulated in the B-cell follicles of draining lymph nodes and stimulated much enhanced germinal center reactions. Strikingly, the enhanced adjuvant activity of the combined vector was absent in B cell–deficient mice, supporting the notion that B cells are important for the adjuvant effects of the combined CTA1-DD/ISCOMs vector.

In vitro studies have shown that ISCOM particles themselves are taken up and presented preferentially by DCs [121], a finding we have extended by showing that ISCOMs accumulate in DCs in vitro and in the DC-rich T cell–dependent areas of lymph nodes in vivo. Interestingly, the pattern of uptake in vivo was consistent with localization in the fibroreticular conduits of T-cell areas, which others have shown to be important sites of accumulation of Ag-loaded DCs and initial interactions between DCs and T cells [3,78,122,123]. In contrast, ISCOMs containing CTA1-DD were taken up very efficiently by B cells as well as by DCs in vitro and were presented by both APCs to CD4$^+$ T cells, presumably reflecting the ability of the DD portion to bind to B cells selectively via their surface Ig [120]. Furthermore, subcutaneously injected CTA1-DD/ISCOMs had a unique ability to accumulate in the B-cell follicles of lymph nodes, where they were retained for at least 24 hours, by which time conventional ISCOMs were virtually undetectable in the lymph node. Of most importance, the augmented adjuvant properties of CTA1-DD/ISCOMs were largely absent in the B-cell KO mice. For these reasons we propose that this novel combined vector is so effective because it targets both DCs and B cells in vivo. Altogether these features highlight the potential usefulness of CTA1-DD/ISCOMs as practical mucosal vaccine vectors which will provide flexible and stable means of inducing protective immunity against a variety of pathogens.

Apparently, mechanistically complementing mucosal adjuvants may provide novel strategies for the generation of potent and safe mucosal vaccines. The use of our combined vector is now being tested for mucosal vaccines against tuberculosis, and preliminary results have demonstrated the feasibility to use the combined vector in an intranasal vaccine (Figure 3.5).

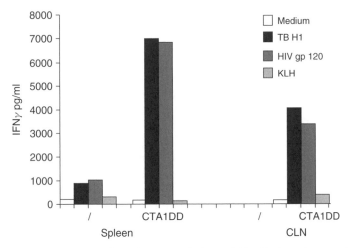

**Figure 3.5** CTA1DD strongly enhances priming of T cells in spleen and cervical lymph nodes (CLN) for recall IFN-γ production following intranasal immunizations with the hybrid H1 antigens (ESAT-6/Ag85) given together with HIV gp120 and the combined CTA1DD/ISCOM adjuvant vector. An unrelated protein, KLH, was added to the cultures as a negative control.

## 3.17  CONCLUDING REMARKS

A better understanding of the underlying mode of action of adjuvants in vivo will help improve their efficacy in achieving the desired immunomodulatory effects. We already have extensive information about some of the mechanisms that adjuvants employ, but it is understanding the fine tuning and the sequential expression of critical factors by APCs in vivo that will aid in developing powerful vaccine adjuvants. There is no doubt that future vaccine adjuvants will be better targeted and more effective, such that less antigen and fewer side effects will be reported. Exploiting single or combined mechanisms such as the TLR and NOD pathways or the mutant or engineered toxins in adjuvant formulations for depot effects or for mucosal administration appears especially interesting. In our own research we have designed two promising adjuvant candidates for future mucosal vaccines. Both CTA1-DD or CTA1-DD/ISCOMS represent conceptually important steps toward the construction of novel, rationally designed, safe, and efficacious mucosal vaccine adjuvants. Indeed, control over the priming process leading to tolerance or an active IgA response by using or not using targeted ADP-ribosylating enzymes appears to be a most powerful strategy for future vaccine adjuvant developments. Such a strategy would greatly facilitate the construction of anti-infectious vaccines as well as vaccines inducing tolerance against autoimmune diseases.

## ACKNOWLEDGMENTS

I would like to thank members of my group and former colleagues for invaluable contributions to the work on adjuvants. In particular, I would like to mention my co-inventor Dr. Björn Löwenadler, Arexis AB; Prof. Allan Mowat, Glasgow; Prof. John Nedrud, Cleveland; Prof. Paula Ricciardi-Castagnoli, Milan; Prof. Peter Andersen, Copenhagen; and Dr Karin Lövgren, ISCONOVA AB. I am also grateful to Dubravka Grdic, Anja Helgeby, Anna Eriksson, Annemarie Hasselgren, Anneli Stensson, Karin Schön, Lena Ågren, Lena Ekman, Anna Sjöblom-Hallén, and Johan Mattson. These studies were not possible without the generous grant support from the European Commission (grants QLK2-CT-2001-01702, QLK2-CT-1999-00228, and LSHP-CT-2003-503240), Vetenskapsrådet, SAREC, LUA/ALF, and Cancerfonden.

## REFERENCES

1. Steinman, R.M., Bonifaz, L., Fujii, S., Liu, K., Bonnyay, D., Yamazaki, S., Pack, M., Hawiger, D., Iyoda, T., Inaba, K. & Nussenzweig, M.C. The innate functions of dendritic cells in peripheral lymphoid tissues. *Adv Exp Med Biol* 2005, 560, 83–97.

2. Adams, C.L., Kobets, N., Meiklejohn, G.R., Millington, O.R., Morton, A.M., Rush, C.M., Smith, K.M. & Garside, P. Tracking lymphocytes in vivo. *Arch Immunol Ther Exp (Warsz)* 2004, 52, 173–187.

3. Lindquist, R.L., Shakhar, G., Dudziak, D., Wardemann, H., Eisenreich, T., Dustin, M.L. & Nussenzweig, M.C. Visualizing dendritic cell networks in vivo. *Nat Immunol* 2005, 5, 1243–1250.

4. Bukreyev, A. & Belyakov, I.M. Expression of immunomodulating molecules by recombinant viruses: Can the immunogenicity of live virus vaccines be improved? *Expert Rev Vaccines* 2002, 1, 233–245.

5. Vogel, F.R. Immunologic adjuvants for modern vaccine formulations. *Ann NY Acad Sci* 1995, 754, 153–160.

6. Greenwald, R.J., Freeman, G.J. & Sharpe, A.H. The B7 family revisited. *Annu Rev Immunol* 2005, 23, 515–548.

7. Crane, I.J. & Forrester, J.V. Th1 and Th2 lymphocytes in autoimmune disease. *Crit Rev Immunol* 2005, 25, 75–102.

8. Raine, T., Zaccone, P., Dunne, D.W. & Cooke, A. Can helminth antigens be exploited therapeutically to downregulate pathological Th1 responses? *Curr Opin Invest Drugs* 2004, 5, 1184–1191.

9. Bluestone, J.A. & Abbas, A.K. Natural versus adaptive regulatory T cells. *Nat Rev Immunol* 2003, 3, 253–257.

10. Baecher-Allan, C. & Anderson, D.E. Immune regulation in tumor-bearing hosts. *Curr Opin Immunol* 2006, 18, 214–219.

11. Berrih-Aknin, S., Fuchs, S. & Souroujon, M.C. Vaccines against myasthenia gravis. *Expert Opin Biol Ther* 2005, 5, 983–995.

12. Matzinger, P. & Guerder, S. Does T-cell tolerance require a dedicated antigen-presenting cell? *Nature* 1989, **338**, 74–76.

13. Underhill, D.M. & Ozinsky, A. Toll-like receptors: key mediators of microbe detection. *Curr Opin Immunol* 2002, **14**, 103–110.

14. Barton, G.M. & Medzhitov, R. Control of adaptive immune responses by Toll-like receptors. *Curr Opin Immunol* 2002, **14**, 380–338.

15. Lauw, F.N., Caffrey, D.R. & Golenbock, D.T. Of mice and man: TLR11 (finally) finds profilin. *Trends Immunol* 2005, **26**, 509–511.

16. Inohara, N. & Nunez, G. NODs: intracellular proteins involved in inflammation and apoptosis. *Nat Rev Immunol* 2003, **3**, 371–382.

17. Strober, W., Murray, P.J., Kitani, A. & Watanabe, T. Signalling pathways and molecular interactions of NOD1 and NOD2. *Nat Rev Immunol* 2006, **6**, 9–20.

18. Murray, P.J. NOD proteins: an intracellular pathogen-recognition system or signal transduction modifiers? *Curr Opin Immunol* 2005, **17**, 352–358.

19. Ricciardi-Castagnoli, P. & Granucci, F. Opinion: interpretation of the complexity of innate immune responses by functional genomics. *Nat Rev Immunol* 2002, **2**, 881–889.

20. Sharpe, A. Analysis of costimulation in vivo using transgenic and knockout mice. *Curr Opin Immunol* 1995, **7**, 389–395.

21. Liang, Y., Zhou, Y. & Shen, P. NF-kappaB and its regulation on the immune system. *Cell Mol Immunol* 2004, **1**, 343–350.

22. Holmgren, J., Czerkinsky, C., Lycke, N. & Svennerholm, A.-M. Strategies for the induction of immune responses at mucosal surfaces making use of cholera toxin B subunit as immunogen, carrier and adjuvant. *Am J Med Hyg* 1994, **50**, 42–54.

23. Lycke, N. Targeted vaccine adjuvants based on modified cholera toxin. *Curr Mol Med* 2005, **5**, 591–597.

24. Duverger, A., Jackson, R.J., van Ginkel, F.W., Fischer, R., Tafaro, A., Leppla, S.H., Fujihashi, K., Kiyono, H., McGhee, J.R. & Boyaka, P.N. Bacillus anthracis edema toxin acts as an adjuvant for mucosal immune responses to nasally administered vaccine antigens. *J Immunol* 2006, **176**, 1776–1783.

25. He, J., Gurunathan, S., Iwasaki, A., Ash-Shaheed, B. & Kelsall, B.L. Primary role for Gi protein signaling in the regulation of interleukin 12 production and the induction of T helper cell type 1 responses. *J Exp Med* 2000, **191**, 1605–1610.

26. Ryan, M., McCarthy, L., Rappuoli, R., Mahon, B.P., & Mills, K.H. Pertussis toxin potentiates Th1 and Th2 responses to co-injected antigen: adjuvant action is associated with enhanced regulatory cytokine production and expression of the co-stimulatory molecules B7-1, B7-2 and CD28. *Int Immunol* 1998, **10**, 651–662.

27. Hou, W., Wu, Y., Sun, S., Shi, M., Sun, Y., Yang, C., Pei, G., Gu, Y., Zhong, C. & Sun, B. Pertussis toxin enhances Th1 responses by stimulation of dendritic cells. *J Immunol* 2003, **170**, 1728–1736.

28. Hornquist, E. & Lycke, N. Cholera toxin adjuvant greatly promotes antigen priming of T cells. *Eur J Immunol* 1993, **23**, 2136–2143.

29. Yamamoto, M., Kiyono, H., Kweon, M.N., Yamamoto, S., Fujihashi, K., Kurazono, H., Imaoka, K., Bluethmann, H., Takahashi, I., Takeda, Y., Azuma, M.

& McGhee, J.R. Enterotoxin adjuvants have direct effects on T cells and antigen-presenting cells that result in either interleukin-4-dependent or -independent immune responses. *J Infect Dis* 2000, **182**, 180–190.

30. Bonifaz, L.C., Bonnyay, D.P., Charalambous, A., Darguste, D.I., Fujii, S., Soares, H., Brimnes, M.K., Moltedo, B., Moran, T.M. & Steinman, R.M. In vivo targeting of antigens to maturing dendritic cells via the DEC-205 receptor improves T cell vaccination. *J Exp Med* 2004, **199**, 815–824.

31. Getahun, A. & Heyman, B. How antibodies act as natural adjuvants. *Immunol Lett* 2006, **104**, 38–45.

32. Lehner, T. Innate and adaptive mucosal immunity in protection against HIV infection. *Vaccine* 2003, **21**(Supp 2), S68–S76.

33. Cohen-Sfady, M., Nussbaum, G., Pevsner-Fischer, M., Mor, F., Carmi, P., Zanin-Zhorov, A., Lider, O. & Cohen, I.R. Heat shock protein 60 activates B cells via the TLR4-MyD88 pathway. *J Immunol* 2005, **175**, 3594–3602.

34. Lee, Y., Haas, K.M., Gor, D.O., Ding, X., Karp, D.R., Greenspan, N.S., Poe, J.C. & Tedder, T.F. Complement component C3d-antigen complexes can either augment or inhibit B lymphocyte activation and humoral immunity in mice depending on the degree of CD21/CD19 complex engagement. *J Immunol* 2005, **175**, 8011–8023.

35. Edelman, R. Adjuvants for the future. In *New Generation Vaccines*, 2nd ed., Woodrow, G.C., Levine, M.M., Kaper, J.B. & Cobon, G.S., Eds., Marcel Dekker, New York, 1997, pp. 173–192.

36. Jackson, D.C., Lau, Y.F., Le, T., Suhrbier, A., Deliyannis, G., Cheers, C., Smith, C., Zeng, W. & Brown, L.E. A totally synthetic vaccine of generic structure that targets Toll-like receptor 2 on dendritic cells and promotes antibody or cytotoxic T cell responses. *Proc Natl Acad Sci USA* 2004, **101**, 15440–15445.

37. Bergfors, E., Bjorkelund, C. & Trollfors, B. Nineteen cases of persistent pruritic nodules and contact allergy to aluminium after injection of commonly used aluminium-adsorbed vaccines. *Eur J Pediatr* 2005, **164**, 691–697.

38. Gupta, R.K. Aluminum compounds as vaccine adjuvants. *Adv Drug Deliv Rev* 1998, **32**, 155–172.

39. Germain, R.N. & Jenkins, M.K. In vivo antigen presentation. *Curr Opin Immunol* 2004, **16**, 120–125.

40. Perrier, P., Martinez, F.O., Locati, M., Bianchi, G., Nebuloni, M., Vago, G., Bazzoni, F., Sozzani, S., Allavena, P. & Mantovani, A. Distinct transcriptional programs activated by interleukin-10 with or without lipopolysaccharide in dendritic cells: induction of the B cell–activating chemokine, CXC chemokine ligand 13. *J Immunol* 2004, **172**, 7031–7042.

41. Moynagh, P.N. TLR signalling and activation of IRFs: revisiting old friends from the NF-kappaB pathway. *Trends Immunol* 2005, **26**, 469–476.

42. Moynagh, P.N. The NF-kappaB pathway. *J Cell Sci* 2005, **118**, 4589–4592.

43. Ulrich, J.T. MPLr immunostimulant: adjuvant formulations. In *Vaccine Adjuvants: Preparation Methods and Research Protocols*, O'Hagan, D.T., Ed., Humana Press, Totowa, NJ, 2000, pp. 273–282.

44. Krieg, A.M. CpG motifs in bacterial DNA and their immune effects. *Annu Rev Immunol* 2002, **20**, 709–760.

45. Wagner, H. The immunobiology of the TLR9 subfamily. *Trends Immunol* 2004, **25**, 381–386.

46. Chu, N.R., DeBenedette, M.A., Stiernholm, B.J.N., Barber, B.H. & Watts, T.H. Role of IL-12 and 4-1 BB ligand in cytokine production by CD28⁺ and CD28⁻ T cells. *J Immunol* 1997, **158**, 3081–3089.

47. van Duin, D., Medzhitov, R. & Shaw, A.C. Triggering TLR signaling in vaccination. *Trends Immunol* 2006, **27**, 49–55.

48. Xie, H., Gursel, I., Ivins, B.E., Singh, M., O'Hagan, D.T., Ulmer, J.B. & Klinman, D. M. CpG oligodeoxynucleotides adsorbed onto polylactide-co-glycolide microparticles improve the immunogenicity and protective activity of the licensed anthrax vaccine. *Infect Immun* 2005, **73**, 828–833.

49. Adamsson, J., Lindblad, M., Lundqvist, A., Kelly, D., Holmgren, J. & Harandi, A. Novel immunostimulatory agent based on CpG oligodeoxynucleotide linked to the non-toxic B subunit of cholera toxin. *J Immunol* 2006, **15**, 4902–4913.

50. Moschos, S.A., Bramwell, V.W., Somavarapu, S. & Alpar, H.O. Modulating the adjuvanticity of alum by co-administration of muramyl di-peptide (MDP) or Quil-A. *Vaccine* 2006, **24**, 1081–1086.

51. Gobec, S., Sollner-Dolenc, M., Urleb, U., Wraber, B., Simcic, S. & Filipic, M. Modulation of cytokine production by some phthalimido-desmuramyl dipeptides and their cytotoxicity. *Farmaco* 2004, **59**, 345–352.

52. Tada, H., Aiba, S., Shibata, K., Ohteki, T. & Takada, H. Synergistic effect of NOD1 and NOD2 agonists with toll-like receptor agonists on human dendritic cells to generate interleukin-12 and T helper type 1 cells. *Infect Immun* 2005, **73**, 7967–7976.

53. Fritz, J.H., Girardin, S.E., Fitting, C., Werts, C., Mengin-Lecreulx, D., Caroff, M., Cavaillon, J.M., Philpott, D.J. & Adib-Conquy, M. Synergistic stimulation of human monocytes and dendritic cells by Toll-like receptor 4 and NOD1- and NOD2-activating agonists. *Eur J Immunol* 2005, **35**, 2459–2470.

54. Kersten, G.F. & Crommelin, D.J. Liposomes and ISCOMs. *Vaccine* 2003, **21**, 915–920.

55. Morein, B., Sundquist, B., Höglund, S., Dalsgaard, K. & Osterhaus, A. Iscom, a novel structure for antigenic presentation of membrane proteins from enveloped viruses. *Nature* 1984, **308**, 457–462.

56. Sanders, M.T., Brown, L.E., Deliyannis, G. & Pearse, M.J. ISCOM-based vaccines: the second decade. *Immunol Cell Biol* 2005, **83**, 119–128.

57. Mowat, A.M. Dendritic cells and immune responses to orally administered antigens. *Vaccine* 2005, **23**, 1797–1799.

58. Takahashi, H., Takeshita, T., Morein, B., Putney, S., Germain, R.N. & Berzofsky, J.A. Induction of CD8⁺ cytotoxic T cells by immunization with purified HIV-1 envelope protein in ISCOMs. *Nature* 1990, **344**, 873–875.

59. Smith, R.E., Donachie, A.M., Grdic, D., Lycke, N. & Mowat, A.M. Immune-stimulating complexes induce an IL-12-dependent cascade of innate immune responses. *J Immunol* 1999, **162**, 5536–5546.

60. Davidsen, J., Rosenkrands, I., Christensen, D., Vangala, A., Kirby, D., Perrie, Y., Agger, E.M. & Andersen, P. Characterization of cationic liposomes based on

dimethyldioctadecylammonium and synthetic cord factor from *M. tuberculosis* (trehalose 6,6'-dibehenate) – a novel adjuvant inducing both strong CMI and antibody responses. *Biochim Biophys Acta* 2005, **1718**, 22–31.

61. Siegrist, C.A., Pihlgren, M., Tougne, C., Efler, S.M., Morris, M.L., Al Adhami, M.J., Cameron, D.W., Cooper, C.L., Heathcote, J., Davis, H.L. & Lambert, P.H. Co-administration of CpG oligonucleotides enhances the late affinity maturation process of human anti-hepatitis B vaccine response. *Vaccine* 2004, **23**, 615–622.

62. Kim, S.K., Ragupathi, G., Cappello, S., Kagan, E. & Livingston, P.O. Effect of immunological adjuvant combinations on the antibody and T-cell response to vaccination with MUC1-KLH and GD3-KLH conjugates. *Vaccine* 2000, **19**, 530–537.

63. Olszewska, W., Partidos, C.D. & Steward, M.W. Antipeptide antibody responses following intranasal immunization: effectiveness of mucosal adjuvants. *Infect Immun* 2000, **68**, 4923–4929.

64. Moschos, S.A., Bramwell, V.W., Somavarapu, S. & Alpar, H.O. Comparative immunomodulatory properties of a chitosan-MDP adjuvant combination following intranasal or intramuscular immunisation. *Vaccine* 2005, **23**, 1923–1930.

65. Berry, L.J., Hickey, D.K., Skelding, K.A., Bao, S., Rendina, A.M., Hansbro, P.M., Gockel, C.M. & Beagley, K.W. Transcutaneous immunization with combined cholera toxin and CpG adjuvant protects against *Chlamydia muridarum* genital tract infection. *Infect Immun* 2004, **72**, 1019–1028.

66. Baudner, B.C., Giuliani, M.M., Verhoef, J.C., Rappuoli, R., Junginger, H.E. & Giudice, G.D. The concomitant use of the LTK63 mucosal adjuvant and of chitosan-based delivery system enhances the immunogenicity and efficacy of intranasally administered vaccines. *Vaccine* 2003, **21**, 3837–3844.

67. Mowat, A.M., Parker, L.A., Beacock-Sharp, H., Millington, O.R. & Chirdo, F. Oral tolerance: overview and historical perspectives. *Ann N Y Acad Sci* 2004, **1029**, 1–8.

68. Lycke, N. From toxin to adjuvant: the rational design of a vaccine adjuvant vector, CTA1-DD/ISCOM. *Cell Microbiol* 2004, **6**, 23–32.

69. Salmond, R.J., Pitman, R.S., Jimi, E., Soriani, M., Hirst, T.R., Ghosh, S., Rincon, M. & Williams, N.A. CD8+ T cell apoptosis induced by *Escherichia coli* heat-labile enterotoxin B subunit occurs via a novel pathway involving NF-kappaB-dependent caspase activation. *Eur J Immunol* 2002, **32**, 1737–1747.

70. Lavelle, E.C., Jarnicki, A., McNeela, E., Armstrong, M.E., Higgins, S.C., Leavy, O. & Mills, K.H. Effects of cholera toxin on innate and adaptive immunity and its application as an immunomodulatory agent. *J Leukoc Biol* 2004, **75**, 756–763.

71. Stevceva, L. & Ferrari, M.G. Mucosal adjuvants. *Curr Pharm Des* 2005, **11**, 801–811.

72. Childers, N.K., Miller, K.L., Tong, G., Llarena, J.C., Greenway, T., Ulrich, J.T. & Michalek, S.M. Adjuvant activity of monophosphoryl lipid A for nasal and oral immunization with soluble or liposome-associated antigen. *Infect Immun* 2000, **68**, 5509–5516.

73. Harandi, A.M. & Holmgren, J. CpG DNA as a potent inducer of mucosal immunity: implications for immunoprophylaxis and immunotherapy of mucosal infections. *Curr Opin Investig Drugs* 2004, **5**, 141–145.

74. Mowat, A.M. Anatomical basis of tolerance and immunity to intestinal antigens. *Nat Rev Immunol* 2003, **3**, 331–341.

75. O'Hagan, D.T., MacKichan, M.L., & Singh, M. Recent developments in adjuvants for vaccines against infectious diseases. *Biomol Eng* 2001, **18**, 69–85.

76. Eriksson, A.M., Schon, K.M. & Lycke, N.Y. The cholera toxin – derived CTA1-DD vaccine adjuvant administered intranasally does not cause inflammation or accumulate in the nervous tissues. *J Immunol* 2004, **173**, 3310–3319.

77. Iwasaki, A. & Medzhitov, R. Toll-like receptor control of the adaptive immune responses. *Nat Immunol* 2004, **5**, 987–995.

78. Itano, A.A., McSorley, S.J., Reinhardt, R.L., Ehst, B.D., Ingulli, E., Rudensky, A.Y. & Jenkins, M.K. Distinct dendritic cell populations sequentially present antigen to CD4 T cells and stimulate different aspects of cell-mediated immunity. *Immunity* 2003, **19**, 47–57.

79. Pulendran, B., Banchereau, J., Maraskovsky, E. & Maliszewski, C. Modulating the immune response with dendritic cells and their growth factors. *Trends Immunol* 2001, **22**, 41–47.

80. Hawiger, D., Inaba, K., Dorsett, Y., Guo, M., Mahnke, K., Rivera, M., Ravetch, J.V., Steinman, R.M., & Nussenzweig, M.C. Dendritic cells induce peripheral T cell unresponsivenes under steady state conditions in vivo. *J Exp Med* 2001, **194**, 769–779.

81. Mempel, T.R., Henrickson, S.E. & Von Andrian, U.H. T-cell priming by dendritic cells in lymph nodes occurs in three distinct phases. *Nature* 2004, **427**, 154–159.

82. Fan, E., O'Neal, C.J., Mitchell, D.D., Robien, M.A., Zhang, Z., Pickens, J.C., Tan, X.J., Korotkov, K., Roach, C., Krumm, B., Verlinde, C.L., Merritt, E.A. & Hol, W.G. Structural biology and structure-based inhibitor design of cholera toxin and heat-labile enterotoxin. *Int J Med Microbiol* 2004, **294**, 217–223.

83. De Haan, L. & Hirst, T.R. Cholera toxin: a paradigm for multifunctional engagement of cellular mechanisms (review). *Mol Membr Biol* 2004, **21**, 77–92.

84. Miller, C.E., Majewski, J., Faller, R., Satija, S. & Kuhl, T.L. Cholera toxin assault on lipid monolayers containing ganglioside GM1. *Biophys J* 2004, **86**, 3700–3708.

85. Soriani, M., Bailey, L. & Hirst, T.R. Contribution of the ADP-ribosylating and receptor-binding properties of cholera-like enterotoxins in modulating cytokine secretion by human intestinal epithelial cells. *Microbiology* 2002, **148**, 667–676.

86. Rappuoli, R., Pizza, M., Douce, G. & Dougan, G. Structure and mucosal adjuvanticity of cholera and *Escherichia coli* heat-labile enterotoxins. *Immunol Today* 1999, **20**, 493–500.

87. Kawamura, Y.I., Kawashima, R., Shirai, Y., Kato, R., Hamabata, T., Yamamoto, M., Furukawa, K., Fujihashi, K., McGhee, J.R., Hayashi, H. & Dohi, T. Cholera toxin activates dendritic cells through dependence on GM1-ganglioside which is mediated by NF-kappaB translocation. *Eur J Immunol* 2003, **33**, 3205–3512.

88. Freytag, L.C. & Clements, J.D. Mucosal adjuvants. *Vaccine* 2005, **23**, 1804–1813.

89. Giuliani, M.M., Del Giudice, G., Giannelli, V., Dougan, G., Douce, G., Rappuoli, R. & Pizza, M. Mucosal adjuvanticity and immunogenicity of LTR72, a novel mutant of *Escherichia coli* heat-labile enterotoxin with partial knockout of ADP- ribosyltransferase activity [in process citation]. *J Exp Med* 1998, **187**, 1123–1132.

90. Sandvig, K., Spilsberg, B., Lauvrak, S.U., Torgersen, M.L., Iversen, T.G. & van Deurs, B. Pathways followed by protein toxins into cells. *Int J Med Microbiol* 2004, **293**, 483–490.

91. Zhu, X., Kim, E., Boman, A.L., Hodel, A., Cieplak, W. & Kahn, R.A. ARF binds the C-terminal region of the *Escherichia coli* heat-labile toxin (LTA1) and competes for the binding of LTA2. *Biochemistry* 2001, **40**, 4560–4568.

92. Massol, R.H., Larsen, J.E., Fujinaga, Y., Lencer, W.I. & Kirchhausen, T. Cholera toxin toxicity does not require functional Arf6- and dynamin-dependent endocytic pathways. *Mol Biol Cell* 2004, **15**, 3631–3641.

93. Stevens, L.A., Moss, J., Vaughan, M., Pizza, M. & Rappuoli, R. Effects of site-directed mutagenesis of *Escherichia coli* heat-labile enterotoxin on ADP-ribosyltransferase activity and interaction with ADP-ribosylation factors. *Infect Immun* 1999, **67**, 259–265.

94. Tsai, B., Rodighiero, C., Lencer, W. I. & Rapoport, T.A. Protein disulfide isomerase acts as a redox-dependent chaperone to unfold cholera toxin. *Cell* 2001, **104**, 937–948.

95. Gardby, E., Wrammert, J., Schon, K., Ekman, L., Leanderson, T. Lycke, N. Strong differential regulation of serum and mucosal IgA responses as revealed in CD28-deficient mice using cholera toxin adjuvant. *J Immunol* 2003, **170**, 55–63.

96. Chen, X., Winkler-Pickett, R.T., Carbonetti, N.H., Ortaldo, J.R., Oppenheim, J.J. & Howard, O.M. Pertussis toxin as an adjuvant suppresses the number and function of CD4(+)CD25(+) T regulatory cells. *Eur J Immunol* 2006, **36**, 671–680.

97. Gagliardi, M.C., Sallusto, F., Marinaro, M., Langenkamp, A., Lanzavecchia, A. & De Magistris, M.T. Cholera toxin induces maturation of human dendritic cells and licences them for Th2 priming. *Eur J Immunol* 2000, **30**, 2394–2403.

98. Gagliardi, M.C. & De Magistris, M.T. Maturation of human dendritic cells induced by the adjuvant cholera toxin: role of cAMP on chemokine receptor expression. *Vaccine* 2003, **21**, 856–861.

99. Holmgren, J., Harandi, A.M. & Czerkinsky, C. Mucosal adjuvants and anti-infection and anti-immunopathology vaccines based on cholera toxin, cholera toxin B subunit and CpG DNA. *Expert Rev Vaccines* 2003, **2**, 205–217.

100. Grdic, D., Ekman, L., Schon, K., Lindgren, K., Mattsson, J., Magnusson, K.E., Ricciardi-Castagnoli, P. & Lycke, N. Splenic marginal zone dendritic cells mediate the cholera toxin adjuvant effect: dependence on the ADP-ribosyltransferase activity of the holotoxin. *J Immunol* 2005, **175**, 5192–5202.

101. Holmgren, J. & Czerkinsky, C. Mucosal immunity and vaccines. *Nat Med* 2005, **11**, S45–S53.

102. Kaper, J.B., Lockman, H., Baldini, M.M. & Levine, M.M. A recombinant live oral cholera vaccine. *Nature* 1984, **308**, 655–658.

103. Fujihashi, K., Koga, T., van Ginkel, F.W., Hagiwara, Y. & McGhee, J.R. A dilemma for mucosal vaccination: efficacy versus toxicity using enterotoxin-based adjuvants. *Vaccine* 2002, **20**, 2431–2438.

104. Mutsch, M., Zhou, W., Rhodes, P., Bopp, M., Chen, R.T., Linder, T., Spyr, C. & Steffen, R. Use of the inactivated intranasal influenza vaccine and the risk of Bell's palsy in Switzerland. *N Engl J Med* 2004, **350**, 896–903.

105. Glueck, R. Pre-clinical and clinical investigation of the safety of a novel adjuvant for intranasal immunization. *Vaccine* 2001, **20**, (Suppl 1), S42–S44.

106. Douce, G., Giuliani, M.M., Giannelli, V., Pizza, M.G., Rappuoli, R. & Dougan, G. Mucosal immunogenicity of genetically detoxified derivatives of heat labile toxin from *Escherichia coli* [in process citation]. *Vaccine* 1998, **16**, 1065–1073.

107. Kweon, M.N., Yamamoto, M., Watanabe, F., Tamura, S., Van Ginkel, F.W., Miyauchi, A., Takagi, H., Takeda, Y., Hamabata, T., Fujihashi, K., McGhee, J.R. & Kiyono, H. A nontoxic chimeric enterotoxin adjuvant induces protective immunity in both mucosal and systemic compartments with reduced IgE antibodies. *J Infect Dis* 2002, **186**, 1261–1269.

108. Agren, L.C., Ekman, L., Lowenadler, B. & Lycke, N.Y. Genetically engineered nontoxic vaccine adjuvant that combines B cell targeting with immunomodulation by cholera toxin A1 subunit. *J Immunol* 1997, **158**, 3936–3946.

109. Uhlen, M., Guss, B., Nilsson, B., Gatenbeck, S., Philipson, L. & Lindberg, M. Complete sequence of the staphylococcal gene encoding protein A. A gene evolved through multiple duplications. *J Biol Chem* 1984, **259**, 1695–1702.

110. Agren, L., Lowenadler, B. & Lycke, N. A novel concept in mucosal adjuvanticity: the CTA1-DD adjuvant is a B cell–targeted fusion protein that incorporates the enzymatically active cholera toxin A1 subunit. *Immunol Cell Biol* 1998, **76**, 280–287.

111. Levine, M.M., Black, R.E., Clements, M.L., Lanata, C., Sears, S., Honda, T., Young, C.R. & Finkelstein, R.A. Evaluation in humans of attenuated *Vibrio cholerae* El Tor Ogawa strain Texas Star-SR as a live oral vaccine. *Infect Immun* 1984, **43**, 515–522.

112. Agren, L.C., Ekman, L., Lowenadler, B., Nedrud, J.G. & Lycke, N.Y. Adjuvanticity of the cholera toxin A1-based gene fusion protein, CTA1-DD, is critically dependent on the ADP-ribosyltransferase and Ig-binding activity. *J Immunol* 1999, **162**, 2432–2440.

113. Agren, L., Sverremark, E., Ekman, L., Schon, K., Lowenadler, B., Fernandez, C. & Lycke, N. The ADP-ribosylating CTA1-DD adjuvant enhances T cell–dependent and independent responses by direct action on B cells involving anti-apoptotic Bcl-2- and germinal center-promoting effects. *J Immunol* 2000, **164**, 6276–6286.

114. van Ginkel, F.W., Jackson, R.J., Yuki, Y. & McGhee, J.R. Cutting edge: the mucosal adjuvant cholera toxin redirects vaccine proteins into olfactory tissues. *J Immunol* 2000, **165**, 4778–4782.

115. Choi, A.H., McNeal, M.M., Flint, J.A., Basu, M., Lycke, N.Y., Clements, J.D., Bean, J.A., Davis, H.L., McCluskie, M.J., VanCott, J.L. & Ward, R.L. The level of protection against rotavirus shedding in mice following immunization with a chimeric VP6 protein is dependent on the route and the coadministered adjuvant. *Vaccine* 2002, **20**, 1733–1740.

116. Akhiani, A.A., Stensson, A., Schon, K. & Lycke, N. The nontoxic CTA1-DD adjuvant enhances protective immunity against helicobacter pylori infection following mucosal immunization. *Scand J Immunol* 2006, **63**, 97–105.

117. De Filette, M., Ramne, A., Birkett, A., Lycke, N., Lowenadler, B., Min Jou, W., Saelens, X. & Fiers, W. The universal influenza vaccine M2e-HBc administered intranasally in combination with the adjuvant CTA1-DD provides complete protection. *Vaccine* 2006, **24**, 544–551.

118. Cong, Y., Weaver, C. & Elson, C. The mucosal adjuvanticity of cholera toxin involves enhancement of co-stimulatory activity by selective up-regulation of B7.2 expression. *J Immunol* 1997, **159**, 5301.

119. Mowat, A.M., Donachie, A.M., Jagewall, S., Schon, K., Lowenadler, B., Dalsgaard, K., Kaastrup, P. & Lycke, N. CTA1-DD-immune stimulating complexes: a novel, rationally designed combined mucosal vaccine adjuvant effective with nanogram doses of antigen. *J Immunol* 2001, **167**, 3398–3405.

120. Helgeby, A., Robson, N.C., Donachie, A.M., Beackock-Sharp, H., Lovgren, K., Schon, K., Mowat, A. & Lycke, N.Y. The combined CTA1-DD/ISCOM adjuvant vector promotes priming of mucosal and systemic immunity to incorporated antigens by specific targeting of B cells. *J Immunol* 2006, **176**, 3697–3706.

121. Robson, N.C., Beacock-Sharp, H., Donachie, A.M. & Mowat, A.M. The role of antigen-presenting cells and interleukin-12 in the priming of antigen-specific CD4+ T cells by immune stimulating complexes. *Immunology* 2003, **110**, 95–104.

122. Katakai, T., Hara, T., Lee, J.H., Gonda, H., Sugai, M. & Shimizu, A. A novel reticular stromal structure in lymph node cortex: an immuno-platform for interactions among dendritic cells, T cells and B cells. *Int Immunol* 2004, **16**, 1133–1142.

123. Galibert, L., Diemer, G.S., Liu, Z., Johnson, R.S., Smith, J.L., Walzer, T., Comeau, M.R., Rauch, C.T., Wolfson, M.F., Sorensen, R.A., Van der Vuurst de Vries, A.R., Branstetter, D.G., Koelling, R.M., Scholler, J., Fanslow, W.C., Baum, P.R., Derry, J.M. & Yan, W. Nectin-like protein 2 defines a subset of T-cell zone dendritic cells and is a ligand for class-I-restricted T-cell-associated molecule. *J Biol Chem* 2005, **280**, 21955–21964.

# 4

# ALUMINUM-CONTAINING ADJUVANTS: PROPERTIES, FORMULATION, AND USE

Stanley L. Hem and Harm HogenEsch

Aluminum-containing adjuvants have become an important component of many vaccines since their discovery by Glenny, Pope, Waddington, and Wallace in 1926 [1]. They safely potentiate the immune response, thereby enabling effective vaccines to be produced. The search for agents to potentiate the immune response began in 1925 when Ramon [2] added such agents as agar, tapioca, lecithin, starch oil, and bread crumbs to tetanus and diphtheria vaccine. The search for improved adjuvants continues today, but aluminum-containing adjuvants remain the mainstay of vaccine formulations. Therefore, it is important to understand their properties and mechanism of immunopotentiation in order to formulate the most effective vaccines. In this chapter we present the current understanding of the mechanism of immunopotentiation; structure and properties; stability; adsorption and elution mechanisms; elimination; formulation; and safety of aluminum-containing adjuvants.

## 4.1 MECHANISM OF IMMUNOPOTENTIATION

Although aluminum-containing adjuvants have been administered to human beings and animals in millions of doses of vaccines, surprisingly little is known about the mechanisms by which these adjuvants enhance the immune response. Aluminum-adjuvanted vaccines are usually administered intramuscularly or subcutaneously. Vaccine antigens need to reach the draining lymph node,

*Vaccine Adjuvants and Delivery Systems,* Edited by Manmohan Singh
Copyright © 2007 John Wiley & Sons, Inc.

where they are recognized by naive T and B cells. Soluble antigens can reach the lymph node via the afferent lymphatics and can interact directly with the immunoglobulin receptors on B cells, or they are taken up and processed by resident dendritic cells for presentation to T cells. Alternatively, antigens are taken up by immature dendritic cells at the site of injection and carried via the afferent lymph vessels to the lymph node for presentation to T cells. Recent experiments show that soluble antigens reach the lymph node within minutes, whereas dendritic cells with antigen arrive after eight to 12 hours [3,4]. Whereas activated T cells can be detected following the arrival of soluble antigen in the lymph node, sustained T-cell activation may only occur after the second wave of antigen arrives.

It is generally thought that the adsorption of antigens to aluminum salts is critical to the adjuvant effect [5]. The traditional view is that antigens adsorbed onto aluminum adjuvants are slowly released from an injection site, resulting in exposure of the immune system to antigens over a prolonged period of time and activation of a sufficient number of lymphocytes. This is known as the *depot hypothesis* and is based on the observation that a significant portion of aluminum adjuvant–adsorbed diphtheria toxoid was retained at the injection site for days or weeks, whereas soluble toxoid disappeared rapidly [6,7]. Subsequent studies showed that surgical removal of the injection site four days after injection impaired the immune response to diphtheria toxoid, whereas removal after seven days had no effect [8]. This suggests that it is critical for the immune response to diphtheria toxoid that the antigen be released and reach the lymph nodes over a period of four to seven days following injection. Aluminum-containing adjuvants cause a mild inflammatory response at the injection site, resulting in the recruitment of neutrophils, eosinophils, and macrophages [9,10]. It is likely that inflammation also increases the number of dendritic cells at the injection site, similar to inflammatory responses in the respiratory tract [11]. Aluminum-containing adjuvants are particulate in nature and facilitate the uptake of antigens by antigen-presenting cells [12,13]. Aluminum phosphate adjuvant is composed of aggregates approximately 3 $\mu$m in diameter and was taken up more efficiently than the larger aluminum hydroxide adjuvant aggregates [13]. Thus, the physical association of antigens with aluminum-containing adjuvants delays the diffusion of antigens from the injection site, allowing more time for uptake by newly recruited dendritic cells and directly enhances the uptake of antigens by phagocytosis.

Aluminum-containing adjuvants also enhance the immune response by mechanisms that do not rely on the adsorption of antigens, but rather, have a direct or indirect effect on antigen-presenting cells. This is perhaps most clearly demonstrated by the increased immune response induced by DNA vaccines when mixed with aluminum phosphate adjuvants [14,15]. Aluminum hydroxide adjuvant was not effective, as it tightly adsorbed the DNA plasmids and did not allow gene expression. Aluminum phosphate adjuvant did not bind the DNA and did not affect expression of the DNA plasmid. The aluminum phosphate adjuvant enhanced the immune response even when administered three days before or after the injection of the DNA vaccine [14]. It is unlikely that the newly synthesized and released antigens are adsorbed onto the aluminum salts in vivo, because many interstitial proteins have a high affinity for aluminum salts and prevent adsorption of antigens [16].

There is some evidence that aluminum-containing adjuvants activate dendritic cells directly. Human peripheral blood monocytes pulsed with aluminum hydroxide adjuvant–adsorbed tetanus toxoid induced a much better proliferative response by autologous T cells than did monocytes pulsed with soluble tetanus toxoid [17]. This correlated with an increased uptake of aluminum hydroxide adjuvant–adsorbed tetanus toxoid and induction of IL-1 secretion. Incubation of peripheral blood mononuclear cells, containing both monocytes and lymphocytes, with aluminum-containing adjuvants in vitro resulted in increased expression of CD86 and MHC (major histocompatibility complex) class II on monocytes [18,19]. Many cells acquired a dendritic morphology and expressed CD83, a marker of mature dendritic cells [18]. However, the effect of aluminum adjuvants on MHC class II expression was not seen with purified monocytes or following neutralization of IL-4, suggesting that this effect was indirect [18]. In contrast, aluminum adjuvants directly increased the expression of MHC class II, CD86, and CD83 on mature macrophages [20].

The activation of dendritic cells is believed to be critical to the mechanism of action of many experimental adjuvants that contain microbial products such as lipopolysaccharides, muramyl dipeptide, and bacterial (CpG) DNA. The activation by microbial products occurs via interaction with Toll-like receptors (TLRs), resulting in activation of signaling pathways, including the NFκB signaling pathway, and increased expression of co-stimulatory molecules and secretion of cytokines. The adaptor molecule MyD88 serves as a link between TLRs and the NFκB signaling pathway. The immunopotentiation effect of aluminum hydroxide adjuvant was not diminished in MyD88-deficient mice compared with wild-type control mice [17], suggesting that TLRs are not involved in the mechanism by which aluminum hydroxide adjuvant enhances the immune response. However, recent studies have identified several other adaptor molecules that can compensate for the loss of MyD88 (reviewed in [21]), and therefore interaction between aluminum-containing adjuvants and TLRs cannot be excluded.

Following activation by antigen-presenting dendritic cells, CD4[+] T cells can differentiate into cytokine-secreting effector cells. Th1 cells are characterized by the expression of IFN-γ and induce cell-mediated immune responses and opsonizing antibodies. Th2 cells express IL-4, IL-5, and IL-13 and induce primarily humoral immune responses. The control of CD4[+] T-cell differentiation is complex and includes genetic factors, antigen dose, and the nature of the infectious agent. Adjuvants exert a strong influence on the type of immune response. Many adjuvants, including complete Freund's adjuvant, saponin-containing adjuvants, and immunostimulating complexes, stimulate Th1-biased immune responses, whereas aluminum-containing adjuvants stimulate a Th2-biased immune response [22–24]. IL-12 plays an important role in stimulating the differentiation of Th1 cells. It induces the expression of IFN-$\gamma$ by T cells and NK (natural killer) cells and is produced primarily by dendritic cells and macrophages [25,26]. Microbial stimuli that signal through TLRs are potent signals for the production of IL-12 [27,28] and skew the CD4 T-cell differentiation toward Th1. Activated dendritic cells that do not secrete IL-12 can induce CD4$^+$ differentiation toward Th2. Several factors have been identified that inhibit IL-12 expression by dendritic cells, including prostaglandin E2 and complement products iC3b and C5a [29–31]. Interestingly, aluminum hydroxide adjuvant activates the complement system in a manner independent of the classical or alternative pathway [32]. The complement products may play a role in Th2 differentiation in the immune response to aluminum adjuvanted vaccines.

IL-4 is not only produced by Th2 cells but also promotes the differentiation of CD4$^+$ T cells into Th2 cells. Functional deletion of the IL-4 gene did not interfere with the adjuvant effect of aluminum hydroxide adjuvant and did not affect the differentiation of Th2 cells as demonstrated by unchanged IL-5 secretion [33]. However, in the absence of IL-4, an aluminum-containing adjuvant induced the production of antigen-specific IgG2a antibodies and IFN-$\gamma$-secreting T cells, indicating a Th1-mediated immune response. Similar results were obtained with IL-4R$\alpha$-deficient and Stat-6-deficient mice, demonstrating that both IL-4 and IL-13 are dispensable for the adjuvant effect of aluminum-containing adjuvants, but inhibit Th1 differentiation [34].

## 4.2   STRUCTURE AND PROPERTIES

### 4.2.1   Aluminum Hydroxide Adjuvant

An x-ray diffraction pattern of aluminum hydroxide adjuvant is shown in Figure 4.1. The x-ray diffraction patterns of the two major forms of aluminum hydroxide [Al(OH)$_3$], bayerite and gibbsite, are also presented in Figure 4.1. The x-ray diffraction pattern of aluminum hydroxide adjuvant is significantly different from those of aluminum hydroxide. Thus, aluminum hydroxide adjuvant is not chemically Al(OH)$_3$ as its name implies. The broad reflections by aluminum hydroxide adjuvant at 12.6, 27.5, 38.2, 48.4, and 64.4 °$2\theta$ correspond to $d$-spacings of 6.46, 3.18, 2.35, 1.86, and 1.44 Å. These $d$-spacings identify aluminum hydroxide adjuvant as poorly crystalline boehmite, an aluminum oxyhydroxide with the formula AlO(OH). Poorly crystalline boehmite is widely used in industrial applications as an adsorbent or catalyst.

The infrared spectrum of aluminum hydroxide adjuvant also identifies it as poorly crystalline boehmite [35]. The absorption band at 1070 cm$^{-1}$ and the strong shoulder at 3100 cm$^{-1}$ are characteristic of poorly crystalline boehmite. Aluminum hydroxide adjuvant is composed of fibrous primary particles that average 4.5 × 2.2 × 10 nm along the $a$, $b$, and $c$ axes [36] (Figure 4.2). The primary crystallite size can be characterized by the line broadening of the

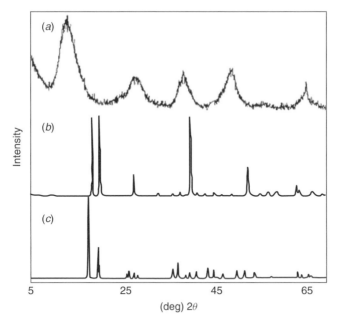

**Figure 4.1** Reference x-ray diffraction patterns using CuKα radiation of (a) aluminum hydroxide adjuvant; (b) bayerite; (c) gibbsite.

**Figure 4.2** Transmission electron micrograph of aluminum hydroxide adjuvant. (From Ref. 35.)

(020) x-ray reflection, which occurs at 12.6 °2θ. The width of the (020) band at half height (WHH) is directly related to the protein adsorption capacity [37]. The primary particles form loose, irregular aggregates that range from 1 to 10 μm. The aggregates are easily dissociated by the shear associated with mixing but readily re-form when the shear is removed.

The small primary particles are responsible for the use of poorly crystalline boehmite as an adsorbent. Traditional techniques for measuring surface area require complete desiccation of the sample. The drying process causes the very small primary particles to fuse irreversibly. Thus, traditional techniques underestimate the surface area of aluminum hydroxide adjuvant. The surface area of aluminum hydroxide adjuvant has been determined by a gravimetric Fourier-transformed infrared (FTIR) spectrophotometer that measured the FTIR spectra and water sorption isotherm simultaneously. The surface area determined by this method was approximately 500 m²/g [36]. Such a high surface area is unusual for a crystalline material and approaches the surface area values reported for expandable clay minerals of 600 to 800 m²/g.

Aluminum hydroxide adjuvant exhibits very low solubility in aqueous media, but the solubility increases in acidic or alkaline media [38] (Figure 4.3). The adsorbent properties of aluminum hydroxide adjuvant are related to the surface area as well as the surface groups. The surface is composed of hydroxyl groups coordinated to aluminum (Figure 4.4a). Such hydroxyl groups, known as metallic hydroxyls, can accept a proton to produce a positive site or can donate a proton to produce a negative site at the surface [39]. Thus, the surface

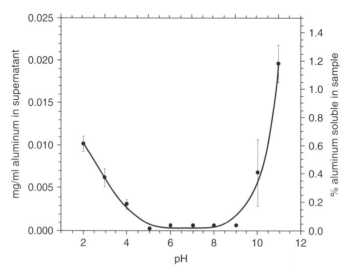

**Figure 4.3** Effect of pH on the solubility of aluminum hydroxide adjuvant after a 45-minute exposure period. The error bars represent 95% confidence intervals of the mean. (From Ref. 38.)

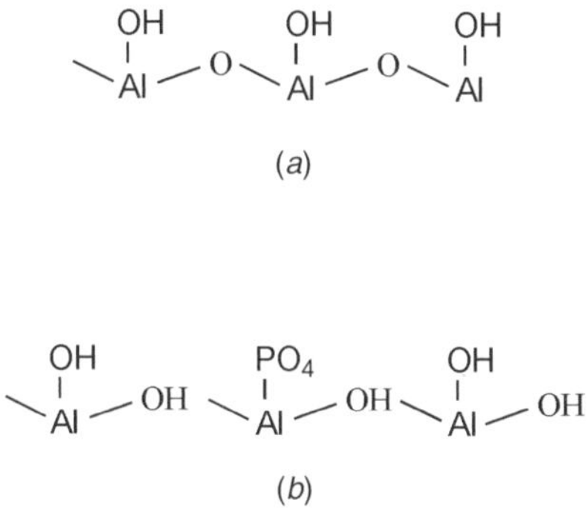

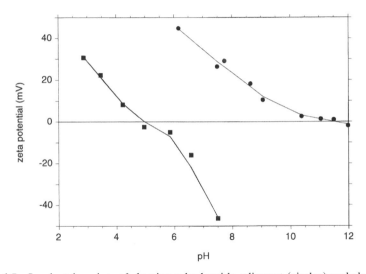

**Figure 4.4** Schematic of surface of (*a*) aluminum hydroxide adjuvant and (*b*) aluminum phosphate adjuvant.

**Figure 4.5** Isoelectric points of aluminum hydroxide adjuvant (circles) and aluminum phosphate adjuvant (squares). (From Ref. 38.)

charge of aluminum hydroxide adjuvant is pH dependent. The isoelectric point of aluminum hydroxide adjuvant is 11.4 (Figure 4.5). The hydroxyl groups at the surface of aluminum hydroxide adjuvant can be displaced by phosphate or fluoride ions, a process known as *ligand exchange* [39], which is an important adsorption mechanism for phosphorylated antigens.

### 4.2.2 Aluminum Phosphate Adjuvant

Aluminum phosphate adjuvant is amorphous to x-rays and therefore cannot be identified by x-ray diffraction. The FTIR spectrum [35] exhibits an absorption band at $1100\,cm^{-1}$ which is characteristic of phosphate. There is a broad hydroxyl-stretching band around $3400\,cm^{-1}$ due to adsorbed water and/ or structural hydroxyls. When the sample was heated to $200°C$, a prominent band at $3164\,cm^{-1}$ was apparent that is characteristic of structural hydroxyls. Thus, aluminum phosphate adjuvant is chemically amorphous aluminum hydroxyphosphate and not $AlPO_4$, as its name implies. Aluminum hydroxy-phosphate is not a stoichiometric compound, and the hydroxyl and phosphate composition depends on the precipitation reactants and conditions.

The primary particles are plates having a diameter of approximately 50 nm (Figure 4.6). The primary particles form loose, irregular aggregates of 1 to 10 µm. The aggregates are easily dissociated by the shear associated with mixing but re-form readily when the shear is removed. Although the surface area of aluminum phosphate adjuvant has not been determined, the very small primary particles seen in Figure 4.6 are likely to have a high surface area. Aluminum phosphate adjuvant is more soluble than aluminum hydroxide adjuvant, and like aluminum hydroxide adjuvant, it is more soluble in acidic or alkaline media (Figure 4.7).

A schematic drawing of the surface of aluminum phosphate adjuvant is seen in Figure 4.4b. The surface is composed of both hydroxyl and phosphate groups. The relative proportion of hydroxyl and phosphate groups is deter-mined during precipitation. The hydroxyl groups are metallic hydroxyls and exhibit a pH-dependent surface charge [39]. The isoelectric point of aluminum phosphate adjuvant is inversely related to the P/Al molar ratio (Figure 4.8).

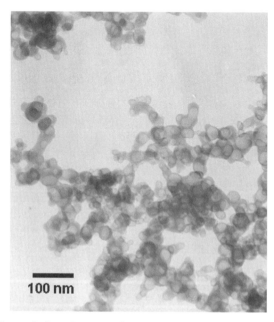

100 nm

**Figure 4.6**   Transmission electron micrograph of aluminum phosphate adjuvant.

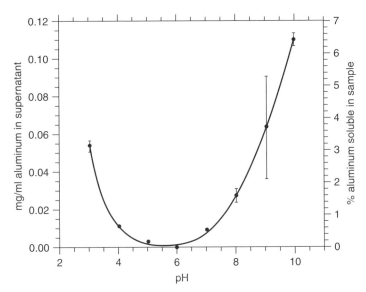

**Figure 4.7** Effect of pH on the solubility of aluminum phosphate adjuvant after a 45-minute exposure period. The error bars represent 95% confidence intervals of the mean. (From Ref. 38.)

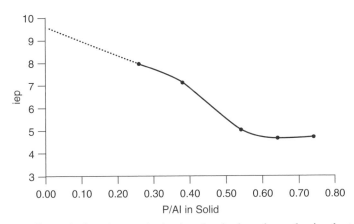

**Figure 4.8** Effect of phosphate substitution for hydroxyl on the isoelectric point (PZC) of aluminum phosphate adjuvant. (From Ref. 40.)

Substitution of phosphate anions for hydroxyl anions lowers the isoelectric point. It is interesting to note that the isoelectric point in Figure 4.8 extrapolates to 9.6, the isoelectric point of $Al(OH)_3$ [41], at a P/Al molar ratio of zero. Aluminum phosphate adjuvant can also undergo ligand exchange with phosphorylated antigens, although there are fewer hydroxyl groups at the surface than are present in aluminum hydroxide adjuvant.

### 4.2.3   In Situ Precipitation

The first use of aluminum-containing adjuvants by Glenny et al. [1] precipitated the aluminum-containing adjuvant in the presence of diphtheria toxoid. Alum, a soluble aluminum compound having the formula $KAl(SO_4)_2 \cdot 12H_2O$, was the source of the aluminum cation. The adjuvant was precipitated by the addition of a base such as sodium hydroxide. Such vaccines are known as *alum-precipitated vaccines* [42]. Vaccines continue to be prepared by the in situ precipitation of the adjuvant.

When vaccines are prepared by in situ precipitation, the antigen is usually dissolved in a phosphate buffer which is mixed with the alum solution. When the antigen–phosphate–alum solution is precipitated by the addition of a solution of sodium hydroxide, the precipitate is amorphous aluminum hydroxyphosphate sulfate [35], as both phosphate anions from the buffer and sulfate anions from the alum substitute for hydroxyls in the precipitate. The morphology is similar to that of aluminum phosphate adjuvant shown in Figure 4.6. Because aluminum hydroxyphosphate sulfate is not a stoichiometric compound, the relative proportions of hydroxyl, phosphate, and sulfate depend on the precipitation recipe and conditions.

There are two views regarding the location of the antigen in a vaccine prepared by in situ precipitation. A World Health Organization report [43] on immunological adjuvants states that "sometimes the aluminum salt is formed in the presence of antigen, securing occlusion of the antigen in the adjuvant." Another view is that the antigen is adsorbed on the surface of the adjuvant when a vaccine is prepared by in situ precipitation. The adsorption of antigen by the adjuvant is supported by a study [44] that found that lysozyme in a vaccine prepared by in situ precipitation could be eluted completely by treatment with sodium dodecyl sulfate. Thus, lysozyme was accessible to the liquid media and was not occluded within the aluminum-containing adjuvant. This result suggests that the in situ precipitation of alum involves two separate processes: precipitation of amorphous aluminum hydroxyphosphate sulfate and adsorption of the antigen by the solid phase.

## 4.3   STABILITY

### 4.3.1   Aging

Although aluminum hydroxide adjuvant is crystalline, its x-ray diffraction bands are broader than those of the crystalline forms of aluminum hydroxide (Figure 4.1). The degree of order in crystalline solids is characterized by the WHH of the major diffraction band. The WHH is inversely related to the degree of order in the crystalline material. The relatively large WHH values for aluminum hydroxide adjuvant suggest that the solid phase may become more highly ordered during aging. Table 4.1 shows the changes in the degree of order of two samples of aluminum hydroxide adjuvant that were aged at

**TABLE 4.1 Effect of Aging at Room Temperature on the Properties of Two Samples of Aluminum Hydroxide Adjuvant**

| | A | | | B | | |
|---|---|---|---|---|---|---|
| | WHH (°2θ) | pH | BSA[a] Adsorption Capacity (mg/mg Al) | WHH (°2θ) | pH | BSA Adsorption Capacity (mg/mg Al) |
| Initial | 4.20 | 6.30 | 2.9 | 3.90 | 6.50 | 2.7 |
| 6 months | 3.75 | 6.24 | 2.8 | 3.72 | 6.48 | 2.3 |
| Diluted, 6 months | 3.62 | 6.15 | 2.8 | 3.53 | 6.39 | 2.3 |
| 15 months | 3.47 | 6.24 | 2.6 | 3.25 | 6.40 | 2.4 |
| Diluted, 15 months | — | 5.98 | 2.1 | — | 6.15 | 1.7 |

*Source:* Data from Burrell, L.S., White, J.L. & Hem, S.L. Stability of aluminum-containing adjuvants during aging at room temperature. *Vaccine* 2000, **18**, 2188–2192.
[a]BSA, bovine serum albumin.

room temperature. The samples were aged as received (10.6 mg Al/mL) and then diluted to 1.7 mg Al/mL with doubly distilled water. The WHH of both samples of aluminum hydroxide adjuvant decreased during aging, indicating that the adjuvants were becoming more highly ordered. The decrease in WHH during aging was accompanied by a decrease in pH (Table 4.1). The development of order occurs by the formation of double hydroxide bridges by sequential deprotonation and dehydration reactions:

$$AL(OH_2)_6^{3+} \rightleftharpoons AL(OH)(OH_2)_5^{2+} + H^+ \tag{4.1}$$

$$2Al(OH)(OH_2)_5^{2+} \rightleftharpoons Al_2(OH)_2(OH_2)_8^{4+} + 2H_2O \tag{4.2}$$

The deprotonation reaction results in a decrease in the pH. The decrease in pH seen during the aging of the aluminum hydroxide adjuvants is confirmation that order developed during aging.

The protein adsorption capacity of the aluminum hydroxide adjuvants also decreased during aging (Table 4.1). The diluted samples exhibited a greater decrease in bovine serum albumin adsorption capacity than that of the undiluted samples. Thus, even though aluminum hydroxide adjuvant is crystalline, further order develops during aging, which results in a decrease in the protein adsorption capacity.

An aluminum phosphate adjuvant containing 4.4 mg Al/mL and having a P/Al molar ratio of 1.0 was aged at room temperature for 15 months. The adjuvant was amorphous to x-rays initially and during the 15-month aging period. However, order was developing during aging as the pH decreased, indicating the formation of double hydroxide bridges, and the lysozyme adsorption capacity decreased (Table 4.2).

**TABLE 4.2    Effect of Aging at Room Temperature on the Properties of an Aluminum Phosphate Adjuvant Having a P/Al Molar Ratio of 1.0**

|           | pH   | Lysozyme Adsorption Capacity (mg/mg Al) |
|-----------|------|-----------------------------------------|
| Initial   | 6.50 | 0.80                                    |
| 6 months  | 6.34 | 0.72                                    |
| 15 months | 6.32 | 0.54                                    |

*Source:* Data from Burrell, L.S., White, J.L. & Hem, S.L. Stability of aluminum-containing adjuvants during aging at room temperature. *Vaccine* 2000, **18**, 2188–2192.

Stability studies of a widely used antacid, aluminum hydroxycarbonate, have shown that carbonate substitution for hydroxyl improves the stability by blocking the formation of double hydroxide bridges [45]. It is likely that phosphate acts in a similar manner to slow the development of order in amorphous aluminum hydroxyphosphate. This hypothesis was confirmed by a study of the stability of five aluminum phosphate adjuvants having P/Al molar ratios ranging from 0.26 to 0.74 [40]. It was found that the rate of aging was greatest for the adjuvant having a P/Al molar ratio of 0.26 and that a P/Al molar ratio of at least 0.50 was needed to minimize the rate of aging.

### 4.3.2    Autoclaving

Autoclaving aluminum hydroxide adjuvant at 121°C increased the degree of order in the crystalline solid, as evidenced by a decrease in the WHH of the 020 reflection [46]. The bovine serum albumin (BSA) adsorption capacity decreased during autoclaving, although the change was not statistically significant at the 95% confidence level. The pH and viscosity decreased due to autoclaving. The viscosity of suspensions is usually inversely related to particle size.

Aluminum phosphate adjuvant also demonstrated an increase in order during autoclaving, although the x-ray diffraction pattern remained amorphous. Both the pH and lysozyme adsorption capacity decreased during autoclaving at 121°C for 30 or 60 minutes. These changes are consistent with an increase in order.

Aluminum hydroxide adjuvant and aluminum phosphate adjuvant both exhibited an increase in order during autoclaving, which resulted in a decrease in protein adsorption capacity. Autoclaving conditions should be selected that minimize exposure time to elevated temperatures. Procedures requiring repeated autoclaving of the same samples should be avoided.

### 4.3.3    Freezing

It is well recognized that freezing can adversely affect the properties of suspensions [47]. The monograph for aluminum hydroxide adjuvant in the *European Pharmacopoeia* states "Do not allow to freeze." The monographs for aluminum hydroxide adjuvant and aluminum phosphate adjuvant in *Pharmaceutical Excipients 2004* both state that the adjuvant "must not be allowed to freeze as the hydrated colloid structure will be irreversibly damaged" [48]. The two major commercial manufacturers of aluminum-containing adjuvants both warn that freezing must be avoided.

When vaccines formulated with aluminum-containing adjuvants are frozen, the physical properties as well as the immunological properties may be adversely affected. The World Health Organization sponsored a collaborative study [49] in which six laboratories studied the effect of freezing diphtheria, pertussis, tetanus (DPT) vaccine. Of the eight adsorbed vaccines tested, six exhibited irreversible physical changes, including the appearance of particulate matter and an increased rate of sedimentation. Three vaccines suffered losses in pertussis potency and two vaccines exhibited a decrease in tetanus potency, but no change was observed in the two vaccines in which the diphtheria potency was tested. The inconsistent effect of freezing on the physical properties and immunogenicity of the vaccines reported in this study suggests that factors that were not controlled contribute to the freeze–thaw stability of aluminum-adjuvanted vaccines. The authors concluded that "although some of the adsorbed DPT vaccines tested did not show evidence of decreased potency or increased toxicity, we recommend that adsorbed vaccines submitted to freezing are discarded."

A study [50] of the storage of pertussis vaccines at $-3°C$ hypothesized that in addition to the physical changes noted for aluminum-containing adjuvants that freezing may cause changes "in the secondary, tertiary or quaternary structure or even fragmentation of the protein antigens."

In summary, it is widely agreed, as stated in the World Health Organization documents *Safe Vaccine Handling, Cold Chain and Immunizations* [51] and *Thermostability of Vaccines* [52], that aluminum-adjuvanted vaccines should not be used if they were exposed to freezing conditions.

## 4.4  ADSORPTION OF ANTIGEN

### 4.4.1  Adsorption Isotherm

Adsorption isotherms of the adsorption of antigens by aluminum-containing adjuvants provide useful information for the formulation of vaccines. They are constructed by adding increasing concentrations of the antigen to a series of samples of the adjuvant at a controlled pH. Each sample is composed of the same concentration of the aluminum-containing adjuvant and a different concentration of the antigen. The samples are allowed to equilibrate and the concentration of antigen in solution is determined by analyzing the supernatant. This value, concentration of antigen in solution at equilibrium, is plotted on the $x$-axis. The amount of antigen that was adsorbed, determined by difference, divided by the weight of the adjuvant is plotted on the $y$-axis.

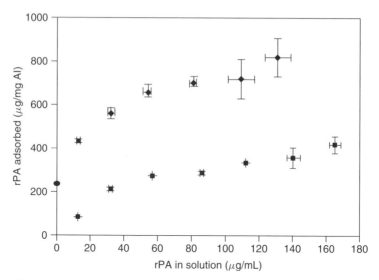

**Figure 4.9** Adsorption isotherms of recombinant protective antigen and aluminum hydroxide adjuvant at pH 7.4, 25°C: in water (crosses) and Dulbeccos phosphate-buffered saline (equates). (From Ref. 53.)

The adsorption isotherm of recombinant protective antigen (rPA) and aluminum hydroxide adjuvant is shown in Figure 4.9. Adsorption was determined in both water and Dulbeccos phosphate-buffered saline. The adsorption isotherms contain three phases. For example, in water the adsorption increased as the equilibrium concentration of rPA in solution increased up to 52 μg/mL. The adsorption then remained constant as the equilibrium concentration of rPA in solution increased from 52 μg/mL to 79 μg/mL. The amount of rPA adsorbed increased as the concentration of rPA in solution at equilibrium increased from 79 μg/mL to 123 μg/mL.

The rate of increase in the first phase is related to the strength of the adsorption force and is characterized by the adsorption coefficient. The adsorption of antigens by aluminum-containing adjuvants is based on an equilibrium between the adsorbed antigen and antigen in solution:

$$\text{antigen adsorbed} \overset{k_f}{\underset{k_r}{\gtrless}} \text{antigen in solution} \qquad (4.3)$$

The adsorption coefficient is the constant that describes the equilibrium. Mathematically, it is the rate of the forward reaction ($k_f$) divided by the rate of the reverse reaction ($k_r$).

The relatively constant amount of rPA that was adsorbed in the second phase indicates that rPA has formed a monolayer on the adjuvant surface. In this region no additional adsorption occurred even though the concentration of antigen in solution at equilibrium increased. The amount of antigen that can be adsorbed as a monolayer is known as the *adsorption capacity*.

As the concentration of antigen in solution at equilibrium increases, a second layer of antigen may adsorb on the monolayer of antigen that represents the adsorption capacity. This multilayer adsorption may occur in some systems and is seen in the third phase of the rPA adsorption isotherms.

A mathematical description of adsorption was developed by Langmuir that includes the adsorption capacity and adsorption coefficient. A test for whether the data from an adsorption isotherm can be analyzed by the Langmuir equation is to plot the data according to the linear form of the Langmuir equation. When the data from Figure 4.9 were plotted according to the linear form of the Langmuir equation, the best-fit lines had correlation coefficients of 0.9850 for water and 0.9376 for Dulbeccos phosphate-buffered saline. The adsorption coefficients were 100 and 4.6 mL/mg, respectively, indicating stronger adsorption forces in water than in Dulbeccos buffered saline.

Adsorption isotherms are frequently characterized as having either high or low affinity. Comparison of the adsorption isotherms of hepatitis B surface antigen (HBsAg) by aluminum hydroxide adjuvant and rPA by aluminum hydroxide adjuvant illustrates this characterization. As shown in Figure 4.10, the adsorption capacity of HBsAg was reached even though the concentration of HBsAg in solution at equilibrium was below the detection point of the assay. Thus, the equilibrium shown in equation (4.3) strongly favors adsorbed antigen. This is a high-affinity adsorption isotherm. In contrast, at least 52 μg rPA/mL in solution at equilibrium was required for monolayer coverage by rPA to be completed. The equilibrium for this system is not as strongly in favor

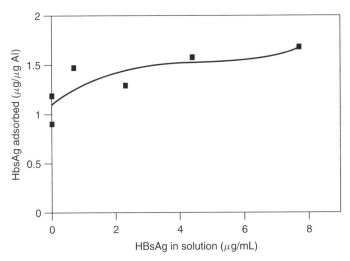

**Figure 4.10** Adsorption isotherm of hepatitis B surface antigen by aluminum hydroxide adjuvant in doubly distilled water at pH 7.4, 25°C. (From Ref. 54.)

of adsorbed antigen as for the HBsAg system. The rPA adsorption isotherm is referred to as a low-affinity adsorption isotherm.

The adsorption coefficient can be used to predict the displacement of an adsorbed antigen by an added antigen or protein. Heimlich et al. [16] studied the in vitro displacement of adsorbed proteins and found that a protein having a larger adsorption coefficient was able to displace a protein with a smaller adsorption coefficient.

### 4.4.2   Adsorption Mechanisms

The major mechanisms by which aluminum-containing adjuvants adsorb antigens are electrostatic attraction, hydrophobic forces, and ligand exchange. Electrostatic attraction is the most frequently encountered mechanism of adsorption. It takes advantage of the high isoelectric point (11.4) of aluminum hydroxide adjuvant and the low isoelectric point (5.0) of commercial aluminum phosphate adjuvant. Once the isoelectric point of the antigen is known, the adjuvant having the opposite surface charge is selected. Seeber et al. [55] illustrated this when they showed that lysozyme (isoelectric point 11.0) was adsorbed extensively by aluminum phosphate adjuvant at pH 7.4 but was poorly adsorbed by aluminum hydroxide adjuvant.

As expected with any electrostatic interaction, the ionic strength of the medium affects adsorption. Al-Shakhshir et al. [56] showed that lysozyme was adsorbed by aluminum phosphate adjuvant at pH 7.4 in the presence of 0.06 M NaCl. Adsorption was greatly reduced at isotonic ionic strength (0.15 M NaCl) and was not observed in 0.25 M NaCl. The ionic strength of vaccines should be kept as low as possible when the antigen is adsorbed by electrostatic attraction. Polyols may be used to adjust the tonicity, rather than salts.

It is possible to change the iep of either aluminum hydroxide adjuvant or aluminum phosphate adjuvant by pretreatment with phosphate anion. Aluminum hydroxide adjuvant was pretreated with five concentrations of phosphate anion [57]. The iep was reduced from 11.4 to 4.6 by the phosphate pretreatment. As shown in Figure 4.11, positively charged lysozyme was adsorbed only when the phosphate pretreatment produced a negative surface charge. Thus, if either aluminum hydroxide adjuvant or aluminum phosphate adjuvant is desired for a vaccine formulation, the surface charge may be reversed in the case of aluminum hydroxide adjuvant by pretreatment with phosphate or by reducing the phosphate content of aluminum phosphate adjuvant (Figure 4.8).

An interesting example of complex electrostatic adsorption is presented by human respiratory syncytial virus vaccine [58]. The antigen, BBG2Na, is a recombinant chimeric protein that has a bipolar two-domain structure. The BB domain has an isoelectric point of 5.5 and the G2Na domain has an isoelectric point of 10.0. The BBG2Na antigen is adsorbed strongly by both aluminum hydroxide adjuvant and aluminum phosphate adjuvant. It was concluded that when BBG2Na was adsorbed by aluminum hydroxide adjuvant,

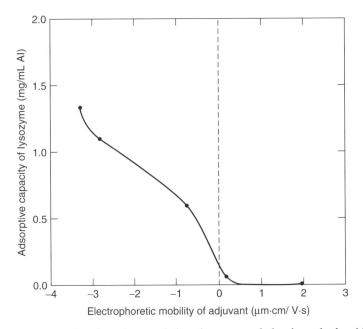

**Figure 4.11**    Effect of surface charge of phosphate-treated aluminum hydroxide adjuvants on the adsorption capacity of lysozyme at pH 7.4 and 25°C. (From Ref. 57.)

negatively charged BB domain was oriented to the adjuvant surface. Conversely, the positively charged G2Na domain was responsible for adsorption by aluminum phosphate adjuvant.

An antigen having exposed hydrophobic regions may be adsorbed by aluminum-containing adjuvants by hydrophobic attractive forces. The contribution of hydrophobic forces to antigen adsorption can be tested by the effect of ethylene glycol on adsorption. Ethylene glycol stabilizes the hydration layer of proteins, which renders hydrophobic interactions thermodynamically unfavorable. The adsorption capacity of lysozyme by aluminum phosphate adjuvant was reduced when ethylene glycol was added [56]. In contrast, the adsorption isotherm of bovine serum albumin by aluminum hydroxide adjuvant was not affected by the addition of ethylene glycol. The observation that hydrophobic interactions contribute to the adsorption of lysozyme but not bovine serum albumin is consistent with lysozyme's longer retention time during hydrophobic interaction chromatography than that of bovine serum albumin.

Ligand exchange occurs with phosphorylated antigens and is the strongest adsorption force. Phosphate forms an inner sphere surface complex with aluminum that is the inorganic equivalent of a covalent band [39]. Phosphate binds more strongly to aluminum than hydroxyl and displaces surface hydroxyl from aluminum hydroxide adjuvant and aluminum phosphate adjuvant.

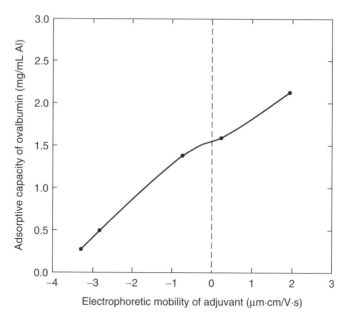

**Figure 4.12**   Effect of surface charge of phosphate-treated aluminum hydroxide adjuvant on the adsorption capacity of ovalbumin at pH 7.4 and 25°C. (From Ref. 57.)

Ligand exchange may occur even when an electrostatic repulsive force is present. The adsorption of ovalbumin (OVA) by the five phosphate-treated aluminum hydroxide adjuvants that were studied with lysozyme (Figure 4.11) is shown in Figure 4.12.

Ovalbumin has an isoelectric point of 5 and contains up to two phosphate groups. Thus, it can adsorb to positively charged aluminum hydroxide adjuvant by both electrostatic attraction and ligand exchange. An electrostatic repulsive force was present when the surface charge was negative. However, as shown in Figure 4.12, extensive adsorption of ovalbumin occurred when the phosphate-treated aluminum hydroxide adjuvant was negatively charged. Adsorption under these conditions is due to ligand exchange. The reason that the adsorption capacity decreased with the higher concentration of phosphate pretreatment is because the number of surface hydroxyl groups available for ligand exchange was reduced as the concentration of phosphate anion used for the pretreatment increased.

Adsorption by ligand exchange is also affected by the degree of phosphorylation of the antigen. Four ovalbumin samples were prepared having different degrees of phosphorylation [59]. The four ovalbumin samples were negatively charged and were electrostatically repelled by aluminum phosphate adjuvant. However, adsorption occurred that was related directly to the degree of phosphorylation of ovalbumin (Table 4.3). The greater the degree of phosphorylation, the greater the opportunity for adsorption by ligand exchange. Thus, the

**TABLE 4.3   Adsorption of Ovalbumin Samples by Aluminum Phosphate Adjuvant at pH 7.4, 25°C**

| Ovalbumin Sample | $PO_4$/OVA Molar Ratio | % Adsorption | % Elution Upon Exposure to Interstitial Fluid for 21 Hours |
|---|---|---|---|
| POVA | 3.2 | 99 | 26 |
| OVA | 1.8 | 38 | 100 |
| DPOVA 1 | 1.2 | 26 | 100 |
| DPOVA 2 | 0.14 | 0 | 100 |

*Source:* Data from Ref. 59.

ovalbumin sample containing $3.2\,mol\ PO_4$/mol ovalbumin was completely adsorbed by ligand exchange, but the sample having only $0.14\,mol\ PO_4$/mol ovalbumin was not adsorbed, as little, if any, ligand exchange was possible, and electrostatic repulsive forces prevented adsorption.

## 4.5   FORMULATION OF THE VACCINE

### 4.5.1   Selection of the Buffer

The potential interaction of the buffer species with aluminum-containing adjuvants should be considered when formulating a vaccine. Two buffer systems require special consideration: phosphate and citrate. Phosphate anions are adsorbed by both aluminum hydroxide adjuvant and aluminum phosphate adjuvant by ligand exchange. Figure 4.13 shows that the isoelectric point of

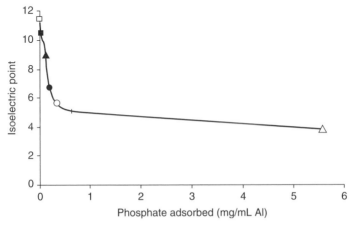

**Figure 4.13**   Relationship between phosphate adsorption by aluminum hydroxide adjuvant and the isoelectric point. (From Ref. 60.)

aluminum hydroxide adjuvant can be decreased from 11.4 to approximately 4 by adsorption of phosphate anion. Similarly, the isoelectric point of aluminum phosphate adjuvant is decreased by phosphate substitution for surface hydroxyls. The isoelectric point of aluminum phosphate adjuvant ranged from approximately 8 to 4, depending on phosphate adsorption (Figure 4.8).

The adsorption of phosphate anion present in a vaccine formulation as buffer can affect the adsorption of an antigen by affecting both the surface charge and the number of surface hydroxyls available for ligand exchange with phosphate groups of the antigen. Figure 4.12 shows that the adsorption capacity of aluminum hydroxide adjuvant for ovalbumin, which may adsorb by electrostatic attraction and/or ligand exchange, decreases due to exposure to phosphate anion. Citrate buffers may lead to an increase in the soluble aluminum concentration. The citrate anion is an α-hydroxycarboxylic acid that adsorbs to an aluminum-containing adjuvant and solubilizes the aluminum by formation of a soluble aluminum–citrate complex [61].

Buffers that do not alter the properties of aluminum-containing adjuvants are acetate and TRIS. However, phosphate buffers may be used as long as their effect on the aluminum-containing adjuvant is understood. For example, adsorption of the experimental malaria antigen R32tet32 was enhanced when the vaccine was formulated with a phosphate buffer. R32tet32 has an isoelectric point of 12.8 and was electrostatically repelled by aluminum hydroxide adjuvant [62]. However, adsorption of R32tet32 increased as the concentration of the phosphate buffer was increased from 3 mM to 20 mM. The buffer ions were adsorbing to the aluminum hydroxide adjuvant and reducing the isoelectric point as shown in Figure 4.13. Lowering the isoelectric point of aluminum hydroxide adjuvant led to the development of electrostatic attractive forces between R32tet32 and the adjuvant.

The use of a phosphate buffer can also decrease the adsorption of an antigen. Recombinant protective antigen has an isoelectric point of 5.6 and is adsorbed electrostatically by aluminum hydroxide adjuvant [53]. Figure 4.9 shows that the adsorption capacity of recombinant protective antigen was decreased when formulated with Dulbeccos phosphate-buffered saline in comparison to water. The decrease in the adsorption capacity is probably caused by a decrease in the isoelectric point of the aluminum hydroxide adjuvant due to adsorption of phosphate anion and an increase in the ionic strength that reduces electrostatic forces.

### 4.5.2 Microenvironment pH

It is a basic principle of colloid chemistry that charged particles electrostatically attract ions of opposite charge (counterions) to form the Gouy–Chapman double layer. Negatively charged surfaces attract cations including protons to form the double layer. Thus, the pH of the double-layer region (i.e., microenvironment pH) will be lower than the bulk pH which is routinely measured by a pH electrode. Similarly, the double layer surrounding positively charged particles will be rich in anions, including hydroxyls, and the microenvironment pH will be higher than the bulk pH.

Antigens that are adsorbed to aluminum-containing adjuvants may be exposed to a pH that is different from the bulk pH, especially if the adjuvant has a surface charge. A recent study [63] used glucose-1-phosphate as a model antigen to study the microenvironment pH of aluminum hydroxide adjuvant. Glucose-1-phosphate adsorbs strongly to aluminum hydroxide adjuvant by ligand exchange of the phosphate group. The rate of hydrolysis of glucose-1-phosphate adsorbed to positively charged aluminum hydroxide adjuvant at pH 5.4 was significantly slower than the rate of hydrolysis of a solution of glucose-1-phosphate at the same pH. The positively charged adjuvant was electrostatically attracting anions to create a microenvironment pH that was higher than 5.4, thus reducing the rate of acid-catalyzed hydrolysis. The adsorbed glucose-1-phosphate at pH 5.4 hydrolyzed at the same rate as a solution of glucose-1-phosphate at pH 7.3. Thus, it was concluded that the microenvironment pH of the aluminum hydroxide adjuvant was approximately 2 pH units higher than the bulk pH.

The chemical stability of antigens that degrade by a pH-dependent mechanism can be optimized by modifying the surface charge of the aluminum-containing adjuvant to produce the pH of maximum stability in the microenvironment of the adjuvant. Antigens that have been reported to be sensitive to pH changes in terms of immunogenicity and stability include foot-and-mouth disease virus, tetanus toxoid, influenza A virus, and *Mycoplasma hyopneumoniae* [63]. The surface charge of aluminum hydroxide adjuvant may be modified by pretreatment with phosphate anion (Figure 4.13). The isoelectric point of aluminum phosphate adjuvant can range from 9.6 to 4.0, depending on the phosphate to hydroxyl ratio (Figure 4.8). Thus, it is possible to formulate an aluminum-adjuvanted vaccine that will have the necessary surface charge to produce the microenvironment pH required to optimize the chemical stability of an antigen that degrades by a pH-dependent mechanism.

### 4.5.3 Content Uniformity

Obtaining an accurate dose is always a concern when administering suspensions. It is also a concern for vaccines, as the dose of most antigens is in micrograms, whereas the quantity of the aluminum-containing adjuvant may be up to 0.85 mg Al per dose. Fortunately, the particle morphology of aluminum-containing adjuvant leads to uniform distribution of the antigen as long as adequate mixing procedures are followed. Aluminum-containing adjuvants exist as primary particles of approximately 50 nm. The high surface free energy of such small particles causes them to aggregate. Thus, the functioning particles are aggregates of 2 to 20 µm. The aggregates disperse and reaggregate readily when mixed. This behavior is shown in Figure 4.14.

Green and red fluorescent probes were conjugated to bovine serum albumin and adsorbed to aluminum hydroxide adjuvant. The suspensions were exam-

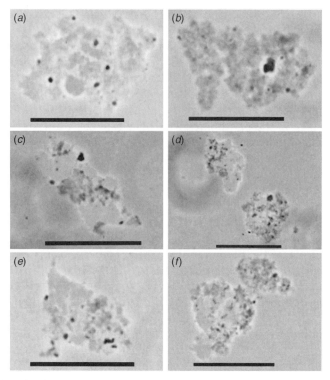

**Figure 4.14**  Distribution of BODIPY FL-labeled bovine serum albumin adsorbed to aluminum hydroxide adjuvant and BODIPY TR-labeled bovine serum albumin adsorbed to aluminum hydroxide adjuvant in a combination vaccine: (*a*) BODIPY FL-labeled bovine serum albumin adsorbed to aluminum hydroxide adjuvant prior to combination; (*b*) BODIPY TR-labeled bovine serum albumin adsorbed to aluminum hydroxide adjuvant prior to combination; (*c*) 15 minutes after combination with mixing; (*d*) 30 minutes of mixing; (*e*) 45 minutes of mixing; (*f*) 60 minutes of mixing; The bars represent 5 μm. (From Ref. 64.) (*See insert for color representation of figure.*)

ined by fluorescence microscopy. Figure 4.14*a* and *b* show green BODIPY FL-labeled bovine serum albumin adsorbed to aluminum hydroxide adjuvant and red BODIPY TR-labeled bovine serum albumin adsorbed to aluminum hydroxide adjuvant, respectively. Equal volumes of the two suspensions were combined and mixed with a magnetic stir bar. Following combination, the aggregates were composed of green fluorescent regions and red fluorescent regions (Figure 4.14*c*). The large, well-defined fluorescent regions became smaller during the 60 minutes of mixing (Figure 4.14*d*–*f*). The yellow color was caused by superposition of green and red fluorescent regions using imaging software. The fluorescent images indicate that mixing causes the adjuvant aggregates to deaggregate and then reaggregate by combining fragments of green fluorescent adjuvant with fragments of red fluorescent adjuvant. This process of deaggregation and reaggregation led to uniform distribution of the red- and green-labeled bovine serum albumin throughout the aluminum hydroxide adjuvant. Thus, the morphology of aluminum-containing adjuvants is favorable for the uniform distribution of antigen as long as adequate mixing occurs.

### 4.5.4 Detoxification of Lipopolysaccharide

Endotoxin or lipopolysaccharide is a component of the outer membrane of gram-negative bacteria. It is released when the bacteria undergo autolysis. Endotoxin causes inflammatory responses or endotoxic shock due to the stimulation of cytokines such as tumor necrosis factor-$\alpha$ (TNF$\alpha$) and IL-6. Thus, it is important to eliminate or detoxify endotoxin during the manufacture of vaccines. Aluminum hydroxide adjuvant, but not aluminum phosphate adjuvant, detoxifies endotoxin.

Adsorption isotherms of endotoxin with aluminum hydroxide adjuvant or aluminum phosphate adjuvant reveal a large difference in the adsorption coefficient [65]. The adsorption coefficients of endotoxin with aluminum hydroxide adjuvant or aluminum phosphate adjuvant are 13,000 and 0.2 mL/$\mu$g, respectively. The strong adsorption of endotoxin by aluminum hydroxide adjuvant is due to ligand exchange with the two phosphate groups in the lipid A part of endotoxin. Adsorption of endotoxin by aluminum phosphate adjuvant is much weaker, as phosphate has substituted for hydroxyl at many surface sites. Thus, ligand exchange is inhibited.

The toxicity of an endotoxin solution, endotoxin and aluminum hydroxide adjuvant or endotoxin and aluminum phosphate adjuvant, was tested in rats [65]. TNF$\alpha$ and IL-6 were detected in the serum of the rats that received the endotoxin solution or endotoxin and aluminum phosphate adjuvant. No TNF$\alpha$ or IL-6 was detected in the serum of the rats that received endotoxin and aluminum hydroxide adjuvant.

Aluminum hydroxide adjuvant but not aluminum phosphate adjuvant protected rats from the toxic effects of endotoxin. The detoxification of endotoxin by aluminum hydroxide adjuvant is due to irreversible adsorption of endotoxin by ligand exchange. Thus, the use of aluminum hydroxide adjuvant provides an extra margin of safety in the event that endotoxin was introduced into the vaccine inadvertently.

## 4.6 ELUTION OF ANTIGEN

### 4.6.1 In Vaccine

It may be desirable during in vitro testing of aluminum adjuvanted vaccines to elute the antigen to determine directly the amount of adsorbed antigen or to examine the conformation of the antigen. Although this is a difficult problem, limited success has been achieved by the use of surfactants, dissolution of the adjuvant, or addition of competing ions.

There are few published studies on the elution of antigens from aluminum-containing adjuvants, but a number of publications describe the use of surfactants to elute proteins from polymer surfaces. For example, 3% sodium dodecyl sulfate has been used to elute fibrinogen from polyurethane surfaces [66]. An aging effect was noted as less fibrinogen was eluted with increasing residence time. It was concluded that changes in the conformation of the adsorbed fibrinogen occurred which resulted in stronger adsorption. Triton X-100, a nonionic surfactant, has been used to elute insulin that was adsorbed to poly(vinyl chloride) bags [67].

A series of anionic, cationic, and nonionic surfactants were screened for their ability to elute ovalbumin from aluminum hydroxide adjuvant or lysozyme from aluminum phosphate adjuvant [68]. The surfactants studied were Triton X-100 and lauryl maltoside (nonionic), lauryl sulfobetaine (zwitterionic), sodium dodecyl sulfate (anionic), cetylpyridinium chloride, and dodecyltrimethylammonium chloride (cationic). Cetylpyridinium chloride produced the greatest degree of elution (60%) of ovalbumin from aluminum hydroxide adjuvant. Sodium dodecyl sulfate completely eluted lysozyme from aluminum phosphate adjuvant. The effectiveness of the surfactants in eluting either ovalbumin or lysozyme was related directly to their ability to denature the protein. Thus, the disadvantage of using surfactants to elute antigens is the potential to denature the antigen.

Another approach to elute antigens during the in vitro testing of vaccines is to dissolve the aluminum-containing adjuvant. Dissolution can be achieved by adjusting the pH, adding citrate anion, or a combination of citrate anion and pH. The pH of minimum solubility of aluminum hydroxide adjuvant is 5.0 [38]. Elution of ovalbumin was observed when the pH was adjusted to values below 4.1 or above 7.4. Maximum elution was observed at pH 2.9 or 11.0. This elution can be explained by the dissolution of the aluminum hydroxide adjuvant.

$\alpha$-Hydroxycarboxylic acids such as citric acid, lactic acid, or malic acid are good chelators of metal ions, due to the formation of stable five- or six-membered rings. Citrate anion is well known for its solubilizing effect on aluminum-containing minerals. Seeber et al. [69] demonstrated that aluminum hydroxide adjuvant and aluminum phosphate adjuvant can be solubilized by a sodium citrate solution at pH 7.4. Aluminum phosphate adjuvant is significantly more soluble than aluminum hydroxide adjuvant.

If the properties of the antigen permit, the best approach to eluting antigens by dissolving the adjuvant is to combine dissolution at low pH with solubilization by chelation with citrate anion. Thus, the exposure of a vaccine to a solution of sodium citrate adjusted to pH 4 is frequently an effective technique to elute antigens.

Antigens that are adsorbed by electrostatic attraction may be eluted by reversing the surface charge of the aluminum-containing adjuvant. For example, electrostatic attractive forces play a role in the adsorption of ovalbumin (iso-electric point 5.0) by aluminum hydroxide adjuvant (isoelectric point 11.4) at pH 7.4. The addition of 4 mM phosphate anion caused 45% of the adsorbed ovalbumin to be eluted [70]. The addition of 4 mM phosphate anion to aluminum hydroxide adjuvant caused the isoelectric point to change from 11.4 to 6.7. Thus, the surface charge of the aluminum hydroxide adjuvant became negative and a repulsive electrostatic force operated between the ovalbumin and the adjuvant. Similar to the elution of fibrinogen with surfactant, an aging effect was observed when ovalbumin was eluted by exposure to phosphate anion. The elution of ovalbumin by 4 mM phosphate anion decreased from 45% to 27% when the vaccine aged for 17 days.

### 4.6.2   In Interstitial Fluid

The importance of the degree of adsorption of the antigen by aluminum-containing adjuvants following intramuscular or subcutaneous administration was recognized by Chang and co-workers [71], who tested three lysozyme vaccines in rabbits. All the vaccines contained the same concentration of lyso-zyme, but the degree of adsorption by the aluminum-containing adjuvant was 3, 35 or 85%. In contrast to a lysozyme solution, all three vaccines produced the same immunopotentiation. However, when mixed in vitro with sheep lymph fluid, which is identical to interstitial fluid, the degree of adsorption of lysozyme was 40% for all three vaccines. It was concluded that the degree of adsorption in interstitial fluid, not the degree of adsorption on the vaccine, correlated with the immune response. It is not surprising that the degree of adsorption of an antigen by an aluminum-containing adjuvant can change upon intramuscular or subcutaneous administration as key components of interstitial fluid such as phosphate anion [70], citrate [69], fibrinogen [16], and albumin [16] have been found to affect the degree of adsorption.

The contributions of electrostatic attraction and ligand exchange to the degree of adsorption of antigens upon exposure to interstitial fluid were dem-onstrated by Morefield et al. [59]. Ovalbumin (OVA) was either conjugated with phosphoserine to increase its phosphorylation or treated with potato acid phosphatase to reduce the degree of phosphorylation. Four ovalbumin samples containing from 3.2 to 0.14 mol $PO_4$/mol ovalbumin were prepared. Ovalbumin has an isoelectric point of 4.6 and is attracted electrostatically to aluminum hydroxide adjuvant (isoelectric point 11.4). In addition, the phosphate groups of ovalbumin are available to undergo ligand exchange with surface hydroxyls of aluminum hydroxide adjuvant. As shown in Table 4.4, these two adsorption mechanisms led to virtually complete adsorption of all four ovalbumin samples.

However, the adsorbed ovalbumin vaccines behaved differently when exposed in vitro to interstitial fluid. When mixed with interstitial fluid the percent elution was inversely related to the $PO_4$/OVA molar ratio (Figure 4.15). This behavior suggests that ovalbumin molecules that are not

**TABLE 4.4   Adsorption of Ovalbumin Samples by Aluminum Hydroxide Adjuvant and Elution in Interstitial Fluid**

| Ovalbumin Sample | $PO_4$/OVA Molar Ratio | % Adsorption | % Elution Upon Exposure to Interstitial Fluid for 21 Hours |
|---|---|---|---|
| POVA | 3.2 | 100 | 3 |
| OVA | 1.8 | 99 | 36 |
| DPOVA 1 | 1.2 | 97 | 73 |
| DPOVA 2 | 0.14 | 95 | 100 |

*Source:* Data from Ref. 59.

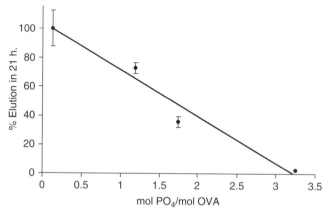

**Figure 4.15**   Effect of phosphate content of ovalbumin samples on elution from aluminum hydroxide adjuvant upon exposure to interstitial fluid at 37°C. (From Ref. 59.)

phosphorylated and are therefore adsorbed by electrostatic attractive forces are readily eluted when exposed to interstitial fluid. Conversely, ovalbumin samples that are phosphorylated adsorb by both ligand exchange and electrostatic attraction. The contribution of ligand exchange to the adsorption mechanism increased as the $PO_4$/OVA molar ratio increased. Examination of Table 4.4 and Figure 4.15 lead to the conclusion that antigens adsorbed by ligand exchange do not readily elute upon exposure to interstitial fluid. Thus, the degree of adsorption in interstitial fluid, following administration, can be controlled through the degree of phosphorylation of the antigen.

The degree of adsorption in interstitial fluid can also be controlled by the degree of phosphorylation of the adjuvant. Aluminum phosphate adjuvant has an isoelectric point of approximately 5. Thus, it is electrostatically repelled by the negatively charged ovalbumin samples and can only be adsorbed by ligand exchange. However, the surface of aluminum phosphate adjuvant is composed of both hydroxyl and phosphate groups. Only the hydroxyl groups can undergo ligand exchange with phosphate groups of ovalbumin. Ligand exchange is responsible for the direct relationship shown in Table 4.3 between the $PO_4$/OVA molar ratio and the percent adsorption by aluminum phosphate adjuvant. The complete elution of ovalbumin samples containing 1.8 mol of phosphate per mol of ovalbumin or less is due to the limited opportunity for ligand exchange with aluminum phosphate adjuvant due to the limited number of surface hydroxyl groups. Thus, elution following administration can also be controlled by modifying the degree of phosphorylation of the adjuvant.

## 4.7   ELIMINATION

Interstitial fluid contains significant concentrations of citric acid, lactic acid, and malic acid. These $\alpha$-hydroxycarboxylic acids are capable of chelating aluminum and solubilizing aluminum-containing compounds. In vitro dissolution studies using citrate solutions at pH 7.4 that mimic the concentration of $\alpha$-hydroxycarboxylic acids in interstitial fluid revealed that both aluminum hydroxide adjuvant and aluminum phosphate adjuvant are solubilized, although the amorphous aluminum phosphate adjuvant is more readily solubilized. Thus, it is likely that aluminum-containing adjuvants dissolve in interstitial fluid, and the aluminum is transported in the lymph to the blood, where it is excreted by the kidney.

Aluminum-containing adjuvants were used for many decades before a technique was available that could detect changes in the normal aluminum plasma concentration (5 μg Al/L) caused by the small amount of aluminum allowed in vaccines (<0.85 mg Al per dose). Accelerator mass spectrometry can accurately measure very small amounts of aluminum, $10^{-17}$ g in blood or urine [72,73]. This technique enabled the absorption, tissue distribution, and elimination of aluminum-containing adjuvants to be studied.

Aluminum phosphate adjuvant and aluminum hydroxide adjuvants were prepared by substituting a small amount of $^{26}AlCl_3$ for the naturally occurring $^{27}AlCl_3$ during the precipitation process. $^{26}Al$ does not occur naturally, so any $^{26}Al$ in blood, tissues, or urine came from the aluminum-containing adjuvant.

Two female New Zealand white rabbits received an intramuscular injection of $^{26}Al$ aluminum phosphate adjuvant and two rabbits received a similar injection of $^{26}Al$ aluminum hydroxide adjuvant. One rabbit received an equivalent intravenous injection of $^{26}Al$-labeled aluminum citrate solution, and another rabbit received an equivalent intramuscular dose of aluminum phosphate adjuvant containing no $^{26}Al$ as a cross-contamination monitor. Blood and urine samples were collected for 28 days, after which the animals were euthanized and the kidney, spleen, liver, heart, lymph node, and brain were analyzed for $^{27}Al$.

Figure 4.16 shows the concentration of aluminum in the blood following intramuscular administration of the adjuvants. Al was detected in the blood of all four rabbits at the first sampling point (one hour). Thus, dissolution of both adjuvants begins upon administration. Table 4.5 shows that approximately

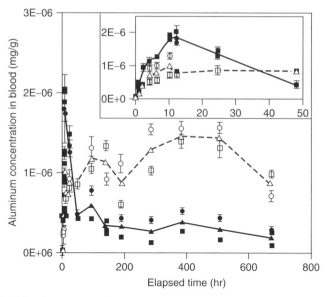

**Figure 4.16** Blood concentration profile after intramuscular administration of $^{26}$Al-labeled aluminum hydroxide adjuvant: (■) rabbit 1; (●) rabbit 2; (▲) mean, or aluminum phosphate adjuvant; (□) rabbit 3; (○) rabbit 4; (△) mean; the solid line represents aluminum hydroxide adjuvant, and the dashed line, aluminum phosphate adjuvant. (From Ref. 72.)

**TABLE 4.5   Pharmacokinetic Parameters After Intramuscular Injection of**
**$^{26}$Al-Containing Aluminum Hydroxide and Aluminum Phosphate Adjuvants**

| Adjuvant | % Adsorbed in 28 Days | Cumulative Al in Urine After 28 Days (%) |
|---|---|---|
| Aluminum hydroxide | | |
| Rabbit 1 | 13 | 5.0 |
| Rabbit 2 | 22 | 6.2 |
| Average | 17 | 5.6 |
| Aluminum phosphate | | |
| Rabbit 3 | 47 | 10 |
| Rabbit 4 | 55 | 33 |
| Average | 51 | 22 |

*Source:* Data from Ref. 72.

three times more aluminum appeared in the blood from aluminum phosphate adjuvant than from aluminum hydroxide adjuvant during the 28-day study.

Similarly, the cumulative urinary excretion of aluminum phosphate adjuvant was four times greater than that of aluminum hydroxide adjuvant. The average tissue distribution of $^{26}$Al was approximately three times greater for aluminum phosphate adjuvant than for aluminum hydroxide adjuvant. Thus, the higher concentration of $^{26}$Al in the blood from aluminum phosphate adjuvant appears to be responsible for the greater distribution to the tissues.

It can be concluded from the in vitro and in vivo studies that interstitial fluid is capable of dissolving both aluminum phosphate adjuvant and aluminum hydroxide adjuvant following intramuscular administration. Amorphous aluminum phosphate adjuvant exhibits a more rapid rate of dissolution and absorption than that of crystalline aluminum hydroxide adjuvant. The ability of the body to eliminate aluminum-containing adjuvants may be partly responsible for their excellent safety record.

## 4.8  SAFETY

It is difficult to assess the contribution of the aluminum-containing adjuvant when adverse reactions to a vaccine occur, as clinical trials rarely include treatment groups that receive the vaccine with and without the adjuvant [74]. When one considers the number of doses of aluminum-adjuvanted vaccines that have been administered since aluminum-containing adjuvants were first included in vaccines in 1926 and the incidence of adverse effects, it is easy to conclude that aluminum-containing adjuvants pose a very small safety risk [42]. They are not associated with any immune complex disorders [75]. Severe local reactions such as erythema, subcutaneous nodules, and contact hypersensitivity are associated with aluminum-containing adjuvants [74]. These local reactions are noted more frequently when aluminum-adjuvanted vaccines are administered subcutaneously than when administered intramuscularly [76]. A study of the effect of the amount of aluminum-containing adjuvant and local swelling reaction found an inconsistent relationship [77].

A new safety issue was recently raised when Gherardi et al. [78–80] reported a focal histological lesion observed in biopsy samples which they termed *macrophagic myofaciitis* (MMF). However, due to the limited number of cases and the lack of controls, the causal relationship between MMF and aluminum-adjuvanted vaccines is in dispute. A recent study in Cynomolgus monkeys concluded that aluminum-adjuvanted vaccines cause histopathological changes restricted to a region within 20 mm from the injection site but that these changes are not associated with abnormal clinical signs [81].

A statement by Clements and Griffiths [75] appears to summarize the current understanding of the safety of aluminum-containing adjuvants: "Regarding aluminum-based adjuvants, we cannot say categorically that they cause no problems. We can say they have been used in vaccines for 70 years and have played a part in making the classical vaccines as successful as they have been. With such a phenomenal number of doses administered over this period of time, we can be confident that if there is a potential for causing adverse reactions to aluminum in these vaccines, the rate is inordinately small. Anything more would have been detected years ago."

## ACKNOWLEDGMENTS

We are grateful to Merck Research Laboratories for their long-standing support of our adjuvant research.

## REFERENCES

1. Glenny, A.T., Pope, C.G., Waddington, H. & Wallace, U. The antigenic value of toxoid precipitated by potassium alum. *J Pathol Bacteriol* 1926, **29**, 31–40.
2. Ramon, G. Sur l'augmentation de l'antitoxine chez chevaux producteurs de serum antidiphtherique. *Bull Soc Cent Med Vet* 1925, **101**, 227–234.
3. Itano, A.A., McSorley, S.J., Reinhardt, R.L. et al. Distinct dendritic cell populations sequentially present antigen to CD4 T cells and stimulate different aspects of cell-mediated immunity. *Immunity* 2003, **19**(1), 47–57.
4. Sixt, M., Kanazawa, N., Selg, M. et al. The conduit system transports soluble antigens from the afferent lymph to resident dendritic cells in the T cell area of the lymph node. *Immunity* 2005, **22**(1), 19–29.
5. Gupta, R.K., Rost, B.E., Relyveld, E. & Siber, G.R. Adjuvant properties of aluminum and calcium compounds. In *Vaccine Design: The Subunit and Adjuvant Approach*, Powell, M.F. & M.J. Newman, Eds., Plenum Press, New York, 1995, pp. 229–248.
6. Glenny, A.T., Buttle, A.H. & Stevens, M.F. Rate of disappearance of diphtheria toxoid injected into rabbits and guinea-pigs: toxoid precipitated with alum. *J Pathol Bacteriol* 1931, **34**, 267–275.
7. Harrison, W.T. Some observations on the use of alum-precipitated diphtheria toxoid. *Am J Public Health* 1935, **25**, 298–300.
8. Holt, L.B. *Developments in Diphtheria Prophlaxis*, W. Heinemann Medical Books, London, 1950, p. 99.
9. White, R.G., Coons, A.H. & Connolly, J.M. Studies on antibody production: III. The alum granuloma. *J Exp Med* 1955, **102**(1), 73–82.
10. Walls, R.S. Eosinophil response to alum adjuvants: involvement of T cells in non-antigen-dependent mechanisms. *Proc Soc Exp Biol Med* 1977, **156**(3), 431–435.
11. McWilliam, A.S., Napoli, S., Marsh, A.M. et al. Dendritic cells are recruited into the airway epithelium during the inflammatory response to a broad spectrum of stimuli. *J Exp Med* 1996, **184**(6), 2429–2432.
12. Mannhalter, J.W., Neychev, H.O., Zlabinger, G.J., Ahmad, R. & Eibl, M.M. Modulation of the human immune response by the non-toxic and non-pyrogenic adjuvant aluminium hydroxide: effect on antigen uptake and antigen presentation. *Clin Exp Immunol* 1985, **61**(1), 143–151.
13. Morefield, G.L., Sokolovska, A., Jiang, D., HogenEsch, H., Robinson, J.P. & Hem, S.L. Role of aluminum-containing adjuvants in antigen internalization by dendritic cells in vitro. *Vaccine* 2005, **23**(13), 1588–1595.
14. Ulmer, J.B., DeWitt, C.M., Chastain, M. et al. Enhancement of DNA vaccine potency using conventional aluminum adjuvants. *Vaccine* 2000, **18**(1–2), 18–28.

15. Wang, S., Liu, X., Fisher, K. et al. Enhanced type I immune response to a hepatitis B DNA vaccine by formulation with calcium- or aluminum phosphate. *Vaccine* 2000, **18**(13), 1227–1235.

16. Heimlich, J.M., Regnier, F.E., White, J.L. & Hem, S.L. The in vitro displacement of adsorbed model antigens from aluminium-containing adjuvants by interstitial proteins. *Vaccine* 1999, **17**(22), 2873–2881.

17. Schnare, M., Barton, G.M., Holt, A.C., Takeda, K., Akira, S. & Medzhitov, R. Toll-like receptors control activation of adaptive immune responses. *Nat Immunol* 2001, **2**(10), 947–950.

18. Ulanova, M., Tarkowski, A., Hahn-Zoric, M., Hanson, L.A. & Moingeon, P. The common vaccine adjuvant aluminum hydroxide up-regulates accessory properties of human monocytes via an interleukin-4-dependent mechanism. *Infect Immun* 2001, **69**(2), 1151–1159.

19. Wilcock, L.K., Francis, J.N. & Durham, S.R. Aluminium hydroxide down-regulates T helper 2 responses by allergen-stimulated human peripheral blood mononuclear cells. *Clin Exp Allergy* 2004, **34**(9), 1373–1378.

20. Rimaniol, A.C., Gras, G., Verdier, F. et al. Aluminum hydroxide adjuvant induces macrophage differentiation towards a specialized antigen-presenting cell type. *Vaccine* 2004, **22**(23–24), 3127–3135.

21. Takeda, K. & Akira, S. Toll-like receptors in innate immunity. *Int Immunol* 2005, **17**(1), 1–14.

22. Bomford, R. The comparative selectivity of adjuvants for humoral and cell-mediated immunity. *Clin Exp Immunol* 1980, **39**(2), 435–441.

23. Comoy, E.E., Capron, A. & Thyphronitis, G. In vivo induction of type 1 and 2 immune responses against protein antigens. *Int Immunol* 1997, **9**(4), 523–531.

24. Bomford, R., Stapleton, M., Winsor, S., McKnight, A. & Andronova, T. The control of the antibody isotype response to recombinant human immunodeficiency virus gp120 antigen by adjuvants. *AIDS Res Hum Retroviruses* 1992, **8**, 1765–1771.

25. Trinchieri, G. Interleukin-12 and the regulation of innate resistance and adaptive immunity. *Nat Rev Immunol* 2003, **3**(2), 133–146.

26. Heufler, C., Koch, F., Stanzl, U. et al. Interleukin-12 is produced by dendritic cells and mediates T helper 1 development as well as interferon-gamma production by T helper 1 cells. *Eur J Immunol* 1996, **26**(3), 659–668.

27. Cella, M., Scheidegger, D., Palmer-Lehmann, K., Lane, P., Lanzavecchia, A. & Alber, G. Ligation of CD40 on dendritic cells triggers production of high levels of interleukin-12 and enhances T cell stimulatory capacity: T-T help via APC activation. *J Exp Med* 1996, **184**(2), 747–752.

28. Cella, M., Salio, M., Sakakibara, Y., Langen, H., Julkunen, I. & Lanzavecchia, A. Maturation, activation, and protection of dendritic cells induced by double-stranded RNA. *J Exp Med* 1999, **189**(5), 821–829.

29. Kalinski, P., Schuitemaker, J.H., Hilkens, C.M. & Kapsenberg, M.L. Prostaglandin E2 induces the final maturation of IL-12-deficient CD1a+CD83+ dendritic cells: the levels of IL-12 are determined during the final dendritic cell maturation and are resistant to further modulation. *J Immunol* 1998, **161**(6), 2804–2809.

30. Wittmann, M., Zwirner, J., Larsson, V.A. et al. C5a suppresses the production of IL-12 by IFN-gamma-primed and lipopolysaccharide-challenged human monocytes. *J Immunol* 1999, **162**(11), 6763–6769.

31. Marth, T. & Kelsall, B.L. Regulation of interleukin-12 by complement receptor 3 signaling. *J Exp Med* 1997, **185**(11), 1987–1995.

32. Ramanathan, V.D., Badenoch-Jones, P. & Turk, J.L. Complement activation by aluminium and zirconium compounds. *Immunology* 1979, **37**(4), 881–888.

33. Brewer, J.M., Conacher, M., Satoskar, A., Bluethmann, H. & Alexander, J. In interleukin-4-deficient mice, alum not only generates T helper 1 responses equivalent to Freund's complete adjuvant, but continues to induce T helper 2 cytokine production. *J Immunol* 1996, **26**(9), 2062–2066.

34. Brewer, J.M., Conacher, M., Hunter, C.A., Mohrs, M., Brombacher, F. & Alexander, J. Aluminium hydroxide adjuvant initiates strong antigen-specific Th2 responses in the absence of IL-4- or IL-13-mediated signaling. *J Immunol* 1999, **163**(12), 6448–6454.

35. Shirodkar, S., Hutchinson, R.L., Perry, D.L., White, J.L. & Hem, S.L. Aluminum compounds used as adjuvants in vaccines. *Pharm Res* 1990, **7**(12), 1282–1288.

36. Johnston, C.T., Wang, S.L. & Hem, S.L. Measuring the surface area of aluminum hydroxide adjuvant. *J Pharm Sci* 2002, **91**(7), 1702–1706.

37. Masood, H., White, J.L. & Hem, S.L. Relationship between protein adsorptive capacity and the x-ray diffraction pattern of aluminium hydroxide adjuvants. *Vaccine* 1994, **12**(2), 187–189.

38. Rinella, J.V., White, J.L. & Hem, S.L. Effect of pH on the elution of model antigens from aluminum-containing adjuvants. *J Colloid Interface Sci* 1998, **205**(1), 161–165.

39. Stumm, W. *Chemistry of the Solid–Water Interface.* Wiley, New York, 1992, pp. 13–38.

40. Jiang, D., Johnston, C.T. & Hem, S.L. Using rate of acid neutralization to characterize aluminum phosphate adjuvant. *Pharm Dev Technol* 2003, **8**(4), 349–356.

41. Parks, G.A. The isoelectric points of solid oxides, solid hydroxides, and aqueous hydroxo complex systems. *Chem Rev* 1965, **65**, 177–198.

42. Lindblad, E.B. Aluminium adjuvants: in retrospect and prospect. *Vaccine* 2004, **22**(27–28), 3658–3668.

43. Immunological Adjuvants. Report of a WHO scientific group. *World Health Organ Tech Rep Ser* **1976**(595), 8.

44. Hem, K.J., Dandashli, E.A., White, J.L. & Hem, S.L. Accessibility of antigen in vaccines produced by in-situ alum precipitation. *Vaccine Res* 1996, **5**(4), 187–191.

45. Serna, C.J., Lyons, J.C., White, J.L. & Hem, S.L. Stabilization of aluminum hydroxide gel by specifically adsorbed carbonate. *J Pharm Sci* 1983, **72**(7), 769–771.

46. Burrell, L.S., Lindblad, E.B., White, J.L. & Hem, S.L. Stability of aluminium-containing adjuvants to autoclaving. *Vaccine* 1999, **17**(20–21), 2599–2603.

47. Zapata, M.I., Feldkamp, J.R., Peck, G.E., White, J.L. & Hem, S.L. Mechanism of freeze–thaw instability of aluminum hydroxycarbonate and magnesium hydroxide gels. *J Pharm Sci* 1984, **73**(1), 3–8.

48. Rowe, R.C. *Pharmaceutical Excipients, 2004,* Pharmaceutical Press, London, 2004.

49. The effects of freezing on the appearance, potency and toxicity of adsorbed and unadsorbed DPT vaccines. *Weekly Epidemiol Rec* 1980, **55**, 385–392.

50. Boros, C.A., Hanlon, M., Gold, M.S. & Roberton, D.M. Storage at −3 degrees C for 24 h alters the immunogenicity of pertussis vaccines. *Vaccine* 2001, **19**(25–26), 3537–3542.

51. WHO. *Safe Vaccine Handling, Cold Chain and Immunizations*, World Health Organization, Geneva, 1998.

52. Galazka, A., Milstien, J. & Zaffran, M. *Thermostability of Vaccines*, World Health Organization, Geneva, 1998.

53. Jendrek, S., Little, S.F., Hem, S., Mitra, G. & Giardina, S. Evaluation of the compatibility of a second generation recombinant anthrax vaccine with aluminum-containing adjuvants. *Vaccine* 2003, **21**, 3011–3018.

54. Iyer, S., Robinett, R.S., HogenEsch, H. & Hem, S.L. Mechanism of adsorption of hepatitis B surface antigen by aluminum hydroxide adjuvant. *Vaccine* 2004, **22**(11–12), 1475–1479.

55. Seeber, S.J., White, J.L. & Hem, S.L. Predicting the adsorption of proteins by aluminium-containing adjuvants. *Vaccine* 1991, **9**(3), 201–203.

56. Al-Shakhshir, R.H., Regnier, F.E., White, J.L. & Hem, S.L. Contribution of electrostatic and hydrophobic interactions to the adsorption of proteins by aluminium-containing adjuvants. *Vaccine* 1995, **13**(1), 41–44.

57. Chang, M.F., White, J.L., Nail, S.L. & Hem, S.L. Role of the electrostatic attractive force in the adsorption of proteins by aluminum hydroxide adjuvant. *PDA J Pharm Sci Technol* 1997, **51**(1), 25–29.

58. Dagouassat, N., Robillard, V., Haeuw, J.F. et al. A novel bipolar mode of attachment to aluminium-containing adjuvants by BBG2Na, a recombinant subunit hRSV vaccine. *Vaccine* 2001, **19**(30), 4143–4152.

59. Morefield, G.L., Jiang, D., Romero-Mendez, I.Z., Geahlen, R.L., Hogenesch, H. & Hem, S.L. Effect of phosphorylation of ovalbumin on adsorption by aluminum-containing adjuvants and elution upon exposure to interstitial fluid. *Vaccine* 2005, **23**(12), 1502–1506.

60. Iyer, S., HogenEsch, H. & Hem, S.L. Effect of the degree of phosphate substitution in aluminum hydroxide adjuvant on the adsorption of phosphorylated proteins. *Pharm Dev Technol* 2003, **8**(1), 81–86.

61. Wang, M.K., White, J.L. & Hem, S.L. Effect of polybasic acids on structure of aluminum hydroxycarbonate gel. *J Pharm Sci* 1980, **69**(6), 668–671.

62. Callahan, P.M., Shorter, A.L. & Hem, S.L. The importance of surface charge in the optimization of antigen–adjuvant interactions. *Pharm Res* 1991, **8**(7), 851–858.

63. Wittayanukulluk, A., Jiang, D., Regnier, F.E. & Hem, S.L. Effect of microenvironment pH of aluminum hydroxide adjuvant on the chemical stability of adsorbed antigen. *Vaccine* 2004, **22**(9–10), 1172–1176.

64. Morefield, G.L., HogenEsch, H., Robinson, J.P. & Hem, S.L. Distribution of adsorbed antigen in mono-valent and combination vaccines. *Vaccine* 2004, **22**(15–16), 1973–1984.

65. Shi, Y., HogenEsch, H., Regnier, F.E. & Hem, S.L. Detoxification of endotoxin by aluminum hydroxide adjuvant. *Vaccine* 2001, **19**(13–14), 1747–1752.

66. Chinn, J.A., Posso, S.E., Horbett, T.A. & Ratner, B.D. Postadsorptive transitions in fibrinogen adsorbed to polyurethanes: changes in antibody binding and sodium dodecyl sulfate elutability. *J Biomed Mater Res* 1992, **26**(6), 757–778.

67. Twardowski, Z.J., Nolph, K.D., McGary, T.J. & Moore, H.L. Nature of insulin binding to plastic bags. *Am J Hosp Pharm* 1983, **40**(4), 579–582.

68. Rinella, J.V., Workman, R.F., Hermodson, M.A., White, J.L. & Hem, S.L. Elutability of proteins from aluminum-containing vaccine adjuvants by treatment with surfactants. *J Colloid Interface Sci* 1998, **197**(1), 48–56.

69. Seeber, S.J., White, J.L. & Hem, S.L. Solubilization of aluminum-containing adjuvants by constituents of interstitial fluid. *J Parenter Sci Technol* 1991, **45**(3), 156–159.

70. Rinella, J.V., White, J.L. & Hem, S.L. Effect of anions on model aluminum-adjuvant-containing vaccines. *J Colloid Interface Sci* 1995, **172**, 121–130.

71. Chang, M., Shi, Y., Nail, S.L. et al. Degree of antigen adsorption in the vaccine or interstitial fluid and its effect on the antibody response in rabbits. *Vaccine* 2001, **19**(20–22), 2884–2889.

72. Flarend, R.E., Hem, S.L., White, J.L. et al. In vivo absorption of aluminium-containing vaccine adjuvants using $^{26}$Al. *Vaccine* 1997, **15**(12–13), 1314–1318.

73. Hem, S.L. Elimination of aluminum adjuvants. *Vaccine* 2002, **20**(Suppl 3), S40–S43.

74. Baylor, N.W., Egan, W. & Richman, P. Aluminum salts in vaccines: US perspective. *Vaccine* 2002, **20**(Suppl 3), S18–S23.

75. Clements, C.J. & Griffiths, E. The global impact of vaccines containing aluminium adjuvants. *Vaccine* 2002, **20**(Suppl 3), S24–S33.

76. Pittman, P.R. Aluminum-containing vaccine associated adverse events: role of route of administration and gender. *Vaccine* 2002, **20**(Suppl 3), S48–S50.

77. Rennels, M.B., Deloria, M.A., Pichichero, M.E. et al. Lack of consistent relationship between quantity of aluminum in diphtheria–tetanus–acellular pertussis vaccines and rates of extensive swelling reactions. *Vaccine* 2002, **20**(Suppl 3), S44–S47.

78. Gherardi, R.K., Coquet, M., Cherin, P. et al. Macrophagic myofasciitis: an emerging entity. Groupe d'Études et Recherche sur les Maladies Musculaires Acquises et Dysimmunitaires (GERMMAD) de l'Association Française contre les Myopathies (AFM). *Lancet* 1998, **352**, 347–352.

79. Gherardi, R.K., Coquet, M., Cherin, P. et al. Macrophagic myofasciitis lesions assess long-term persistence of vaccine-derived aluminium hydroxide in muscle. *Brain* 2001, **124**(Pt 9), 1821–1831.

80. Cherin, P., Laforet, P., Gherardi, R.K. et al. [Macrophagic myofasciitis: description and etiopathogenic hypotheses. Study and Research Group on Acquired and Dysimmunity-Related Muscular Diseases (GERMMAD) of the French Association Against Myopathies (AFM)]. *Rev Med Interne* 1999, **20**(6), 483–489.

81. Verdier, F., Burnett, R., Michelet-Habchi, C., Moretto, P., Fievet-Groyne, F. & Sauzeat, E. Aluminium assay and evaluation of the local reaction at several time points after intramuscular administration of aluminium containing vaccines in the *Cynomolgus* monkey. *Vaccine* 2005, **23**(11), 1359–1367.

# 5

# MF59: A SAFE AND POTENT OIL-IN-WATER EMULSION ADJUVANT

DEREK T. O'HAGAN AND MANMOHAN SINGH

## 5.1 EMULSIONS AS ADJUVANTS

Emulsions are liquid dispersions of two immiscible phases, usually an oil and water, either of which may comprise the dispersed phase or the continuous phase to provide water-in-oil or oil-in-water emulsions. Emulsions are generally unstable and need to be stabilized by surfactants, which lower interfacial tension and prevent coalescence of the dispersed droplets. Stable emulsions can be prepared through the use of surfactants, which orientate themselves at the interface between the two phases since they comprise both hydrophobic and hydrophilic components. Although charged surfactants are excellent stabilizers, nonionic surfactants are widely used in pharmaceutical emulsions due to their lower toxicity and their lower sensitivity to formulation additives. Surfactants can be defined by their ratio of hydrophilic to hydrophobic components, called their *hydrophile-to-lipophile balance* (HLB), which gives information on their relative affinity for water and oil phases. At the high end of the scale, the surfactants are predominantly hydrophilic and can be used to stabilize oil-in-water (o/w) emulsions. In contrast, oil-soluble surfactants at the lower end of the scale are used primarily to stabilize water-in-oil (w/o) emulsions. The commonly used polysorbate-based surfactants (Tweens) have HLB values in the range 9 to 16, while sorbitan esters (Spans) have HLBs in the lower range 2 to 9. Extensive pharmaceutical experience with emulsions has established that a mixture of surfactants offers maximum stability, probably due to the formation of more rigid films at the interface. The physicochemical

*Vaccine Adjuvants and Delivery Systems*, Edited by Manmohan Singh

characteristics of emulsions, including droplet size and viscosity, are controlled by a variety of factors, including the choice of surfactants, the ratio of continuous to disperse phases, and the method of preparation used. For an emulsion to be used in an injectable product, stability and viscosity are important parameters. In general, stability is enhanced by having smaller droplets, and viscosity is decreased by having a higher volume of the continuous phase.

Emulsions have a long history of use as vaccine adjuvants in both human and veterinary products. Almost 70 years ago, Freund demonstrated the adjuvant effect of mineral (paraffin) oil combined with mycobacterial cells, called *Freund's complete adjuvant* (FCA) [1]. The w/o emulsion, without bacterial cells, known as *Freund's incomplete adjuvant* (FIA), has subsequently been used in several veterinary vaccines [2]. Recent studies have explored the structure–activity relationships for w/o emulsion adjuvants of the FIA type [3]. Although w/o emulsions containing mineral oils such as FIA have been used extensively as vaccine adjuvants in human subjects, particularly for influenza vaccination [4], they are generally considered to be too reactogenic for widespread use [5]. Nevertheless, long-term follow-up of individuals immunized with FIA have established that there are no significant long-term adverse events, although local reactogenicity was common during the initial immunizations [6]. More recently, w/o emulsions with high oil content, based on mineral oils and non-mineral oils, have been evaluated as vaccine adjuvants for malaria and HIV vaccines [7]. Ongoing clinical trials continue to demonstrate that these newer generation w/o emulsions (Montanide) induce potent immune responses but also show a significant occurrence of local reactions, which can occasionally be severe [8]. Due to the reactogenicity of w/o emulsions, o/w approaches were evaluated as alternatives and were initially promoted as delivery systems for immunopotentiators [9].

## 5.2 INITIAL DEVELOPMENT OF MF59 OIL-IN-WATER ADJUVANT

In the 1980s a number of groups were working on the development of novel adjuvant formulations, including emulsions, ISCOMs (immunostimulating complexes), liposomes, and microparticles [10]. All of these approaches had the potential to be more potent and effective than the established aluminium-based adjuvants, which were the only adjuvants available in licensed human vaccines at the time. Often, the novel adjuvant formulations contained immunopotentiators of natural or synthetic origin, which were included to enhance potency. However, the inclusion of immunopotentiators often raised concerns about the overall safety of the adjuvant formulation.

Based on the long history of the use of emulsions as adjuvants, including the widely used FIA, several groups investigated the development of improved emulsion formulations, which might prove acceptable for use in humans.

Scientists at Syntex developed an o/w emulsion adjuvant using biodegradable squalene as oil and used this formulation as a delivery system for a synthetic immunopotentiator called $N$-acetymuramyl-L-threonyl-D-isoglutamine (threonyl-MDP) [9]. The Syntex adjuvant formulation (SAF) was designed to be as potent as FCA, but with greater potential to prove acceptable for widespread human use. A closely related immunopotentiator, $N$-acetyl-L-alanyl-D-isoglutamine (MDP), had originally been identified in 1974 as the minimal structure isolated from the peptidoglycan of mycobacterial cell walls, which had adjuvant activity [11]. However, MDP was pyrogenic and induced uveitis in rabbits [12], making it unacceptable as an adjuvant for human vaccines. Therefore, various derivatives of MDP were synthesized in an effort to identify a potent molecule with an acceptable toxicology profile, including threonyl-MDP. Unlike lipopolysaccharide (LPS) and its synthetic derivatives, which are also used as adjuvants [e.g., monophosphorye lipid A (MPL)], MDP does not activate immune cells through TLR2 or TLR4 [13]. Recently, it was discovered that MDP activates immune cells through interaction with the nucleotide-binding domain (NOD), which acts as an intracellular recognition system for bacterial components [14]. In addition to threonyl-MDP, SAF also contained a pluronic polymer surfactant (L121), which was included to help bind antigens to the surface of the emulsion droplets. Unfortunately, clinical evaluations of SAF as an adjuvant for a HIV vaccine showed it to have an unacceptable profile of reactogenicity [15]. As an alternative to SAF, Chiron scientists developed a squalene-based o/w emulsion as a delivery system for an alternative synthetic MDP derivative, muramyltripeptide phosphatidylethanolamine (MTP-PE). MTP-PE was lipidated to allow it to be more easily incorporated into lipid-based adjuvant formulations and to reduce toxicity [16]. Unfortunately, clinical testing showed that emulsions of MTP-PE also showed an unacceptable level of reactogenicity, which made them unsuitable for routine clinical use [17,18]. Nevertheless, these studies highlighted that the squalene-based emulsion alone, without an additional immunopotentiator, was well tolerated and had comparable immunogenicity to the formulation containing MTP-PE [18,19]. These observations resulted in the development of the MF59 o/w emulsion as an injectable adjuvant. The composition of MF59 is shown in Figure 5.1.

The small droplet size of MF59, generated through the use of a microfluid-izer in the emulsion preparation process, was crucial to potency but also enhanced stability and allowed the formulation to be sterile filtered. MF59 alone, without the use of additional immunopotentiators, was used in subsequent studies and proved sufficiently potent and safe to allow successful product development [20]. Hence, our early clinical experience with o/w emulsions served to highlight the need for careful selection of immunopotentiators if they need to be included in adjuvant formulations. In addition, the experience with MF59 showed that o/w emulsions can be highly effective adjuvants, with an acceptable safety profile.

Appearance: milky white oil in water (o/w emulsion)

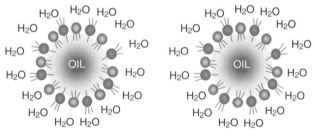

Composition:       0.5% Tween 80 - water-soluble surfactant
0.5% Span 85 - oil-soluble surfactant
4.3% Squalene oil
Water for injection
10 nM Na-citrate buffer

Density: 0.9963 g/mL                    Emulsion droplet size: ~165 nm

Viscosity: close to water, easy to inject

**Figure 5.1**  Submicron MF59 emulsion composition. (*See insert for color representation of figure.*)

## 5.3  MECHANISM OF ACTION OF MF59 ADJUVANT

Early studies designed to determine the mechanism of action of MF59 focused on the possibility of a depot effect for coadministered antigen, since there had been suggestions that the emulsion may retain antigen at the injection site and promote sustained antigen presentation. However, it was shown that an antigen depot was not established at the injection site and that the emulsion was cleared rapidly [21]. The lack of an antigen depot at the injection site with MF59 was confirmed in later studies [22], which also established that MF59 and antigen were cleared independently. Subsequently, it was thought that perhaps the emulsion acted as a direct delivery system and was responsible for promoting the uptake of antigen into antigen-presenting cells (APCs). This theory was linked to earlier observations with SAF emulsion, which contained a pluronic surfactant that was thought to be capable of binding antigen to the emulsion droplets [9]. However, studies with recombinant antigens showed that MF59 was an effective adjuvant, despite no evidence of binding of the antigens to the oil droplets. Moreover, an adjuvant effect was still observed if MF59 was injected up to 24 hours before the antigen and up to one hour after, confirming that direct association was not required [21]. Nevertheless, administration of MF59 24 hours after the antigen resulted in a much reduced adjuvant effect, suggesting that the emulsion was activating immune cells, which were then able to better process and present the coadministered antigen. A direct effect on cytokine levels in vivo has also been observed [23]. Moreover, more recent studies have confirmed the ability of MF59 to have a direct effect on immune cells, triggering the release of chemokines and other factors responsible for recruitment and maturation of immune cells.

Hence, although the exact mechanism of action of MF59 adjuvant remains to be defined, it appears to function predominantly as a delivery system and promotes the uptake of coadministered vaccine antigens into APC [24,25]. However, for them to be taken up, there does not appear to be a need for the antigen to be directly associated with the emulsion droplets. Rather, it seems likely that MF59 recruits and activates APCs at the injection site, which are then better able to take up, transport, and process coadministered antigens. Nevertheless, further studies are necessary to better define the mechanism of action of MF59. Observations on the mechanism of action of MF59 can be summarized as follows:

- Minimal retention of antigen and adjuvant at the injection site.
- MF59 and antigen are cleared independently from the injection site.
- Early studies showed that herpes simplex virus (HSV) gD2 did not associate with droplets.
- HSV gB2 associates (~10%), but does not increase, immunogenicity.
- Administration of MF59 24 hours before antigen works well.
- Administration of MF59 24 hours after antigen is ineffective.
- Enhanced cytokines can be detected in local lymph nodes or serum after MF59 injection.
- MF59 recruits APCs (MØ and dendritic cells) to the injection site.
- MF59 is taken up by APCs and transported to local lymph nodes.
- MF59 enhances uptake of antigen into APCs.
- MF59 triggers apoptosis of MØ in draining lymph nodes.

## 5.4 COMPOSITION OF MF59

MF59 is a low-oil-content o/w emulsion. The oil used for MF59 is squalene, which is a naturally occurring substance found in plants and in the livers of a range of species, including humans. Squalene is an intermediate in the human steroid hormone biosynthetic pathway and is a direct synthetic precursor to cholesterol. Therefore, squalene is a biodegradable and biocompatible naturally occurring oil. Eighty percent of shark liver oil is squalene, and shark liver provides the original natural source of the squalene that is used to prepare MF59. MF59 also contains two nonionic surfactants, Tween 80 and Span 85, which are designed to optimally stabilize the emulsion droplets. Citrate buffer is also used in MF59 to stabilize pH and the squalene content, since early formulations in water were not sufficiently stable for further development. Although single vial formulations can be developed with vaccine antigens dispersed directly in MF59, MF59 can also be added to antigens immediately prior to their administration. Although a less favorable option, combination prior to administration may be necessary to ensure optimal antigen stability.

## 5.5   MANUFACTURING OF MF59

Details of the manufacturing process for MF59 at the 50-L scale have previously been described [26]. The process involves dispersing Span 85 in the squalene phase and Tween 80 in the aqueous phase, before high-speed mixing to form a coarse emulsion. The coarse emulsion is then passed repeatedly through a microfluidizer to produce an emulsion of uniform small droplet size (165 nm) which can be sterile-filtered and filled into vials. Methods have also been published previously to allow the preparation of MF59 at a small scale, for use in research studies, but a microfluidizer is always required [27]. MF59 is characterized extensively by various physicochemical criteria after preparation and some of the parameters that are characterized are:

- Visual inspection
- Mean particle size
- Number of large particles
- Surfactant and squalene content
- Carbonyl content
- pH
- Endotoxin evaluation

## 5.6   PRECLINICAL EXPERIENCE WITH MF59

Preclinical experience with MF59 is extensive and has been reviewed on several previous occasions [26] and [28]. MF59 has been shown to be a potent adjuvant in a diverse range of species, in combination with a broad range of vaccine antigens, to include recombinant proteins, isolated viral membrane antigens, bacterial toxoids, protein polysaccharide conjugates, peptides, and viruslike particles. MF59 is particularly effective for inducing high levels of antibodies, including functional titers (neutralizing, bactericidal, and opsonophagocytic titers) and is generally more potent than alum (Table 5.1).

In a recent preclinical study, we compared MF59 and alum directly for several different vaccines and confirmed that MF59 was more potent, although alum performed well for bacterial toxoids [29]. MF59 has also shown enhanced potency over alum when compared directly in nonhuman primates with protein polysaccharide conjugate vaccines [30] (Table 5.2) and with a recombinant viral antigen [27]. In preclinical studies, compared to various alternative adjuvants, MF59 is the most potent adjuvant for flu vaccines (Figure 5.2).

In addition to immunogenicity studies, extensive preclinical toxicology studies have been undertaken with MF59 in combination with a range of

**TABLE 5.1    MF59 Represents a Workable Alternative to Alum for a Range of Traditional and New-Generation Vaccines**

| Formulation | DT ELISA Titer | Toxin Neutral (EU/mL) | TT ELISA Titer | MenC ELISA Titer | MenC BCA Titer | HBsAgM (IU/ML Titer) | MenB ELISA Titer | MenB BCA Titer |
|---|---|---|---|---|---|---|---|---|
| Alum | 8,568 | 14.1 | 31,028 | 11,117 | 19,766 | 9,118 | 8,143 | 512 |
| MF59 | 4,625 | 4.1 | 84,922 | 29,526 | 32,768 | 41,211 | 107,638 | 4,096 |

**TABLE 5.2    Comparison of MF59 with MenC and Hib Conjugate Vaccines in Infant Baboons**

| | ELISA Antibody Responses Hib (µg/mL) | | | MenC (IgG Titer) | | |
|---|---|---|---|---|---|---|
| Adjuvant | 1 Dose | 2 Doses | 3 Doses | 1 Dose | 2 Doses | 3 Doses |
| Alum | 8,568 | 14.1 | 31,028 | 11,117 | 19,766 | 9,118 |
| MF59 | 4,625 | 4.1 | 84,922 | 29,526 | 32,768 | 41,211 |

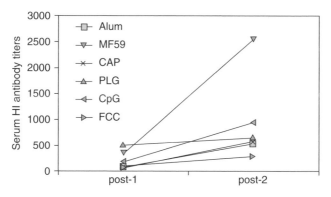

**Figure 5.2**    Evaluation of various delivery systems with flu cell culture antigen (FCC). MF59 emulsion exhibits the most potent response after two immunizations.

different antigens in a number of species. In these studies it has been shown that MF59 is neither mutagenic nor teratogenic, and did not induce sensitization in an established guinea pig model to assess contact hypersensitivity. The favorable toxicological profile established for MF59 allowed extensive clinical testing for MF59 with a number of different vaccine candidates. Observations from preclinical toxicological studies and clinical safety studies are summarized as follows:

- Neither mutagenic nor teratogenic
- Does not include hypersensitization
- No significant systemic adverse effects
- Minimal local effects at the injection site; reversible short-term reactions

## 5.7  CLINICAL EXPERI ENCE WITH MF59 ADJUVANT

The largest clinical experience with MF59 has been obtained with the adjuvanted influenza vaccine Fluad, which was licensed initially in Italy in 1997 and is now licensed in more than 20 countries. More than 20 million doses of this product have now been used in humans. The adjuvanted influenza vaccine was initially targeted for vaccination of the elderly, since conventional vaccines do not provide optimal protection in this age group [31]. For this reason, most of the clinical trials with MF59-adjuvanted influenza vaccines have been performed in elderly subjects, in which a significant adjuvant effect has consistently been observed [32]. The increased immunogenicity of MF59-adjuvanted influenza was shown to be particularly important in subsets of the elderly population, which have a higher risk of developing influenza and its most severe complications, including subjects with a low preimmunization titer and subjects affected by chronic diseases [32,33]. Additionally, immunogenicity against heterovariant flu viruses was enhanced by MF59, a feature that is particularly beneficial when the vaccine antigens do not match completely those of the circulating viruses [32,34,35]. The addition of MF59 to influenza vaccine did not affect the safety profile of the vaccine, which was very well tolerated [32]. MF59 was also evaluated as a potential adjuvant for pandemic influenza vaccines and was shown to induce a highly significant enhancement of antibody titers (Figure 5.3) [36,37].

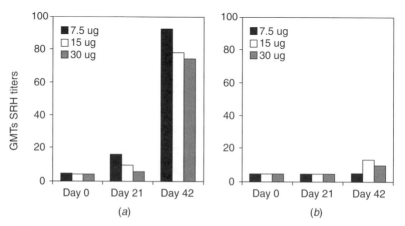

**Figure 5.3**  Enhanced efficacy of a pandemic influenza vaccine with MF59 adjuvant emulsion at three dose levels: (*a*) with MF59 and (*b*) vaccine alone.

Importantly, MF59 also allowed a significant reduction in the antigen dose, an observation that might be very important to increase the vaccine production capacity when a real pandemic occurs [36]. Notably, 7.5 μg of MF59-adjuvanted H5 hemagglutinin was significantly more immunogenic than 30 μg of plain H5 hemagglutinin [36]. As already shown for the interpandemic vaccine [32,34,35], broader cross-neutralization against heterovariant pandemic strains was also an additional benefit of an MF59-adjuvanted vaccine [38]. This is an important observation, which might favor the use of MF59-adjuvanted pandemic vaccines for stockpiling purposes.

Clinical testing of MF59 with other vaccine candidates, including the herpes simplex (HSV) and hepititus B (HBV) viruses, has provided additional evidence of the safety, tolerability, and potency of MF59 in adults [39–41]). These data have been reviewed recently [28]. Clinical data on the use of MF59 as an adjuvant for pediatric vaccines has also been obtained, with cytomegalovirus (CMV) and HIV vaccines. Seronegative toddlers immunized with an MF59-adjuvanted CMV gB vaccine showed antibody titers that were higher than those found in adults naturally infected with CMV. Moreover, the MF59-adjuvanted vaccine was well tolerated in this age group [42]. Additionally, an MF59-adjuvanted HIV vaccine was evaluated in newborns born to HIV-positive mothers [43–45]. The vaccine was very well tolerated and despite the presence of maternal antibodies, induced an antibody response in 87% of the infants immunized [44,45]. Moreover, the MF59 vaccine was significantly more potent than alum for the induction of cell-mediated immune responses (proliferative T-cell responses) against homologous and heterologous strains of HIV [43].

In summary, clinical testing of MF59 adjuvant has resulted in the registration in more than 20 countries of an effective and well-tolerated influenza vaccine for use in the elderly population. However, it has also been demonstrated that MF59 can be administered safely with a range of antigens, to diverse age groups, including the pediatric population.

## 5.8   COMBINATION OF MF59 WITH IMMUNOPOTENTIATORS

Although MF59 is generally a more potent adjuvant than alum [46], it cannot be expected to be suitable for all vaccines. MF59 is particularly effective for enhancing antibody and T-cell proliferative responses [26,46]. However, it is not a potent adjuvant for the induction of Th1 cellular immune responses, which may be required to provide protective immunity against some viruses and additional intracellular pathogens. Nevertheless, Th1 immunopotentiators, including CpG oligonucleotides [47], have been added successfully to MF59 to improve its potency and to alter the type of response induced [48]. Although the formulation of MF59 can be modified to promote the association of CpG with the oil droplets [48], more recent studies suggest that this may not be

**TABLE 5.3    MF59 in Combination with an Immunopotentiator: CpG and E1E2 (HCV) in Mouse**

| Adjuvant | Total IgG | IgG1 | IgG2a |
|---|---|---|---|
| MF59 | 3,475 | 31,262 | 3,801 |
| MF59 + CpG | 5,260 | 3,846 | 46,747 |

necessary, and simple addition of CpG to MF59 may be sufficient in some situations (Table 5.3).

However, careful choice is needed in considering which immunopotentiators to add to MF59 emulsion and how best to formulate them. Our early experience in the clinic showed that MTP-PE added to MF59 gave an unacceptable level of reactogenicity [17,18]. Fortunately, subsequent data showed that the inclusion of MTP-PE was not necessary to enhance the immunogenicity of antigens combined with MF59. Although preclinical studies showed that the potency of MF59 was enhanced by the inclusion of MTP-PE [49], animal models were not able to predict the poor tolerability of MTP-PE in humans.

In addition to immunopotentiators, alternative delivery systems, including microparticles, can also be added to MF59 to enhance potency [50]. However, the level of enhancement achieved would need to be highly significant and probably enabling for vaccine efficacy, to justify development of such a complex formulation approach.

## 5.9    USE OF MF59 IN PRIME–BOOST SETTINGS

As an alternative to the inclusion of immunopotentiators into MF59 to promote a Th1 response, MF59 can be used as a booster vaccine with recombinant proteins once a Th1 response has already been established by immunization with DNA [51]. Recently, this strategy has been shown to be highly promising for the development of a vaccine against HIV, since all arms of the immune response, including CTL responses, T-helper responses, and neutralizing antibodies, are induced by this combination immunization approach [52–55]. A similar approach of DNA prime and protein boost in MF59 has also shown significant promise in nonhuman primates as a vaccine strategy against hepitis C virus (HCV) [56]. Alternatively, protein in MF59 can also be used to boost Th1 responses primed by immunization with attenuated viral vectors. The concept of an attenuated viral vector prime followed by MF59 boost has been established in the clinic using canarypox vectors, as a strategy for both HIV [57] and CMV [58]. Studies are also showing very encouraging preclinical data with alternative viral vectors, including alphaviruses and adenoviruses (unpublished data).

## 5.10  FUTURE PERSPECTIVES ON THE USE OF MF59

In an extensive range of clinical studies, MF59 has proven to be a safe and potent vaccine adjuvant, resulting in the licensure of an MF59-adjuvanted influenza vaccine in more than 20 countries. The ability of MF59 to induce significantly enhanced titers against potential pandemic flu strains at low antigen doses appears highly promising, as does the ability of MF59 to offer neutralization against heterologous strains. The clinical data obtained previously with a H5N3 flu strain was recently reproduced with a H9N2 strain, and studies with a H5N1 strain are currently ongoing. The dose-sparing aspect of MF59 may be particularly attractive given the current limited capacity worldwide for influenza vaccine production, which is not sufficient to deal with a pandemic, given the high dose requirements for unadjuvanted influenza vaccine [35].

The encouraging safety and tolerability profile of MF59, in combination with immunogenicity data, suggest that MF59 is an appropriate adjuvant for use in pediatric populations. The stronger adjuvant effect of MF59 compared to alum in newborn infants immunized with an HIV vaccine has established the basis for further use of MF59 in this population. Moreover, preclinical data have firmly established that MF59 is a more potent adjuvant than alum for a wide range of vaccines, including recombinant proteins and protein polysaccharide conjugates. Finally, if necessary, MF59 may be combined with immunopotentiators to enable the development of more complex vaccines (e.g., against HCV and/or HIV), which may also require the use of a prime with DNA or viral vectors.

## ACKNOWLEDGMENTS

We would like to recognize the contributions of many colleagues in Novartis Vaccines and Diagnostics, Inc. to the data and ideas contained in this chapter, particularly the vaccine delivery group. In addition, we would like to acknowledge the excellent assistance of Nelle Cronen in manuscript preparation and formatting.

## REFERENCES

1. Freund, J., Casals, J. & Hosmer, E.P. Sensitization and antibody formation after injection of turbecle bacili and paraffin oil. *Proc Soc Exp Biol Med* 1937, **37**, 509–513.

2. Hilleman, M.R. Critical appraisal of emulsified oil adjuvants applied to viral vaccines. *Prog Med Virol* 1966, **8**, 131–182.

3. Jansen, T., Hofmans, M.P., Theelen, M.J. & Schijns, V.E. Structure–activity relations of water-in-oil vaccine formulations and induced antigen-specific antibody responses. *Vaccine* 2005, **23**(8), 1053–1060.

4. Salk, J.E., Laurent, A.M. & Bailey, M.L. Direction of research on vaccination against influenza; new studies with immunologic adjuvants. *Am J Public Health* 1951, **41**(6), 669–677.

5. Edelman, R. Vaccine adjuvants. *Rev Infect Dis* 1980, **2**(3), 370–383.

6. Page, W. Long-term followup of army recruits immunized with Freund's incomplete adjuvanted vaccine. *Vaccine Res* 1993, **2**, 141–149.

7. Aucouturier, J., Dupuis, L., Deville, S., Ascarateil, S. & Ganne, V. Montanide ISA 720 and 51: a new generation of water in oil emulsions as adjuvants for human vaccines. *Expert Rev Vaccines* 2002, **1**(1), 111–118.

8. Audran, R., Cachat, M., Lurati, F., Soe, S., Leroy, O., Corradin, G. et al. Phase I malaria vaccine trial with a long synthetic peptide derived from the merozoite surface protein 3 antigen. *Infect Immun* 2005, **73**(12), 8017–8026.

9. Allison, A.C. & Byars, N.E. An adjuvant formulation that selectively elicits the formation of antibodies of protective isotypes and of cell-mediated immunity. *J Immunol Methods* 1986, **95**(2), 157–168.

10. Vogel, F.R. & Powell, M.F. In *Vaccine Design: The Subunit and Adjuvant Approach*, Powell, M.F. & Newman, M.J., Eds., Plenum Press, New York, 1995, pp. 141–228.

11. Ellouz, F., Adam, A., Ciorbaru, R. & Lederer, E. Minimal structural requirements for adjuvant activity of bacterial peptidoglycan derivatives. *Biochem Biophys Res Commun* 1974, **59**(4), 1317–1325.

12. Waters, R.V., Terrell, T.G. & Jones, G.H. Uveitis induction in the rabbit by muramyl dipeptides. *Infect Immun* 1986, **51**(3), 816–825.

13. Vidal, V.F., Casteran, N., Riendeau, C.J., Kornfeld, H., Darcissac, E.C., Capron, A. et al. Macrophage stimulation with Murabutide, an HIV-suppressive muramyl peptide derivative, selectively activates extracellular signal-regulated kinases 1 and 2, C/EBPbeta and STAT1: role of CD14 and Toll-like receptors 2 and 4. *Eur J Immunol* 2001, **31**(7), 1962–1971.

14. Phillpotts, R.J., Venugopal, K. & Brooks, T. Immunisation with DNA polynucleotides protects mice against lethal challenge with St. Louis encephalitis virus. *Arch Virol* 1996, **141**(3–4), 743–749.

15. Kenney, R.T. & Edelman, R. *New Generation Vaccines*, Marcel Dekker, New York, 2004.

16. Wintsch, J., Chaignat, C.L., Braun, D.G., Jeannet, M., Stalder, H., Abrignani, S. et al. Safety and immunogenicity of a genetically engineered human immunodeficiency virus vaccine. *J Infect Dis* 1991, **163**(2), 219–225.

17. Keitel, W., Couch, R., Bond, N., Adair, S., Van Nest, G. & Dekker, C. Pilot evaluation of influenza virus vaccine (IVV) combined with adjuvant. *Vaccine* 1993, **11**(9), 909–913.

18. Keefer, M.C., Graham, B.S., McElrath, M.J., Matthews, T.J., Stablein, D.M., Corey, L. et al. Safety and immunogenicity of Env 2-3, a human immunodeficiency virus type 1 candidate vaccine, in combination with a novel adjuvant, MTP-PE/MF59. NIAID AIDS Vaccine Evaluation Group. *AIDS Res Hum Retroviruses* 1996, **12**(8), 683–693.

19. Kahn, J.O., Sinangil, F., Baenziger, J., Murcar, N., Wynne, D., Coleman, R.L. et al. Clinical and immunologic responses to human immunodeficiency virus (HIV) type 1SF2 gp120 subunit vaccine combined with MF59 adjuvant with or without muramyl

tripeptide dipalmitoyl phosphatidylethanolamine in non-HIV-infected human volunteers. *J Infect Dis* 1994, **170**(5), 1288–1291.

20. Ott, G., Radhakrishnan, R., Fang, J.-H. & Hora, M. The adjuvant MF59: a ten year perspective. In *Vaccine Adjuvants: Preparation Methods and Research Protocols*, O'Hagan, D., Ed., Humana Press, Totowa, NJ, 2001, pp. 211–228.

21. Ott, G., Barchfeld, G.L., Chernoff, D., Radhakrishnan, R., van Hoogevest, P. & Van Nest, G. MF59: design and evaluation of a safe and potent adjuvant for human vaccines. In *Vaccine Design: The Subunit and Adjuvant Approach*, Powell, M.F. & Newman, M.J., Eds., Plenum Press, New York, 1995, pp. 277–296.

22. Dupuis, M., McDonald, D.M. & Ott, G. Distribution of adjuvant MF59 and antigen gD2 after intramuscular injection in mice. *Vaccine* 1999, **18**(5–6), 434–439.

23. Valensi, J.P., Carlson, J.R. & Van Nest, G.A. Systemic cytokine profiles in BALB/c mice immunized with trivalent influenza vaccine containing MF59 oil emulsion and other advanced adjuvants. *J Immunol* 1994, **153**(9), 4029–4039.

24. Dupuis, M., Murphy, T.J., Higgins, D., Ugozzoli, M., Van Nest, G., Ott, G. et al. Dendritic cells internalize vaccine adjuvant after intramuscular injection. *Cell Immunol* 1998, **186**(1), 18–27.

25. Dupuis, M., Denis-Mize, K., LaBarbara, A., Peters, W., Charo, I.F., McDonald, D.M. et al. Immunization with the adjuvant MF59 induces macrophage trafficking and apoptosis. *Eur J Immunol* 2001, **31**(10), 2910–2918.

26. Ott, G. Vaccine adjuvants: preparation methods and research protocols. In *Vaccine Adjuvants: Preparation Methods and Research Protocols*, O'Hagan, D., Ed., Humana Press, Totowa, NJ, 2000.

27. Traquina, P., Morandi, M., Contorni, M. & Van Nest, G. MF59 adjuvant enhances the antibody response to recombinant hepatitis B surface antigen vaccine in primates. *J Infect Dis* 1996, **174**(6), 1168–1175.

28. Podda, A. & Del Giudice, G. MF59-adjuvanted vaccines: increased immunogenicity with an optimal safety profile. *Expert Rev Vaccines* 2003, **2**(2), 197–203.

29. Singh, M., Ugozzoli, M., Kazzaz, J., Chesko, J., Soenawan, E., Mannucci, D. et al. A preliminary evaluation of alternative adjuvants to alum using a range of established and new generation vaccine antigens. *Vaccine* 2006, **24**(10), 1680–1686.

30. Granoff, D.M., McHugh, Y.E., Raff, H.V., Mokatrin, A.S. & Van Nest, G.A. MF59 adjuvant enhances antibody responses of infant baboons immunized with *Haemophilus influenzae* type b and *Neisseria meningitidis* group C oligosaccharide–CRM197 conjugate vaccine. *Infect Immun* 1997, **65**(5), 1710–1715.

31. Strassburg, M.A., Greenland, S., Sorvillo, F.J., Lieb, L.E. & Habel, L.A. Influenza in the elderly: report of an outbreak and a review of vaccine effectiveness reports. *Vaccine* 1986, **4**(1), 38–44.

32. Podda, A. The adjuvanted influenza vaccines with novel adjuvants: experience with the MF59-adjuvanted vaccine. *Vaccine* 2001, **19**(17–19), 2673–2680.

33. Banzhoff, A., Nacci, P. & Podda, A. A new MF59-adjuvanted influenza vaccine enhances the immune response in the elderly with chronic diseases: results from an immunogenicity meta-analysis. *Gerontology* 2003, **49**(3), 177–184.

34. De Donato, S., Granoff, D., Minutello, M., Lecchi, G., Faccini, M., Agnello, M. et al. Safety and immunogenicity of MF59-adjuvanted influenza vaccine in the elderly. *Vaccine* 1999, **17**(23–24), 3094–3101.

35. Daems, R., Del Giudice, G. & Rappuoli, R. Anticipating crisis: towards a pandemic flu vaccination strategy through alignment of public health and industrial policy. *Vaccine* 2005, **23**(50), 5732–5742.

36. Nicholson, K., Colegate, A., Podda, A., Stephenson, I., Wood, J., Ypma, E. et al. Confronting a potential H5N1 pandemic: a randomised controlled trial of conventional and MF59 adjuvanted influenza A/Duck/Singapore/97 (H5N3) surface antigen vaccine. *Lancet* 2001, **9272**, 357.

37. Stephenson, I., Nicholson, K.G., Colegate, A., Podda, A., Wood, J., Ypma, E. et al. Boosting immunity to influenza H5N1 with MF59-adjuvanted H5N3 A/Duck/Singapore/97 vaccine in a primed human population. *Vaccine* 2003, **21**(15), 1687–1693.

38. Stephenson, I., Nicholson, K.G., Bugarini, R., Podda, A., Wood, J., Zambon, M. et al. Cross reactivity to highly pathogenic avian influenza H5N1 viruses following vaccination with non-adjuvanted and MF-59-adjuvanted influenza A/Duck/Singapore/97 (H5N3) vaccine: a potential priming strategy. *J Infect Dis* 2005, **191**.

39. Heineman, T.C., Clements-Mann, M.L., Poland, G.A., Jacobson, R.M., Izu, A.E., Sakamoto, D. et al. A randomized, controlled study in adults of the immunogenicity of a novel hepatitis B vaccine containing MF59 adjuvant. *Vaccine* 1999, **17**(22), 2769–2778.

40. Langenberg, A.G., Burke, R.L., Adair, S.F., Sekulovich, R., Tigges, M., Dekker, C.L. et al. A recombinant glycoprotein vaccine for herpes simplex virus type 2: safety and immunogenicity [corrected] [published erratum appears in *Ann Intern Med* 1995 Sep 1, **123**(5), 395]. *Ann Intern Med* 1995, **122**(12), 889–898.

41. Corey, L., Langenberg, A.G., Ashley, R., Sekulovich, R.E., Izu, A.E., Douglas, J.M., Jr. et al. Recombinant glycoprotein vaccine for the prevention of genital HSV-2 infection: two randomized controlled trials. Chiron HSV Vaccine Study Group [see comments]. *JAMA* 1999, **282**(4), 331–340.

42. Mitchell, D.K., Holmes, S.J., Burke, R.L., Duliege, A.M. & Adler, S.P. Immunogenicity of a recombinant human cytomegalovirus gB vaccine in seronegative toddlers. *Pediatr Infect Dis J* 2002, **21**(2), 133–138.

43. Borkowsky, W., Wara, D., Fenton, T., McNamara, J., Kang, M., Mofenson, L. et al. Lymphoproliferative responses to recombinant HIV-1 envelope antigens in neonates and infants receiving gp120 vaccines. AIDS Clinical Trial Group 230 Collaborators. *J Infect Dis* 2000, **181**(3), 890–896.

44. Cunningham, C.K., Wara, D.W., Kang, M., Fenton, T., Hawkins, E., McNamara, J. et al. Safety of 2 recombinant human immunodeficiency virus type 1 (HIV-1) envelope vaccines in neonates born to HIV-1-infected women. *Clin Infect Dis* 2001, **32**(5), 801–807.

45. McFarland, E.J., Borkowsky, W., Fenton, T., Wara, D., McNamara, J., Samson, P. et al. Human immunodeficiency virus type 1 (HIV-1) gp120-specific antibodies in neonates receiving an HIV-1 recombinant gp120 vaccine. *J Infect Dis* 2001, **184**(10), 1331–1335 (Epub 2001 Oct 1310).

46. Ott, G. In *Vaccine Design: The Subunit and Adjuvant Approach*, Powell, M.F. & Newman, M.J., Eds., Plenum Press, New York, 1995.

47. Klinman, D.M. Use of CpG oligodeoxynucleotides as immunoprotective agents. *Expert Opin Biol Ther* 2004, **4**(6), 937–946.

48. O'Hagan, D.T., Singh, M., Kazzaz, J., Ugozzoli, M., Briones, M., Donnelly, J. et al. Synergistic adjuvant activity of immunostimulatory DNA and oil/water emulsions for immunization with HIV p55 gag antigen. *Vaccine* 2002, **20**(27–28), 3389–3398.

49. Burke, R.L., Goldbeck, C., Ng, P., Stanberry, L., Ott, G. & Van Nest, G. The influence of adjuvant on the therapeutic efficacy of a recombinant genital herpes vaccine. *J Infect Dis* 1994, **170**(5), 1110–1119.

50. O'Hagan, D.T., Ugozzoli, M., Barackman, J., Singh, M., Kazzaz, J., Higgins, K. et al. Microparticles in MF59, a potent adjuvant combination for a recombinant protein vaccine against HIV-1. *Vaccine* 2000, **18**(17), 1793–1801.

51. Cherpelis, S., Shrivastava, I., Gettie, A., Jin, X., Ho, D.D., Barnett, S.W. et al. DNA vaccination with the human immunodeficiency virus type 1 SF162DeltaV2 envelope elicits immune responses that offer partial protection from simian/human immunodeficiency virus infection to CD8(+) T-cell-depleted rhesus macaques. *J Virol* 2001, **75**(3), 1547–1550.

52. Otten, G.R., Schaefer, M., Greer, C., Calderon-Cacia, M., Coit, D., Kazzaz, J. et al. Induction of broad and potent anti-HIV immune responses in rhesus macaques by priming with a DNA vaccine and boosting with protein-adsorbed PLG microparticles. *J Virol* 2003, **77**, 6087–6092.

53. Otten, G., Schaefer, M., Doe, B., Liu, H., Srivastava, I., zur Megede, J. et al. Enhancement of DNA vaccine potency in rhesus macaques by electroporation. *Vaccine* 2004, **22**(19), 2489–2493.

54. Otten, G.R., Schaefer, M., Doe, B., Liu, H., Megede, J.Z., Donnelly, J. et al. Potent immunogenicity of an HIV-1 gag-pol fusion DNA vaccine delivered by in vivo electroporation. *Vaccine* 2005 (Epub ahead of print, Sep 20).

55. Otten, G.R., Schaefer, M., Doe, B., Liu, H., Srivastava, I., Megede, J. et al. Enhanced potency of plasmid DNA microparticle human immunodeficiency virus vaccines in rhesus macaques by using a priming-boosting regimen with recombinant proteins. *J Virol* 2005, **79**(13), 8189–8200.

56. O'Hagan, D.T., Singh, M., Dong, C., Ugozzoli, M., Berger, K., Glazer, E. et al. Cationic microparticles are a potent delivery system for a HCV DNA vaccine. *Vaccine* 2004, **23**(5), 672–680.

57. AIDS Vaccine Evaluation Group 022 Protocol Team: cellular and humoral immune responses to a canarypox vaccine containing human immunodeficiency virus type 1 Env, Gag, and Pro in combination with rgp120. *J Infect Dis* 2001, **183**(4), 563–570.

58. Bernstein, D.I., Schleiss, M.R., Berencsi, K., Gonczol, E., Dickey, M., Khoury, P. et al. Effect of previous or simultaneous immunization with canarypox expressing cytomegalovirus (CMV) glycoprotein B (gB) on response to subunit gB vaccine plus MF59 in healthy CMV-seronegative adults. *J Infect Dis* 2002, **185**(5), 686–690 (Epub 2002 Feb 2006).

# 6

# TLR4 AGONISTS AS VACCINE ADJUVANTS

DAVID A. JOHNSON AND JORY R. BALDRIDGE

## 6.1 INTRODUCTION

Most vaccines in use today employ killed or attenuated microbes or microbial fragments to stimulate a protective immune response against the cognate infectious agent. However, problems associated with the manufacture of conventional vaccines and the presence of nonessential components and antigens have resulted in considerable effort to refine vaccines as well as to develop well-defined synthetic antigens using chemical and recombinant techniques [1]. But the refinement and simplification of microbial vaccines has led to a concomitant loss in potency. In addition, low-molecular-weight synthetic antigens, although devoid of potentially harmful contaminants, are themselves not very immunogenic [1,2]. These observations have led to investigations on coadministering *additives* known as *adjuvants* with vaccine antigens to potentiate the activity of vaccines and the weak immunogenicity of synthetic epitopes [2].

Currently, the only adjuvant licensed for human use in the United States is a group of aluminum salts known as alum [1]. But alum, which acts by a depot mechanism, is not without side effects and enhances humoral (Th2) immunity principally [1,3]. The recognition that cell-mediated immune responses, particularly the induction of cytotoxic T lymphocytes (CTLs), are crucial for generating protective immunity in the case of many intracellular pathogens and cancers has prompted efforts to develop new vaccine adjuvants that enhance both antibody and T-cell responses [1]. For certain diseases it may also be desirable to stimulate predominantly a Th1 or Th2 immune response

*Vaccine Adjuvants and Delivery Systems*, Edited by Manmohan Singh
Copyright © 2007 John Wiley & Sons, Inc.

or stimulate both mucosal and systemic immunity. Thus, new adjuvants are needed that help control the magnitude, direction, and duration of the immune response against antigens. The discovery of mammalian Toll-like receptors (TLRs) and other pattern-recognition receptors (PRRs) on innate immune cells, and dendritic cells (DCs) in particular, not only has demonstrated the integral role that stimulation of the innate immune system plays in triggering adaptive immune responses to vaccine antigens but has also provided natural product leads that are useful in the design of new adjuvants.

Although the addition of microbial components to vaccines has long been known to enhance adaptive immune responses, the molecular mechanisms involved have not been well understood. Only recently were PRRs on cells of the immune system shown to engage some of these microbial products. For example, the main cell surface component of gram-negative bacteria, such as lipopolysaccharide (LPS, endotoxin) and its active principle, lipid A, are ligands for Toll-like receptor 4 (TLR4) and accessory molecules such as MD-2 on dendritic and other immune cells. As a result, many lipid A molecules are potent adjuvants for protein and carbohydrate antigens and can markedly enhance both humoral and cell-mediated responses. However, the profound pyrogenicity and lethal toxicity of LPS and lipid A have limited their medicinal use. Thus, considerable effort has been directed toward the development of semisynthetic and synthetic lipid A mimetics with simplified structures and improved toxicity and activity profiles for use as vaccine adjuvants as well as stand-alone therapeutics for enhancing host resistance to infectious disease and cancer.

## 6.2  TLR4-MEDIATED INDUCTION OF ADAPTIVE IMMUNITY

The TLR family consists of at least 11 known members that collectively recognize conserved pathogen-associated molecular patterns (PAMPs) common to most or all known pathogens. These evolutionarily conserved receptors are type 1 membrane proteins with extracellular, transmembrane, and cytoplasmic regions. The extracellular or amino-terminus portion has leucine-rich motifs that are thought to be involved with PAMP recognition. Of the 11 known family members, five TLRs appear to specialize in recognition of structural molecules from bacteria: TLR4 detects LPS; TLRs 1, 2, and 6 cooperate in the recognition of bacterial lipopeptides, and TLR5 detects bacterial flagellin [4]. In contrast, TLRs 3, 7, 8, and 9 focus on the detection of viruses and nucleic acids: TLR3 detects double-stranded RNA; TLRs 7 and 8 recognize single-stranded RNA; and TLR9 detects unmethylated DNA, which is more common to bacteria and viruses [4]. Accordingly, the cellular expression of TLRs 3, 7, 8, and 9 is restricted to cytoplasmic compartments, where they are more likely to encounter pathogen-associated nucleic acids [4]. The most recent TLR discovery, TLR11, is found in mice and is associated with the detection of protease-sensitive molecules from urogenic bacteria [5] and a profillin-like protein from the protozoan parasite, *Toxoplasma gondii* [6]. Thus, each TLR specializes in the recognition of different microbial ligands, and collectively the TLRs detect most if not all classes of pathogens.

All TLRs have conserved cytoplasmic tails, which share amino acid homology with the IL-1 receptor (IL-1R). This conserved Toll/IL-1R (TIR) region is critical for the induction of intracellular signaling pathways that lead to gene expression. The TIR domain regulates signaling via association with intracellular adaptor molecules, including MyD88, TRIF, TIRAP, and TRAM [7,8]. All TLRs utilize the MyD88 adaptor protein with the exception of TLR3, which associates with TRIF. Following PAMP-mediated activation, TLRs recruit and activate MyD88, which subsequently induces recruitment of members of the IL-1 receptor-associated kinase family (IRAK) and TRAF6. The majority of TLRs appear to associate directly with MyD88, but TLR4 and TLR2 work through the intermediate adapter molecule, TIRAP. By either mechanism this MyD88-dependent pathway leads to expression of inflammatory cytokine and chemokine genes, including TNF$\alpha$, IL-1, and IL-6. TLR3 does not associate with MyD88 but instead employs the TIR-containing adaptor molecule TRIF to regulate intracellular signaling. Activation of TRIF in this MyD88-independent pathway induces the recruitment of IRF-3, which leads to the expression of type I interferons (IFN$\alpha$/$\beta$). TLR4 activation can also lead to the expression of IFN$\alpha$/$\beta$ through the engagement of TRIF; however, this MyD88-independent TLR4 pathway uses the intermediate adapter molecule TRAM. Thus, the stimulation of distinct TLR family members can differentially regulate the expression of inflammatory cytokines, chemokines, and type I interferons through utilization of a variety of intracellular adaptor molecules and their associated pathways.

TLRs are expressed strategically in tissues and cells that maximize their capacity for immune surveillance and immune cell recruitment in the event of infection. Consequently, TLRs are expressed on epithelial cells lining the respiratory and digestive tracts, on resident leukocytes of the mucosa, and on leukocytes in circulation. Some differential expression of TLRs on antigen-presenting cells has been noted. Monocytes and macrophages express most TLRs with the exception of TLR3. Dendritic cell (DC) subsets, key cells for the uptake and presentation of antigens to T lymphocytes, appear to have distinct TLR expression patterns with myeloid DCs (mDCs) expressing predominantly TLRs 1, 2, 3, 4, 5, 6, and 8, while plasmacytoid DCs (pDCs) express TLRs 7 and 9 [4,9]. Interestingly, freshly isolated human DCs do not appear to express TLR4 or to respond to LPS (TLR4 ligand) stimulation. In addition to their role in allergic reactions, mast cells are capable of phagocytosis, presentation of antigen, and secretion of inflammatory cytokines. Human mast cells express TLRs 2, 4, 6, and 8. Distributed in this way, TLRs are positioned to recognize invading pathogens at the most likely portals of entry, to activate phagocytic cells to control the initial spread of infection, and to recruit and activate lymphocytes that provide antigen-specific resistance as well as immunologic memory.

The realization that TLR molecules are key mediators of innate and adaptive immune responses has led to increased evaluation of TLR agonists as vaccine adjuvants and as stand-alone immunomodulators. In this chapter, we focus predominantly on agonists of Toll-like receptor 4.

LPS, the natural ligand for TLR4, is a potent immunostimulant that activates immunocompetent cells and provokes the release of inflammatory cytokines and chemokines. However, cellular recognition of LPS is complex, and TLR4 alone is insufficient for detection and signaling. At least three other proteins aid in the detection, binding, and transfer of LPS to the cell surface receptor complex: LPS-binding protein (LBP), CD14, and MD-2 [10]. LBP, a secreted serum protein, binds gram-negative bacteria, facilitating the extrication and transfer of monomeric units of LPS from the bacterial cell wall to CD14, which is present as a soluble serum protein and as a glycosylphosphatidylinositol-anchored protein on PMNs, monocytes, and macrophages. CD14 has long been recognized as an LPS receptor but does not have a cytoplasmic tail that would enable it to induce signal transduction. Instead, CD14 transfers LPS to MD-2, a critical component of the membrane-bound TLR4/MD-2 heterodimer that forms the ultimate LPS receptor complex. MD-2, a secreted glycoprotein, possesses a putative lipid-binding pocket within its hydrophobic core capable of sequestering the acyl chains of LPS and is considered to be the LPS-binding component of the TLR4/MD-2 receptor complex. In support of this concept, MD-2 can discriminate between lipid A molecules of varying levels of acylation and subsequently affect the response to these agonists [11]. MD-2 also possesses conserved amino acid sequences that are critical for association with the extracellular portion of TLR4, the signaling component of the receptor complex.

Understanding where TLR4 is expressed, how it interacts with other extracellular proteins, and how it triggers intracellular signaling pathways should help in the design of agonists that are safe and beneficial for use as vaccine adjuvants. As mentioned above, in vitro studies have determined that MD-2 with its lipid-binding pocket is a critical component of the TLR4 receptor complex, but exactly how variations in lipid A acylation patterns are accommodated and their effect on signaling remains unclear. Furthermore, how the phosphorylated glucosamine backbone interacts with TLR4/MD-2 and how it affects signaling is not understood. Studies with natural lipid A variants and synthetic lipid A mimetics are now under way which may shed light on these unknowns. There is already convincing evidence that TLR4 agonists can be produced that have reduced toxicity profiles while maintaining beneficial vaccine adjuvant capabilities.

## 6.3  NATURALLY DERIVED LIPID A DERIVATIVES AND SYNTHETIC LIPID A MIMETICS

### 6.3.1  Disaccharides

Several years ago, it was discovered that the toxic effects of *Salmonella minnesota* R595 lipid A (structure **1**) could be ameliorated by selective hydrolysis

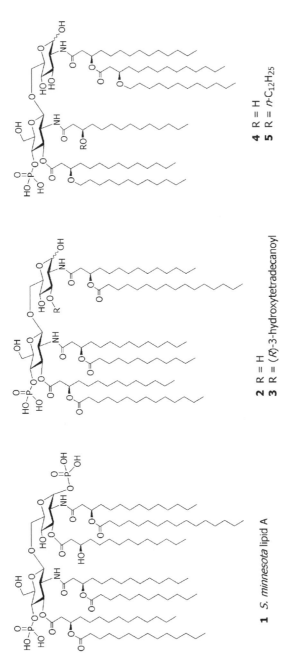

**1** *S. minnesota* lipid A

**2** R = H
**3** R = (*R*)-3-hydroxytetradecanoyl

**4** R = H
**5** R = *n*-C₁₂H₂₅

**Figure 6.1** Synthetic and naturally derived derivatives of *S. minnesota* lipid A.

135

of the 1-$O$-phosphono and ($R$)-3-hydroxytetradecanoyl groups [12]. The resulting chemically modified natural lipid A product, monophosphoryl lipid A (MPL) adjuvant, is an effective adjuvant in prophylactic and therapeutic vaccines and shows an excellent safety profile in humans [13]. The first TLR agonist–containing vaccine (Fendrix vaccine for hepatitis B containing MPL) was recently approved in Europe, and a second MPL-adjuvanted vaccine (Cervarix for human papilloma virus) is in phase III clinical trials in Europe. In addition to the major hexaacyl component (**2**), MPL adjuvant comprises several less highly acylated compounds, due primarily to inherent heterogeneity in the LPS from which it is manufactured.

Another detoxified lipid A product closely related to MPL adjuvant and containing heptaacyl derivative (**3**) as a major component is being employed as an adjuvant in therapeutic cancer vaccines against non–small cell lung (NSCL) and prostate cancer [14]. A phase III clinical trial with an NSCL liposomal-based vaccine that elicits a cellular immune response to the tumor-associated antigen mucin MUC-1 began in 2007. Biomira, the developer of this and other therapeutic cancer vaccines, recently reported that the synthetic 3-desacyl monophosphoryl lipid A (MLA) derivatives **4** and **5**, which possess hydrolytically more stable alkoxy-substituted acyl residues and a unique trilipid moiety, induce T-cell proliferation and levels of IFN-$\gamma$ comparable to those induced by the detoxified lipid A (**3**) and its congeners in experimental MUC-1 vaccines in murine models [15]. MPL adjuvant has also been shown to possess activity similar to that of synthetic MLA derivatives **4** and **5**, and this detoxified lipid A in experimental models for NSCL and prostate cancer [14,15].

Preclinical and clinical results with these monophosphoryl lipid A derivatives suggest that the structural requirements for adjuvanticity are less stringent than for other endotoxicities. The effect of fatty acid structure on endotoxicity in the MLA series has also been evaluated with a series of chain-length homologs of MLA (**2**) [16]. Although the MLA normal (secondary) fatty acid chain length was found to play an indispensable role in the expression of various endotoxic activities, including TLR4-mediated cytokine induction in peripheral-blood monocytic cells (PBMCs), prepared synthetic **2** as well as synthetic MLA derivative **6**, which contains 14-carbon secondary acyl chains and the same overall number of acyl carbon atoms as **2**, induced significantly higher tetanus toxoid–specific antibodies of all classes of immunoglobulins tested, including complement-fixing IgG2a and IgG2b isotypes than those of the highly endotoxic 10-carbon homolog **7** in murine vaccine models.

These data suggest that fatty acid structure, which is known to influence molecular conformation as well as solution aggregate structure, may be more important than the presence or absence of the 1-phosphate and 3-hydroxytetradecanoyl groups (both absent in MPL) in determining the biological activity of lipid A molecules. For example, the reduced toxicity of MPL component **2** relative to that of parent lipid A (**1**) cannot be reconciled with the high pyro-

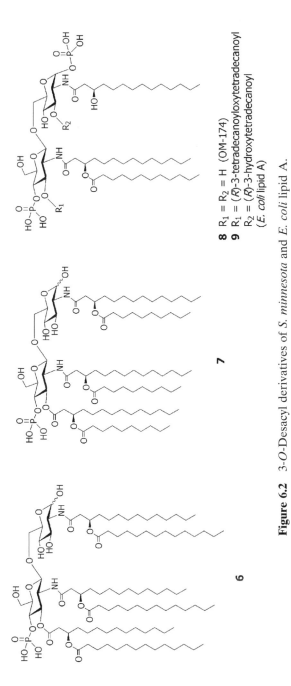

**8**  R₁ = R₂ = H  (OM-174)
**9**  R₁ = (R)-3-tetradecanoyloxytetradecanoyl
        R₂ = (R)-3-hydroxytetradecanoyl
        (E. coli lipid A)

**7**

**6**

**Figure 6.2**  3-O-Desacyl derivatives of S. minnesota and E. coli lipid A.

genicity and lethal toxicity of MLA derivative **7** on the basis of chemical modification (deacylation and dephosphorylation) alone. Further, OM Pharma recently reported that the triacyl diphosphoryl lipid A (**8**; OM-174), derived from *Escherichia coli* lipid A (**9**), exhibits strong TLR4-dependent adjuvant activity (humoral, mucosal, and cell-mediated immunity) in vaccine models but is more than 10,000 times less pyrogenic than its lipid A progenitor (**9**) despite the presence of the anomeric phosphate in **8** [17,18]. Triacyl derivative **8**, which induces high levels of IL-6 and nitric oxide (NO) in human mononuclear cells and murine macrophages, respectively, also activates dendritic cells in vivo via TLR4 [19]. Nonetheless, direct comparison of the adjuvant activity of the underacylated diphosphate **8**, which may also signal via TLR2, with that of related disaccharides has not been reported, making it difficult to draw definitive conclusions about the importance of acyl chain number and pattern and the presence of anomeric phosphate in **8** to its adjuvanticity and reduced toxicity. In general, increasing or decreasing the number of lipid chains present in disaccharide lipid A derivatives from an optimum of six reduces TLR4 agonist activity and toxicity [20,21]. The low endotoxicity observed for certain helicobacter, pseudomonas, and other lipopolysaccharides has been attributed to the presence of underacylated lipid A components [22,23]. But factors besides TLR4 agonism, such as aqueous solubility and stability, solution aggregation, and other physicochemical properties, are also important to adjuvanticity.

Biophysical characterization of the triacyl derivative **8** shows that individual molecules adopt an inverted conical shape under nearly physiological conditions and a micellular $H_I$ aggregate structure [24]. This molecular shape and supramolecular structure are both very different from those of the potent TLR4 agonist **9** (conical/inverted cubic) and the TLR4 antagonist lipid IVa (cylindrical/lamellar) [25]. This shows that physicochemical characterization of lipid A–like molecules can provide insight into the molecular requirements for different endotoxic activities, including adjuvanticity, and improved toxicity/bioactivity profiles. Nevertheless, for certain underacylated LPS variants it has been reported that solution aggregation properties are not important determinants of the TLR4 agonist activity and that the monomeric LPS:MD-2 complexes formed from underacylated variants are less adept at activating TLR4 and triggering oligomerization [26].

### 6.3.2   Monosaccharides and Acyclic Derivatives

Numerous subunit derivatives of bacterial lipid A have also been prepared with the aim of separating toxic properties from beneficial immunostimulatory effects. These molecules are typically synthetic analogs of either the reducing or nonreducing glucosamine moieties of lipid A or molecules in which one of the saccharide units or the entire disaccharide backbone has been replaced with an acyclic scaffold. The availability of synthetic lipid A subunit analogs makes it possible to carry out precise investigations on the relationship between chemical structure and biological activity without ambiguity due to heterogeneity in naturally derived materials and/or contamination by other bioactive substances.

Synthetic analogs of either the reducing or nonreducing glucosamine moieties of lipid A containing up to five fatty acids and lacking an aglycon unit typically exhibit low biological activity. However, certain monosaccharides— particularly nonreducing subunit analogs possessing three acyl residues— exhibit broad endotoxic activities but low pyrogenicity and lethal toxicity compared to E. coli lipid A (9). For example, GLA-60 (10), a compound possessing nearly the same fatty acid substituents in the same relative positions as in the disaccharide OM-174 (8), showed the strongest B-cell activation and adjuvant activities among various nonreducing subunit analogs that have been examined [27]. Structure–activity relationship (SAR) investigations showed the importance of normal fatty acid chain length, stereochemistry, and the number, type, and relative position of the fatty acyl moieties in this series. However, despite the low pyrogenicity of GLA-60 (10), and in marked contrast to disaccharide (8), compound (10) was two orders of magnitude less active than E. coli lipid A (9) with respect to its ability to induce TNFα production in human mononuclear cells. Biophysical characterization of 10 showed that it adopts a slightly conical molecular conformation and mainly unilamellar aggregate structures, which is consistent with its reduced bioactivity [28]. Adding a fatty acid onto the side-chain hydroxyl group of 10 to form tetra-acyl derivative 11 (GLA-47) abolishes TNFα and IL-6 induction in human U937 cells and PBMCs [27].

Interestingly, the triacyl sulfoglycolipid 12 (ONO-4007), which is structurally related to GLA-60 and GLA-47, exhibited strong antitumor activity in several animal models via intratumoral production of TNFα [29]. However, this compound activates human monocytes to release TNFα only in GM-CSF-primed monocytes, not monocytes from human whole blood [30]. Thus, this sulfo lipid A analog may activate human monocytes by a pathway different from that of lipid A.

It is important to note, however, that certain lipid A–like compounds, such as the monosaccharide GLA-47 (11) and the disaccharide lipid IVa, the biosynthetic precursor of E. coli lipid A (9), are TLR4 agonists in murine models but antagonists in human cells. Thus, cells of the human immune system exhibit much stricter structural requirements for TLR4 agonism than do murine macrophages [27,31]. Conversely, the structural requirements for antagonism are stricter in murine cells than in human cells. This clear difference between the activation of murine and human immune cells underscores the importance of using human cell systems for the discovery of clinically useful vaccine adjuvants that activate TLR4.

In contrast to nonreducing subunit analogs of lipid A, monosaccharide derivatives of the reducing glucosamine are considerably less immunostimulatory and often exhibit TLR4 antagonist properties [32,33]. Despite seminal studies showing that the biosynthetic precursor to the reducing sugar of lipid

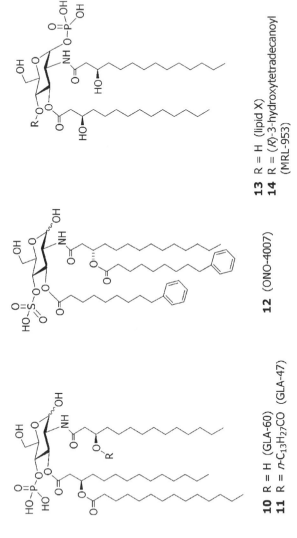

**10** R = H (GLA-60)
**11** R = n-C₁₃H₂₇CO (GLA-47)

**12** (ONO-4007)

**13** R = H (lipid X)
**14** R = (R)-3-hydroxytetradecanoyl
(MRL-953)

**Figure 6.3** Synthetic analogs of the reducing and nonreducing glucosamine units of lipid A.

A (lipid X, **13**) possesses immunostimulatory activity [34], highly purified synthetic lipid X was shown to be devoid of activity and to actually antagonize the effects of LPS in vitro [35]. In comparison, the synthetic triacyl derivative **14** (MRL-953) was found to be highly protective in experimental models of microbial infection, presumably via TLR4 activation of monocytes [36]. While the immunostimulatory activity of **14** and other triacyl monosaccharides speaks to the importance of having an optimum of three acyl residues in monosaccharides lipid A derivatives, the ability of 1-phosphate intermediates to act as glycosyl donors during chemical synthesis, potentially forming small amounts of biologically active acylated β-(1→6)glucosamine disaccharides [35], brings into question the biological activity of certain monosaccharide analogs. The chemical instability of the anomeric phosphate (as well as the acyloxyacyl moieties of natural lipid A) has engendered studies on the design of lipid A analogs with improved chemical stability. This has been accomplished in part by replacing the labile anomeric phosphate with "bioisosteric" acidic groups in both monosaccharides and disaccharides as well as replacing one of the sugars (particularly the structurally less conserved reducing sugar) and/or entire β-(1→6)diglucosamine moiety with stable acyclic mimetics. Such compounds also provide a structural motif more amenable to systematic investigations of structure–activity relationships.

Since the spatial arrangement, chain length, and number of fatty acyl residues in lipid A and other bacterial amphiphiles appeared important for immunostimulant activity [37,38], we designed a new class of stable acyclic lipid A mimetics known as aminoalkyl glucosaminide 4-phosphates (AGPs) [39]. The AGPs are synthetic mimetics of MLA derivative **2** in which the reducing sugar has been replaced with a conformationally flexible N-acylaminoalkyl (aglycon) unit. Several members of this class, including prototypical AGPs RC-529 (**15**) and CRX-527 (**16**), have been shown to improve humoral and cell-mediated immune responses to a variety of antigens in mice [39,40], as well as to enhance nonspecific resistance (NSR) in mice to *Listeria monocytogenes* and influenza infections [41]. RC-529, which can be synthesized in high yield and purity on a large scale, was recently shown to be a safe and effective adjuvant in a pivotal phase III vaccine trial with a recombinant hepatitis B antigen [42]. CRX-527, which contains an N-acylated serine aglycon unit that is structurally similar to the TLR4-active lipoamino acid flavolipin [43], shows promise as a stand-alone therapeutic for enhancing host resistance to bacterial and viral infection [44]. CRX-527 also induced higher titers of tetanus toxoid–specific antibodies, including complement-fixing IgG2a and IgG2b antibodies, and CTL responses, than those of both synthetic **2** and naturally derived MPL adjuvant in murine models [39]. The immunostimulatory activity of the AGP class of lipid A mimetics has been shown to be strictly dependent on TLR4 and MD-2, but not CD14, in HeLa cell transfectants [45].

As further testimony to the feasibility of this approach to adjuvant design, the lipid A mimetics **17** and **18** possessing an N-acylated pentaerythritol aglycon unit and the same acyl residues as RC-529 (**15**) were found to induce

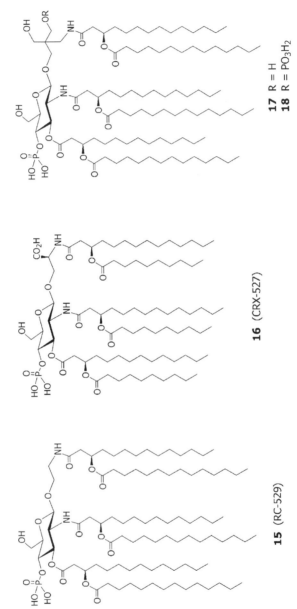

**15** (RC-529)

**16** (CRX-527)

**17** R = H
**18** R = PO₃H₂

**Figure 6.4** Aminoalkyl glucosamine 4-phosphate (AGP) class of monosaccharide lipid A mimetics.

142

high levels of cytokines in human adherent antigen-presenting cells (APCs) and antigen-specific T-cell proliferation in an experimental MUC-1 liposomal vaccine [46]. Interestingly, the monophosphate **17** induced higher levels of TNFα, IL-6, and IL-8 in human APCs than the corresponding diphosphate (**18**). This observation corroborates other studies showing that the distance between anionic groups and/or the spatial relationship of the aglycon phosphate or phosphate bioisostere to the aglycon lipid moiety are important to TLR4 agonist activity [47]. OM Pharma recently described synthetic acyclic versions of the triacyl diphosphoryl lipid A (**8**; OM-174) in which the disaccharide backbone of **8** has been replaced with a pseudodipeptide bearing three fatty acids and one or two phosphate moieties or other acidic groups [18,48]. Two of these diseco derivatives, OM-294-DP (**19**) and OM-197-MP-AC (**20**), which comprise homoserine- and ornithine-derived amino acids as well as a functionalized side chain in **20** suitable for conjugation, were reported to exhibit strong adjuvant activity in several vaccine models. Pseudopeptides **19** and **20** showed TLR4-dependent up-regulation of T-cell co-stimulatory molecules such as CD40 and CD86 and TLR4-dependent induction of TNFα and IL-12 (p40) in murine macrophages. Compound **20** was able to activate both monocyte-derived dendritic cells and leukemia-derived DCs. Like disaccharide predecessor **8**, these new acyclic lipid A mimetics are more than 10,000 times less pyrogenic than *E. coli* LPS in rabbits.

Eisai has also described a series of acyclic lipid A analogs that are potent vaccine adjuvants when administered with an antigen either subcutaneously or intranasaly [49,50]. These dimeric compounds, which comprise two trilipid serinol units connected to a stable linker via phosphodiester groups, activate NFκB signaling through the TLR4 receptor like other structural mimetics of lipid A [49,51]. The ability of these symmetric phospholipids to stimulate TNFα and IL-6 in human cell systems as well as to enhance the generation of antigen-specific antibodies in vaccine models against tetanus toxoid and other antigens was shown to be highly dependent on the stereochemistry of the lipid moieties as well as the length of the chain connecting the serinol units [49]. Both the stereochemistry and spatial arrangement of the lipids in these dimeric compounds probably affect their overall molecular shape and thus their ability to interact with the TLR4 receptor and accessory molecules.

Some of the most potent compounds in this series, exemplified by urea derivative **21** (ER803022) and malonamide **22** (ER804058), possess the same stereochemistry as that of natural lipid A (all of *R* configuration) and one- or three-carbon spacer units between the monomeric units; longer linker units reportedly led to diminished activity [49]. The high TNFα-inducing ability of **22** in human monocytes is somewhat surprising in view of the fact that this analog possesses the same (keto and unsaturated) fatty acid moieties as those of the nontoxic lipid A from *Rhodobacter capsulatus* as well as the closely related synthetic TLR4 antagonist E5531 [52]. Nonetheless, these acyclic derivatives were found to adopt a conical shape and nonlamellar, inverted hexagonal $H_{II}$ aggregate structures in solution, which is consistent with their high

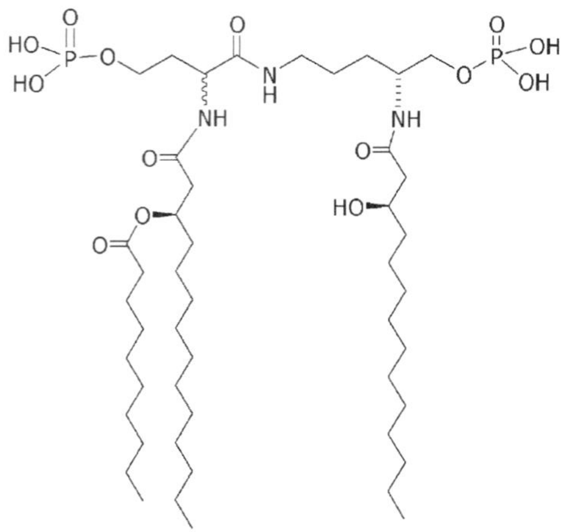

144

**19** (OM-294-DP)

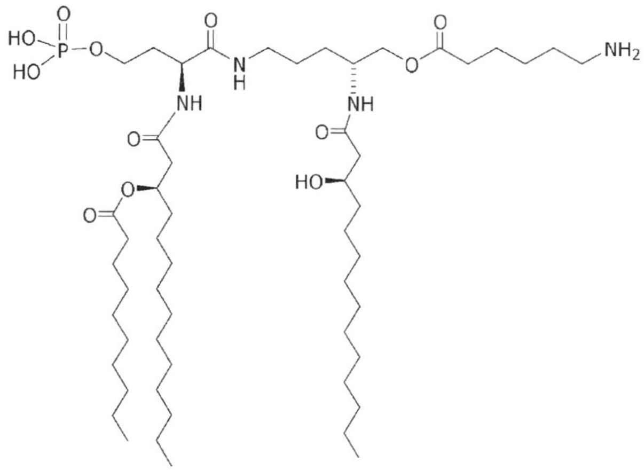

**20** (OM-197-MP-AC)

**Figure 6.5** Synthetic acyclic triacyl analogs of naturally derived OM-174.

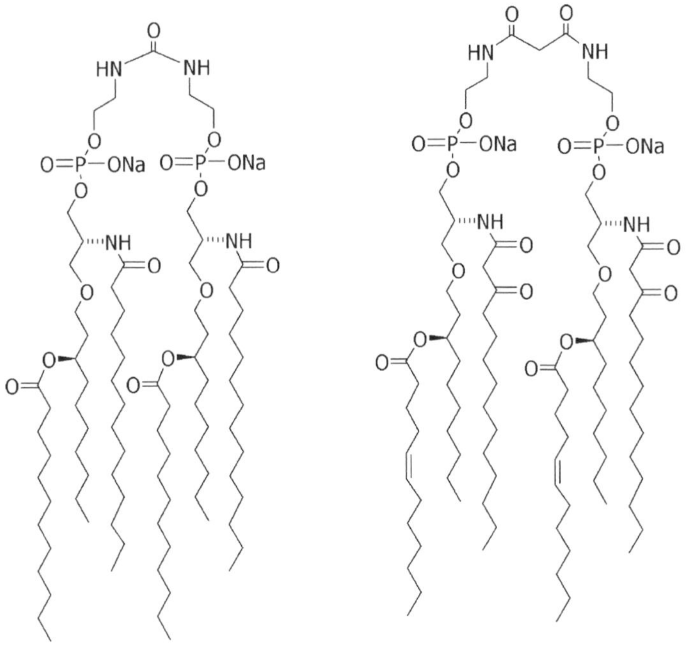

**21** (ER803022)                              **22** (ER804058)

**Figure 6.6**   Acyclic lipid A analogs containing two trilipid serinol units.

agonist activity [53]. In marked contrast to other subunit lipid A analogs possessing three lipid moieties [e.g., GLA-60 (**10**)], the monomeric units of these compounds were reported to be inactive.

## 6.4   SELF-ADJUVANTED VACCINES: TLR4 AGONIST–ANTIGEN CONSTRUCTS

Although covalently linking endogenous TLR2 ligands such as tripalmitoyl-*S*-glycerol cysteine (Pam$_3$Cys) to various peptide, glycopeptide, and carbohydrate antigens often enhances humoral and antigen-specific CTL responses [54], the construction of synthetic vaccines or self-adjuvanted vaccines with TLR4 agonists has not been investigated as extensively. A few experimental synthetic vaccines incorporating the triacyl lipid A structure **23** have been designed against HIV, melanoma, and leukemia using Tn (α-N-acetylgalactos-amine-O-serine), sialyl Tn, or peptide antigens present on cancer cells or the HIV-1 viral envelope (compounds **24** to **27**) [55–57]. Many of these constructs exhibit mitogenic activity on C3H/He mouse splenocytes in vitro that is greater than that of the lipid A analog **23** alone or conjugated to a linker, suggesting the possibility of antigen-specific immune responses. However, in some cases the mitogenic responses were weaker with the synthetic construct than with an admixture of antigen and **23** containing a β-alanine linker [57].

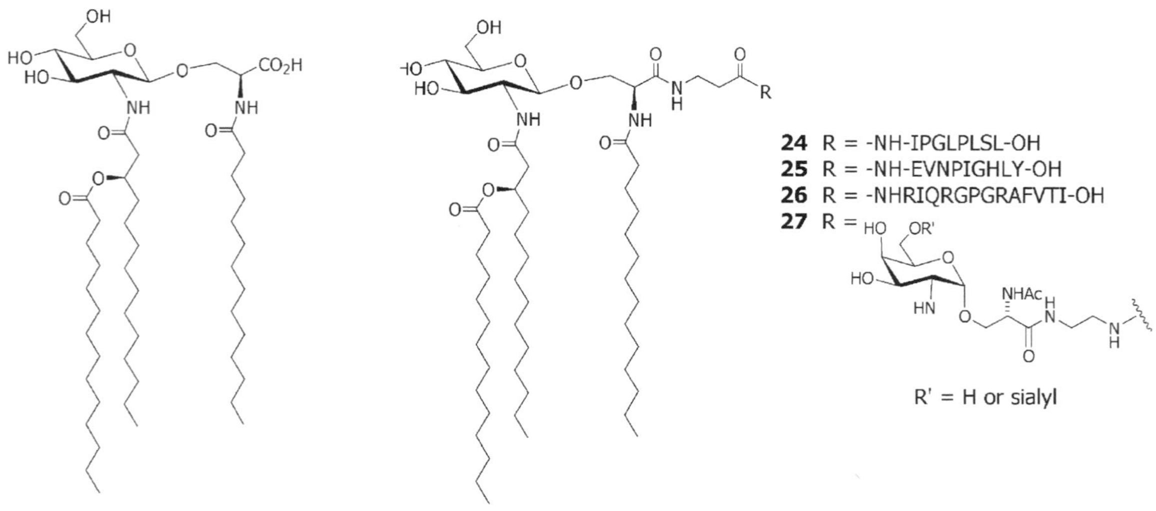

23

24 R = -NH-IPGLPLSL-OH
25 R = -NH-EVNPIGHLY-OH
26 R = -NHRIQRGPGRAFVTI-OH
27 R = H or sialyl

R' = H or sialyl

**Figure 6.7** Experimental synthetic vaccines containing covalently bound triacyl monosaccharide adjuvant.

Similarly, antigen–adjuvant conjugates of TLR4 agonist OM-197-MP (**28**) have been generated via reductive amination of a side-chain aldehyde with peptide antigens with or without a protein carrier present [58]. In the case of the antigen $(NANP)_6P_2P_{30}$, which comprises a recurring peptide sequence from *Plasmodium falciparum* (NANP) and two peptide sequences from tetanus toxoid ($P_2$ and $P_{30}$), immunization with monoconjugate **29** (one adjuvant molecule per antigen construct) led to slightly greater IgG responses than with the antigen–adjuvant admixture. Higher-order conjugates containing a mixture of mono-, di-, and tri-conjugates or coadministering additional adjuvant with the monoconjugate did not improve humoral responses. In the case of a synthetic peptide corresponding to the T-cell epitope of circumsporozoite protein of *Plasmodium yoelli*, the corresponding adjuvant–antigen tetraconjugate (four molecules of **28** per antigen) elicited moderate but significant antibody titers and CTL responses compared to the admixture or monoconjugate. As has been observed with certain other self-adjuvanted vaccines exhibiting weak immune responses, a stronger CTL response was elicited by the monoconjugate **30** when coadministered with additional OM-197-MP (**28**). Enhancement of humoral responses (IgM) in mice with the OM-197-H1N1 hemagglutinin conjugate **31** in comparison to the analogous admixtures appears to be equivocal with or without added adjuvant.

Despite the moderate improvement in serological responses with TLR4 agonist–peptide conjugates, it is not clear whether these conjugates actually interact with TLR4 and co-stimulatory receptors on cells of the immune system. These constructs may help target the bound antigen into APCs much like conventional delivery systems promote antigen uptake. The amphipathic nature of TLR4 agonists in general and their tendency to form liposomelike aggregates in solution may facilitate endocytosis by APCs via a "microdelivery" mechanism. Clearly, the development of effective synthetic vaccines using TLR4 agonists will require much more work to better define the cell types, specific receptors, signaling pathways, and physicochemical processes involved.

## 6.5 SYNERGISM BETWEEN TLR4 AND OTHER TLR AGONISTS OR VACCINE ADJUVANTS

Another promising approach toward optimizing immune responses, particularly as our understanding of the immunobiology of TLR4 and other TLR receptors and of the importance of Th1 versus Th2 responses expands, is the combination of TLR4 agonists with other TLR agonists or molecular adjuvants to stimulate multiple TLRs and other immune receptors simultaneously. The potential synergistic effect of using multiple immune response modifiers was explored initially by Ribi, who noted enhanced responses when MPL was combined with mycobacterial cell wall skeleton (CWS), which is now thought to act on TLR2 [59]. These early studies led to the development of Detox (later Enhanzen) adjuvant, a clinical-grade adjuvant containing MPL and CWS

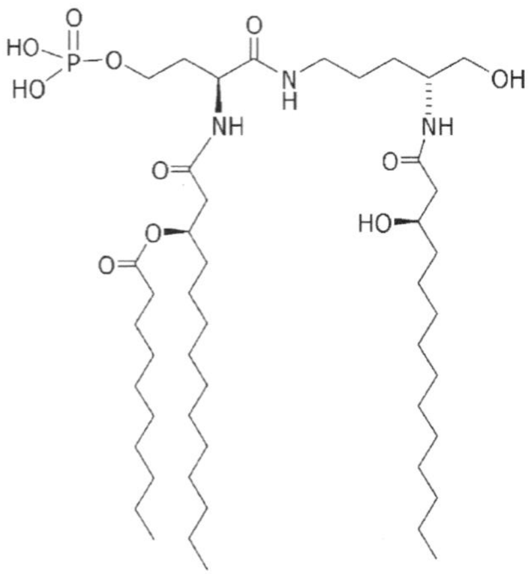

**28** (OM-197-MP)

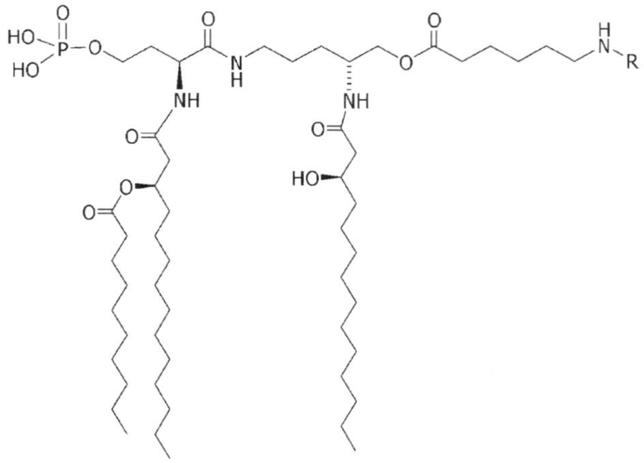

**29** R = -(NANP)$_6$P$_2$P$_{30}$
**30** R = -NH-SERSYVPSAEQI-OH
**31** R = -H1N1

**Figure 6.8** Experimental synthetic vaccines containing covalently bound triacyl acyclic adjuvant.

formulated as an oil-in-water emulsion. This combination adjuvant, which was more potent then MPL alone, was evaluated in a number of cancer vaccines and eventually approved as a component of a melanoma vaccine [60]. Further evidence in support of this concept is found when the TLR9 ligand CpG is combined with the TLR4 agonist MPL, resulting in a strongly skewed Th1 type of immune response to vaccination [61]. However, the best example of the potential synergy mediated by TLR4 agonists and other immunomodifiers is exemplified in the combination of MPL and QS21.

Through its action on TLR4, MPL is known to stimulate inflammatory cytokines, including IL-12, to up-regulate accessory molecules, including B7 proteins, all of which are important in the induction of Th1 immunity. QS21, a purified triterpine saponin, is a strong adjuvant for which the mode of action is still under debate, but it is not known to interact with any identified TLR. Numerous preclinical studies indicated that the combination of MPL and QS21 provided a much stronger adjuvant for the induction of humoral and cell-mediated immune responses than did either adjuvant alone. Based on this information, a novel oil-in-water formulation containing MPL and QS21, called SBAS2 (now AS02A), was developed and evaluated in candidate malarial vaccine studies [62]. *Plasmodium falciparum* malaria is responsible for significant morbidity and mortality in endemic regions of the world, causing in excess of 2 million deaths per year. Despite considerable effort, prophylactic vaccines targeting this pathogen are currently unavailable. An open-label clinical trial was conducted in a laboratory setting to evaluate the safety and efficacy of a vaccine containing recombinant *P. falciparum* circumsporozoite antigen RTS,S and SBAS2 [63]. Remarkably, six of seven immunized subjects did not develop parasitemia upon experimental challenge with *P. falciparum*. These results were confirmed in subsequent studies and complemented by demonstrated protection in a field study [64,65]. Considering that no other malaria vaccine to date has been found to elicit this degree of protection in humans, these findings are highly significant. Follow-on studies demonstrated that protection correlated with antigen-specific, interferon-$\gamma$-producing $CD4^+$ and $CD8^+$ cells, which were induced reliably only when the stronger SBAS2 adjuvant was used [66]. For disease indications such as malaria, HIV, and cancer that have proven resistant to traditional vaccination strategies, the complementary combination of TLR4 agonists with other immune-response modifiers may facilitate the induction of improved immune responses.

## 6.6 RAPID-ACTING VACCINES

The full potential of TLR4 agonists as adjuvants and stand-alone immunomodulators is only now beginning to be realized. Armed with an increased understanding of TLR4 signaling, we can now evaluate novel vaccine approaches that take full advantage of both the innate and the adaptive immune responses. For example, preclinical studies have shown that in their capacity as stimulators of innate immunity, TLR4 agonists are able to provide short-term, nonspecific resistance or protection against bacterial and viral challenge within hours of administration [41,44]. These same agonists act as mucosal adjuvants that mediate strong adaptive immune responses to coadministered vaccine antigens, including influenza [67]. Exploitation of both TLR4 agonist characteristics may yield vaccines that provide (1) short-term, nonspecific protection against pathogens and (2) durable, antigen-specific acquired immunity. A rapid-acting vaccine approach such as this may provide much appreciated relief in such critical times as during a pandemic influenza outbreak.

## 6.7  TLR4-ACTIVE VACCINE ADJUVANTS: CONCLUDING REMARKS

It is widely noted that vaccines provide one of the most cost-effective and successful health care tools for preventing disease and mortality. TLR4 agonists have shown potential in a number of human vaccine studies with infectious disease, cancer, and allergy vaccines. The results from clinical trials with TLR4-active adjuvants indicate that we can expect continued growth in the already impressive results of vaccination by using this class of adjuvants in new and improved vaccines (Table 6.1). The clinical evidence indicates that

TABLE 6.1  Selected Clinical Trials with TLR4 Agonists as Vaccine Adjuvants

| Clinical Indication | Adjuvant | Trial Highlights | Refs. |
|---|---|---|---|
| Hepatitis B | MPL + alum (SBAS4) | Enhanced seroconversion; higher GMT; enhanced cell-mediated immunity | 68, 70, 73, 74 |
| Malaria | MPL + QS21 (SBAS2) | Resistance to parasitemia; enhanced humoral and cell-mediated immunity | 63–65 |
| Herpes type 2 | MPL + alum (SBAS4) | Enhanced binding and neutralizing antibody; enhanced cell proliferation; enhanced IFN-γ | 75 |
| *Streptococcus pneumoniae* | MPL ± alum | Neonate patient population; enhanced cell proliferation; enhanced IFN-γ | 73 |
| Melanoma | MPL + CWS (Detox) | Extended survival | 60 |
| Grass pollen allergy | MPL + tyrosine | Reduced nasal symptoms; reduced skin-prick sensitivity | 71, 77 |
| Hepatitis B | RC-529 + alum | Enhanced seroconversion; higher GMT; synthetic TLR4 agonist | 69 |

the TLR4 agonists MPL and RC-529 induce protective responses to hepatitis vaccines faster and with fewer administrations than do alum-adjuvanted vaccines [68,76]. Persons who had low or no response to traditional hepatitis vaccines were stimulated to sero-protective levels of immunity following vaccination with MPL-adjuvanted vaccines [70]. TLR4 and TLR9 agonists evaluated in clinical trials with allergy vaccines were shown to bias antigen-specific responses toward a protective Th1 type of immunity [74,77]. As discussed previously, the complementary activity induced by MPL and QS21 has led to the most effective malaria vaccine candidate developed to date. In summary, the discovery of TLRs has greatly increased our understanding of how adjuvants function and has opened up exciting new possibilities, leading us to believe that TLR4-active adjuvants can improve on the already impressive results of vaccines.

# REFERENCES

1. Arnon, R. & Van Regenmortel, M.H. Structural basis of antigenic specificity and design of new vaccines. *Faseb J* 1992, **6**(14), 3265–3274.
2. Johnson, A.G. Molecular adjuvants and immunomodulators: new approaches to immunization. *Clin Microbiol Rev* 1994, **7**(3), 277–289.
3. Edelman, R. Vaccine adjuvants. *Rev Infect Dis* 1980, **2**(3), 370–383.
4. Iwasaki, A. & Medzhitov, R. Toll-like receptor control of the adaptive immune responses. *Nat Immunol* 2004, **5**(10), 987–995.
5. Zhang, D., Zhang, G., Hayden, M.S. et al. A Toll-like receptor that prevents infection by uropathogenic bacteria. *Science* 2004, **303**(5663), 1522–1526.
6. Yarovinsky, F., Zhang, D., Andersen, J.F. et al. TLR11 activation of dendritic cells by a protozoan profilin-like protein. *Science* 2005, **308**(5728), 1626–1629.
7. Akira, S. & Takeda, K. Toll-like receptor signalling. *Nat Rev Immunol* 2004, **4**(7), 499–511.
8. Takeda, K. & Akira, S. TLR signaling pathways. *Semin Immunol* 2004, **16**(1), 3–9.
9. Takeda, K., Kaisho, T. & Akira, S. Toll-like receptors. *Annu Rev Immunol* 2003, **21**, 335–376.
10. Miyake, K. Innate recognition of lipopolysaccharide by Toll-like receptor 4-MD-2. *Trends Microbiol* 2004, **12**(4), 186–192.
11. Gangloff, M. & Gay, N.J. MD-2: the Toll "gatekeeper" in endotoxin signalling. *Trends Biochem Sci* 2004, **29**(6), 296–300.
12. Myers, K.R., Truchot, A., Ward, J.I., Hudson, Y. & Ulrich, J.T. A critical determinant of lipid A entotoxic activity. In *Cellular and Molecular Aspects of Endotoxin Reactions*, Nowotny, A., Spitzer, J.J. & Ziegler, E.J., Eds., Elsevier Science, Amsterdam, 1990, pp. 145–156.
13. Ulrich, J.T. & Myers, K.R. Monophosphoryl lipid A as an adjuvant: past experiences and new directions. In *Vaccine Design: the Subunit and Adjuvant Approach*, Powell, J.D. & Newman, J., Eds., Plenum Press, New York, 1995, pp. 495–524.

14. North, S. & Butts, C. Vaccination with BLP25 liposome vaccine to treat non–small cell lung and prostate cancers. *Expert Rev Vaccines* 2005, **4**(3), 249–257.

15. Jiang, Z.H., Bach, M.V., Budzynski, W.A., Krantz, M.J., Koganty, R.R. & Longenecker, B.M. Lipid A structures containing novel lipid moieties: synthesis and adjuvant properties. *Bioorg Med Chem Lett* 2002, **12**(16), 2193–2196.

16. Johnson, D.A., Keegan, D.S., Sowell, C.G. et al. 3-*O*-Desacyl monophosphoryl lipid A derivatives: synthesis and immunostimulant activities. *J Med Chem* 1999, **42**(22), 4640–4649.

17. Meraldi, V., Audran, R., Romero, J.F. et al. OM-174, a new adjuvant with a potential for human use, induces a protective response when administered with the synthetic C-terminal fragment 242–310 from the circumsporozoite protein of *Plasmodium berghei*. *Vaccine* 2003, **21**(19–20), 2485–2491.

18. Davies, G. OM-triacyl adjuvants: activators of innate immunity for use as vaccine adjuvants or in cellular immunotherapy. www.who.int/entity/vaccine_research/about/2003_novel_adjuvants/en/02_davies.pdf, 2003.

19. Pajak, B., Garze, V., Davies, G., Bauer, J., Moser, M. & Chiavaroli, C. The adjuvant OM-174 induces both the migration and maturation of murine dendritic cells in vivo. *Vaccine* 2003, **21**(9–10), 836–842.

20. Qureshi, N. & Takayama, K. Structure and function of lipid A. In *The Bacteria*, Vol. XI, Iglewski, B.H. & Clark, V.L., Eds., Academic Press, San Diego, CA, 1990, pp. 319–338.

21. Funatogawa, K., Matsuura, M., Nakano, M., Kiso, M. & Hasegawa, A. Relationship of structure and biological activity of monosaccharide lipid A analogues to induction of nitric oxide production by murine macrophage RAW264.7 cells. *Infect Immun* 1998, **66**(12), 5792–5798.

22. Moran, A.P., Lindner, B. & Walsh, E.J. Structural characterization of the lipid A component of *Helicobacter pylori* rough- and smooth-form lipopolysaccharides. *J Bacteriol* 1997, **179**(20), 6453–6463.

23. Kulshin, V.A., Zahringer, U., Lindner, B., Jager, K.E., Dmitriev, B.A. & Rietschel, E.T. Structural characterization of the lipid A component of *Pseudomonas aeruginosa* wild-type and rough mutant lipopolysaccharides. *Eur J Biochem* 1991, **198**(3), 697–704.

24. Brandenburg, K., Lindner, B., Schromm, A. et al. Physicochemical characteristics of triacyl lipid A partial structure OM-174 in relation to biological activity. *Eur J Biochem* 2000, **267**(11), 3370–3377.

25. Brandenburg, K. & Wiese, A. Endotoxins: relationships between structure, function, and activity. *Curr Top Med Chem* 2004, **4**(11), 1127–1146.

26. Teghanemt, A., Zhang, D., Levis, E.N., Weiss, J.P. & Gioannini, T.L. Molecular basis of reduced potency of underacylated endotoxins. *J Immunol* 2005, **175**(7), 4669–4676.

27. Matsuura, M., Kiso, M. & Hasegawa, A. Activity of monosaccharide lipid A analogues in human monocytic cells as agonists or antagonists of bacterial lipopolysaccharide. *Infect Immun* 1999, **67**(12), 6286–6292.

28. Brandenburg, K., Matsuura, M., Heine, H. et al. Biophysical characterization of triacyl monosaccharide lipid A partial structures in relation to bioactivity. *Biophys J* 2002, **83**(1), 322–333.

29. Yang, D., Satoh, M., Ueda, H., Tsukagoshi, S. & Yamazaki, M. Activation of tumor-infiltrating macrophages by a synthetic lipid A analog (ONO-4007) and its implication in antitumor effects. *Cancer Immunol Immunother* 1994, **38**(5), 287–293.

30. Matsumoto, N., Aze, Y., Akimoto, A. & Fujita, T. ONO-4007, an antitumor lipid A analog, induces tumor necrosis factor-alpha production by human monocytes only under primed state: different effects of ONO-4007 and lipopolysaccharide on cytokine production. *J Pharmacol Exp Ther* 1998, **284**(1), 189–195.

31. Tamai, R., Asai, Y., Hashimoto, M. et al. Cell activation by monosaccharide lipid A analogues utilizing Toll-like receptor 4. *Immunology* 2003, **110**(1), 66–72.

32. Perera, P.Y., Manthey, C.L., Stutz, P.L., Hildebrandt, J. & Vogel, S.N. Induction of early gene expression in murine macrophages by synthetic lipid A analogs with differing endotoxic potentials. *Infect Immun* 1993, **61**(5), 2015–2023.

33. Hawkins, L.D., Christ, W.J. & Rossignol, D.P. Inhibition of endotoxin response by synthetic TLR4 antagonists. *Curr Top Med Chem* 2004, **4**(11), 1147–1171.

34. Proctor, R.A., Will, J.A., Burhop, K.E. & Raetz, C.R. Protection of mice against lethal endotoxemia by a lipid A precursor. *Infect Immun* 1986, **52**(3), 905–907.

35. Aschauer, H., Grob, A., Hildebrandt, J., Schuetze, E. & Stuetz, P. Highly purified lipid X is devoid of immunostimulatory activity: isolation and characterization of immunostimulating contaminants in a batch of synthetic lipid X. *J Biol Chem* 1990, **265**(16), 9159–9164.

36. Lam, C., Schutze, E., Hildebrandt, J. et al. SDZ MRL 953, a novel immunostimulatory monosaccharidic lipid A analog with an improved therapeutic window in experimental sepsis. *Antimicrob Agents Chemother* 1991, **35**(3), 500–505.

37. Werner, G.H. & Jolles, P. Immunostimulating agents: What next? A review of their present and potential medical applications. *Eur J Biochem* 1996, **242**(1), 1–19.

38. Seydel, U., Labischinski, H., Kastowsky, M. & Brandenburg, K. Phase behavior, supramolecular structure, and molecular conformation of lipopolysaccharide. *Immunobiology* 1993, **187**(3–5), 191–211.

39. Johnson, D.A., Sowell, C.G., Johnson, C.L. et al. Synthesis and biological evaluation of a new class of vaccine adjuvants: aminoalkyl glucosaminide 4-phosphates (AGPs). *Bioorg Med Chem Lett* 1999, **9**(15), 2273–2278.

40. Baldridge, J.R., McGowan, P., Evans, J.T. et al. Taking a Toll on human disease: Toll-like receptor 4 agonists as vaccine adjuvants and monotherapeutic agents. *Expert Opin Biol Ther* 2004, **4**(7), 1129–1138.

41. Baldridge, J.R., Cluff, C.W., Evans, J.T. et al. Immunostimulatory activity of aminoalkyl glucosaminide 4-phosphates (AGPs): induction of protective innate immune responses by RC-524 and RC-529. *J Endotoxin Res* 2002, **8**(6), 453–458.

42. Engers, H., Kieny, M.P., Malhotra, P. & Pink, J.R. Third meeting on Novel Adjuvants Currently in or Close to Clinical Testing World Health Organization: Organisation Mondiale de la Sante, Fondation Merieux, Annecy, France, Jan. 7–9, 2002. *Vaccine* 2003, **21**(25–26), 3503–3524.

43. Gomi, K., Kawasaki, K., Kawai, Y., Shiozaki, M. & Nishijima, M. Toll-like receptor 4-MD-2 complex mediates the signal transduction induced by flavolipin, an amino acid–containing lipid unique to *Flavobacterium meningosepticum*. *J Immunol* 2002, **168**(6), 2939–2943.

44. Cluff, C.W., Baldridge, J.R., Stover, A.G. et al. Synthetic Toll-like receptor 4 agonists stimulate innate resistance to infectious challenge. *Infect Immun* 2005, **73**(5), 3044–3052.

45. Stover, A.G., Da Silva Correia, J., Evans, J.T. et al. Structure–activity relationship of synthetic Toll-like receptor 4 agonists. *J Biol Chem* 2004, **279**(6), 4440–4449.

46. Jiang, Z.H., Budzynski, W.A., Skeels, L.N., Krantz, M.J. & Koganty, R.R. Novel lipid A mimetics derived from pentaerythritol: synthesis and their potent agonistic activity. *Tetrahedron* 2002, **58**, 8833–8842.

47. Pedron, T., Girard, R., Eustache, J. et al. New synthetic analogs of lipid A as lipopolysaccharide agonists or antagonists of B lymphocyte activation. *Int Immunol* 1992, **4**(4), 533–540.

48. Corradin, G., Rivier, D., Martin, O., Davies, G. & Bauer, J. The new synthetic compounds OM-294-MP and OM-294-DP induce cellular and humoral responses in 2 murine vaccine models. Presented at the 4th Annual Conference on Vaccine Research, Arlington, VA, 2001.

49. Hawkins, L.D., Ishizaka, S.T., McGuinness, P. et al. A novel class of endotoxin receptor agonists with simplified structure, toll-like receptor 4-dependent immunostimulatory action, and adjuvant activity. *J Pharmacol Exp Ther* 2002, **300**(2), 655–661.

50. Przetak, M., Chow, J., Cheng, H., Rose, J., Hawkins, L.D. & Ishizaka, S.T. Novel synthetic LPS receptor agonists boost systemic and mucosal antibody responses in mice. *Vaccine* 2003, **21**(9–10), 961–970.

51. Lien, E., Chow, J.C., Hawkins, L.D. et al. A novel synthetic acyclic lipid A–like agonist activates cells via the lipopolysaccharide/Toll-like receptor 4 signaling pathway. *J Biol Chem* 2001, **276**(3), 1873–1880.

52. Christ, W.J., Asano, O., Robidoux, A.L. et al. E5531, a pure endotoxin antagonist of high potency. *Science* 1995, **268**(5207), 80–83.

53. Brandenburg, K., Hawkins, L., Garidel, P. et al. Structural polymorphism and endotoxic activity of synthetic phospholipid-like amphiphiles. *Biochemistry* 2004, **43**(13), 4039–4046.

54. Jiang, Z.H. & Koganty, R.R. Synthetic vaccines: the role of adjuvants in immune targeting. *Curr Microbiol* 2003, **10**(15), 1423–1429.

55. Miyajima, K., Nekado, T., Ikeda, K. & Achiwa, K. Synthesis of Tn and sialyl Tn antigen-lipid A analog conjugates for synthetic vaccines. *Chem Pharm Bull (Tokyo)* 1997, **45**(9), 1544–1546.

56. Miyajima, K., Nekado, T., Ikeda, K. & Achiwa, K. Synthesis of Tn, sialyl Tn and HIV-1-derived peptide antigen conjugates having a lipid A analog as an immunoadjuvant for synthetic vaccines. *Chem Pharm Bull (Tokyo)* 1998, **46**(11), 1676–1682.

57. Ikeda, K., Miyajima, K., Maruyama, Y. & Achiwa, K. Synthesis of cancer peptide antigen-lipid A analog conjugates for synthetic vaccines. *Chem Pharm Bull (Tokyo)* 1999, **47**(4), 563–568.

58. Bauer, Jacques, Martin, Richard, O. & Rodriguez, S. Novel acyl-dipeptide-like compounds bearing an accessory functional side chain spacer, a method for preparing the same and pharmaceutical compositions containing such products. Filed Apr. 22, 2005, US 2005/0192232 A1 Sept. 1, 2005, 113, 443.

59. Rudbach, J.A., Johnson, D.A. & Ulrich, T. Ribi adjuvants: chemistry, biology and utility in vaccines for human and veterinary medicine. In *The Theory and Practical Application of Adjuvants*, Stewart-Tull, D.E.S., Ed., Wiley, New York, 1995, pp. 287–313.

60. Mitchell, M.S. & Von Eschen, K.B. Phase III trial of Melacine melanoma theraccine versus combination chemotherapy in the treatment of stage IV melanoma. *Proc Am Soc Clin Oncol* 1997, **16**, 494a.

61. Weeratna, R.D., McCluskie, M.J., Xu, Y. & Davis, H.L. CpG DNA induces stronger immune responses with less toxicity than other adjuvants. *Vaccine* 2000, **18**(17), 1755–1762.

62. Garcon, N., Heppner, D.G. & Cohen, J. Development of RTS,S/AS02: a purified subunit-based malaria vaccine candidate formulated with a novel adjuvant. *Expert Rev Vaccines* 2003, **2**(2), 231–238.

63. Stoute, J.A., Slaoui, M., Heppner, D.G. et al. A preliminary evaluation of a recombinant circumsporozoite protein vaccine against *Plasmodium falciparum* malaria. RTS,S Malaria Vaccine Evaluation Group. *N Engl J Med* 1997, **336**(2), 86–91.

64. Kester, K.E., McKinney, D.A., Tornieporth, N. et al. Efficacy of recombinant circumsporozoite protein vaccine regimens against experimental *Plasmodium falciparum* malaria. *J Infect Dis* 2001, **183**(4), 640–647.

65. Bojang, K.A., Milligan, P.J., Pinder, M. et al. Efficacy of RTS,S/AS02 malaria vaccine against *Plasmodium falciparum* infection in semi-immune adult men in The Gambia: a randomised trial. *Lancet* 2001, **358**(9297), 1927–1934.

66. Sun, P., Schwenk, R., White, K. et al. Protective immunity induced with malaria vaccine, RTS,S, is linked to *Plasmodium falciparum* circumsporozoite protein-specific CD4+ and CD8+ T cells producing IFN-gamma. *J Immunol* 2003, **171**(12), 6961–6967.

67. Baldridge, J.R., Yorgensen, Y., Ward, J.R. & Ulrich, J.T. Monophosphoryl lipid A enhances mucosal and systemic immunity to vaccine antigens following intranasal administration. *Vaccine* 2000, **18**(22), 2416–2425.

68. Thoelen, S., Van Damme, P., Mathei, C. et al. Safety and immunogenicity of a hepatitis B vaccine formulated with a novel adjuvant system. *Vaccine* 1998, **16**(7), 708–714.

69. Thoelen, S., De Clercq, N. & Tornieporth, N. A prophylactic hepatitis B vaccine with a novel adjuvant system. *Vaccine* 2001, **19**(17–19), 2400–2403.

70. Desombere, I., Van der Wielen, M., Van Damme, P. et al. Immune response of HLA DQ2 positive subjects, vaccinated with HBsAg/AS04, a hepatitis B vaccine with a novel adjuvant. *Vaccine* 2002, **20**(19–20), 2597–2602.

71. Jacques, P., Moens, G., Desombere, I. et al. The immunogenicity and reactogenicity profile of a candidate hepatitis B vaccine in an adult vaccine non-responder population. *Vaccine* 2002, **20**(31–32), 3644–3649.

72. Stanberry, L.R., Spruance, S.L., Cunningham, A.L. et al. Glycoprotein-D-adjuvant vaccine to prevent genital herpes. *N Engl J Med* 2002, **347**(21), 1652–1661.

73. Vernacchio, L., Bernstein, H., Pelton, S. et al. Effect of monophosphoryl lipid A (MPL) on T-helper cells when administered as an adjuvant with pneumocococcal-CRM197 conjugate vaccine in healthy toddlers. *Vaccine* 2002, **20**(31–32), 3658–3667.

74. Drachenberg, K.J., Wheeler, A.W., Stuebner, P. & Horak, F. A well-tolerated grass pollen–specific allergy vaccine containing a novel adjuvant, monophosphoryl lipid A, reduces allergic symptoms after only four preseasonal injections. *Allergy* 2001, **56**(6), 498–505.

75. Mothes, N., Heinzkill, M., Drachenberg, K.J. et al. Allergen-specific immunotherapy with a monophosphoryl lipid A–adjuvanted vaccine: reduced seasonally boosted immunoglobulin E production and inhibition of basophil histamine release by therapy-induced blocking antibodies. *Clin Exp Allergy* 2003, **33**(9), 1198–1208.

76. Dupont, J.-C., Altclas, J., Sigelchifer, M., Von Eschen, E.B., Timmermans, I. & Wagener, A. Efficacy and safety of AgB/RC529: a novel two dose adjuvant vaccine against hepatitis B. Presented at the 42nd Interscience Conference on Antimicrobial Agents and Chemotherapy, San Diego, CA, 2002.

77. Simons, F.E., Shikishima, Y., Van Nest, G., Eiden, J.J. & HayGlass, K.T. Selective immune redirection in humans with ragweed allergy by injecting Amb a 1 linked to immunostimulatory DNA. *J Allergy Clin Immunol* 2004, **113**(6), 1144–1151.

# 7

# IMMUNOSTIMULATORY CpG OLIGODEOXYNUCLEOTIDES AS VACCINE ADJUVANTS*

DENNIS M. KLINMAN, DEBBIE CURRIE, AND HIDEKAZU SHIROTA

## 7.1  INTRODUCTION

The mammalian immune system utilizes two general strategies to combat infectious diseases. Pathogen exposure rapidly triggers an "innate" immune response characterized by the production of immunostimulatory cytokines, chemokines, and polyreactive IgM antibodies (reviewed in [1–3]). This innate immune response is elicited by exposure to *pathogen-associated molecular patterns* (PAMPs) expressed by a diverse group of infectious microorganisms [3] and serves to limit the early proliferation and spread of infectious organisms in vivo. Subsequently, the host mounts an "adaptive" immune response directed against determinants uniquely expressed by the infectious pathogen. This results in pathogen-specific immunity characterized by the production of high-affinity antibodies and the generation of cytotoxic T cells that provide long-lasting protection against subsequent exposure to the same organism [4].

The recognition of PAMPs is mediated by members of the Toll-like family of receptors (TLRs) [5,6]. Bacterial DNA represents one example of a PAMP. Due to differences in the frequency of utilization and the methylation pattern of CpG dinucleotides by prokaryotes versus eukaryotes, unmethylated CpG motifs are present at a much higher frequency in the genomes of bacteria than

---

*The assertions herein are those of the authors and are not to be construed as official or as reflecting the views of the U.S. Food and Drug Administration at large.

---

*Vaccine Adjuvants and Delivery Systems*, Edited by Manmohan Singh
Copyright © 2007 John Wiley & Sons, Inc.

in those of mammals [7,8]. Toll-like receptor 9 detects unmethylated CpG motifs [9–11]. Thus, exposure to unmethylated CpG DNA released during an infection provides a "danger signal" that triggers an innate immune response that helps the host eliminate the pathogen [12].

Synthetic oligodeoxynucleotides (ODNs) expressing CpG motifs similar to those found in bacterial DNA stimulate the same type of innate immune response [12–15]. The potential utility of immunomodulatory ODNs for the treatment of cancer and allergic disorders and to improve host resistance to infection are being widely evaluated. This review focuses on efforts to harness CpG ODNs as vaccine adjuvants designed to improve antigen-presenting cell function and to promote the induction of adaptive immune responses.

## 7.2   IMMUNOMODULATORY PROPERTIES OF CpG ODNS

CpG ODNs are rapidly internalized by immune cells, where they interact with intracellular TLR9 [9,16]. This interaction is exquisitely sensitive to modifications of the CpG motif [9,10]. Cells lacking TLR9 do not respond to CpG DNA, but can be made responsive if transfected to express that receptor [9,10].

B cells and plasmacytoid dendritic cells (pDCs) are the primary human cell types that express TLR9 and respond to CpG stimulation [9,10,17,18]. Activation of these cells by CpG DNA initiates an immunostimulatory cascade that culminates in the maturation, differentiation, and/or proliferation of natural killer (NK) cells, T cells, and monocytes or macrophages [15,19,20]. Together, these secrete cytokines and chemokines that create a pro-inflammatory [IL-1, IL-6, IL-18, and tumor necrosis factor (TNF)] and Th1-biased (IFN$\alpha$ and IL-12) immune milieu [9,10,12,15,16,21,22].

Due to evolutionary divergence among the TLR9 molecules expressed by different species, the sequence motif (unmethylated CpG dinucleotide plus flanking regions) that optimally stimulates cells from one species may be ineffective in another species [23]. For example, the amino acid sequences of murine and human TLR9 differ by 24% [9]. Whereas the optimal sequence motif in mice consists of two 5' purines, unmethylated CpG, and then two 3' pyrimidines [12,15,22,24], the optimal motif in humans is T<u>C</u>GTT and/or T<u>C</u>GTA [11,21,23,25–27]. In addition, the cell populations that express TLR9 may differ among species. In mice, immune cells of the myeloid lineage [including monocytes, macrophages, and myeloid denderitic cells (DCs)] express TLR9 and respond to CpG stimulation, whereas in humans these cell types do not express TLR9 and are not directly activated by CpG ODNs [28–30].

At least three structurally distinct classes of synthetic CpG ODNs have been described that are capable of stimulating cells that express human TLR9 [26,27,31,32] K-type ODNs (also referred to as B type) encode multiple CpG motifs on a phosphorothioate backbone. K ODNs trigger pDCs to differentiate and produce TNF$\alpha$, and B cells to proliferate and secrete Ig [26,27,33] (Table 7.1).

**TABLE 7.1   Comparison of D-, K-, and C-Type ODNs**

| | |
|---|---|
| ODN type | D, also referred to as A |
| Example | GGTGCATCGATGCAGGGGGG |
| Structural characteristics | Mixed *phosphodiester*/phosphorothioate backbone |
| | Single CpG motif (underline) |
| | CpG flanking region forms a palindrome (**bold**) |
| | Poly-G tail at the 3′ end |
| Immunomodulatory activity | APC maturation |
| | Preferentially stimulates pDCs to secrete IFNα |
| ODN type | K, also referred to as B |
| Example | TCCATGGACGTTCCTGAGCGTT |
| Structural characteristics | Phosphorothioate backbone |
| | Multiple CpG motifs (underline) |
| | 5′ motif most stimulatory |
| Immunomodulatory activity | pDC maturation |
| | Preferentially supports the production of TNFα and IL-6 |
| | Triggers B-cell activation, including the production of gM |
| ODN type | C |
| Example | TCGTCGTTCGAACGACGTTGAT |
| Structural characteristics | Phosphorothioate backbone |
| | Multiple CpG motifs (underline) |
| | TCG dimer at the 5′ end |
| | CpG motif embedded in a central palindrome |
| Immunomodulatory activity | Stimulates B cells and pDCs |
| | Induces production of IL-6 and IFNα |

Stimulation of these cells is initiated by the binding of K ODNs to TLR9 in lysosomal vesicles [34], and proceeds through an IRF5-mediated pathway. D-type ODNs (also referred to as A type) are constructed of a mixed phosphodiester–phosphorothioate backbone and contain a single hexameric purine–pyrimidine–CG–purine–pyrimidine motif flanked by self-complementary bases that form a stem-loop structure capped at the 3′ end by a poly G tail [26]. This poly-G tail interacts with CXCL16 expressed on the surface of pDCs which increases their uptake and directs them into early endosomes [34,35]. TLR9-mediated CpG stimulation proceeds through IRF7 in these lysosomes, culminating in the production of IFNα [35]. Thus, the poly-G tail of D ODN accounts for their ability to trigger pDC to secrete IFNα rather than TNFα, and their lack of activity on cells that do not express CXCL16 [26,27,35]. C-type ODNs resemble the K type in being composed entirely of phosphorothioate nucleotides. C-type ODNs were originally described as expressing a TCGTCG at the 5′ end, and commonly contain an internal K-type motif (such as GTCGTT) embedded in a palindromic sequence [36]. This class of ODNs is capable of stimulating B cells to secrete IL-6 and pDCs to produce IFNα (thus combining some of the stimulatory properties of D- and K-type ODNs) [31,32].

Studies examining the activity of peripheral-blood mononuclear cells (PBMCs) from humans, macaques, chimpanzees, and orangutans indicate that primates respond to the same broad classes of CpG ODN [37–40]. In contrast, rodents respond poorly to some [9,10] but not all [25,32] ODNs, which are highly active in primates. Thus, studies designed to examine the therapeutic potential of CpG ODNs are typically initiated in mice and then followed by studies in which the specific ODNs planned for clinical use are evaluated in nonhuman primates.

## 7.3 CpG ODNS AS VACCINE ADJUVANTS

The ability of CpG DNA to promote the production of Th1 and pro-inflammatory cytokines and induce the maturation and activation of professional antigen-presenting cells suggests that they might improve the immunogenicity of coadministered antigens. Consistent with such a possibility, studies involving K-type ODNs in mice established that CpG-mediated stimulation boosted both humoral and cell-mediated responses to proteins such as ovalbumin and keyhole limpet hemocyanin [41,42].

These adjuvant properties were improved significantly by maintaining the ODNs in close physical and temporal proximity to the antigen. Physically binding ODNs to antigen, cross-linking them with alum, or co-incorporating them into lipid emulsions or vesicles generated IgG responses 10- to 1000-fold greater than induced by antigen alone [42–44]. Subsequent studies suggested that this adjuvant effect had three components: (1) a CpG-induced enhancement in antigen-presenting cells (APC) function, (2) a CpG-dependent induction of a cytokine or chemokine microenvironment supportive of antigen-specific immunity, and (3) an improvement in antigen uptake mediated by DNA-binding receptors on APCs (this final effect being CpG independent) [45].

Further study established that CpG ODNs could boost the immune response to coadministered vaccines. Adding K-type ODNs to a variety of vaccines [including those targeting influenza virus, measles virus, hepatitis B virus surface antigen (HBsAg), tetanus, and anthrax] increased antigen-specific antibody titers in mice by up to orders orders of magnitude [44,46–52] Consistent with their effects on Th1 cytokine production, CpG ODN preferentially induced the production of IFNα, which supported the secretion of IgG2a antibodies and facilitated the development of antigen-specific cytotoxic T lymphocytes (CTLs) [44,46–49,53].

## 7.4 CpG ODNS IMPROVE THE RESPONSE TO MUCOSAL VACCINES

Many pathogens gain access to the host through the respiratory, gastrointestinal, vaginal, or rectal mucosa. Thus, the ability of CpG ODNs to boost mucosal

**TABLE 7.2   CpG ODNs as Immune Adjuvants$^a$**

| ODN Type | Antigenic Target | Species | Route | Refs. |
|---|---|---|---|---|
| Toxins | | | | |
| K | Tetanus toxoid | Mouse | Systemic, oral | 50, 58 |
| K | Diptheria toxin | Mouse | Systemic | 58 |
| Viruses | | | | |
| K | Measles | Neonatal mice | Systemic | 47 |
| K | Hepatitis B | Mouse, monkey, human | Intranasal, systemic | 59–62 |
| K | Cytomegalovirus | Mouse | Systemic | 63 |
| K | Influenza | Mouse, human | | 46, 55, 64 |
| Bacteria | | | | |
| K | *Brucella abortus* | Mouse | Systemic | 65 |
| K, D | Anthrax | Mouse, monkey, human | Systemic | 52, 66 |
| K | Tuberculosis | Guinea pig | Systemic | 67 |
| Parasites | | | | |
| K | Malaria | Mouse | Systemic | 68, 69 |
| K, D | Leishmania | Mouse, monkey | Systemic, intradermal | 37, 61, 70 |
| K | Trypanosome | Mouse | Systemic | 71 |

$^a$This table provides representative examples of the type of immune response elicited when CpG ODNs are coadministered as vaccine adjuvants.

immunity was examined. Administering CpG ODNs with formalin-inactivated influenza virus intranasally significantly increased flu-specific antibody levels in the serum, saliva, and genital tract of mice [46]. Similarly, CpG ODNs delivered with vaccine or β-galactosidase stimulated strong antigen-specific IgA responses throughout the mucosal immune system and in the serum [54–56] (Table 7.2). Spleen cells from intranasally immunized mice preferentially produced IFNγ rather than IL-4 when reexposed to antigen in vitro. Animals immunized intravaginally with an HSV-2 vaccine combined with CpG ODNs rapidly developed a strong mucosal and systemic Th1 immune response that protected against lethal HSV-2 infection [57]. They also generated MHC-restricted, antigen-specific CTL, replicating the effects of parenterally injected CpG ODNs plus antigen [54].

## 7.5   CpG ODNS AUGMENT THE IMMUNE RESPONSE OF IMMUNOCOMPROMIZED ANIMALS

Vaccination is most effective when "herd immunity" is achieved and the number of individuals remaining susceptible to infection is minimized. Individuals whose immune system is compromised (due to age, disease, or immunosuppressive therapy) are more resistant to vaccination and more susceptible to infection. Such individuals thus place the broader community at risk. Studies were performed to examine whether CpG ODNs, by stimulating the innate immune system, could overcome defects in the adaptive immune response of animals with various types of immunodeficiency.

Due to the immaturity of the neonatal immune system, newborns frequently mount an inadequate response to foreign pathogens [72,73]. For example, newborns respond poorly to immunization with HBsAg, attenuated measles virus, and tetanus toxoid [47,49]. Studies in mice indicate that both antibody and CTL responses by young mice are enhanced by co-delivery of CpG ODNs with antigen [47,49]. Interpreting these findings is complicated, however, since the young animals were immunized repeatedly, obscuring the effect of a single early dose of CpG ODNs on subsequent immune responsiveness. At the other end of the age spectrum, very old animals develop defects in cell-mediated and humoral immunity [74,75]. Vaccines for the elderly must compensate for these alterations in immune function. Several recent reports indicate that adding CpG ODNs significantly boosts vaccine immunogenicity in geriatric mice and/or restores IgG and CD4 T-cell priming to young adult levels [76–78].

Coexisting infection can also reduce vaccine responsiveness. For example, HIV-infected patients experience a progressive deterioration in the number and functional activity of $CD4^+$ T cells, and respond suboptimally to vaccination [79–81]. Yet retroviral infection has less effect on the innate than on the adaptive immune system [61]. Thus, PBMCs from HIV-infected subjects (and SIV-infected macaques) can respond to CpG stimulation despite declines in antigen-specific immunity [61]. The ability of CpG ODNs to boost the immune response of SIV-infected macaques to a hepatitis B vaccine was therefore investigated. Unlike healthy macaques, SIV-infected animals were unable to mount a protective antibody response when vaccinated repeatedly with Engerix-B (HIV patients show a similar loss of vaccine responsiveness) [79,80]. Only 20% of the SIV-infected macaques ever developed protective titers of antibody. By comparison, the addition of K- or D-type ODNs to the vaccine boosted Ab titers to protective levels in all animals with viral loads < $10^7$ copies/mL [61]. Although the antibody levels achieved were significantly lower than that of similarly immunized uninfected animals, these findings indicate that inclusion of CpG ODNs can boost the immunogenicity of vaccines in both normal and immunocompromized hosts. Similarly, the innate hyporesponsiveness of orangutans to hepatitis B vaccination was overcome by coadministration of CpG ODNs [39] (Table 7.2).

## 7.6  ADJUVANT ACTIVITY OF CpG ODNS IN NONHUMAN PRIMATES

Building on the observation that nonhuman primates respond to the same CpG ODNs that stimulate human cells and that CpG ODNs boost immunity to the hepatitis B vaccine [82], studies evaluating their effect with other vaccines were initiated. When K-type ODNs were coadministered with a peptide-based malaria vaccine, a significant increase in the antigen-specific serum IgG response of Aotus monkeys was observed [68]. In contrast, inclusion of control ODN had no effect on vaccine immunogenicity [68].

K-type ODNs not only increase the magnitude but also accelerate the development of antigen-specific immunity. For example, when coadministered with AVA (the licensed anthrax vaccine) K ODNs induced a sixfold-higher antibody response in rhesus macaques than does AVA alone [52]. This enhanced antibody response arose in less than two weeks and resulted in significantly greater protection against anthrax infection as monitored using a serum transfer system. Upon boosting, the avidity of the anti-anthrax antibodies elicited by K-type ODNs plus AVA was significantly higher than that induced by AVA alone [52]. Of note, despite the ability of K-type ODNs to improve humoral immune responses in vivo, their ability to significantly improve T-cell responses in primates remains uncertain.

In another model, rhesus macaques were immunized with a candidate leishmania vaccine [heat-killed leishmania vaccine (HKLV)] plus either K or D ODNs. Animals vaccinated with HKLV alone and then challenged with *Leishmania major* developed large cutaneous lesions [37]. Monkeys vaccinated with HKLV plus K ODNs also developed large lesions, although the lesions developed somewhat more slowly than in controls. This result confirmed that not all CpG ODNs improve vaccine-induced immunity. By comparison, animals immunized with HKLV plus D-type ODN had significantly smaller lesions, consistent with a reduced parasite burden (Figure 7.1) [37,83]. PBMCs from these animals also had a higher proportion of cells that were stimulated by leishmania antigens to secrete IFN in vitro than did those immunized with HKLV alone [37].

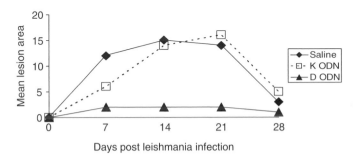

Days post leishmania infection

**Figure 7.1** Six rhesus macaques/group were treated with 500 ug of a mixture of D ODN, K ODN or saline i.d. 3 days before and 3 days after infectious challenge with 10,000,000 L. amonzensis metacyclic promastigotes. Data represent the average area of the lesions. Note that macaques treated with D ODN had significantly smaller lesions (p < 0.05) than animals treated with K ODN or saline.

## 7.7 HUMAN CLINICAL EXPERIENCE

CpG ODNs have been administered to hundreds of subjects in more than a dozen clinical trials. Many of these studies explored the safety and immuno-modulatory properties of CpG ODNs delivered alone or in combination with various vaccines, antibodies, or allergens. Three clinical trials in which CpG ODNs were used as vaccine adjuvants have been described. The first was a double-blind study in which CpG ODNs were coadministered multiple times with Engerix-B (the licensed hepatitis B virus vaccine). Healthy adult volunteers immunized with the vaccine plus CpG ODNs developed serum IgG antibody responses more rapidly than did those immunized with vaccine alone. The mean antibody titer in subjects treated with CpG ODNs plus Engerix-B was 13- to 45-fold higher than in recipients of vaccine alone after both primary and secondary immunization [62,64].

In the second double-blind study, CpG ODNs was coadministered with the Fluarix influenza vaccine. Inclusion of CpG ODNs did not increase the antibody response of naive recipients compared to Fluarix alone, but did increase antibody titers among subjects with preexisting antiflu antibodies. PBMCs from CpG ODNs vaccinated subjects responded to in vitro restimulation by secreting significantly higher levels of IFN$\alpha$ than did PBMCs from control vaccines [64]. No serious adverse events attributed to the use of CpG ODNs were observed. None of the subjects exposed to CpG ODNs developed signs or symptoms of autoimmune disease [84].

In the third recently completed study, CpG ODNs were administered in combination with AVA, the approved anthrax vaccine. Inclusion of the ODNs significantly accelerated the induction of a strong humoral immune response while boosting the peak anti-protective antigen response by six- to eightfold (nearly identical to the effects observed in rhesus macaques) [66].

## 7.8 CpG ODN SAFETY

Several safety concerns have been raised by the clinical use of CpG ODNs. These include the possibility that ODNs might (1) enhance the immunogenicity of self proteins at the site of delivery, thereby triggering the development of organ-specific or systemic autoimmune disease; (2) increase the susceptibility of the host's immune system to pathogenic agents that cause toxic shock; or (3) induce abnormalities in host tissues such as those of the liver, lymphoid, or hematopoietic system [85].

The ability of CpG DNA to affect the development of autoimmune disease is supported by studies showing that high doses of bacterial DNA elicit the production of autoantibodies against double-stranded DNA in normal mice and accelerate autoantibody production in lupus-prone animals [86–88]. Bacterial DNA also stimulates the production of IL-6 and blocks the apoptotic death of activated lymphocytes, functions that predispose to the development of autoimmune disease by facilitating the persistence of self-reactive lymphocytes [89–92].

To clarify the magnitude of this safety concern, in vivo experiments were conducted in which mice were injected repeatedly with immunostimulatory doses of CpG DNA. The number of IgG anti-DNA secreting B cells rose two-to threefold [93], and serum IgG anti-DNA antibody titers rose by up to 60%. The magnitude of these effects was insufficient to induce or worsen systemic autoimmunity in some models while promoting disease development in others [93–97].

The situation was even more complex for organ-specific autoimmune diseases that are typically promoted by the type of Th1 response preferentially elicited by CpG DNA. In an IL-12-dependent model of experimental allergic encephalomyelitis (similar to multiple sclerosis), animals treated with CpG DNA and then challenged with autoantigen developed autoreactive Th1 effector cells that caused disease, whereas mice injected with autoantigen alone remained disease-free [98,99]. In a molecular mimicry model, CpG DNA coadministered with *Chlamydia*-derived antigens promoted the induction of autoimmune myocarditis [100]. CpG ODNs also increased the susceptibility of mice to arthritis [101,102]. These findings indicate that CpG motifs can promote the development of deleterious autoimmune reactions under certain circumstances.

Another concern was the possibility that CpG ODNs might facilitate the development of toxic shock. Agents such as lipopolysaccharide and D-galactosamine promote toxic shock by triggering the overproduction of TNFα, a cytokine also elicited by CpG DNA. Indeed, CpG ODNs enhanced the mortality and morbidity induced by subpathologic doses of lipopolysaccharide and D-galactosamine [103–105]. Finally, CpG ODNs at high concentration, or when administered repeatedly at short intervals, can induce abnormalities in the liver, spleen, and in the production of platelets and blood cells [106]. To examine whether these toxicities are likely to occur under normal circumstances, CpG ODNs at doses equal to or exceeding those typically used as immune adjuvants were injected weekly for 4 months into normal BALB/c mice. All of the animals treated remained physically fit and none showed macroscopic or microscopic evidence of tissue damage or inflammation [107]. Similarly, no adverse health effects were reported in studies involving the delivery of CpG ODNs to nonhuman primates when hundreds of milligrams of antisense ODNs were administered repeatedly to patients with active Crohn's disease [108], or when DNA vaccines composed of bacterial plasmids that contain CpG motifs were administered to normal volunteers [61,109].

Adverse events (AEs) were observed among recipients of CpG ODNs in combination with various protein-based vaccines. These were predominantly injection-site reactions (such as pain and erythema) and flulike symptoms. Vaccinees reported that these events were typically short-lived and did not interfere with the activities of daily living. The intensity of these AEs was similar in recipients of vaccine alone versus recipients of vaccine plus ODNs, although the frequency of AEs was higher among those co-vaccinated with CpG ODNs versus vaccine alone [62,64,66]. Only in the trial involving the anthrax vaccine were any serious (grade 3) AEs observed, with preliminary data showing no significant increase in their frequency due to coadministration of CpG ODNs [66]. There were no clinically relevant changes in hematocrit or white blood cell count among immunized volunteers, nor were there any changes in liver or renal function. None of the subjects exposed to CpG ODNs developed signs or symptoms of autoimmune disease [84].

Thus, while concerns remain that CpG ODNs might have adverse effects under certain conditions, it does not appear that these agents are toxic to normal animals or humans when used as vaccine adjuvants. Clinical studies are thus proceeding, and subjects continue to be closely monitored for their development of adverse reactions.

It should also be appreciated that CpG ODNs are not universally successful as vaccine adjuvants. Failures of both D- and K-type ODNs to promote immunity in nonhuman primates were described above. Of greater concern is recent evidence that CpG-induced increases in immune activity do not necessarily correlate with improved vaccine efficacy. For example, rodents treated with CpG ODNs plus a respiratory syncytial virus (RSV) protein-based vaccine were only modestly protected from viral challenge and were at increased risk of pulmonary pathology [110]. Similarly, other immune adjuvants may be superior to CpG ODNs for use with specific vaccine antigens [111].

## 7.9  SUMMARY

CpG ODNs stimulate cells that express TLR9, initiating an immunomodulatory cascade that culminates in the production of Th1 and pro-inflammatory cytokines and chemokines. There is consistent evidence that CpG ODNs function as adjuvants when coadministered with conventional protein-based vaccines, boosting antigen-specific antibody and cell-mediated immune responses. CpG ODNs can both accelerate and magnify vaccine-specific immunity. These effects could be of considerable benefit when the rapid induction of a protective immune response is required (e.g., when confronted by the release of a biopathogen) [112]. Yet the adjuvant effect of CpG ODNs is strongest when the amount of immunogen being administered is suboptimal. This is usually referred to as an *antigen-sparing effect*. When high doses of vaccine are administered, the magnitude of the immunologic boosting attributable to CpG ODNs is modest.

The utility of CpG ODNs is enhanced by their ability to promote mucosal as well as systemic immunity. This is of considerable importance for pathogens that gain access to the host through the respiratory, gastrointestinal, and reproductive tracts. Several studies document that coadministering CpG ODNs with vaccines significantly increases antigen-specific IgA levels at mucosal sites and IgG levels systemically [46,48,54]. An additional benefit of CpG ODNs is their ability to boost immunity in groups with reduced immune function that typically respond poorly to vaccination, such as newborns, the elderly, and the immunosuppressed. Eliminating reservoirs of susceptible individuals reduces the likelihood of pathogen transmission and thus enhances the community's resistance to infection.

Preclinical studies involving nonhuman primates confirm the expectation that CpG ODNs selected for their ability to stimulate human immune cells are active in vivo. Although the magnitude of this effect varies with the type of antigen and ODN utilized, studies involving HKLV indicate that CpG ODNs can convert an otherwise ineffective vaccine to one that provides significant protection from infection [37], while studies with Engerix-B document that CpG ODNs facilitate the induction of protective immunity in immunocompromised hosts [61]. Experiments involving AVA demonstrate that CpG ODNs improve both the magnitude and rapidity with which protection is elicited [52].

## 7.10  FUTURE PROSPECTS

Clinical studies designed to evaluate the safety and activity of CpG ODNs in humans are ongoing. Available results suggest that these agents are generally safe, and in some cases boost the immunogenicity of coadministered vaccines significantly [64,66]. The recent discovery of a cell surface receptor that selectively promotes the uptake of D-class CpG ODNs and triggers the production of IFNγ should allow the tailoring of these adjuvants to enhance those immune responses that provide optimal control of target pathogens. In this context, progress will require further research to identify ODNs of different classes that are optimally active in humans when coadministered with specific vaccines (and vaccine types). Supporting such product development will be efforts to determine how each class of ODN regulates discrete elements of the immune system. It will also be important to establish the optimal dose, duration, and site(s) of vaccine/ODN delivery. Finally, the issue of long-term CpG ODN safety will require continued exploration. We expect these efforts to improve the utility of CpG-augmented vaccines for the induction of protective immunity against infectious pathogens.

## REFERENCES

1. Marrack, P. & Kappler, J. Subversion of the immune system by pathogens. *Cell* 1994, **76**(2), 323–332.
2. Medzhitov, R. & Janeway, C.A., Jr. Innate immunity: impact on the adaptive immune response. *Curr Opin Immunol* 1997, **9**(1), 4–9.
3. Medzhitov, R. & Janeway, C.A., Jr. Innate immunity: the virtues of a nonclonal system of recognition. *Cell* 1997, **91**(3), 295–298.

4. Paul, W.E. *Fundamental Immunology*, Raven Press, New York, 1993.

5. Underhill, D.M. & Ozinsky, A. Toll-like receptors: key mediators of microbe detection. *Curr Opin Immunol* 2002, **14**(1), 103–110.

6. Vasselon, T. & Detmers, P.A. Toll receptors: a central element in innate immune responses. *Infect Immun* 2002, **70**(3), 1033–1041.

7. Razin, A. & Friedman, J. DNA methylation and its possible biological roles. *Prog Nucleic Acid Res Mol Biol* 1981, **25**, 33–52.

8. Cardon, L.R., Burge, C., Clayton, D.A. & Karlin, S. Pervasive CpG suppression in animal mitochondrial genomes. *Proc Natl Acad Sci USA* 1994, **91**(9), 3799–3803.

9. Hemmi, H., Takeuchi, O., Kawai, T. et al. A Toll-like receptor recognizes bacterial DNA. *Nature* 2000, **408**(6813), 740–745.

10. Takeshita, F., Leifer, C.A., Gursel, I. et al. Cutting edge: Role of Toll-like receptor 9 in CpG DNA-induced activation of human cells. *J Immunol* 2001, **167**(7), 3555–3558.

11. Bauer, S., Kirschning, C.J., Hacker, H. et al. Human TLR9 confers responsiveness to bacterial DNA via species-specific CpG motif recognition. *Proc Natl Acad Sci USA* 2001, **98**(16), 9237–9242.

12. Krieg, A.M., Yi, A.K., Matson, S. et al. CpG motifs in bacterial DNA trigger direct B-cell activation. *Nature* 1995, **374**(6522), 546–549.

13. Yamamoto, S., Katoaka, T., Yano, O., Kuramoto, E., Shimada, S. & Tokunaga, T. Antitumor Effect of Nucleic Acid Fraction from Bacteria. *Proc Jpn Soc Immunol* 1989(48), 272–281.

14. Yamamoto, S., Yamamoto, T., Kataoka, T., Kuramoto, E., Yano, O. & Tokunaga, T. Unique palindromic sequences in synthetic oligonucleotides are required to induce IFN [correction of INF] and augment IFN-mediated [correction of INF] natural killer activity. *J Immunol* 1992, **148**(12), 4072–4076.

15. Klinman, D.M., Yi, A.K., Beaucage, S.L., Conover, J. & Krieg, A.M. CpG motifs expressed by bacterial DNA rapidly induce lymphocytes to secrete IL-6, IL-12, and IFNg. *Proc Natl Acad Sci USA* 1996, **93**(7), 2879–2883.

16. Ishii, K.J., Takeshita, F., Gursel, I. et al. Potential role of phosphatidylinositol 3 kinase, rather than DNA-dependent protein kinase, in CpG DNA-induced immune activation. *J Exp Med* 2002, **196**(2), 269–274.

17. Gursel, M., Verthelyi, D., Gursel, I., Ishii, K.J. & Klinman, D.M. Differential and competitive activation of human immune cells by distinct classes of CpG oligodeoxynucleotide. *J Leukoc Biol* 2002, **71**(5), 813–820.

18. Hornung, V., Rothenfusser, S., Britsch, S. et al. Quantitative expression of Toll-like receptor 1–10 mRNA in cellular subsets of human peripheral blood mononuclear cells and sensitivity to CpG oligodeoxynucleotides. *J Immunol* 2002, **168**(9), 4531–4537.

19. Sun, S., Zhang, X., Tough, D.F. & Sprent, J. Type I interferon-mediated stimulation of T cells by CpG DNA. *J Exp Med* 1998, **188**(12), 2335–2342.

20. Stacey, K.J., Sweet, M.J. & Hume, D.A. Macrophages ingest and are activated by bacterial DNA. *J Immunol* 1996, **157**(5), 2116–2122.

21. Ballas, Z.K., Rasmussen, W.L. & Krieg, A.M. Induction of NK activity in murine and human cells by CpG motifs in oligodeoxynucleotides and bacterial DNA. *J Immunol* 1996, **157**(5), 1840–1847.

22. Halpern, M.D., Kurlander, R.J. & Pisetsky, D.S. Bacterial-DNA induces murine interferon-gamma production by stimulation of interleukin-12 and tumor-necrosis-factor-alpha. *Cell Immunol* 1996, **167**(1), 72–78.

23. Rankin, R., Pontarollo, R., Ioannou, X. et al. CpG motif identification for veterinary and laboratory species demonstrates that sequence recognition is highly conserved. *Antisense Nucleic Acid Drug Dev* 2001, **11**(5), 333–340.

24. Roman, M., Martin-Orozco, E., Goodman, J.S. et al. Immunostimulatory DNA sequences function as T helper-1-promoting adjuvants [see comments]. *Nat Med* 1997, **3**(8), 849–854.

25. Broide, D.H., Stachnick, G., Castaneda, D. et al. Systemic administration of immunostimulatory DNA sequences mediates reversible inhibition of Th2 responses in a mouse model of asthma. *J Clin Immunol* 2001, **21**(3), 175–182.

26. Verthelyi, D., Ishii, K., Gursel, M., Takeshita, F. & Klinman, D. Human peripheral blood cells differentially recognize and respond to two distinct CPG motifs. *J Immunol* 2001, **166**(4), 2372–2377.

27. Krug, A., Rothenfusser, S., Hornung, V. et al. Identification of CpG oligonucleotide sequences with high induction of IFN-alpha/beta in plasmacytoid dendritic cells. *Eur J Immunol* 2001, **31**(7), 2154–2163.

28. Kadowaki, N., Ho, S., Antonenko, S. et al. Subsets of human dendritic cell precursors express different Toll-like receptors and respond to different microbial antigens. *J Exp Med* 2001, **194**(6), 863–869.

29. Krug, A., Towarowski, A., Britsch, S. et al. Toll-like receptor expression reveals CpG DNA as a unique microbial stimulus for plasmacytoid dendritic cells which synergizes with CD40 ligand to induce high amounts of IL-12. *Eur J Immunol* 2001, **31**(10), 3026–3037.

30. Bauer, M., Redecke, V., Ellwart, J.W. et al. Bacterial CpG-DNA triggers activation and maturation of human CD11c–, CD123+ dendritic cells. *J Immunol* 2001, **166**(8), 5000–5007.

31. Hartmann, G., Battiany, J., Poeck, H. et al. Rational design of new CpG oligonucleotides that combine B cell activation with high IFN-alpha induction in plasmacytoid dendritic cells. *Eur J Immunol* 2003, **33**(6), 1633–1641.

32. Marshall, J.D., Fearon, K., Abbate, C. et al. Identification of a novel CpG DNA class and motif that optimally stimulate B cell and plasmacytoid dendritic cell functions. *J Leukoc Biol* 2003, **73**(6), 781–792.

33. Hartmann, G. & Krieg, A.M. Mechanism and function of a newly identified CpG DNA motif in human primary B cells. *J Immunol* 2000, **164**(2), 944–953.

34. Honda, K., Ohba, Y., Yanai, H. et al. Spatiotemporal regulation of MyD88-IRF-7 signalling for robust type-I interferon induction. *Nature* 2005, **434**(7036), 1035–1040.

35. Gursel, M., Gursel, I., Mostowski, H.S. & Klinman, D.M. CXCL16 influences the nature and specificity of CpG-induced immune activation. *J Immunol* 2006, **177**(3), 1575–1580.

36. Hartmann, G., Weeratna, R.D., Ballas, Z.K. et al. Delineation of a CpG phosphorothioate oligodeoxynucleotide for activating primate immune responses in vitro and in vivo. *J Immunol* 2000, **164**(3), 1617–1624.

37. Verthelyi, D., Kenney, R.T., Seder, R.A., Gam, A.A., Friedag, B. & Klinman, D.M. CpG oligodeoxynucleotides as vaccine adjuvants in primates. *J Immunol* 2002, **168**(4), 1659–1663.

38. Kanellos, T.S., Sylvester, I.D., Butler, V.L. et al. Mammalian granulocyte-macrophage colony-stimulating factor and some CpG motifs have an effect on the immunogenicity of DNA and subunit vaccines in fish. *Immunology* 1999, **96**(4), 507–510.

39. Davis, H.L., Suparto, I.I., Weeratna, R.R. et al. CpG DNA overcomes hyporesponsiveness to hepatitis B vaccine in orangutans. *Vaccine* 2000, **18**(18), 1920–1924.

40. Verthelyi, D. & Klinman, D.M. Immunoregulatory activity of CpG oligonucleotides in humans and nonhuman primates. *Clin Immunol* 2003, **109**(1), 64–71.

41. Klinman, D.M. Therapeutic applications of CpG-containing oligodeoxynucleotides. *Antisense Nucleic Acid Drug Dev* 1998, **8**(2), 181–184.

42. Klinman, D.M., Barnhart, K.M. & Conover, J. CpG motifs as immune adjuvants. *Vaccine* 1999, **17**(1), 19–25.

43. Gursel, I., Gursel, M., Ishii, K.J. & Klinman, D.M. Sterically stabilized cationic liposomes improve the uptake and immunostimulatory activity of CpG oligonucleotides. *J Immunol* 2001, **167**(6), 3324–3328.

44. Davis, H.L., Weeranta, R., Waldschmidt, T.J., Tygrett, L., Schorr, J. & Krieg, A.M. CpG DNA is a potent enhancer of specific immunity in mice immunized with recombinant hepatitis B surface antigen. *J Immunol* 1998, **160**(2), 870–876.

45. Shirota, H., Sano, K., Hirasawa, N. et al. Novel roles of CpG oligodeoxynucleotides as a leader for the sampling and presentation of CpG-tagged antigen by dendritic cells. *J Immunol* 2001, **167**(1), 66–74.

46. Moldoveanu, Z., Love-Homan, L., Huang, W.Q. & Krieg, A.M. CpG DNA, a novel immune enhancer for systemic and mucosal immunization with influenza virus. *Vaccine* 1998, **16**(11–12), 1216–1224.

47. Kovarik, J., Bozzotti, P., Love-Homan, L. et al. CpG oligodeoxynucleotides can circumvent the Th2 polarization of neonatal responses to vaccines but may fail to fully redirect Th2 responses established by neonatal priming. *J Immunol* 1999, **162**(3), 1611–1617.

48. McCluskie, M.J. & Davis, H.L. CpG DNA is a potent enhancer of systemic and mucosal immune responses against hepatitis B surface antigen with intranasal administration to mice. *J Immunol* 1998, **161**(9), 4463–4466.

49. Brazolot Millan, C.L., Weeratna, R., Krieg, A.M., Siegrist, C.A. & Davis, H.L. CpG DNA can induce strong Th1 humoral and cell-mediated immune responses against hepatitis B surface antigen in young mice. *Proc Natl Acad Sci USA* 1998, **95**(26), 15553–15558.

50. Eastcott, J.W., Holmberg, C.J., Dewhirst, F.E., Esch, T.R., Smith, D.J. & Taubman, M.A. Oligonucleotide containing CpG motifs enhances immune response to mucosally or systemically administered tetanus toxoid. *Vaccine* 2001, **19**(13–14), 1636–1642.

51. Xie, H., Gursel, I., Ivins, B.E. et al. CpG oligodeoxynucleotides adsorbed onto polylactide-co-glycolide microparticles improve the immunogenicity and protective activity of the licensed anthrax vaccine. *Infect Immun* 2005, **73**(2), 828–833.

52. Klinman, D.M., Xie, H., Little, S.F., Currie, D. & Ivins, B.E. CpG oligonucleotides improve the protective immune response induced by the anthrax vaccination of rhesus macaques. *Vaccine* 2004, **22**(21–22), 2881–2886.

53. Branda, R.F., Moore, A.L., Lafayette, A.R. et al. Amplification of antibody production by phosphorothioate oligodeoxynucleotides. *J Lab Clin Med* 1996, **128**(3), 329–338.

54. Horner, A.A., Ronaghy, A., Cheng, P.M. et al. Immunostimulatory DNA is a potent mucosal adjuvant. *Cell Immunol* 1998, **190**(1), 77–82.

55. McCluskie, M.J. & Davis, H.L. Oral, intrarectal and intranasal immunizations using CpG and non-CpG oligodeoxynucleotides as adjuvants. *Vaccine* 2000, **19**(4–5), 413–422.

56. Hernandez, H.M., Figueredo, M., Garrido, N., Sanchez, L. & Sarracent, J. Intranasal immunisation with a 62 kDa proteinase combined with cholera toxin or CpG adjuvant protects against *Trichomonas vaginalis* genital tract infections in mice. *Int J Parasitol* 2005, **35**(13), 1333–1337.

57. Tengvall, S., Lundqvist, A., Eisenberg, R.J., Cohen, G.H. & Harandi, A.M. Mucosal administration of CpG oligodeoxynucleotide elicits strong CC and CXC chemokine responses in the vagina and serves as a potent Th1-tilting adjuvant for recombinant gD2 protein vaccination against genital herpes. *J Virol* 2006, **80**(11), 5283–5291.

58. von Hunolstein, C., Mariotti, S., Teloni, R. et al. The adjuvant effect of synthetic oligodeoxynucleotide containing CpG motif converts the anti–*Haemophilus influenzae* type b glycoconjugates into efficient anti-polysaccharide and anti-carrier polyvalent vaccines. *Vaccine* 2001, **19**(23–24), 3058–3066.

59. Kowalczyk, D.W., Wlazlo, A.P., Shane, S. & Ertl, H.C. Vaccine regimen for prevention of sexually transmitted infections with human papillomavirus type 16. *Vaccine* 2001, **19**(25–26), 3583–3590.

60. McCluskie, M.J., Weeratna, R.D., Krieg, A.M. & Davis, H.L. CpG DNA is an effective oral adjuvant to protein antigens in mice. *Vaccine* 2000, **19**(7–8), 950–957.

61. Verthelyi, D., Gursel, M., Kenney, R.T. et al. CpG oligodeoxynucleotides protect normal and SIV-infected macaques from leishmania infection. *J Immunol* 2003, **170**(9), 4717–4723.

62. Halperin, S.A., Van Nest, G., Smith, B., Abtahi, S., Whiley, H. & Eiden, J.J. A phase I study of the safety and immunogenicity of recombinant hepatitis B surface antigen coadministered with an immunostimulatory phosphorothioate oligonucleotide adjuvant. *Vaccine* 2003, **21**(19–20), 2461–2467.

63. Temperton, N.J., Quenelle, D.C., Lawson, K.M. et al. Enhancement of humoral immune responses to a human cytomegalovirus DNA vaccine: adjuvant effects of aluminum phosphate and CpG oligodeoxynucleotides. *J Med Virol* 2003, **70**(1), 86–90.

64. Klinman, D.M., Ishii, K.J., Gursel, M., Gursel, I., Takeshita, S. & Takeshita, F. Immunotherapeutic applications of CpG-containing oligodeoxynucleotides. *Drug News Perspect* 2000, **13**(5), 289–296.

65. Al-Mariri, A., Tibor, A., Mertens, P. et al. Protection of BALB/c mice against *Brucella abortus* 544 challenge by vaccination with bacterioferritin or P39 recombinant proteins with CpG oligodeoxynucleotides as adjuvant. *Infect Immun* 2001, **69**(8), 4816–4822.

66. Rynkiewicz, D., Rathkopf, M., Ransom, J. et al. Marked enhancement of antibody response to anthrax vaccine adsorbed with CpG 7909 in healthy volunteers. *ICAAC Abstr* 2005, LB-25.

67. Hogarth, P.J., Jahans, K.J., Hecker, R., Hewinson, R.G. & Chambers, M.A. Evaluation of adjuvants for protein vaccines against tuberculosis in guinea pigs. *Vaccine* 2003, **21**(9–10), 977–982.

68. Jones, T.R., Obaldia, N., 3rd, Gramzinski, R.A. et al. Synthetic oligodeoxynucleotides containing CpG motifs enhance immunogenicity of a peptide malaria vaccine in *Aotus* monkeys. *Vaccine* 1999, **17**(23–24), 3065–3071.

69. Su, Z., Tam, M.F., Jankovic, D. & Stevenson, M.M. Vaccination with novel immunostimulatory adjuvants against blood-stage malaria in mice. *Infect Immun* 2003, **71**(9), 5178–5187.

70. Mendez, S., Tabbara, K., Belkaid, Y. et al. Coinjection with CpG-containing immunostimulatory oligodeoxynucleotides reduces the pathogenicity of a live vaccine against cutaneous leishmaniasis but maintains its potency and durability. *Infect Immun* 2003, **71**(9), 5121–5129.

71. Frank, F.M., Petray, P.B., Cazorla, S.I., Munoz, M.C., Corral, R.S. & Malchiodi, E.L. Use of a purified *Trypanosoma cruzi* antigen and CpG oligodeoxynucleotides for immunoprotection against a lethal challenge with trypomastigotes. *Vaccine* 2003, **22**(1), 77–86.

72. Sterzl, J. & Silverstein, A.M. Developmental aspects of immunity. *Adv Immunol* 1967, **6**, 337–459.

73. Hunt, D.W., Huppertz, H.I., Jiang, H.J. & Petty, R.E. Studies of human cord blood dendritic cells: evidence for functional immaturity. *Blood* 1994, **84**(12), 4333–4343.

74. Miller, R.A. The cell biology of aging: immunological models. *J Gerontol* 1989, **44**(1), B4–B8.

75. Thoman, M.L. & Weigle, W.O. The cellular and subcellular bases of immunosenescence. *Adv Immunol* 1989, **46**, 221–261.

76. Qin, W., Jiang, J., Chen, Q. et al. CpG ODN enhances immunization effects of hepatitis B vaccine in aged mice. *Cell Mol Immunol* 2004, **1**(2), 148–152.

77. Siegrist, C.A., Pihlgren, M., Tougne, C. et al. Co-administration of CpG oligonucleotides enhances the late affinity maturation process of human anti-hepatitis B vaccine response. *Vaccine* 2004, **23**(5), 615–622.

78. Sen, G., Chen, Q. & Snapper, C.M. Immunization of aged mice with a pneumococcal conjugate vaccine combined with an unmethylated CpG-containing oligodeoxynucleotide restores defective immunoglobulin G antipolysaccharide responses and specific CD4+-T-cell priming to young adult levels. *Infect Immun* 2006, **74**(4), 2177–2186.

79. Diamant, E.P., Schechter, C., Hodes, D.S. & Peters, V.B. Immunogenicity of hepatitis B vaccine in human immunodeficiency virus-infected children. *Pediatr Infect Dis J* 1993, **12**(10), 877–878.

80. Wong, E.K., Bodsworth, N.J., Slade, M.A., Mulhall, B.P. & Donovan, B. Response to hepatitis B vaccination in a primary care setting: influence of HIV infection, CD4+ lymphocyte count and vaccination schedule. *Int J STD AIDS* 1996, **7**(7), 490–494.

81. Pacanowski, J., Kahi, S., Baillet, M. et al. Reduced blood CD123+ (lymphoid) and CD11c+ (myeloid) dendritic cell numbers in primary HIV-1 infection. *Blood* 2001, **98**(10), 3016–3021.

82. Payette, P.J., Ma, X., Weeratna, R.D. et al. Testing of CpG-optimized protein and DNA vaccines against the hepatitis B virus in chimpanzees for immunogenicity and protection from challenge. *Intervirology* 2006, **49**(3), 144–151.

83. von Stebut, E., Belkaid, Y., Nguyen, B.V., Cushing, M., Sacks, D.L. & Udey, M.C. *Leishmania major*–infected murine Langerhans cell–like dendritic cells from susceptible mice release IL-12 after infection and vaccinate against experimental cutaneous leishmaniasis. *Eur J Immunol* 2000, **30**(12), 3498–3506.

84. Klinman, D.M. CpG DNA as a vaccine adjuvant. *Expert Rev Vaccines* 2003, **2**(2), 305–315.

85. Heikenwalder, M., Polymenidou, M., Junt, T. et al. Lymphoid follicle destruction and immunosuppression after repeated CpG oligodeoxynucleotide administration. *Nat Med* 2004, **10**(2), 187–192.

86. Gilkeson, G.S., Ruiz, P., Howell, D., Lefkowith, J.B. & Pisetsky, D.S. Induction of immune-mediated glomerulonephritis in normal mice immunized with bacterial DNA. *Clin Immunol Immunopathol* 1993, **68**(3), 283–292.

87. Gilkeson, G.S., Pippen, A.M. & Pisetsky, D.S. Induction of cross-reactive anti-dsDNA antibodies in preautoimmune NZB/NZW mice by immunization with bacterial DNA. *J Clin Invest* 1995, **95**(3), 1398–1402.

88. Steinberg, A.D., Krieg, A.M., Gourley, M.F. & Klinman, D.M. Theoretical and experimental approaches to generalized autoimmunity. *Immunol Rev* 1990, **118**, 129–163.

89. Klinman, D.M. Polyclonal B cell activation in lupus-prone mice precedes and predicts the development of autoimmune disease. *J Clin Invest* 1990, **86**(4), 1249–1254.

90. Linker-Israeli, M., Deans, R.J., Wallace, D.J., Prehn, J., Ozeri-Chen, T. & Klinenberg, J.R. Elevated levels of endogenous IL-6 in systemic lupus erythematosus: a putative role in pathogenesis. *J Immunol* 1991, **147**(1), 117–123.

91. Krieg, A.M. CpG DNA: a pathogenic factor in systemic lupus erythematosus? *J Clin Immunol* 1995, **15**(6), 284–292.

92. Yi, A.K., Hornbeck, P., Lafrenz, D.E. & Krieg, A.M. CpG DNA rescue of murine B lymphoma cells from anti-IgM-induced growth arrest and programmed cell death is associated with increased expression of c-myc and bcl-xL. *J Immunol* 1996, **157**(11), 4918–4925.

93. Mor, G., Singla, M., Steinberg, A.D., Hoffman, S.L., Okuda, K. & Klinman, D.M. Do DNA vaccines induce autoimmune disease? *Hum Gene Ther* 1997, **8**(3), 293–300.

94. Katsumi, A., Emi, N., Abe, A., Hasegawa, Y., Ito, M. & Saito, H. Humoral and cellular-immunity to an encoded protein-induced by direct DNA injection. *Hum Gene Ther* 1994, **5**(11), 1335–1339.

95. Gilkeson, G.S., Conover, J., Halpern, M., Pisetsky, D.S., Feagin, A. & Klinman, D.M. Effects of bacterial DNA on cytokine production by (NZB/NZW)F1 mice. *J Immunol* 1998, **161**(8), 3890–3895.

96. Christensen, S.R., Kashgarian, M., Alexopoulou, L., Flavell, R.A., Akira, S. & Shlomchik, M.J. Toll-like receptor 9 controls anti-DNA autoantibody production in murine lupus. *J Exp Med* 2005, **202**(2), 321–331.

97. Viglianti, G.A., Lau, C.M., Hanley, T.M., Miko, B.A., Shlomchik, M.J. & Marshak-Rothstein, A. Activation of autoreactive B cells by CpG dsDNA. *Immunity* 2003, **19**(6), 837–847.

98. Segal, B.M., Klinman, D.M. & Shevach, E.M. Microbial products induce autoimmune disease by an IL-12-dependent pathway. *J Immunol* 1997, **158**(11), 5087–5090.

99. Segal, B.M., Chang, J.T. & Shevach, E.M. CpG oligonucleotides are potent adjuvants for the activation of autoreactive encephalitogenic T cells in vivo. *J Immunol* 2000, **164**(11), 5683–5688.

100. Bachmaier, K., Neu, N., de la Maza, L.M., Pal, S., Hessel, A. & Penninger, J.M. *Chlamydia* infections and heart disease linked through antigenic mimicry. *Science* 1999, **283**(5406), 1335–1339.

101. Zeuner, R.A., Verthelyi, D., Gursel, M., Ishii, K.J. & Klinman, D.M. Influence of stimulatory and suppressive DNA motifs on host susceptibility to inflammatory arthritis. *Arthritis Rheum* 2003, **48**(6), 1701–1707.

102. Deng, G.M., Nilsson, I.M., Verdrengh, M., Collins, L.V. & Tarkowski, A. Intra-articularly localized bacterial DNA containing CpG motifs induces arthritis. *Nat Med* 1999, **5**(6), 702–705.

103. Sparwasser, T., Miethke, T., Lipford, G. et al. Bacterial DNA causes septic shock [letter]. *Nature* 1997, **386**(6623), 336–337.

104. Cowdery, J.S., Chace, J.H., Yi, A.K. & Krieg, A.M. Bacterial DNA induces NK cells to produce IFN-gamma in vivo and increases the toxicity of lipopolysaccharides. *J Immunol* 1996, **156**(12), 4570–4575.

105. Hartmann, G., Krug, A., Waller-Fontaine, K. & Endres, S. Oligodeoxynucleotides enhance lipopolysaccharide-stimulated synthesis of tumor necrosis factor: dependence on phosphorothioate modification and reversal by heparin. *Mol Med* 1996, **2**(4), 429–438.

106. Singh, R.R., Kumar, V., Ebling, F.M. et al. T cell determinants from autoantibodies to DNA can upregulate autoimmunity in murine systemic lupus erythematosus. *J Exp Med* 1995, **181**(6), 2017–2027.

107. Klinman, D.M., Takeno, M., Ichino, M. et al. DNA vaccines: safety and efficacy issues. *Springer Semin Immunopathol* 1997, **19**(2), 245–256.

108. Yacyshyn, B.R., Barish, C., Goff, J. et al. Dose ranging pharmacokinetic trial of high-dose alicaforsen (intercellular adhesion molecule-1 antisense oligodeoxynucleotide) (ISIS 2302) in active Crohn's disease. *Aliment Pharmacol Ther* 2002, **16**(10), 1761–1770.

109. Liu, M.A. DNA vaccines: a review. *J Intern Med* 2003, **253**(4), 402–410.

110. Prince, G.A., Mond, J.J., Porter, D.D., Yim, K.C., Lan, S.J. & Klinman, D.M. Immunoprotective activity and safety of a respiratory syncytial virus vaccine: mucosal delivery of fusion glycoprotein with a CpG oligodeoxynucleotide adjuvant. *J Virol* 2003, **77**(24), 13156–13160.

111. Wille-Reece, U., Flynn, B.J., Lore, K. et al. Toll-like receptor agonists influence the magnitude and quality of memory T cell responses after prime-boost immunization in nonhuman primates. *J Exp Med* 2006, **203**(5), 1249–1258.

112. Klinman, D.M., Verthelyi, D., Takeshita, F. & Ishii, K.J. Immune recognition of foreign DNA: A cure for bioterrorism? *Immunity* 1999, **11**(2), 123–129.

# 8

# SMALL MOLECULE IMMUNOPOTENTIATORS AS VACCINE ADJUVANTS

FENGFENG XU, NICHOLAS M. VALIANTE, AND JEFFREY B. ULMER

## 8.1 CONTROL OF INNATE AND ADAPTIVE IMMUNITY BY TLR SIGNALING

The innate immune system functions as a first line of defense against infectious microorganisms: to either eradicate or contain them. The system also facilitates the induction of appropriate adaptive immune responses to eliminate the infection and prevent its recurrence. Three families of receptors found in innate immune cells and known to detect microbes are Toll-like receptors (TLRs), nucleotide-binding oligomerization domain (NOD)-like receptors (NLRs), and retinoic acid-inducible gene (RIG)-I-like receptors (RLRs). TLRs recognize bacteria, viruses, fungi, and protozoans [1,2], NLRs detect intracellular bacteria [3,4], and RLRs react to the presence of viruses [5–7]. TLR recognition and subsequent triggering of downstream reactions have been investigated extensively for their broad impact on controlling innate as well as adaptive immunity [8,9].

TLRs are highly conserved transmembrane proteins consisting of a leucine-rich repeat (LRR), a single transmembrane domain, and a cytoplasmic tail. The latter contains the canonical Toll–interleukin 1 (IL-1) receptor (TIR) domain, which provides the structural basis for forming an active signaling complex. The LRR domain is thought to contain the ligand-binding site, as well as binding sites for coreceptors. Although TLRs share a high degree of structural similarity, their recognition is selective for structurally diverse microbial ligands. This specificity has been shown for TLR2 [peptidoglycan,

*Vaccine Adjuvants and Delivery Systems*, Edited by Manmohan Singh
Copyright © 2007 John Wiley & Sons, Inc.

bacterial lipopeptides, and certain types of lipopolysaccharide (LPS)], TLR3 (double-stranded RNA), TLR4 (LPS), TLR5 (bacterial flagellin), TLR7 (ssRNA), and TLR9 (bacterial DNA). After TLR stimulation, adaptor proteins, including MyD88, TRIF, TRAM, and TIRAP (also known as Mal), are thought to be selectively recruited to different TLRs (see Figure 8.1).

A complex of IL-1 receptor–associated kinase 1 (IRAK-1), IRAK-4, and other proteins, such as Toll-interacting protein (Tollip), may also be recruited to certain TLRs. TNF receptor–associated factor (TRAF6) is crucial in transducing the signal downstream because it interacts with the receptor complex as well as with downstream complexes containing proteins such as the kinase TAK-1 and its coactivators TAB-1 and TAB-2. TAK-1 then triggers the activation of IKKκ, which phosphorylates IκB proteins that normally sequester NFκB proteins in the cytoplasm. TAK-1 also activates the JNK and p38 MAPK pathways, leading to the activation of the AP-1 transcription factor complex. TRAF6 signal transduction also results in the activation of another trans-

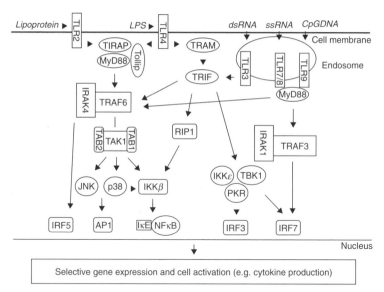

**Figure 8.1** TLR signaling pathways (see the text for details). A presumed interaction between a TLR and its ligand initiates a signaling cascade of kinase reactions and leads to activation of specific transcriptional factors (AP1, NFκB, and IRFs), which transcribe genes involved in immune cell activation and differentiation, including production of cytokines. Abbreviations: AP-1, activating protein; IκB, inhibitor of NFκB; IKK, IκB kinase; IRAK, IL-1R-associated kinase; IRF, interferon regulatory factor; PKR, double-stranded RNA-dependent protein kinase; TAB, TAK1-binding protein; TAK1, TGF-β-activating kinase; TBK1, TRAF-family-membrane-associated NFκB activator-binding kinase 1; TIRAP, TIR-domain-containing adaptor protein; TRAF, TNF-receptor-associated factor; TRAM, TRIF-related adaptor molecule; TRIF, TIR-domain-containing adaptor protein inducing IFN-β; TIRAP, TIR domain-containing adapter protein; JNK, Jun kinase; RIP1, receptor-interacting protein.

cription factor, IRF5. These proteins then enter the nucleus to activate genes involved in cell activation and differentiation, including pro-inflammatory cytokine production. The TRIF–TRAM pathway, signaled by TLR3 receptor, branches into three distinct pathways involving (1) TRAF6-mediated downstream reactions; (2) RIP-1 protein, which leads to final activation of NFκB; and (3) activation of IKK-ε and TBK1, followed by phosphorylation of IRF3 or IRF7 transcription factor, thereby promoting expression of type I interferon (IFN)-inducible genes. IRF7 can also be activated through TLR7 and TLR9 receptor-mediated pathways (Figure 8.1). TLR signaling can also lead to increased activity of negative regulators such as SOCS1 [10], SIGIRR [11], PI3K [12], CYLD [13], β-arresin [14], and A20 [15], thereby providing a feedback mechanism [16].

Certain TLRs are expressed differentially in macrophages, dentritic cells (DCs), B cells, and some T-cell subsets, thereby enabling activation of a subset of specific cells in response to stimulation of a particular TLR. For example, TLR7 is expressed primarily in B and plasmacytoid dendritic cells (pDCs), a subset of DCs that produces most cellular IFN. The cellular patterns of TLR expression also vary among species. For example, mice differ from primates in that they express TLR9 not only in pDCs and B cells but also in monocytes and myeloid DCs. In addition, human cells but not mouse cells express TLR8 [17]. Besides hematopoietic cells, mouse leukocyte progenitor cells and quiescent stem cells can respond to activation of TLR2 and TLR4 to augment cell proliferation and development of certain cell lineages. Therefore, the link between TLR signaling and specific immunity extends from secondary lymphoid tissues to sites of primary leucopoiesis [18].

During an inflammatory response, TLR activation results in local production of inflammatory cytokines and chemokines, which can activate and mobilize other leukocytes from the circulation. It also programs DC maturation and activation, leading to effective antigen presentation to naive T helper cells located in lymph nodes [9,19]. Activated monocytes or DCs can release IL-6 and other factors to mitigate suppression by T regulatory cells (Tregs) [20]. TLR signaling can act alone or synergize with B-cell receptor (BCR) stimulation and T-cell help to stimulate human naive B cells to proliferate and differentiate (e.g., antibody class switch and plasma cell development) [8,21]. As a result of such TLR-mediated stimulation, a polarized Th1-type adaptive immune response toward a coadministered antigen can be triggered, characterized by generation of antigen-specific IFN-γ-producing T cells and IgG2 antibodies. Therefore, activation of the innate immune system, and TLR in particular appears to be an important component of vaccine-induced immunity.

However, independent of TLR or TIR ligation, a Th1 type of response against antigens expressed by apoptotic cells can also be developed in mice defective in both *Myd88* and *Trif* genes. The response appeared to be triggered indirectly by IFN produced by precursor lymphoid DCs [22]. Hence, the precise role of TLR signaling in controlling specific adaptive immunity against different types of antigens has yet to be fully elucidated [23,24].

## 8.2    EFFECTS OF SMIPS ON CELLULAR FUNCTIONS

The discovery of interferon in 1957 [25] ignited the search for small molecules that could induce IFN in humans to combat viral infection and cancer. By the mid-1980s, a number of such compounds were reported, such as tilorone [26], BL-20803 [27], ataburine [28], CP-20961 [29], U25166 [30], DRB [31], 10-carboxymethyl-9-acridone [32], and bropirimine [33]. Most of these compounds were active only on murine cells, and those few tested in humans yielded only marginal beneficial effects [34,35]. In the ensuing 20 years, several relatively potent base or nucleoside analogs were described. Among the guanosine derivatives, 7-thio-8-oxoguanosine (TOG, Figure 8.2) was shown to stimulate mouse macrophages and NK (natural killer) and B cells, and to suppress replication of a broad spectrum of viruses in mouse models [36–39]. Two other active analogs included 7-deazaguanosine [40,41] and 7-allyl-8-oxoguanosine (loxoribine, [42–46]) (Figure 8.2).

In evaluating various adenine analogs as inhibitors of herpes simplex virus (HSV) DNA replication, Gerster et al. [47] noted that the suppressing effect exhibited by some analogs to a viral replication cycle inhibitor was attributed solely to its activity of inducing interferon in animal cells. Subsequent structure–activity relationship (SAR) studies identified an active inducer, imiquimod (R837, S-26308, 1$H$-imidazo[4,5-$c$]quinolines) [47], which has been developed into a licensed immunotherapeutic treatment of human genital warts, actinic keratosis, and superficial basal cell carcinoma. A more potent imidazoquanoline analog, resiquimod (R848, S-28463, 4-amino-2-ethoxymethyl-$\alpha$,$\alpha$-dimethyl-1$H$-imidazo[4,5-$c$]quinolin-1-ethanol), has also been reported (Figure 8.2 [48]). Other adenine derivatives with different structural scaffolds were also described [49–52], such as SM360320 (9-benzyl-2-methoxyethoxy-8-oxoadenine [51] and SM-276001 [2-butylamino-8-hydroxy-9-(6-methylpyridine-3-ylmethyl) adenine] [52]) (Figure 8.2). The latter was shown to be of potency similar to that of resiquimod in vitro at inducing IFN, but more potent and tolerable when delivered orally to animals [51].

Most of the above-mentioned nucleotide- or base-derived small molecule immunopotentiators (SMIPs) were shown to activate immune cells in ways other than inducing IFN [39,40,43,48,53–56]. For example, acting in a manner similar to CD40 ligand (CD154), resiquimod activates both NF$\kappa$B and MAP kinase pathways in B cells [57], thereby promoting production of antibodies and up-regulation of MHC class II surface receptors [58]. The responses of immune cells to most of these SMIPs are now known to be TLR7 and/or TLR8 dependent [59–63]. Their potency at stimulating cytokine production on immune cells correlates with their activity to enhance TLR- and NF$\kappa$B-dependent reporter gene expression in HEK293 cells (our unpublished data).

Activation of immune cells and their subsets by SMIPs can be profiled by the nature of their TLR dependency. For example, TLR7 SMIPs preferentially stimulate pDCs to produce IFN and its regulated cytokines, while TLR8 functions in monocytes and myeloid DCs to generate highly proinflammatory cytokines such as TNF$\alpha$. TLR8 SMIPs are thus helpful in driving cell-mediated

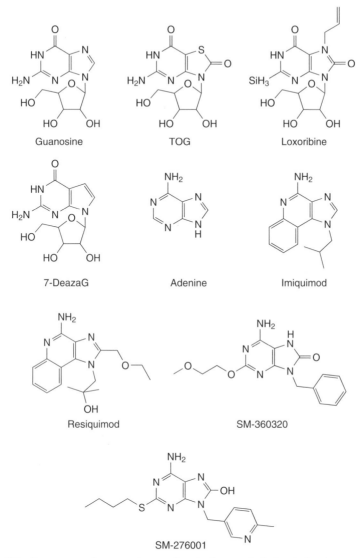

**Figure 8.2** Structure of guanosine and adenine and their derivatives as potent SMIPs.

immune responses, whereas TLR7 SMIPs may be used to promote antibody production [64]. TLR2 and TLR3 agonists appear to activate NK cells directly, whereas TLR7, TLR8, and TLR9 agonists can do so indirectly with the help of cytokines produced by other cells [65]. TLR-mediated B-cell stimulation seems to be regulated differently. TLR7 signaling requires IFN, which sensitizes B cells to highly express TLR7, leading to cell expansion and differentiation, as seen with TLR9 signaling. In contrast, though, even with IFN TLR2, TLR4, and TLR6, ligands are ineffective at activating human B cells. Of note, TLR7 or TLR9-stimulated B-cell expansion and differentiation occur independently of T-cell help and BCR stimulation [66]. Activation of Tregs with TLR8 ligands directly reverses the suppressive function of these cells, a finding that appears to be unique to TLR8 [67]. On neonatal blood cells, TLR8 signaling results in pro-inflammatory responses far in excess of those induced by TLR2, TLR4, or TLR7 agonists. Thus, TLR8-directed adjuvants may potentially be effective in newborns [68]. While TLR specificity in controlling innate immunity has been observed, plasticity in agonist specificity for TLR7 and TLR8 has been noted. For example, the TLR7/TLR8 agonist resiquimod and TLR7 agonist loxoribine, when copresented with thymidine homopolymer oligodeoxynucleotides (ODNs), became more TLR8-like. They induced more TNFα and less IFN on hPBMC, and consistently stimulated a TLR8- and NFκB-dependent reporter gene expression in HEK293 cells [69].

In addition to signaling TLR, imiquimod (but not resiquimod) antagonizes $A_1$ and $A_{2A}$ adenosine receptors and inhibits adenylyl cyclase in cell-based assays. This dual activity synergistically suppresses production of cAMP, which is up-regulated as a feedback mechanism in inflammation. The positive induction of pro-inflammatory cytokines by imiquimod together with its suppressive activity toward the production of the negative regulator cAMP may explain its high potency for inducing inflammation [70]. Like some $A_{2A}$ adenosine receptor–specific compounds, imiquimod can also stimulate transformed human keratinocytes to secrete cytokines [71,72]. Imiquimod can also directly induce apoptosis in vitro in human epithelial cells (HeLa S3) and keratinocytes (HaCaT and A431) and in mouse fibroblasts (McCoy). The induction seems not to be controlled by membrane-bound death receptors, but rather, is dependent on Bcl-2 expression as demonstrated by its overexpression in melanoma cells. Presumably, apoptosis is activated through Bcl-2-controlled release of mitochondrial cytochrome $c$ and subsequent activation of caspase-9 [72,73]. Because the imiquimod dose used clinically is sufficiently high to induce apoptosis in vitro, this proapoptotic activity may be important for eliminating virus-infected dysplastic or neoplastic epithelial cells in vivo. In support of this hypothesis, clinical imiquimod treatment modulates the expression of multiple apoptosis- and oncogenesis-associated genes [74].

The inflammation induced by both resiquimod and imiquimod seems partly controlled by a multiprotein complex termed the *inflammasome*, which includes cryopyrin, caspase-1, and apoptosis-associated specklike protein. This complex functions in promoting caspase-1 activation and processing of pro-IL-1β and is also known to be involved in the NLR signaling pathway. Cryopyrin deficiency causes reduced production of IL-1β and IL-18 but not IL-6 and TNFα in cells stimulated with these compounds [75].

In addition to these nucleotide or base-derived compounds, a number of structurally distinct small molecules, such as some antibiotics, have recently been described to have weak immunomodulatory function [76,77]. Of these compounds, amphotericin B [78] and the antifungal compound nystatin [79] were shown to stimulate TLR2/TLR1.

## 8.3   SMIPS AS VACCINE ADJUVANTS

Antigen-specific Th1-type immune responses are widely considered to be important for protection against intracellular microorganisms and tumors. However, current commercially available adjuvants such as alum and MF59 enhance a Th2- or Th0-type response. Several potent SMIPs, besides being investigated as immunotherapeutic treatments for cancer and viral infections [80–83], have also been evaluated as experimental Th1 adjuvants in animal models.

Initial studies with resiquimod demonstrated that it could enhance an immune response directed against chicken ovalbumin (OVA) immunized subcutaneously in mice when the compound was given together with the antigen or separately by oral delivery. The in vivo induction of proinflammatory cytokines, such as TNF$\alpha$, IL-12, and IFN-$\gamma$, was believed to lead to an increased production of IgG2a and a decrease of IgE antibodies. This small molecule, like CpG ODN, is able to modulate immunity in the presence of alum toward a Th1 type of response (IgG2a) and away from a Th2 response (IgE) [84,85]. Because of its ability to stimulate maturation of epidermal Langerhans cells, which in turn induce T-cell proliferation [86], imiquimod was also used as a topical adjuvant (5% cream formulation) applied to skin. Johnston and Bystryn [87] demonstrated that topically administered compound could enhance both anti-OVA antibody and cellular responses in mice. These enhanced responses were polarized toward a Th1 type, marked by an increase in IgG2a and IgG2b antibodies and CD8$^+$ T cells and a decrease in IgM and IgG1 response [87]. Topical administration of an OVA CTL-epitope peptide formulated with imiquimod elicited a peptide-specific CD8$^+$ T cell-response (cytolytic activity and IFN-$\gamma$ induction) in TCR-transgenic and wild-type mice [88]. Topical application of imiquimod was also shown to augment a murine anti-melanoma effect induced by an intravenous-injected live recombinant listeria vaccine, by enhancing vaccine immunity and bypassing peripheral tolerance [89]. Moreover, this approach caused a greater reduction in central nervous system tumor growth compared to a melanoma-associated antigen-pulsed DC vaccination strategy [90] in mice. In a guinea pig model, imiquimod combined with an HSV glycoprotein effectively resolved HSV viral recurrences. The increased duration and extent of viral protection by imiquimod appeared associated with its improved anti HSV-specific immune responses. A Th1 type of response was also seen in mice with a recombinant HIVgp140 env protein administered together with R848 or imiquimod (our unpublished data).

Imidazoquinoline compounds were also tested as DNA vaccine adjuvants. Resiquimod enhanced an HIV gag DNA vaccine-induced T-cell (cell proliferation, and IFN-$\gamma$ production) and Th1-biased antibody responses in mice [91]. In this study the DNA was formulated with bupivacaine, which is thought to increase DNA transfection efficiency and to bring the water-soluble vaccine and adjuvant together in vivo. Imiquimod increased the potency of three separately encoded HIV DNA vaccines for gag-p37, RT, and Nef, delivered topically to mice in a cream mixture. However, the increase was seen only when the DNA vaccines were delivered by a gene gun but not by intramuscular or intradermal injection, possibly due to a localized effect in the superficial layers of the skin. Compared to GM-CSF, a potent DNA vaccine adjuvant, topical application of imiquimod stimulated an equivalent or stronger Th1-polarized T-cell response but a weaker overall humoral response [92]. This SMIP could also be co-delivered with a DNA vaccine-encoding HER-2/neu, an oncoprotein, into mice using a gene gun. A boosted Th1 response was seen, leading to a four- to eightfold increase in the number of tumor-free mice compared with the animals receiving no adjuvant [93]. Furthermore, imiquimod could be administered by subcutaneous injection immediately after an OVA DNA vaccine delivered by a gene gun, resulting in enhanced antigen-specific CD4$^+$ and CD8$^+$ T-cell responses [94].

Several studies have compared the adjuvanticity of CpG oligodeoxy nucleotides (ODN) and resiquimod in animal models. Vasilakos et al. [84] showed that resiquimod and CpG1826 promoted a similar level of increased responses against the soluble antigen OVA in mice after subcutaneous injection. Weeratna et al. used a particulate antigen, hepatitis B surface antigen (HbsAg), to show that CpG7909, which is less potent than CpG1826 in mice, elicited an overall Th1 type of antibody response superior to that of resiquimod [95], seen after either intramuscular or subcutaneous injections. However, a small molecule (resiquimod-like) adjuvant, when chemically conjugated to an HIV-1 gag protein antigen, promoted improved Th1 and CD8$^+$ T-cell responses in mice, similar to levels seen with CpG1826 [96]. In rhesus macaques, a higher magnitude of Th1 T cells was observed with a similar SMIP–antigen conjugate than with a mixture of HIV-1 gag protein and CpG2395 [97]. Finally, quantitative and qualitative differences were observed between CpG- and SMIP-stimulated response after priming with gag protein plus adjuvant followed by boosting with a recombinant adenovirus-gag. The type of adjuvant used in the prime seemed to affect the stability, magnitude, and quality of Th1 and CD8$^+$ T-cell responses differently during the priming and boosting phases. CpG ODN elicited a high number of short-term effector cells in the primary phase, whereas the TLR7 and TLR8 agonists appeared to establish a larger memory pool boosted by the adenovirus vector [98]. Thus, so far, preclinical data in animal models supports the hypothesis that small molecules targeting the innate immune system have potential utility as vaccine adjuvants.

## 8.4  FUTURE OUTLOOK

In the past few years, the advancement in our understanding of the innate immune system, including the identification of agonists of innate immune receptors and the elucidation of their downstream signaling pathways, has provided a new set of targets for developing novel SMIPs. The success of imidazoquanolines as immunomodulators in humans and the effectiveness of TLR agonists (including imidazoquanolines) as vaccine adjuvants in animal models have provided the impetus for using this approach. The chance of success in identifying and developing novel SMIPs as adjuvants is difficult to quantify. However, we can benefit from the track record of success in modern small molecule drug discovery employing high-throughput screening technologies [99,100]. Hence, the field of novel adjuvant discovery is relatively nascent, but the prospects are excellent.

## ACKNOWLEDGMENTS

We thank Nelle Cronen for her excellent effort in preparing this manuscript.

## REFERENCES

1. Takeda, K. & Akira, S. Toll-like receptors in innate immunity. *Int Immunol* 2005 Jan, **17**(1), 1–14.

2. Janssens, S. & Beyaert, R. Role of Toll-like receptors in pathogen recognition. *Clin Microbiol Rev* 2003 Oct, **16**(4), 637–646.

3. Murray, P.J. NOD proteins: An intracellular pathogen-recognition system or signal transduction modifiers? *Curr Opin Immunol* 2005 Aug, **17**(4), 352–358.

4. Inohara, N., Chamaillard, M., McDonald, C. & Nunez, G. NOD-LRR proteins: role in host–microbial interactions and inflammatory disease. *Annu Rev Biochem* 2005, **74**, 355–383.

5. Yoneyama, M., Kikuchi, M., Matsumoto, K., Imaizumi, T., Miyagishi, M., Taira, K. et al. Shared and unique functions of the DExD/H-box helicases RIG-I, MDA5, and LGP2 in antiviral innate immunity. *J Immunol* 2005 Sep 1, **175**(5), 2851–2858.

6. Kato, H., Sato, S., Yoneyama, M., Yamamoto, M., Uematsu, S., Matsui, K. et al. Cell type-specific involvement of RIG-I in antiviral response. *Immunity* 2005 Jul, **23**(1), 19–28.

7. Kato, H., Takeuchi, O., Sato, S., Yoneyama, M., Yamamoto, M., Matsui, K. et al. Differential roles of MDA5 and RIG-I helicases in the recognition of RNA viruses. *Nature* 2006 May 4, **441**(7089), 101–105.

8. Pasare, C. & Medzhitov, R. Control of B-cell responses by Toll-like receptors. *Nature* 2005 Nov 17, **438**(7066), 364–368.

9. Hemmi, H. & Akira, S. TLR signalling and the function of dendritic cells. *Chem Immunol Allergy* 2005, **86**, 120–135.

10. Baetz, A., Frey, M., Heeg, K. & Dalpke, A.H. Suppressor of cytokine signaling (SOCS) proteins indirectly regulate Toll-like receptor signaling in innate immune cells. *J Biol Chem* 2004 Dec 24, **279**(52), 54708–54715.

11. Wald, D., Qin, J., Zhao, Z., Qian, Y., Naramura, M., Tian, L. et al. SIGIRR, a negative regulator of Toll-like receptor-interleukin 1 receptor signaling. *Nat Immunol* 2003 Sep, **4**(9), 920–927.

12. Fukao, T. & Koyasu, S. PI3K and negative regulation of TLR signaling. *Trends Immunol* 2003 Jul, **24**(7), 358–363.

13. Yoshida, H., Jono, H., Kai, H. & Li, J.D. The tumor suppressor cylindromatosis (CYLD) acts as a negative regulator for Toll-like receptor 2 signaling via negative

cross-talk with TRAF6 AND TRAF7. *J Biol Chem* 2005 Dec 9, **280**(49), 41111–41121.

14. Wang, Y., Tang, Y., Teng, L., Wu, Y., Zhao, X. & Pei, G. Association of beta-arrestin and TRAF6 negatively regulates Toll-like receptor-interleukin 1 receptor signaling. *Nat Immunol* 2006 Feb, **7**(2), 139–147.

15. Boone, D.L., Turer, E.E., Lee, E.G., Ahmad, R.C., Wheeler, M.T., Tsui, C. et al. The ubiquitin-modifying enzyme A20 is required for termination of Toll-like receptor responses. *Nat Immunol* 2004 Oct, **5**(10), 1052–1060.

16. Han, J. & Ulevitch, R.J. Limiting inflammatory responses during activation of innate immunity. *Nat Immunol* 2005 Dec, **6**(12), 1198–1205.

17. Rehli, M. Of mice and men: species variations of Toll-like receptor expression. *Trends Immunol* 2002 Aug, **23**(8), 375–378.

18. Nagai, Y., Garrett, K.P., Ohta, S., Bahrun, U., Kouro, T., Akira, S. et al. Toll-like receptors on hematopoietic progenitor cells stimulate innate immune system replenishment. *Immunity* 2006 Jun, **24**(6), 801–812.

19. Re, F. & Strominger, J.L. Heterogeneity of TLR-induced responses in dendritic cells: from innate to adaptive immunity. *Immunobiology* 2004, **209**(1–2), 191–198.

20. Pasare, C. & Medzhitov, R. Toll pathway-dependent blockade of CD4+CD25+ T cell-mediated suppression by dendritic cells. *Science* 2003 Feb 14, **299**(5609), 1033–1036.

21. Ruprecht, C.R. & Lanzavecchia, A. Toll-like receptor stimulation as a third signal required for activation of human naive B cells. *Eur J Immunol* 2006 Apr, **36**(4), 810–816.

22. Janssen, E., Tabeta, K., Barnes, M.J., Rutschmann, S., McBride, S., Bahjat, K.S. et al. Efficient T cell activation via a Toll-interleukin 1 receptor–independent pathway. *Immunity* 2006 Jun, **24**(6), 787–799.

23. Nemazee, D., Gavin, A., Hoebe, K. & Beutler, B. Immunology: Toll-like receptors and antibody responses. *Nature* 2006 May 18, **441**(7091), E4, discussion, E.

24. Pasare, C. & Medzhitov, R. Pasare and Medzhitov reply. *Nature* 2006 18 May, **441**, E4.

25. Isaacs, A. & Lindenmann, J. Virus interference: I. The interferon. *Proc R Soc Lond B Biol Sci* 1957 Sep 12, **147**(927), 258–267.

26. Krueger, R.E. & Mayer, G.D. Tilorone hydrochloride: an orally active antiviral agent. *Science* 1970 Sep 18, **169**(951), 1213–1214.

27. Siminoff, P., Bernard, A.M., Hursky, V.S. & Price, K.E. BL-20803, a new, low-molecular-weight interferon inducer. *Antimicrob Agents Chemother* 1973 Jun, **3**(6), 742–743.

28. Glaz, E.T., Szolgay, E., Stoger, I. & Talas, M. Antiviral activity and induction of interferon-like substance by quinacrine and acranil. *Antimicrob Agents Chemother* 1973 May, **3**(5), 537–541.

29. Hoffman, W.W., Korst, J.J., Niblack, J.F. & Cronin, T.H. *N,N*-Dioctadecyl-*N′,N′*-bis(2-hydroxyethyl)propanediamine: antiviral activity and interferon stimulation in mice. *Antimicrob Agents Chemother* 1973 Apr, **3**(4), 498–502.

30. Nichol, F.R., Weed, S.D. & Underwood, G.E. Stimulation of murine interferon by a substituted pyrimidine. *Antimicrob Agents Chemother* 1976 Mar, **9**(3), 433–439.

31. Tamm, I. & Sehgal, P.B. A comparative study of the effects of certain halogenated benzimidazole ribosides on RNA synthesis, cell proliferation, and interferon production. *J Exp Med* 1977 Feb 1, **145**(2), 344–356.

32. Taylor, J.L., Schoenherr, C.K. & Grossberg, S.E. High-yield interferon induction by 10-carboxymethyl-9-acridanone in mice and hamsters. *Antimicrob Agents Chemother* 1980 Jul, **18**(1), 20–26.

33. Wierenga, W., Skulnick, H.I., Stringfellow, D.A., Weed, S.D., Renis, H.E. & Eidson, E.E. 5-Substituted 2-amino-6-phenyl-4(3*H*)-pyrimidinones. Antiviral- and interferon-inducing agents. *J Med Chem* 1980 Mar, **23**(3), 237–239.

34. Wierenga, W. Antiviral and other bioactivities of pyrimidinones. *Pharmacol Ther* 1985, **30**(1), 67–89.

35. Adam, A., Ed. *Synthetic Adjuvants*, 4th ed., Wiley, New York, 1985.

36. Smee, D.F., Alaghamandan, H.A., Cottam, H.B., Sharma, B.S., Jolley, W.B. & Robins, R.K. Broad-spectrum in vivo antiviral activity of 7-thia-8-oxoguanosine, a novel immunopotentiating agent. *Antimicrob Agents Chemother* 1989 Sep, **33**(9), 1487–1492.

37. Parandoosh, Z., Ojo-Amaize, E., Robins, R.K., Jolley, W.B. & Rubalcava, B. Stimulation of phosphoinositide signaling pathway in murine B lymphocytes by a novel guanosine analog, 7-thia-8-oxoguanosine. *Biochem Biophys Res Commun* 1989 Sep 29, **163**(3), 1306–1311.

38. Jin, A., Mhaskar, S., Jolley, W.B., Robins, R.K. & Ojo-Amaize, E.A. A novel guanosine analog, 7-thia-8-oxoguanosine, enhances macrophage and lymphocyte antibody-dependent cell-mediated cytotoxicity. *Cell Immunol* 1990 Apr 1, **126**(2), 414–419.

39. Sharma, B.S., Mhaskar, S., Balazs, L. & Siaw, M. Immunomodulatory activity of a novel nucleoside, 7-thia-8-oxoguanosine: I. Activation of natural killer cells in mice. *Immunopharmacol Immunotoxicol* 1992, **14**(1–2), 1–19.

40. Smee, D.F., Alaghamandan, H.A., Gilbert, J., Burger, R.A., Jin, A., Sharma, B.S. et al. Immunoenhancing properties and antiviral activity of 7-deazaguanosine in mice. *Antimicrob Agents Chemother* 1991 Jan, **35**(1), 152–157.

41. Smee, D.F., Alaghamandan, H.A., Ramasamy, K. & Revankar, G.R. Broad-spectrum activity of 8-chloro-7-deazaguanosine against RNA virus infections in mice and rats. *Antiviral Res* 1995 Mar, **26**(2), 203–209.

42. Goodman, M.G., Gupta, S., Rosenthale, M.E., Capetola, R.J., Bell, S.C. & Weigle, W.O. Protein kinase C independent restoration of specific immune responsiveness in common variable immunodeficiency. *Clin Immunol Immunopathol* 1991 Apr, **59**(1), 26–36.

43. Goodman, M.G., Reitz, A.B., Chen, R., Bobardt, M.D., Goodman, J.H. & Pope, B.L. Selective modulation of elements of the immune system by low molecular weight nucleosides. *J Pharmacol Exp Ther* 1995 Sep, **274**(3), 1552–1557.

44. Gupta, S., Vayuvegula, B. & Gollapudi, S. Substituted guanine ribonucleosides as B cell activators. *Clin Immunol Immunopathol* 1991 Nov, **61**(2 Pt 2), S21–S27.

45. Pope, B.L., Chourmouzis, E., Sigindere, J., Capetola, R.J. & Lau, C.Y. In vivo enhancement of murine natural killer cell activity by 7-allyl-8-oxoguanosine (loxoribine). *Int J Immunopharmacol* 1992 Nov, **14**(8), 1375–1382.

46. Reitz, A.B., Goodman, M.G., Pope, B.L., Argentieri, D.C., Bell, S.C., Burr, L.E. et al. Small-molecule immunostimulants: synthesis and activity of 7,8-disubstituted guanosines and structurally related compounds. *J Med Chem* 1994 Oct 14, **37**(21), 3561–3578.

47. Gerster, J.F., Lindstrom, K.J., Miller, R.L., Tomai, M.A., Birmachu, W., Bomersine, S.N. et al. Synthesis and structure-activity-relationships of 1*H*-imidazo[4,5-c]quinolines that induce interferon production. *J Med Chem* 2005 May 19, **48**(10), 3481–3491.

48. Tomai, M.A., Gibson, S.J., Imbertson, L.M., Miller, R.L., Myhre, P.E., Reiter, M.J. et al. Immunomodulating and antiviral activities of the imidazoquinoline S-28463. *Antiviral Res* 1995 Nov, **28**(3), 253–264.

49. Hirota, R., Tajima, S., Yoneda, Y., Tamayama, T., Watanabe, M., Ueda, K. et al. Alopecia of IFN-gamma knockout mouse as a model for disturbance of the hair cycle: a unique arrest of the hair cycle at the anagen phase accompanied by mitosis. *J Interferon Cytokine Res* 2002 Sep, **22**(9), 935–945.

50. Kurimoto, A., Ogino, T., Ichii, S., Isobe, Y., Tobe, M., Ogita, H. et al. Synthesis and structure–activity relationships of 2-amino-8-hydroxyadenines as orally active interferon inducing agents. *Bioorg Med Chem* 2003 Dec 1, **11**(24), 5501–5508.

51. Jin, G., Wu, C.C., Tawatao, R.I., Chan, M., Carson, D.A. & Cottam, H.B. Synthesis and immunostimulatory activity of 8-substituted amino 9-benzyladenines as potent Toll-like receptor 7 agonists. *Bioorg Med Chem Lett* 2006 Sep 1, **16**(17), 4559–4563.

52. Isobe, Y., Kurimoto, A., Tobe, M., Hashimoto, K., Nakamura, T., Norimura, K. et al. Synthesis and biological evaluation of novel 9-substituted-8-hydroxyadenine derivatives as potent interferon inducers. *J Med Chem* 2006 Mar 23, **49**(6), 2088–2095.

53. Colic, M., Vucevic, D., Vasilijic, S., Popovic, L., Pejanovic, V., Jandric, D. et al. Proliferation of spleen cells in culture stimulated by 7-thia-8-oxoguanosine: evidence that both B- and T-cells are the targets of its action. *Methods Find Exp Clin Pharmacol* 1999 Nov, **21**(9), 583–590.

54. Pope, B.L., MacIntyre, J.P., Kimball, E., Lee, S., Zhou, L., Taylor, G.R. et al. The immunostimulatory compound 7-allyl-8-oxoguanosine (loxoribine) induces a distinct subset of murine cytokines. *Cell Immunol* 1995 May, **162**(2), 333–339.

55. Ahonen, C.L., Gibson, S.J., Smith, R.M., Pederson, L.K., Lindh, J.M., Tomai, M.A. et al. Dendritic cell maturation and subsequent enhanced T-cell stimulation induced with the novel synthetic immune response modifier R-848. *Cell Immunol* 1999 Oct 10, **197**(1), 62–72.

56. Wagner, T.L., Ahonen, C.L., Couture, A.M., Gibson, S.J., Miller, R.L., Smith, R.M. et al. Modulation of TH1 and TH2 cytokine production with the immune response modifiers, R-848 and imiquimod. *Cell Immunol* 1999 Jan 10, **191**(1), 10–19.

57. Bishop, G.A., Hsing, Y., Hostager, B.S., Jalukar, S.V., Ramirez, L.M. & Tomai, M.A. Molecular mechanisms of B lymphocyte activation by the immune response modifier R-848. *J Immunol* 2000 Nov 15, **165**(10), 5552–5557.

58. Bishop, G.A., Ramirez, L.M., Baccam, M., Busch, L.K., Pederson, L.K. & Tomai, M.A. The immune response modifier resiquimod mimics CD40-induced B cell activation. *Cell Immunol* 2001 Feb 25, **208**(1), 9–17.

59. Lee, J., Wu, C.C., Lee, K.J., Chuang, T.H., Katakura, K., Liu, Y.T. et al. Activation of anti–hepatitis C virus responses via Toll-like receptor 7. *Proc Natl Acad Sci USA* 2006 Feb 7, **103**(6), 1828–1833.

60. Akira, S. & Hemmi, H. Recognition of pathogen-associated molecular patterns by TLR family. *Immunol Lett* 2003 Jan 22, **85**(2), 85–95.

61. Edwards, A.D., Diebold, S.S., Slack, E.M., Tomizawa, H., Hemmi, H., Kaisho, T. et al. Toll-like receptor expression in murine DC subsets: lack of TLR7 expression by CD8 alpha+ DC correlates with unresponsiveness to imidazoquinolines. *Eur J Immunol* 2003 Apr, **33**(4), 827–833.

62. Hemmi, H., Kaisho, T., Takeuchi, O., Sato, S., Sanjo, H., Hoshino, K. et al. Small anti-viral compounds activate immune cells via the TLR7 MyD88-dependent signaling pathway. *Nat Immunol* 2002 Feb, **3**(2), 196–200.

63. Lee, J., Chuang, T.H., Redecke, V., She, L., Pitha, P.M., Carson, D.A. et al. Molecular basis for the immunostimulatory activity of guanine nucleoside analogs: activation of Toll-like receptor 7. *Proc Natl Acad Sci USA* 2003 May 27, **100**(11), 6646–6651.

64. Gorden, K.B., Gorski, K.S., Gibson, S.J., Kedl, R.M., Kieper, W.C., Qiu, X. et al. Synthetic TLR agonists reveal functional differences between human TLR7 and TLR8. *J Immunol* 2005 Feb 1, **174**(3), 1259–1268.

65. Gorski, K.S., Waller, E.L., Bjornton-Severson, J., Hanten, J.A., Riter, C.L., Kieper, W.C. et al. Distinct indirect pathways govern human NK-cell activation by TLR-7 and TLR-8 agonists. *Int Immunol* 2006 Jul, **18**(7), 1115–1126.

66. Bekeredjian-Ding, I.B., Wagner, M., Hornung, V., Giese, T., Schnurr, M., Endres, S. et al. Plasmacytoid dendritic cells control TLR7 sensitivity of naive B cells via type I IFN. *J Immunol* 2005 Apr 1, **174**(7), 4043–4050.

67. Peng, G., Guo, Z., Kiniwa, Y., Voo, K.S., Peng, W., Fu, T. et al. Toll-like receptor 8–mediated reversal of CD4+ regulatory T cell function. *Science* 2005 Aug 26, **309**(5739), 1380–1384.

68. Levy, O., Suter, E.E., Miller, R.L. & Wessels, M.R. Unique efficacy of Toll-like receptor 8 agonists in activating human neonatal antigen-presenting cells. *Blood* 2006 Aug 15, **108**(4), 1284–1290.

69. Jurk, M., Kritzler, A., Schulte, B., Tluk, S., Schetter, C., Krieg, A.M. et al. Modulating responsiveness of human TLR7 and 8 to small molecule ligands with T-rich phosphorothiate oligodeoxynucleotides. *Eur J Immunol* 2006 Jul, **36**(7), 1815–1826.

70. Schon, M.P., Schon, M. & Klotz, K.N. The small antitumoral immune response modifier imiquimod interacts with adenosine receptor signaling in a TLR7- and TLR8-independent fashion. *J Invest Dermatol* 2006 Jun, **126**(6), 1338–1347.

71. Kono, T., Kondo, S., Pastore, S., Shivji, G.M., Tomai, M.A., McKenzie, R.C. et al. Effects of a novel topical immunomodulator, imiquimod, on keratinocyte cytokine gene expression. *Lymphokine Cytokine Res* 1994 Apr, **13**(2), 71–76.

72. Schon, M.P. & Schon, M. Immune modulation and apoptosis induction: two sides of the antitumoral activity of imiquimod. *Apoptosis* 2004 May, **9**(3), 291–298.

73. Schon, M.P., Wienrich, B.G., Drewniok, C., Bong, A.B., Eberle, J., Geilen, C.C. et al. Death receptor-independent apoptosis in malignant melanoma induced by the small-molecule immune response modifier imiquimod. *J Invest Dermatol* 2004 May, **122**(5), 1266–1276.

74. Barnetson, R.S., Satchell, A., Zhuang, L., Slade, H.B. & Halliday, G.M. Imiquimod induced regression of clinically diagnosed superficial basal cell carcinoma is associated with early infiltration by CD4 T cells and dendritic cells. *Clin Exp Dermatol* 2004 Nov, **29**(6), 639–643.

75. Kanneganti, T.D., Ozoren, N., Body-Malapel, M., Amer, A., Park, J.H., Franchi, L. et al. Bacterial RNA and small antiviral compounds activate caspase-1 through cryopyrin/Nalp3. *Nature* 2006 Mar 9, **440**(7081), 233–236.

76. Hamilton-Miller, J.M. Immunopharmacology of antibiotics: direct and indirect immunomodulation of defence mechanisms. *J Chemother* 2001 Apr, **13**(2), 107–111.

77. Dalhoff, A. & Shalit, I. Immunomodulatory effects of quinolones. *Lancet Infect Dis* 2003 Jun, **3**(6), 359–371.

78. Razonable, R.R., Henault, M., Lee, L.N., Laethem, C., Johnston, P.A., Watson, H.L. et al. Secretion of proinflammatory cytokines and chemokines during amphotericin B exposure is mediated by coactivation of Toll-like receptors 1 and 2. *Antimicrob Agents Chemother* 2005 Apr, **49**(4), 1617–1621.

79. Razonable, R.R., Henault, M., Watson, H.L. & Paya, C.V. Nystatin induces secretion of interleukin (IL)-1beta, IL-8, and tumor necrosis factor alpha by a Toll-like receptor-dependent mechanism. *Antimicrob Agents Chemother* 2005 Aug, **49**(8), 3546–3549.

80. Horsmans, Y., Berg, T., Desager, J.P., Mueller, T., Schott, E., Fletcher, S.P. et al. Isatoribine, an agonist of TLR7, reduces plasma virus concentration in chronic hepatitis C infection. *Hepatology* 2005 Sep, **42**(3), 724–731.

81. Dockrell, D.H. & Kinghorn, G.R. Imiquimod and resiquimod as novel immunomodulators. *J Antimicrob Chemother* 2001 Dec, **48**(6), 751–755.

82. Spruance, S.L., Tyring, S.K., Smith, M.H. & Meng, T.C. Application of a topical immune response modifier, resiquimod gel, to modify the recurrence rate of recurrent genital herpes: a pilot study. *J Infect Dis* 2001 Jul 15, **184**(2), 196–200.

83. Harrison, C.J., Miller, R.L. & Bernstein, D.I. Reduction of recurrent HSV disease using imiquimod alone or combined with a glycoprotein vaccine. *Vaccine* 2001 Feb 8, **19**(13–14), 1820–1826.

84. Vasilakos, J.P., Smith, R.M., Gibson, S.J., Lindh, J.M., Pederson, L.K., Reiter, M.J. et al. Adjuvant activities of immune response modifier R-848: comparison with CpG ODN. *Cell Immunol* 2000 Aug 25, **204**(1), 64–74.

85. Tomai, M.A., Imbertson, L.M., Stanczak, T.L., Tygrett, L.T. & Waldschmidt, T.J. The immune response modifiers imiquimod and R-848 are potent activators of B lymphocytes. *Cell Immunol* 2000 Jul 10, **203**(1), 55–65.

86. Burns, R.P., Jr., Ferbel, B., Tomai, M., Miller, R. & Gaspari, A.A. The imidazoquinolines, imiquimod and R-848, induce functional, but not phenotypic, maturation of human epidermal Langerhans' cells. *Clin Immunol* 2000 Jan, **94**(1), 13–23.

87. Johnston, D. & Bystryn, J.C. Topical imiquimod is a potent adjuvant to a weakly-immunogenic protein prototype vaccine. *Vaccine* 2006 Mar 10, **24**(11), 1958–1965.

88. Rechtsteiner, G., Warger, T., Osterloh, P., Schild, H. & Radsak, M.P. Cutting edge: priming of CTL by transcutaneous peptide immunization with imiquimod. *J Immunol* 2005 Mar 1, **174**(5), 2476–2480.

89. Craft, N., Bruhn, K.W., Nguyen, B.D., Prins, R., Lin, J.W., Liau, L.M. et al. The TLR7 agonist imiquimod enhances the anti-melanoma effects of a recombinant *Listeria monocytogenes vaccine. J Immunol* 2005 Aug 1, **175**(3), 1983–1990.

90. Prins, R.M., Craft, N., Bruhn, K.W., Khan-Farooqi, H., Koya, R.C., Stripecke, R. et al. The TLR-7 agonist, imiquimod, enhances dendritic cell survival and promotes tumor antigen-specific T cell priming: relation to central nervous system antitumor immunity. *J Immunol* 2006 Jan 1, **176**(1), 157–164.

91. Otero, M., Calarota, S.A., Felber, B., Laddy, D., Pavlakis, G., Boyer, J.D. et al. Resiquimod is a modest adjuvant for HIV-1 gag-based genetic immunization in a mouse model. *Vaccine* 2004 Apr 16, **22**(13–14), 1782–1790.

92. Zuber, A.K., Brave, A., Engstrom, G., Zuber, B., Ljungberg, K., Fredriksson, M. et al. Topical delivery of imiquimod to a mouse model as a novel adjuvant for human immunodeficiency virus (HIV) DNA. *Vaccine* 2004 Apr 16, **22**(13–14), 1791–1798.

93. Smorlesi, A., Papalini, F., Orlando, F., Donnini, A., Re, F. & Provinciali, M. Imiquimod and S-27609 as adjuvants of DNA vaccination in a transgenic murine model of HER2/neu-positive mammary carcinoma. *Gene Ther* 2005 Sep, **12**(17), 1324–1332.

94. Thomsen, L.L., Topley, P., Daly, M.G., Brett, S.J. & Tite, J.P. Imiquimod and resiquimod in a mouse model: adjuvants for DNA vaccination by particle-mediated immunotherapeutic delivery. *Vaccine* 2004 Apr 16, **22**(13–14), 1799–1809.

95. Weeratna, R.D., Makinen, S.R., McCluskie, M.J. & Davis, H.L. TLR agonists as vaccine adjuvants: comparison of CpG ODN and resiquimod (R-848). *Vaccine* 2005 Nov 1, **23**(45), 5263–5270.

96. Wille-Reece, U., Flynn, B.J., Lore, K., Koup, R.A., Kedl, R.M., Mattapallil, J.J. et al. HIV Gag protein conjugated to a Toll-like receptor 7/8 agonist improves the magnitude and quality of Th1 and CD8+ T cell responses in nonhuman primates. *Proc Natl Acad Sci USA* 2005 Oct 18, **102**(42), 15190–15194.

97. Wille-Reece, U., Wu, C.Y., Flynn, B.J., Kedl, R.M. & Seder, R.A. Immunization with HIV-1 gag protein conjugated to a TLR7/8 agonist results in the generation of HIV-1 gag-specific Th1 and CD8+ T cell responses. *J Immunol* 2005 Jun 15, **174**(12), 7676–7683.

98. Wille-Reece, U., Flynn, B.J., Lore, K., Koup, R.A., Miles, A.P., Saul, A. et al. Toll-like receptor agonists influence the magnitude and quality of memory T cell responses after prime–boost immunization in nonhuman primates. *J Exp Med* 2006 May 15, **203**(5), 1249–1258.

99. Ulevitch, R.J. Therapeutics targeting the innate immune system. *Nat Rev Immunol* 2004 Jul, **4**(7), 512–520.

100. Pashine, A., Valiante, N.M. & Ulmer, J.B. Targeting the innate immune response with improved vaccine adjuvants. *Nat Med* 2005 Apr, **11**(4 Suppl), S63–S68.

# 9

# NEW ISCOMs MEET UNSETTLED VACCINE DEMANDS

BROR MOREIN, KEFEI HU, KARIN LÖVGREN, AND ERIK D'HONDT

## 9.1 INTRODUCTION

The early concept regarding vaccines was to mimic the natural pathogen by using the whole killed or live microorganism converted to be harmless. This concept became one of the most successful assets in human and animal medicines. Although no winning concept should be abandoned, the limitations of the concept have to be set to define targets for new developments, in this case concerning vaccines with complementary properties. Vaccines based on this two-century-old concept have often been efficient against pathogens that cause acute infections and even death, but in general, vaccines are lacking against pathogens that cause chronic and persistent infections. Such pathogens have the capacity to evade immune killing by deviating immune reactions from components in the organism that are essential for the infection process or by steering the immune reaction of the host to allow and even promote persistence of the pathogen. By defining mechanisms for immune evasion, there are prospects to design vaccines to counteract the strategy of the pathogen by identification of protective antigens and epitopes. Moreover, an immunological concept to circumvent evasion mechanisms includes prominent immunological exposure of protective antigens and epitopes in a vaccine formulated with a potent adjuvant with capacity to modulate for protective immune reactions.

*Vaccine Adjuvants and Delivery Systems*, Edited by Manmohan Singh
Copyright © 2007 John Wiley & Sons, Inc.

## 9.2   CONCEPT AND STRUCTURE OF THE IMMUNOSTIMULATING COMPLEX

### 9.2.1   Requirements of a Vaccine Delivery System

From natural infections and from experiences with vaccines we learned that antigens displayed on particles are potent immunogens. Indeed, live viral vaccines, but also nonreplicating vaccines based on antigens in particles such as recHepB vaccine, inactivated polio vaccine (IPV), or inactivated hepatitis A vaccine, are solid particulate immunogens that induce immune protection even at low dosage. In contrast, single antigen units isolated from pathogens or expressed by recombinant techniques are often weak antigens and require proper physical presentation and adjuvantation. Thus, we are back to Jenner's two-century-old concept [1] to mimic the shape of a pathogen (i.e., the particle form). In 1975, Almeida et al. [2] provided the early example of presenting virus subunit antigens in a particle by integrating the envelope antigens from influenza virus into membranes of liposomes, thereby coining the expression *virosome*.

Soon thereafter, the term *protein micelle* was introduced as a vaccine concept. The micellar particle was composed of envelope antigens from Semliki forest virus (SFV) held together by hydrophobic interactions [3]. Later, virus-like particles (VLPs) used the same concept of particulate multimeric presentation mainly to utilize the possibilities of gene technology [4]. The limitation of these particles is that they do not necessarily take into consideration targeting to antigen-presenting cells (APCs) and the immune modulation aspects (e.g., micelle formulation was as immunogenic as both the whole killed virus particle and the virosome formulated with the same antigen). However, to induce immune protection against parainfluenza 3 virus (PIV3), the causative agent of pneumonia, in the lungs of lambs, micelle formulated with the envelope proteins of PIV3 had to be adjuvanted with an oil-in-water adjuvant, reflecting the limitations of the micelle as well as other technologies based solely on the particle [5].

### 9.2.2   Challenges for Formulating the Ideal Vaccine Delivery System

Thus, a lesson learned is that a complete vaccine antigen delivery system should encompass a particulate form as well as devices targeting antigens to APCs and to cellular compartments, including the endosomal pathway leading to CD4 responses. The objective in selecting APCs and their cellular compartments is to utilize their biological properties to achieve strong immune modulatory properties, the latter being characterized by the cytokine profile. For immunity against certain infections, targeting to the cytosolic pathway is essential to achieving CD8 cytotoxic T-cell response. Finally, the mode of intended administration (i.e., mucosal or parenteral modes) requires different targeting devices.

A balance between the intrinsic immune modulatory properties of the vaccine antigens and those of the delivery system is desirable for optimal immune protection, including the capacity to induce broad immunity against, for example, several variants or subtypes of a virus or other pathogens while avoiding possible side effects. Intrinsic immune modulatory effects of vaccine antigens, possibly enhanced by those in an adjuvant system [6], may have additive protective effects but also provoke adverse reactions. An ideal vaccine delivery system has to balance the properties mentioned to combine efficacy and tolerability, and these aspects were considered for the construction of the new immunostimulating complex (ISCOM).

## 9.3   ISCOM TECHNOLOGY

Saponin, the unique component of the ISCOM, was introduced as an adjuvant for commercial animal vaccines in 1951 [7]. The adjuvant activity of saponin has long been recognized in the literature on adjuvants [8,9]. For many years saponin was used as a crude raw material from several plant species. Later, the superior adjuvant properties of saponin from the tree *Quillaja saponaria Molina* became recognized and of great importance for many animal vaccines. It is used especially in foot-and-mouth disease vaccines, but also in a number of other animal vaccines, including equine influenza vaccines. A more defined preparation of Quillaja saponin, Quil A, was introduced by Dalsgaard [10,11] and his work paved the way for further standardized and characterized products, such as QS-21 [12] and the Quil A fractions QHA and QHC [13], which today are important adjuvants for a number of registered animal vaccines and vaccine development projects. QS-21 is traditionally administered as aqueous solutions admixed with antigens in the vaccine and not as an integral part of a construct.

The Quillaja saponin or fractions thereof became a central component in efforts to create an efficient vaccine delivery system that combines multimeric presentation in a particle with an in-built adjuvant [11,14]. The resulting procedure is now referred to as the *ISCOM technology*, and the nanoparticle ISCOM was born. ISCOMS are spherical cagelike complexes about 40 nm in diameter (Figure 9.1).

The ISCOM is formed and held together by the strong affinity between Quillaja saponins and cholesterol, forming small rings of about 12 nm. These rings are further assembled into 40-nm complexes with the aid of phopholipids (e.g., phosphophatidylcholine). If hydrophobic or amphipathic antigens are present during particle formation, the antigens become physically incorporated into the complex [15]. During the past 20 years, ISCOM technology has been refined and developed significantly in order to fulfill the demands for a modern adjuvant technology, from the production point of views as well as regarding safety and efficacy.

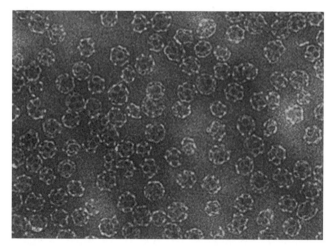

**Figure 9.1** ISCOM particles under electronic microscope.

The first procedure used for ISCOM formation was based on ultracentrifugation. Detergent-solubilized membrane proteins were put on top of a sucrose gradient. The membrane antigens, the necessary molecules of cholesterol and phospholipids, were centrifuged into a Quillaja saponin–containing zone in which the ISCOM particles were formed.

The technology was then expanded to include non-membrane-derived antigens such as recombinant antigens, and a dialysis procedure was found to be more convenient and also to promote new developments. The antigen(s) is (are) mixed with cholesterol and phosphatidylcholine solubilized in a detergent with a high critical micellar concentration, such as β-octylglucoside or MEGA-10; Quillaja saponins are added; and the mixture is dialyzed extensively against phosphate-suffered online (PBS). This procedure was easier to perform and could be brought into a production scale for a commercial vaccine against equine influenza virus produced by Iscotec AB (later named AdVet AB) in Sweden.

Using the dialysis procedure, it was shown that the typical 40-nm structure could be produced from the lipids and Quillaja saponins alone, that is, without antigens present during the formation. The resulting particles, identical in shape and appearance to the ISCOM, were called ISCOM-Matrix. The ISCOM-Matrix was also shown to be a powerful adjuvant admixed with antigens in solution. Hence, there are now three different formulations of Quillaja saponin for use as vaccine adjuvant. The main physical and biological differences of the free saponin extract and ISCOM/ISCOM-Matrix formulations are summarized in Table 9.1.

The ISCOM and ISCOM-Matrix technologies were proven to be potent adjuvant formulations. However, there were needs for further improvements regarding the Quillaja saponin components. Quillaja saponins are heteroge-

**TABLE 9.1   Comparison of Physical and Biological Characteristics of Saponin Extracts and ISCOM/ISCOM-Matrix Formulations**

| Quillaja Saponin Extract | ISCOM/ISCOM-Matrix |
|---|---|
| Mixture of saponin monomers and saponin micelles | Homogeneous preparation of 40-nm particles |
| Highly surface active preparation | Surface activity abolished by complex formation with cholesterol and phospholipid |
| A significant amount of saponin will bind to cells at the injection site, causing local side effects and inflammation | Virtually all saponin material administered will be rapidly taken up by cells |
| Purified peaks of saponin material or fractions can be used | Purified peaks of saponin material or fractions can be used in ISCOM/ISCOM-Matrix; different peaks or fractions can be formulated separately in ISCOM/ISCOM-Matrix for increased efficacy and tolerance |
| Limited chemical stability under storage and physiological conditions; hydrolysis yields deacylated forms of saponins with significantly decreased adjuvant activity | Chemical stability under storage and physiological conditions greatly enhanced $t_{1/2}$ at ambient temperature and neutral pH increased from week(s) to months and higher pH from hours to days ISCOM: stability limited by the stability of the antigen |
| Potent adjuvant | Increased adjuvant effects (magnitude, quality, and duration) ISCOM: CTL more efficiently induced by ISCOMs than with ISCOM-Matrix if the antigen(s) not spontaneously adhere to the matrix by electrostatic interactions, for example |
| Nonrefined or semirefined extracts may need costly purification to remove contaminants to become acceptable for vaccine use | During the production of ISCOM/ISCOM-Matrix only matrix-forming saponins will be retained; all low-molecular-weight (<30 kD) substances will be removed during the process |

neous mixtures of chemically related molecules but with diverse immunological and reactogenic properties.

Two saponin fractions purified by high-performance liquid chromatography, fraction A (QHA) and fraction C (QHC), were selected for further use with ISCOM technology. Both fractions readily formed ISCOM/ISCOM-Matrix particles but exerted different immunological activities [16]. Fraction A was shown to have a very low intrinsic toxicity, with its major immunological activities promoting cellular immune responses. Fraction C is considerably more reactogenic but has strong adjuvant activity, inducing potent antibody responses. A combination of the two fractions denoted ISCOPREP 703, consisting of 70% fraction A and 30% fraction C, and was anticipated to be considerably less reactogenic than the parent saponin extract. The 703 mixture of saponins was chemically and functionally well defined, but the strong reactogenic properties were disappointing. For veterinary vaccine applications the 703 mixture became too expensive with no obvious advantage over the more crude saponin preparations. For human applications, an approach for therapeutic vaccines became based on fraction C alone, and the general perception of ISCOM technology for prophylactic human use was that the reactogenicity of ISCOM formulation should be avoided. Consequently, major efforts were expended to substantially reduce the reactogenicity and to fine-tune the immune modulation in efforts to minimize the dose of fraction C. The resulting improved technology is based on the principle of keeping the various Quillaja saponin fractions in separate 40-nm complexes (PCT/SE03/01180; WO 2004/004762A1; SE2004/001038; WO 2005/002620 A1).

A base level of immunity is supplied by fraction A particles while minute amounts of fraction-C particles suffice to achieve optimal immune responses. In a sensitive BALB/c model, reactogenicity was reduced 8 to 10 times, corresponding to the reduced dose of fraction C. The adjuvant activity was retained or even enhanced; that is, there is no need to increase the total saponin dose, including the major component fraction A, to give optimum immune stimulation compared with the parent composition.

### 9.3.1 Adjuvant Formulations for New Demands

The importance of a favorable antigen formulation in terms of immunogenicity is well accepted to attaining better vaccines. However, formulation of the antigens with an immunomodulator is also essential for nonreplicating vaccines. The urgent need for efficient adjuvant to enhance the efficacy of nonreplicating vaccines is best exemplified by an anticipated threat of an influenza virus pandemic. The immunogenicity of the present influenza virus vaccines is comparatively low, and they are often nonadjuvated because the present registered adjuvants have few additive effects. The capacity in the world for production of flu vaccines is limited, and in the event of the outbreak of a new pandemic with a virus as pathogenic as the recent H5N1 and H7N7 fowl viruses, the vaccine production would suffice for only a fraction of the need. An efficient adjuvant would reduce the dose of antigens fivefold or more, as would be expected with an ISCOM-adjuvanted experimental vaccine [17]. Luckily, so far, these viruses do not spread from person to person.

Using the ISCOM technology to formulate Quillaja saponins into particles has enhanced the biological properties: mainly the adjuvant properties of the saponin but also the chemical stability (see Table 9.1). Playing on that theme, ISCOM formulations with different biological properties have been formulated. For example, ISCOMs with a totally different mucosal immunomodulator, CTA1DD, increased adjuvant properties significantly over a CTA1DD molecule not formulated in an ISCOM [18]. The CTA1DD-ISCOM is particularly efficient for oral application [19]. In similar but yet different ways, the ISCOM technology is combined successfully with many different immunomodulating substances, such as monophosphocyte lipid A (MPL), cholera toxin (CT) recombinant B subunit of cholera toxin rCTB [20], and Al(OH)$_3$. Also, mucosal targeting devices using envelope proteins from respiratory viruses from influenza and respiratory syncytial viruses have been used successfully [21], the overall effect being that doses could be lowered significantly, resulting in increased immune responses and enhanced immunomodulation. In addition, immune responses became balanced in terms of Th1/Th2 features, as discussed below.

### 9.3.2   Classical and new ISCOM Formulations

The ISCOM concept and technology have recently been strengthened by a number of innovations, resulting in improved efficacy and the possibility to make variants that take into consideration the intrinsic immunological features of the vaccine antigen. Aspects on the feasibility to prepare vaccines rationally have also been taken into consideration. The new ISCOM formulations are well tolerated for sensitive individuals, including old and young age groups.

Today, three fundamentally different ISCOM formulations are available:

- ISCOM. The immunostimulating complex, the original formulation, uses a technology that assembles one or more saponin components in the same particle into which the antigen(s) are incorporated.
- ISCOM-Matrix. CSL's trade name for the ISCOM-Matrix is ISCOMATRIX, which contains one or more saponin components in the same matrix structure. Antigen(s) are not incorporated into the particle but are added to the final vaccine formulation. For optimal cytotoxic T-lymphocyte (CTL) induction, antigens are engineered to adsorb to the ISCOM-Matrix. Such formulations, not being ISCOMs with physically incorporated antigens, have been referred to as nonclassical ISCOM formulations [22].
- ISCOM-Matrix M formulations. ISCOM-Matrix M consists of mixtures of two types of particles, ISCOM-Matrix A contains the defined saponin fraction A, and ISCOM-Matrix C contains the defined saponin fraction C. The two ISCOM-Matrix particles are mixed according to a preset proportion prior to adding up with the vaccine antigen(s). The trade name of this formulation for laboratory animal research is ABISCO-100 (Isconova AB).

Targeting devices can be imposed on ISCOM and ISCOM-Matrix. Both structures are versatile in the sense that most components can be replaced (i.e., one or more immune modulators can be used and replace each other). Targeting devices can be changed and selected for a particular purpose. Surface proteins of respiratory viruses can, for example, be used for targeting the local immune system through the mucus of the respiratory tract [21]. The ISCOM or ISCOM-Matrix can incorporate cholera toxin or its B or A subunits for targeting purposes or for enhancing the immunogenicity to these antigens [20]. In the case of the A subunit it was detoxified by a single amino acid mutation [18]. The cholera A subunit formulation showed extraordinary properties to enhance immune responses by the oral route, as described below. Incorpora-tion of vaccine antigens either as complex proteins or as oligopeptides can be done by hydrophobic interactions, electrostatic forces, or chemical conjuga-tion, which virtually facilitates the inclusion of any type of antigen [15,23].

### 9.3.3   Immunological Properties of ISCOM and ISCOM-Matrix Formulations

ISCOMs given either locally or systemically induce a Th1-like cytokine profile, but Th2-type cytokines are also produced; that is, there is a balanced immune response. This was demonstrated not only by the acquired immune reaction but also by the innate response. This cytokine profile is likely to be predeter-mined at early stages when ISCOMs meet the innate cells.

Smith et al. [24] studied the innate response to ovalbumin (OVA) ISCOMs after intraperitoneal injection of ISCOMs in mice and observed intense local inflammation with early recruitment of neutrophils and mast cells followed by macrophages, dendritic cells (DCs), and lymphocytes. An important feature is that the ISCOM formulations target DCs efficiently after both parenteral and mucosal modes of administration [25]. A number of inflammatory mediators, including nitric oxide, reactive oxygen intermediates, TNFα, IL-1, IL-6, IL-8, GM-CSF, IL-12, and IL-18 as well as IFN-γ, are produced. Of the factors, only IL-12 appeared to be essential for the immunogenicity of ISCOMs, whereas IL-6 and inducible nitric oxide synthethase knockout (KO) mice developed normal immune responses to OVA in ISCOMs. These responses were reduced markedly in IL-12 KO mice. Thus, ISCOMs prime antigen-specific immune responses at least in part by activating IL-12-dependent aspects of the innate immune system. This is in sharp contrast to that of the well-known mucosa adjuvant cholera toxin (CT), which is independent of IL-12.

As expected from the innate profile, the acquired immunity was Th1 biased, as demonstrated by Hu et al. [26] using respiratory syncytial virus (RSV) ISCOM containing various fractions of Quillaja saponin. This Th1 profiled immunity is, however, counterbalanced with a simultaneously developed Th2. In most cases it is desirable with a balanced response against both extra- and intracellular pathogens, including viruses and parasites. An extremely biased immune response, whether Th1 or Th2, is usually harmful, as experienced with RSV and many other pathogens. The acquired immune response includes IFN-γ and IL-2, but also IL-4- and CD8-restricted CTLs [27,28]. Both the ISCOM and the ISCOM-Matrix formulations modulate similar types of immune responses: a balanced response but with a stronger focus on Th1

[25,28–30]. The ISCOM formulations target DCs efficiently after both paren-teral and mucosal modes of administration [25].

The importance of strong immune modulatory effects by vaccines, including enhancement of co-stimulatory factors in the innate system, Th1 driving, and prominent stimulation of CD8-restricted CTL and the general stimulation of potent antibody responses, is shown with ISCOM formulations below in various animal species, including low-responding animals [31–34].

### 9.3.4   ISCOM Designs for Mucosal Administration

In view of the fact that most pathogens gain access via mucosal surfaces, ISCOMs were also considered for mucosal administration in various studies. Immune protection against infectious agents attacking via mucosal surfaces is obviously dependent on local mucosal immunity, including mucosal IgA anti-bodies. Classically live vaccines stood for the modus operandi to achieve mucosal immunity by local administration at the mucosal site of infection. Pioneering work using CT as an adjuvant for nonreplicating vaccine antigens showed that strong mucosal immune responses were induced after mucosal administration, including both oral and nasal routes [34,35].

Lövgren et al. [36,37] were the first to use ISCOMs for mucosal application, demonstrating that mice became fully protected against challenge infection after intranasal administration of an influenza ISCOM. Early studies by Jones et al. [38] also revealed that influenza virus ISCOMs after intranasal immuni-zation induced mucosal IgA and a potent CTL response. Later, Claassen et al. [39] demonstrated that oral administration of rabies virus ISCOMs targeted Peyer's patches more efficiently than did killed rabies virus particles, where enterocytes are considered to act as APCs. Intestinal epithelial cells may also be involved in gut-associated mucosal immunity. An in vitro study carried out on a human colon epithelial cell line (Caco-2 cells) by Lazorova et al. [40] showed that ISCOMs containing influenza virus antigen were taken up via the apical surface and the antigens were processed and transported through the basal membrane. These processed peptides were immunogenic and appropri-ate for immunological presentation.

Due to the obvious advantages, including practical ones, Mowat et al. [25,41–43] explored the oral route for administration. His group has used the model antigen–OVA. Like CT and *Escherichia coli* heat-labile enterotoxin (LT), ISCOMs were shown to prevent induction of oral tolerance and to exert adjuvant activity in the digestive tract. Low but repeated oral doses of OVA–ISCOMs induced a wide range of immune responses, including serum antibod-ies and secretion of mucosal IgA, Th1, and Th2 CD4 T-cell responses and MHC class I restricted CTL activity. Oral administration of ISCOMs, as the case is with injection, induces activation of the innate immune system, including neu-trophils, mast cells, macrophages, DCs, and lymphocytes. These cells seem to be activated, as they express high levels of MHC class II, TNFα, IL-1, IL-6, IL-12, and reactive oxygen intermediates and nitric oxide [26–28]. Smith et al.

[24] found in mice that after oral administration, ISCOMs induced an IL-12-dependent cascade of innate immune responses, including IL-2. This IL-12-dependent innate response to orally administered OVA ISCOMs is in agreement with the notion that ISCOMs enhance a Th1 response. In contrast, CT is independent of IL-12 for the induction of mucosal immune responses. Instead, the groups of Mowat and Lycke [25,44] showed that cholera toxin requires production of the cytokines IL-4 for induction of mucosal immunity, which is not a prerequisite for the adjuvant effect of ISCOMs, but ISCOMs depend on IL-12.

Hu et al. [27] showed in a mouse model that the surface proteins of respiratory viruses (RSV), including envelope proteins from RSV [21] have targeting effects when incorporated into ISCOMs. Such ISCOMs were useful for the mucosal mode of administration, inducing potent IgA response and antigen-driven cytokine responses [27]. Synergism between CTB and ISCOMs was seen in the remote genital tract as an adjuvant targeting effect enhancing IgA after intranasal administration. In particular, the membrane proteins of RSV served as efficient targeting device for the passenger antigen gp120 by intranasal application [21]. The IgA responses in the respiratory tract lasted over a long period of time (i.e., 22 weeks after the second immunization with an ISCOM containing human RSV envelope proteins). IgA responses were also detected at the distant genital and intestinal tracts. Moreover, virus-neutralizing antibody was detected at various mucosal sites, including the target organs, upper respiratory tract, and lungs, as well as in serum. These results imply that protective epitopes of RSV were preserved and recognized after intranasal administration [21,45].

ISCOMs have also been used in immunization experiments with bacterial antigens. ISCOMs delivering the membrane antigens from *Mycoplasma mycoides* subsp. *mycoides* (MmmSC), causing a severe chronic disease in the lungs of the bovine species, enhanced in mice by the intranasal route mucosal IgA and IgG and serum antibodies, including subclasses IgG1, IgG2a, and IgG2b [46,47].

Carol et al. [48] demonstrated in dogs that ISCOMs, containing *Echinococcus granulosus* membrane antigens, were effective for nasal administration of parasitic antigens. The intranasal route of immunization induced higher serum IgA but comparatively lower IgG levels than the subcutaneous route, and serum antibodies were of slightly higher avidity. Interestingly, the mucosal route of immunization induced more efficiently antibodies of both IgA and IgG isotypes than did systemic injection directed to carbohydrate epitopes considered as putative protective antigens.

There is a lack of efficient adjuvants for oral administration to enhance local immune response. CT is the exception. However, its safety is strongly questioned. A strong candidate for an efficient generic mucosal adjuvant is the CTA1-DD/ISCOM concept [18]. This construct is rationally designed of three main components, each contributing complementary adjuvant properties. CTA1 is the enzymatically active subunit of cholera toxin that is converted nontoxicly by separation from the A2 and B subunits. It is fused to DD that is a 2× repeat of a synthetic analog from protein A from *Staphylococcus aureus*, which targets B cells. Finally, the construct is formulated into an ISCOM or ISCOM-Matrix construct. This ISCOM vector has strong adjuvant properties by parenteral, nasal, and oral routes of administration, inducing potent cellular and humoral immune responses [19].

Apart from mucosal modes of administration, parenteral administration of ISCOMs has also been tested for the capacity to induce mucosal immune responses. Using an ISCOM model containing sheep erythrocyte membrane proteins in mice, Thapar et al. [49] found that administration to the pelvic presacral space twice resulted in significantly higher antierythrocyte IgA ELISA titers in the vaginal fluid than two administrations of intraperitoneal, subcutaneous, intravaginal, or mixed intraperitoneal–intravaginal immunizations. The notion that presacral space immunization was more effective for IgA induction than direct application to vaginal mucosa confirms that the reproductive organ is an inefficient site for the induction of mucosal immune response.

The results of Thapar, but with a more practical form of vaccine administration, were confirmed by studies in mice by Hu et al. [21,45] and in calves by Hägglund et al. [50], as discussed in detail below. Briefly, these studies show that parenteral immunization with a potent vaccine antigen adjuvant formulation induces mucosal immunity and immune protection exerted primarily by IgG, possibly produced locally.

## 9.4   EXAMPLES ILLUSTRATING THAT ISCOMs INDUCE IMMUNE PROTECTION AGAINST DIFFERENT PATHOGENS IN DIFFERENT SPECIES

Immune protection for experimental or natural infections is a complex biological event. Although a great deal is known about the basic immunology, including the requirements of antibody, Th1, Th2, or CD8 types of responses and their added effects, eventually, the protection against infection has to be tested in a live and, if possible, natural host. In a number of vaccination experiments using ISCOM formulations, protective immunity has been achieved even when other vaccine formulations failed (see Table 9.2). The reason for the potent efficacy observed in so many different vaccine trials and so many species is a result of the balanced recruitment of the various arms of the immune system in combination with the strong immune modulatory effect of ISCOM and ISCOM-Matrix formulations. This balanced response to the antigens included in these formulations also explains the protective effect against experimental infections in primates against HIV-1 (challenge with SHIV), HIV-2, and simian immune deficiency virus (SIV) [6,28,51,52]. The data indicate that the ISCOM formulation can also counterbalance an undesired intrinsic antigen-specific immune modulation, which is a necessity for induction of immune protection against pathogens causing persistent infections (e.g., herpes viruses devote more than 50% of their genes to manipulating the immune system of the host).

**TABLE 9.2   Protective Immunity Induced by Various ISCOMs**

| Antigen | Animal | Disease |
|---|---|---|
| Hemagglutinin, neuramimdase, influenza virus after intranasal–subcutaneous immunization | Mouse | Pneumonia |
| Hemagglutinin measles virus | Mouse | Encephalitis |
| Fusion protein, measles virus | Mouse | Encephalitis |
| Hemagglutinin and fusion protein, phoid distemper virus | Seal | Lethal infection |
| Hemagglutinin and fusion protein, canine distemper virus | Dog | Pneumonia |
| G protein, rabies virus | Mouse | Lethal infection, postexposure immunization |
| Gp120, simian immunodeficiency virus | Monkey | Lethal infection |
| Gp125, HIV-2; Gp120, HIV-1 | Monkey | Viremia |
| Envelope proteins, EHV-2 | Foal | Pneumonia, virus, and secondary bacterial infection |
| Envelope proteins, BRSV | Calf | Pneumonia |
| Envelope proteins, BHV-1 | Cattle | Pneumonia, virus, and secondary bacterial infection |
| Envelope proteins, bovine diarrhea virus | Sheep | Abortion |
| Gp120 and P24 of HIV-1 | Monkey | Viremia |
| Gp70, feline leukemia virus | Cat | Viremia |
| Gp360, Epstein–Barr virus | Tamarin monkey | Lethal tumor |
| Surface antigens, *Toxoplasma gondi* | Mouse | Lethal infection |
| Immunoaffinity purified protein from *Trypanosoma cruzi* | Mouse | Lethal infection |

ISCOMs and ISCOM-Matrix formulations have induced immune protection against a great number of pathogens in a number of species, including viruses from different families and genus. The virus repertoire includes picorna viruses, retrovirus, influenza virus, respiratory syncytial viruses, canine distemper virus, rabies virus, and herpes simplex viruses 1 and 2 respiratory syncytial virus. Antigens from bacterial species and parasites also evoke protective responses when included in ISCOM or formulated with ISCOM-Matrix. The details are provided in Table 9.2 and some examples for protection are discussed below.

### 9.4.1   Horse

The first commercial ISCOM vaccine was against influenza virus causing pneumonia in horses, similar to the disease in humans. This vaccine contains the envelope proteins hemagglutinin (HA) and neuraminidase (NA) and induces a long-lasting immune response (i.e., >15 months), including a cytotoxic T-cell response. The immune response, judged as protective, is broad against variants in the same subserotype.

ISCOMs containing envelope proteins from EHV2, which induced virtually full protection in young foals with maternal antibodies to natural infection, including subsequent clinical disease by a secondary bacterial infection, cause pneumonia with abscesses in the lungs, revealed by nonimmunized controls [53].

### 9.4.2 Ruminants

Bovine herpes virus 1 (BHV-1), which causes bovine infectious rhinotracheitis (IBR), is a common and economically important pathogen of cattle infecting the respiratory and genital tracts. Subsequent bacterial infections often exacerbate the disease. In a Canadian vaccination experiment in cattle [54], ISCOMs containing the BHV-1 envelope proteins administered intramuscularly induced protection to disease and reduced by virus excretion 1000-fold after challenge with virus infection of the respiratory tract. In a similar experiment carried out in Hungary [55], all animals were challenge-infected, two weeks after the second vaccination, by an intranasal application of a virulent BHV-1 strain, predisposing for secondary bacterial infection. A secondary bacterial infection was established four days later by intranasal installation of a virulent *Pasteurella multocida* strain to mimic hard field exposure. All the animals vaccinated with the ISCOMs were fully protected, no virus could be recovered from their nasal secretions, and no clinical symptoms were recorded. In contrast, animals vaccinated with the commercial vaccine responded to challenge with moderate fever and loss of appetite and the virus was isolated from the nasal secretions. The animals in the control group developed severe symptoms. Virus-neutralizing titers of 1/3500 or more were recorded in sera from ISCOM-vaccinated animals.

Bovine virus diarrhea virus (BVDV) is a major concern in cattle and sheep, causing abortion, respiratory distress, and enteric disease. In a sheep model, full protection was achieved with a BVDV ISCOM vaccine against challenge infection causing abortion ([56]; for a review, see [28]).

Two intramuscular injections of ISCOMs containing RSV envelope proteins protected young calves with maternal antibodies against challenge infection in the respiratory tract from clinical disease, and no virus could be isolated from nasal swabs [50,57,58]. For more details, see below.

### 9.4.3 Dog

Dogs are generally given the first vaccine dose at eight weeks of age (i.e., the time of delivery). Earlier vaccination with commercial vaccines fails to induce protective levels of serum antibodies due to maternal antibodies and an immature immune system. Puppies vaccinated with killed parvovirus adjuvanted with ISCOM-Matrix for the first time at three weeks, and given a second dose at six weeks of age had protective antibody levels at eight weeks, while the commercial vaccine failed (not published).

Rabies virus vaccination is obligatory in many countries, and today, live vaccines are not allowed in most countries. Killed rabies virus vaccines have not been as effective as live virus, calling for improvement. After one low dose of $0.7\,\mu g$ of glycoprotein antigens incorporated into ISCOMs, dogs resisted lethal challenge with street virus, and higher virus-neutralizing antibodies were induced than with the live commercial vaccine that was allowed at that time. Similar ISCOM vaccine was shown to be therapeutic in mice when administered after experimental infection, whereas the commercial vaccine failed [59].

### 9.4.4   Cat

Feline leukemia virus (FeLV) causes persistent infection in cat. An experimental ISCOM vaccine containing the envelope protein gp70 and its trans-membrane protein p15E induced protection against experimental infection with FeLV [60]. In a subsequent test in 137 households cats, in contrast to a commercial vaccine, almost all cats responded in three immunodiagnostic tests, including virus neutralization, membrane immunofluorescence, and ELISA.

### 9.4.5   Seal

Young seals vaccinated with canine distemper virus (CDV) envelope proteins incorporated into ISCOMs were protected against natural infection with CDV. This vaccine was used to protect seals during outbreaks of phocide distemper infection ([61]; Osterhaus, personal communications).

### 9.4.6   Poultry

An experimental ISCOM vaccine given as a single low dose $(1\,\mu g)$ of antigen containing membrane proteins including p64 and p56 from *Mycoplasma gallisepticum* protected chickens against experimental infection, as measured by a significantly reduced lesion score (for details, see [62]). The H5N1 subtype of influenza virus occurred in poultry flocks in the Hongkong area and caused deaths in human beings. Later, this virus spread over Asia and to Kazakhstan. This virus has not yet spread from human to human. In light of the threat to human beings, vaccination experiments were carried out in roosters. Experimental ISCOM vaccine containing envelope antigens from the subtype virus induced full protection, whereas a nonadjuvanted classical vaccine failed [63]. It should be noted that very high doses of a conventional nonadjuvanted vaccine were required.

### 9.4.7 Primates

Vaccines tested in primates are mostly experimental vaccines intended for human use. A number of experimental vaccines have been tested in primates, including HIV-1, HIV-2, and SIV. The SIV experiment demonstrated protection against challenge infection with both cell-free and cell-associated virus [52,64]. Of eight vaccine experiments in primates against lentiviruses encompassing simian and human lentiviruses, the protection rate varied between 40 and 100%. Long-standing protection was achieved over the entire experimental period [52,64–67]. The most demanding task is to protect against escape mutants. That problem was addressed with an HIV ISCOM containing antigens from a SF-2 HIV-1 isolate, and protection was obtained against the SF-13 escape. The vaccination strategy was based on priming with recombinant prime with gp120 antigen in ISCOMs and a boost with peptides carried by ISCOMs. This study shows that a prime–boost strategy can refocus immune response to nonimmunodominant epitopes [52,63,67]. Other studies in macaques suggested that the protection induced by ISCOMs against experimental infection resulted from virus-neutralizing antibodies and a balanced Th1 (IFN-$\gamma$,IL-2)–Th2 (IL-4) immune response [6,65]. Other parameters that might correlate to protection are CTL response and production of the chemokines MIP-1$\alpha$, MIP-1h, and RANTES [65]. The latter three chemokines block the second HIV-1 receptors, promoting fusion between the virus and the cell membrane essential for virus entry. Verschoor et al. [66] also showed that both Th1 and Th2, as well as antibody responses, correlated to protection against experimental infection, including virus clearance. These results have a bearing on similar infections in animals with lentiviruses for which vaccines are still lacking.

## 9.5  REGISTERED VACCINES BASED ON ISCOM TECHNOLOGY

To date, ISCOM-formulated vaccines are available commercially. In the 1990s the first ISCOM vaccine, formulated with the envelope proteins from influenza virus, including two subserotypes (H3N8 and H7N7), became approved and widely used in horses. This ISCOM influenza vaccine has a broader capacity to protect against variants within the same subserotype than that of other horse influenza vaccines [68–70]. Another vaccine based on lutenizing hormone releasing hormone (LHRH) to prevent ostrous behavior in female horses is registered and used mainly in racehorses, animals that are carefully observed for side effects that might impair their performance.

Today, a major company is registering a series of vaccines for horses, including a multicomponent vaccine adjuvanted with ISCOM-Matrix prepared with a defined saponin fraction tailored to the species. As with the ISCOM vaccine against equine influenza, some of these other vaccines are also administered at regular intervals. Repeated vaccinations over many years have not led to any changes in the adverse event profile. A vaccine is also registered for cattle against bovine virus diarrhea virus (BVDV) using ISCOM-Matrix as adjuvant.

## 9.6 LESSONS LEARNED FROM ISCOM VACCINES IN ANIMAL EXPERIMENTS

Most results under this heading are from vaccines or experimental vaccines used to protect against pathogens in their natural host that have developed a complex relationship over a long evolutionary period, as in a pathogen–human relationship.

### 9.6.1 Induction of Immune Response and Immune Protection in Early Life and Complications with Secondary Infections

Vaccination in early life is of great interest. However, there are two obstacles to overcome for successful immunization in young infants: first, an immature immune system, and second, interference from passively derived maternal antibodies. ISCOMs have been proven to induce immune response in neonates of several animal species [28] even in the presence of maternal antibodies. Viral infections in newborns, but also at other ages, are often followed by secondary infections (frequently, bacterial) that aggravate the disease.

Cattle are natural hosts for bovine respiratory syncytial virus (BRSV), which is closely related to human RSV (HRSV) and is probably the best animal model for human RSV. It causes high morbidity (i.e., pneumonia) and also mortality in young calves. As in the human situation, effective vaccines are lacking. An ISCOM vaccine containing RSV envelope antigens, most prominently the F protein, induced full protection against respiratory challenge infection after two intramuscular administrations of calves with high levels of maternal antibodies. The protection was against disease and virus shedding [50], as demonstrated with sensitive assays, including virus isolation (VI) and polymerase chain reaction (PCR). Such a sterile immunity is extremely important in preventing viral spread in a population and controlling epidemics. In contrast, a commercial vaccine failed. The calves became ill after challenge and excreted virus comparable to the excretion pattern of the nonvaccinated controls (Table 9.3).

An ISCOM herpes equine virus 2 (EHV-2) vaccine induced virus-neutralizing antibodies after only one immunization of 10-day-old foals with high levels of maternal antibodies. A boost three to four weeks later induced protection against natural infection with EHV-2 and a subsequent severe secondary infection with *Rodococcus equi* [53]. It should be noted that EHV-2 belongs to the same group as Epstein–Barr virus (EBV), an immune-suppressive γ herpesvirus.

As discussed elsewhere in this chapter, cattle vaccinated against the herpesvirus BHV-1 with ISCOMs, containing the envelope proteins of this virus, were protected against experimental infection in the respiratory tract with a

**TABLE 9.3  Comparison of the Efficacy of an ISCOM BRSV Vaccine and a Commercial BRSV Vaccine in Calves with Preexisting Maternal BRSV Antibodies[a]**

| Group | Serum[b] Neut. Titer | Nasal IgG[c] | Positive by VI[d] Day 3 | Day 5 | Positive by PCR[e] Day 4 | Day 5 | Day 10 |
|---|---|---|---|---|---|---|---|
| Control | na[f] | <1 | + | + | + | + | — |
| | na | <1 | + | + | + | + | na |
| | na | <1 | + | + | + | + | — |
| | na | <1 | + | + | + | + | na |
| Commercial | 3.1 | 2.9 | + | — | + | + | — |
| inactivated | 1.8 | 2.0 | + | — | + | + | — |
| vaccine | 1.9 | 2.0 | + | — | + | + | + |
| | 1.5 | <1 | + | — | + | + | — |
| | 2.1 | <1 | + | + | + | + | — |
| ISCOM- | 2.5 | 3.2 | — | — | — | — | — |
| formulated | 2.5 | 3.2 | — | — | — | — | — |
| inactivated | 1.6 | 3.2 | — | — | — | — | — |
| vaccine | 2.1 | 2.6 | — | — | — | — | — |
| | 1.9 | 2.9 | — | — | — | + | — |

[a]The calves were experimentally challenged after the second immunization.
[b]Neutralization titer log 10.
[c]Nasal swabs tested in FBL cells, positive CPE, log 10 titers.
[d]By virus infection.
[e]Real-time PCR on nasal swabs.
[f]Nonanalyzed.

virulent BHV-1 virus and a subsequent secondary experimental bacterial infection.

The hurdles for induction of immune protection in young ages were partly overcome in infant lambs after immunization with rotavirus antigens adjuvanted with ISCOM-Matrix by the oral route. The lambs developed partial protective immune response against experimental infection [71].

An ISCOM-Matrix formulation containing killed parvovirus induced protective levels of neutralizing antibodies in puppies primed at three weeks of age with high levels of maternal antibodies and boosted at six weeks of age as described above. That means that these puppies can be delivered to the customer at eight weeks fully protected by the vaccine. The results show that improved vaccine regimens can be implemented to reduce the endemic spread of virus. In animal setups, this property of a vaccine is of particular value for breeding units. Vaccines presently available do not fulfill such criteria.

The capacity of the ISCOMs to up-regulate the innate immune system is particularly important for individuals with immature or weakened immune systems, including people at both ends of the life span and people with immunodeficiency diseases. Flu and many other frequently occurring infectious diseases establish themselves in these populations, constituting reservoirs for spreading to the healthier general public. This enhanced protection was correlated with the up-regulation of T helper cells Thus, the capacity of ISCOMs to enhance innate as well as acquired immunities, particularly in the young and elderly (see below), give rise to the development of new and more effective antivirus vaccines that would fill these particular needs.

### 9.6.2   Induction of Mucosal Immunity by the Oral Route

The information above from infant lambs teaches that protective immune response against experimental infection can be induced by administration against rotavirus after oral administration of rotavirus antigen adjuvanted with ISCOM-Matrix [70]. In this context the CTA1-DD ISCOM is an interesting concept to be explored for oral administration [18], including infants as well as other age groups.

### 9.6.3   Immune Protection Against Respiratory Viruses

The horse is a natural host for influenza viruses and as such, the best animal model to study phenomena such as immune escape by variants within the same subserotype, a situation relevant to the human influenza setting. The equine influenza virus ISCOM vaccine has a proven safety record, induces long-lasting immune responses, and shows a broader response to new variants in the subsero type than do other influenza vaccines. The immune response includes CTL and production of IFN-$\gamma$ [67–69].

Bovines are exposed to respiratory diseases caused by a number of pathogens being spread among the animals under crowded conditions. Examples of ISCOM vaccines inducing immune protection after parenteral immunization against viruses affecting the respiratory tract include BRSV and BHV-1, as described above [50,54,55]. Both vaccines induced full protection against respiratory challenge infection, the latter including a secondary experimental secondary bacterial infection. In contrast, a commercial vaccine partly failed [50,55].

### 9.6.4   Induction of Immune Protection in the Genital Tract
### Against Abortion

In the world of production animals, genital tract infections and abortions cause substantial economic burdens. One such infectious agent is bovine virus diarrhea virus (BVDV), which infects the respiratory, genital, and digestive tracts in bovine and ovine species. In an experimental model in sheep, BVDV-ISCOM induced full protection against abortion after two intramuscular injections [56].

### 9.6.5   Use of ISCOMs as Adjuvant in Multivalent Vaccines

Vaccines for young animals, including dog, are multivalent with regard to vaccine antigens. For that reason animal vaccines given at an early age contain, in general, live and attenuated vaccine antigens. Killed vaccines that require adjuvant are therefore excluded from the multivalent composition. Adjuvants used today cannot be used with live antigens since they inhibit replication. ISCOM and ISCOM-Matrix formulations are now designed that do not hamper replication of live vaccine components and even enhance the immune response to live vaccine antigens that are included in the formulation. Rabies virus and *Bordetella bronchoseptica* antigens are used as killed antigens and are therefore not included in the canine multicomponent vaccines. However, it would now be possible to combine these antigens with live vaccine antigens, which was shown in the following experiment. Frets were immunized with an experimental multivalent vaccine containing live canine distemper virus, live canine adenoviruses, live parainfluenza 3 viruses, killed canine parvovirus, and killed rabies virus adjuvanted with ISCOM-Matrix. The immune responses were enhanced for all vaccine antigens, including the live ones, with the exception of canine distemper virus, which showed similar responses regardless of whether or not adjuvant was added (patent, not published). Thus, it is possible to mix live and killed vaccine antigens in the same formulation and even enhance the immunogenicity of the live vaccine antigens.

## 9.7 ISCOM VACCINE CANDIDATES IN HUMAN CLINICAL TRIALS

At the time of the introduction of the ISCOM technology as a candidate for human vaccines, the potential toxicity and risks for adverse reactions were discussed. Initially, the Quil A molecule, in a semipurified form, was added to vaccine antigens. In horse, cattle, and pig, these formulations were well tolerated. Quil A then became the cornerstone of ISCOM technology. By formulation of the saponin into the matrix (i.e., ISCOM or ISCOMATRIX), a significantly improved tolerability was achieved compared to free saponin, due primarily to the fact that the cell lytic effect of saponin is avoided in ISCOM formulations. The incorporation in the ISCOM-Matrix structure also increased its accessibility and bioavailability, resulting in increased biological effects or lower doses.

The ISCOM with influenza virus envelope antigens became the first vaccine candidate for evaluation in humans. A further adaptation involved the use of purified Quil A and a combination of components isolated from Quil A to prepare ISCOM-Matrix structures. These formulations are known as Iscoprep (QHC) or Iscoprep 703. To date, ISCOM and ISCOMATRIX based on QHC have been administered to almost 1000 volunteers in clinical trials with antigens from influenza virus, human papilloma virus (HPV-E6E7), melanoma–testicular cancer antigen (NY-ESO-1), and hepatitis C virus (HCV core) [72,73]. Interestingly, the influenza vaccines tested induced significantly higher antibody responses within the first week of vaccination than those of conventional vaccines [74]. Moreover, a general feature was that close to 100% of the vaccines responded to the ISCOM and ISCOMATRIX vaccines as measured by hemagglutination inhibition and virus-neutralizing tests. A high proportion of the vaccines mounted CD4 and CD8 T-cell responses.

Adverse events, although not serious, seem to be associated with two characteristics of the technology used in these clinical trials: There was a high concentration of saponin, of fraction C from saponin, or of a combination of saponin fractions A and C combined in the same particle (Iscoprep 703). The latter combination of A and C fractions in the same particles was shown to be even more toxic than that of fraction C alone. In other clinical trials conducted on vaccine candidates containing QS-21, a dose-related intrinsic toxicity was also observed [75,76].

The added effects of the strong induction of IFN-γ expressing CD8 CTLs and Th1 cells by the ISCOM system and the intrinsic property of influenza antigens to induce IFN-γ probably contributed to the adverse events observed in clinical trials. Indeed, vigorous induction of IFN-γ can induce influenzalike symptoms. In a vaccine, antigen and adjuvant components must be evaluated together to obtain the optimal balance [77,78]. As explained in the following section, observations from clinical trials have been considered in the design of new technology, to prepare it for prophylactic use.

## 9.8   NOVEL ISCOM FORMULATIONS

The characteristics of novel ISCOM formulations build on the experience gained over the years with both ISCOM and ISCOM-Matrix [28,79] formulations in terms of particle size, saponin–cholesterol–lipid interactions, manufacturing process, and basic immunological properties. New developments aim at preserving the important immunological and immune modulatory properties while improving the characterization and addressing understanding of intrinsic toxicity of saponin fractions to obtain formulations for human vaccines.

The entire saponin extract was replaced by two well-defined fractions of saponin, fraction A and fraction C (see further details in Section 3.3). These fractions were selected for potent adjuvant properties and for the retained capacity to formulate ISCOM-Matrix while other fractions of the saponin known for their toxicity were left over. The process was developed according to a new concept. ISCOM-Matrix particles constructed from either fraction A or C were prepared, thus physically separating the two saponin fractions. Both fractions have adjuvant properties with specific relations and interactions with the host cells. Whereas fraction C, encompassing the well-known or QS21 molecule, has a lytic activity on animal and human cells, the fraction A component interacts in a milder, virtually nonlytic and different way.

In an experiment with resting dendritic cells (DCs) in vitro, a high proportion of DCs exposed to fraction C died, whereas DCs exposed to fraction A became transformed into differentiating cells expressing costimulatory molecules on the surface and excreting cytokines in the supernatant (nonpublished data). Those cells that survived exposure to C also induced cytokine production.

Physically separated, the two different fractions can interact independently with the cells they encounter once injected in animal or humans. The strong immunomodulatory effects of fraction A, which has also been observed on the resting DCs, make it possible to reduce the amount of the C component significantly without affecting the potency of the mixed-particle configuration. Moreover, the combination of two separate saponin entities opens up the possibility to vary the formulations and obtain adjuvants with different properties. The immunomodulatory capacity can best be exploited in such a manner.

Thus, the new technology facilitates a significant reduction of the C component, retaining and even enhancing the adjuvant effect while reducing the intrinsic toxicity. This feature is important in view of the proposed use of the adjuvant in prophylactic vaccines designed for human vaccines to be used for a variety of age groups. The reduced toxicity was demonstrated in a sensitive mouse model selected for its discriminatory power for variations in toxicity. BALB/c mice are inoculated subcutaneously, and lethality was the endpoint of the study, reported as follows. Table 9.4 shows that the intrinsic toxicity of ISCOM-Matrix A is low compared to ISCOM-Matrix C, that different combinations of ISCOM-Matrix A particles with ISCOM-Matrix C particles as the basic principle for the ABISCO 100 formulation are well tolerated, and that ISCOM-Matrix A and C in the same particle is comparatively toxic in contrast to the well-tolerated formulation with fractions A and C in different ISCOM-Matrix particles.

**TABLE 9.4 Comparison of Toxic Reactions Caused by Various ISCOM-Matrix Formulations**

| | Amount (μg) | Lethality |
|---|---|---|
| ISCOM-Matrix | | |
| ISCOM-Matrix A | 10 | 0/8 |
| ISCOM-Matrix A | 50 | 0/8 |
| ISCOM-Matrix C | 10 | 0/8 |
| ISCOM-Matrix C | 50 | 8/8 |
| ISCOM-Matrix M | | |
| A/C is 8/2 | 50 | 2/8 |
| A/C is 9/1 | 50 | 0/8 |
| A/C is 9.5/05 | 50 | 0/8 |
| ISCOM-Matrix with A and C in the same single particle | | |
| A/C is 7/3 | 50 | 8/8 |
| A/C is 8/2 | 50 | 8/8 |
| A/C is 9/1 | 10 | 6/8 |
| A/C is 9.5/05 | 50 | 5/8 |

In conclusion, these observations show that when present in the same ISCOM-Matrix particle as fraction A, even low concentrations of saponin fraction C are lethal to mice, whereas the mixed particle technology reduced the intrinsic toxic property significantly because of reduced use of fraction C. It is this mixed particle technology that we now propose for prophylactic human application and for sensible animal species such as cat and mouse.

## 9.9   PROPERTIES OF SPECIAL INTEREST

Overall, a number of experimental results generated with ISCOM-formulated vaccine candidates define properties relevant for designing protective vaccines for humans and animals. Listed below are immunological properties that can be included in vaccine formulations and thereby should be considered for future vaccines.

### 9.9.1   Induction of Nasal IgG After Systemic Vaccination

A general view is that protective mucosal antibodies are of the IgA isotype. Retrospective analyses of the studies by Hu et al. [21] and Hägglund et al. [50] using RSV ISCOMs in a mouse model or in calves indicate that the parenteral route of administration and IgG in the respiratory tract correlated well with immune protection and better than local IgA. Moreover, the neutralizing antibodies in nasal secretion were of the IgG isotype. Whether the parenteral route might not be as efficient as mucosal administration for vaccines aimed at respiratory diseases should be reconsidered, as should the fact that the choice of adjuvant plays an important role. It should be realized that the few adjuvants used in humans have not promoted the development and use of adjuvant in humans and the strong correlation noted between the magnitude of the RSV-specific IgG present in the nasal secretion and protection against the experimental infection merits further studies. It is noteworthy that the protective levels of IgG detected in the nasal secretion were induced by systemic vaccination, which certainly makes the vaccine production easier from the point of view of dosage and the present setsup of vaccine production facilities. Retrospective analyses of several experiments reveal high levels of mucosal IgG antibodies following parenteral administration of ISCOM-formulated vaccines in various animal species, including influenza antigens (Table 9.4).

### 9.9.2   Immune Modulation and Up-Regulation of Co-Stimulatory Factors and CTL in Elderly

Aging is associated with a decline in immune function, and the elderly are therefore more susceptible to infectious diseases and less responsive to vaccination, and a number of immune functions need to be enhanced to approach the capacity of fully immune competent adults. In a murine model, Katz et al. [80] have determined that defects in APC expression of pattern-recognition molecules, co-stimulatory molecules, and cytokine production may play an important role in the reduced clonal expansion of T cells in aging. ISCOMs prominently enhance the antigen targeting, uptake, and activity of antigen-presenting cells, including dendritic and B cells and macrophages, resulting in the production of pro-inflammatory cytokines, above all interleukins IL-1, IL-6, and IL-12. The expression of co-stimulatory molecules, major histocompatibility complex (MHC) class II, B7.1 (CD80) and B7.2 (CD86), is also enhanced. The latter partly explains why the ISCOM is an efficient adjuvant for elderly mice [29,30]. Thus, in aged mice it was clearly demonstrated that influenza-ISCOMs induced significantly higher serum hemagglutination inhibition (HAI) titers than those of a commercial vaccine to the levels obtained in young adult mice that received the split vaccine. Influenza-ISCOMs, but not the commercial vaccine, induced cytotoxic T-lymphocyte (CTL) responses in young and aged mice but not to the same levels as in young adult. In aged mice, influenza-ISCOMs significantly reduced illness and enhanced recovery from viral infection compared with a commercial vaccine [81].

### 9.9.3 Induction of Broad (Cross) Immune Protection

In view of the difficulty to establish vaccines to cover disease situations caused by pathogens with efficient escape mechanisms such as influenza viruses, hepatitis C, and various lentiviruses, there is a need to define conserved protective antigens and epitopes to design devices, presenting those efficiently to APCs.

Influenza-ISCOMs and a monovalent subvirion vaccine prepared with an H1N1 strain of influenza virus were compared in mice for immunogenicity and protection against challenge with homologous and heterotypic influenza viruses. Influenza-ISCOMs, but not subvirion vaccine, fully protected mice against homologous virus challenge after one immunization, as assessed by measurement of virus lung titers. The improved protection induced by influenza-ISCOMs was associated with a 10-fold-higher prechallenge serum hemagglutination inhibition titer. Furthermore, only influenza-ISCOMs fully protected mice against mortality and reduced morbidity following challenge with an influenza virus of the serologically distinct H2N2 subtype. This cross-protection correlated with the induction of virus cross-reactive cytotoxic T lymphocytes that recognized a known MHC class I (H2-Kd)–restricted epitope within the hemagglutinin of influenza virus that is conserved among the H1 and H2 influenza virus subtypes.

Influenza-ISCOMs may offer significant advantages over current commercial formulations as an improved influenza vaccine [82]. In a later study [83], an ISCOM H1N1 vaccine induced protection against H1N1, H2N2, H3N2, and the avian H5 and H9 viruses. Similar observations have been observed with a commercial equine ISCOM vaccine.

The results from laboratory animal models that have nonevolutionary experience with the pathogen studied may not predict a real host–pathogen situation. However, this virgin relation may expose possible immune protective mechanisms: either that protective mechanisms are hidden in the pathogen or that the pathogens evade elimination by manipulating the host defense.

### 9.9.4  Long-Lasting Immune Response (Mouse, Cattle, Horse, and Primates)

ISCOMs containing various antigens induced longer-lasting immune responses than those of conventional vaccines with similar antigens. ISCOMs given by systemic routes gave rise to long-lasting serum antibody response with a higher IgG2a/IgG1 ratio than that of intranasal administration.

Agrawal et al. [84] demonstrated that an ISCOM containing HIV-1 antigens elicited high titer and long-lasting antibody response (60 days and above) along with enhanced IgG2a and 2b responses in mice. Therefore, ISCOMs induce long-lasting immune responses by both mucosal and parenteral routes of administration. By systemic routes, the long-lasting response is characterized as a Th1-driven IgG profile. Osterhaus's group showed that macaques immunized with ISCOM influenza virus vaccine [74] or with measles virus ISCOM vaccine [85] responded with long-lasting (> one year) antibody and cell-mediated immune responses. The commercial horse ISCOM influenza virus vaccine induces a longer immune response measured by HAI or virus-neutralizing tests than do the conventional vaccines [68]. Macaques immunized with ISCOMs carrying gp120 from SIV [86] or gp120 from HIV-2 [52] showed long-lasting immune protection that persisted for two years after a two-dose vaccination regimen measured by a challenge with virulent virus. There exist many more examples not covered here that show ISCOMs with various kinds of antigens inducing long-lasting immune responses, including CTL.

### 9.9.5  Specific Influenza Virus CTL Memory (Human)

There is a need to improve the ability of subunit vaccines to induce CD8+ CTL responses in humans, especially for vaccines used to prevent illness by organisms that undergo antigenic variation at their major neutralizing antibody sites (e.g., influenza A viruses and human immunodeficiency virus).

Murine models have demonstrated the protective role of cross-reactive CTL against influenza A virus antigenic drift. Ennis et al. [87] tested the ability of an adjuvanted carrier (ISCOM-Matrix) to help human antigen-presenting cells to present formalin-killed influenza vaccine to human CD8+ CTL clones in vitro and in vaccinated humans. The results of a randomized double-blind controlled clinical study demonstrate that a single dose of a vaccine formulated into ISCOM particles increased influenza A virus–specific CTL memory in 50 to 60% of recipients, compared to 5% of the recipients of the standard influenza vaccine.

### 9.9.6   Low Responder Turned to High Responder (Mouse and Cattle)

ISCOMs containing gp340 envelope protein of Epstein–Barr virus (EBV) were evaluated in low (H-2d)- and high (H-2k)-responder mice. The high responders reacted with high antibody and T-cell responses with a low dose of the Quillaja saponin component in the ISCOM, whereas the low responders reacted with low antibody and T-cell responses. By addition of extra Quillaja saponin as ISCOM-Matrix, the low responders became high responders. This enhanced response is characterized by a Th1 profile regarding both IgG subclasses and cytokine production [32]. Hübschle et al. [33] turned cattle that were low responders to *Mycoplasma mycoides* to high responders by increasing the dose of the ISCOM adjuvant component from 0.2 mg to 2 mg.

### 9.9.7   Immune Responses in the Presence of Preexisting Specific Antibodies

As described above, ISCOMs induce immune responses and immune protection in newborns with maternal antibodies, which is shown against a number of pathogens in a number of species, including murine, equine, bovine, and canine species, while conventional vaccines have failed [28,50].

It is noteworthy that the immune status of the mothers influences the type of immune response in the offspring induced by active immunization. Blomquist et al. [88] showed that mice born to mothers immunized with Sendai virus adjuvanted with aluminum hydroxide responded with suppressed IgG2a titers compared with those of mice born to nonimmunized mothers. However, significant priming of the IgG2a response was detected in offspring after neonatal priming with ISCOMs followed by a boost when the mice were immunologically mature. At the cytokine level, the maternal immunity from these mothers exerted a significant suppressive effect on the IFN-γ production but not on IL-5 levels after immunization at neonatal age compared with mice born to nonimmunized mothers. In concert with the IgG2a situation, a significant priming effect of neonatal immunization with ISCOMs on IFN-γ levels was recorded after reimmunization at an adult age. Thus, the Th2 bias of newborns can be modulated by a powerful adjuvanted delivery system for vaccine antigens.

## 9.10   CONCLUSIONS

Altogether, the data generated so far in experiments in many animal species, from commercial use in veterinary vaccines and in clinical trials in humans, clearly demonstrates the potency and the immunomodulatory capacity of the ISCOM technology. The novel ISCOM technology further expands its practical use. The results generated in horse and cattle, natural hosts of influenza, RSV, γ-herpesvirus (EBV-like), and herpes simplex type of viruses, provide strong support for using similar formulations in humans against equivalent infections. The capacity to overcome preexisting immunity, to induce a balanced Th1/Th2 response, to induce CD8-positive CTLs, and to increase the IgG titers in the nasal cavity significantly after systemic vaccination, plus a broadened immune protection, are all important features to improve a number of vaccines, including influenza and RSV vaccines including their use at both ends of the life span.

The new developments in the family of ISCOM formulations related to preparing separate ISCOM-Matrix with two different saponin components to optimally exploit specific characteristics of the two components have created extensive possibilities for varying the composition to obtain well-tolerated formulations with improved modulatory effects—relevant properties for all age groups.

The significantly reduced quantity of QHC component in the formulation and the capacity to fine-tune the adjuvant component to the antigens involved have bearings on the use of ISCOM technology for prophylactic use in humans.

## REFERENCES

1. Ford, J.M. Edward Jenner, MD FRS (1749–1823). *J Med Biogr* 2003, **11**(4), 241.
2. Almeida, J.D., Edwards, D.C., Brand, C.M. & Heath, T.D. Formation of virosomes from influenza subunits and liposomes. *Lancet* 1975, **2**(7941), 899–901.
3. Morein, B., Helenius, A., Simons, K., Pettersson, R., Kaariainen, L. & Schirrmacher, V. Effective subunit vaccines against an enveloped animal virus. *Nature* 1978, **276**(5689), 715–718.
4. Petry, H., Goldmann, C., Ast, O. & Luke, W. The use of virus-like particles for gene transfer. *Curr Opin Mol Ther* 2003, **5**(5), 524–528.
5. Morein, B., Sharp, M., Sundquist, B. & Simons, K. Protein subunit vaccines of parainfluenza type 3 virus: immunogenic effect in lambs and mice. *J Gen Virol* 1983, **64**(Pt 7), 1557–1569.
6. Heeney, J.L., van Gils, M.E., van der Meide, P., de Giuli Morghen, C., Ghioni, C., Gimelli, M. et al. The role of type-1 and type-2 T-helper immune responses in HIV-1 vaccine protection. *J Med Primatol* 1998, **27**(2–3), 50–58.
7. Espinet, R.G. Nuevo tipo de vacuna antiaftosa a complejo glucovirico. *Gac Vet* 1951, **74**(74), 1–13.
8. Ramon, G. Sur l'augmentation anormale de l'antitoxine chez les chevaux producteurs de serum antidiptherique. *Bull Soc Centr Med Vet* 1925, **101**, 227–234.
9. Thibault, P.a.R., R. Sur l'accroissement l'immunite antitoxique sous influence de l'addition de diverses substances a l'antigene (anatoxines diphteroque et tetanique). *CR Soc Biol* 1936, **121**(121), 718–721.
10. Dalsgaard, K. Saponin adjuvants. 3. Isolation of a substance from *Quillaja saponaria* Molina with adjuvant activity in food-and-mouth disease vaccines. *Arch Gesamte Virusforsch* 1974, **44**(3), 243–254.

11. Dalsgaard, K. A study of the isolation and characterization of the saponin Quil A: evaluation of its adjuvant activity, with a special reference to the application in the vaccination of cattle against foot-and-mouth disease. *Acta Vet Scand Suppl* 1978(69), 7–40.

12. Kensil, C.R., Patel, U., Lennick, M. & Marciani, D. Separation and characterization of saponins with adjuvant activity from *Quillaja saponaria* Molina cortex. *J Immunol* 1991, **146**(2), 431–437.

13. Ronnberg, B., Fekadu, M. & Morein, B. Adjuvant activity of non-toxic *Quillaja saponaria* Molina components for use in ISCOM matrix. *Vaccine* 1995, **13**(14), 1375–1382.

14. Morein, B., Sundquist, B., Hoglund, S., Dalsgaard, K. & Osterhaus, A. Iscom, a novel structure for antigenic presentation of membrane proteins from enveloped viruses. *Nature* 1984, **308**(5958), 457–460.

15. Lovgren Bengtsson, K. & Morein, B. The ISCOM technology. In *Methods in Molecular Medicine. Vaccine Adjuvants: Preparation Methods and Research Protocols*, Vol. 42, O'Hagan, D.T., Ed., Humana Press, Totowa, NJ, 2000, pp. 239–258.

16. Johansson, M. & Lövgren-Bengtsson, K. ISCOMs with different *Quillaja saponin* components differ in their immunomodulating activities. *Vaccine* 1999, **17**(22), 2894–2900.

17. de Wit, E., Munster, V.J., Spronken, M.I., Bestebroer, T.M., Baas, C., Beyer, W.E. et al. Protection of mice against lethal infection with highly pathogenic H7N7 influenza A virus by using a recombinant low-pathogenicity vaccine strain. *J Virol* 2005, **79**(19), 12401–12407.

18. Lycke, N. From toxin to adjuvant: the rational design of a vaccine adjuvant vector, CTA1-DD/ISCOM. *Cell Microbiol* 2004, **6**(1), 23–32.

19. Mowat, A.M., Donachie, A.M., Jagewall, S., Schon, K., Lowendaler, B., Dalsgaard, K. et al. CTA1-DD-immune stimulating complexes: a novel, rationally designed combined mucosal vaccine adjuvant effective with nanogram doses of antigen. *J Immunol* 2001, **167**(6), 3398–3405.

20. Ekstrom, J., Hu, K.F., Bengtsson, K.L. & Morein, B. ISCOM and ISCOM-Matrix enhance by intranasal route the IgA responses to OVA and rCTB in local and remote mucosal secretions. *Vaccine* 1999, **17**(20–21), 2690–2701.

21. Hu, K.F., Ekstrom, J., Merza, M., Lovgren-Bengtsson, K. & Morein, B. Induction of antibody responses in the common mucosal immune system by respiratory syncytical virus immunostimulating complexes. *Med Microbiol Immunol* (*Berl*) 1999, **187**(4), 191–198.

22. Polakos, N.K., Drane, D., Cox, J., Ng, P., Selby, M.J., Chien, D. et al. Characterization of hepatitis C virus core-specific immune responses primed in rhesus macaques by a nonclassical ISCOM vaccine. *J Immunol* 2001, **166**(5), 3589–3598.

23. Lovgren-Bengtsson, K. Preparation and use of adjuvants. In *Methods in Microbiology*, Vol. 25, Kaufmann, S.H., Ed., Academic Press, Jan Diego, CA, 1998, pp. 471–502.

24. Smith, R.E., Donachie, A.M., Grdic, D., Lycke, N. & Mowat, A.M. Immune-stimulating complexes induce an IL-12-dependent cascade of innate immune responses. *J Immunol* 1999, **162**(9), 5536–5546.

25. Mowat, A.M., Smith, R.E., Donachie, A.M., Furrie, E., Grdic, D. & Lycke, N. Oral vaccination with immune stimulating complexes. *Immunol Lett* 1999, **65**(1–2), 133–140.

26. Hu, K.F., Regner, M., Siegrist, C.A., Lambert, P., Chen, M., Bengtsson, K.L. et al. The immunomodulating properties of human respiratory syncytial virus and immunostimulating complexes containing *Quillaja saponin* components QH-A, QH-C and ISCOPREP703. *FEMS Immunol Med Microbiol* 2005, **43**(2), 269–276.

27. Hu, K.F., Chen, M., Abusugra, I., Monaco, F. & Morein, B. Different respiratory syncytial virus and *Quillaja saponin* formulations induce murine peritoneal cells to express different proinflammatory cytokine profiles. *FEMS Immunol Med Microbiol* 2001, **31**(2), 105–112.

28. Morein, B., Hu, K.F. & Abusugra, I. Current status and potential application of ISCOMs in veterinary medicine. *Adv Drug Deliv Rev* 2004, **56**(10), 1367–1382.

29. Morein, B. & Hu, K.F. Biological aspects and prospects for adjuvants and delivery systems. In *New Vaccine Technology*, Ellis, R.W., Ed., Landes Bioscience, Georgetown, TX, 2001, pp. 274–291.

30. Morein, B. & Bengtsson, K.L. Immunomodulation by ISCOMs, immune stimulating complexes. *Methods* 1999, **19**(1), 94–102.

31. Sambhara, S., Kurichh, A., Miranda, R., Tamane, A., Arpino, R., James, O. et al. Enhanced immune responses and resistance against infection in aged mice conferred by flu-ISCOMs vaccine correlate with up-regulation of costimulatory molecule CD86. *Vaccine* 1998, **16**(18), 1698–1704.

32. Dotsika, E., Karagouni, E., Sundquist, B., Morein, B., Morgan, A. & Villacres-Eriksson, M. Influence of *Quillaja saponaria* triterpenoid content on the immunomodulatory capacity of Epstein–Barr virus ISCOMs. *Scand J Immunol* 1997, **45**(3), 261–268.

33. Hubschle, O.J., Tjipura-Zaire, G., Abusugra, I., di Francesca, G., Mettler, F., Pini, A. et al. Experimental field trial with an immunostimulating complex (ISCOM) vaccine against contagious bovine pleuropneumonia. *J Vet Med B Infect Dis Vet Public Health* 2003, **50**(6), 298–303.

34. Morein, B., Villacres-Eriksson, M., Sjolander, A. & Bengtsson, K.L. Novel adjuvants and vaccine delivery systems. *Vet Immunol Immunopathol* 1996, **54**(1–4), 373–384.

35. Lycke, N. & Holmgren, J. Strong adjuvant properties of cholera toxin on gut mucosal immune responses to orally presented antigens. *Immunology* 1986, **59**(2), 301–308.

36. Lövgren, K., Kaberg, H. & Morein, B. An experimental influenza subunit vaccine (ISCOM): induction of protective immunity to challenge infection in mice after intranasal or subcutaneous administration. *Clin Exp Immunol* 1990, **82**(3), 435–439.

37. Morein, B., Villacres-Eriksson, M. & Lovgren-Bengtsson, K. ISCOM, a delivery system for parenteral and mucosal vaccination. *Dev Biol Stand* 1998, **92**, 33–39.

38. Jones, P.D., Tha Hla, R., Morein, B., Lovgren, K. & Ada, G.L. Cellular immune responses in the murine lung to local immunization with influenza A virus glycoproteins in micelles and immunostimulatory complexes (ISCOMs). *Scand J Immunol* 1988, **27**(6), 645–652.

39. Claassen, I.J., Osterhaus, A.D., Poelen, M., Van Rooijen, N. & Claassen, E. Antigen detection in vivo after immunization with different presentation forms of rabies virus antigen: II. Cellular, but not humoral, systemic immune responses against rabies virus immune-stimulating complexes are macrophage dependent. *Immunology* 1998, **94**(4), 455–460.

40. Lazorova, L., Artursson, P., Engstrom, A. & Sjolander, A. Transport of an influenza virus vaccine formulation (ISCOM) in Caco-2 cells. *Am J Physiol* 1996, **270**(4 Pt 1), G554–G564.

41. Mowat, A.M. & Donachie, A.M. ISCOMs: A novel strategy for mucosal immunization? *Immunol Today* 1991, **12**(11), 383–385.

42. Mowat, A.M., Maloy, K.J. & Donachie, A.M. Immune-stimulating complexes as adjuvants for inducing local and systemic immunity after oral immunization with protein antigens. *Immunology* 1993, **80**(4), 527–534.

43. Mowat, A.M. & Maloy, K.J. *Immune Stimulating Complexes as Vectors for Oral Immunization*, CRC Press, Boca Raton, FL, 1994.

44. Grdic, D., Smith, R., Donachie, A., Kjerrulf, M., Hornquist, E., Mowat, A. et al. The mucosal adjuvant effects of cholera toxin and immune-stimulating complexes differ in their requirement for IL-12, indicating different pathways of action. *Eur J Immunol* 1999, **29**(6), 1774–1784.

45. Hu, K.F., Elvander, M., Merza, M., Akerblom, L., Brandenburg, A. & Morein, B. The immunostimulating complex (ISCOM) is an efficient mucosal delivery system for respiratory syncytial virus (RSV) envelope antigens inducing high local and systemic antibody responses. *Clin Exp Immunol* 1998, **113**(2), 235–243.

46. Abusugra, I., Wolf, G., Bolske, G., Thiaucourt, F. & Morein, B. ISCOM vaccine against contagious bovine pleuropneumonia (CBPP): 1. Biochemical and immunological characterization. *Vet Immunol Immunopathol* 1997, **59**(1–2), 31–48.

47. Abusugra, I. & Morein, B. Iscom is an efficient mucosal delivery system for *Mycoplasma mycoides* subsp. *mycoides* (MmmSC) antigens inducing high mucosal and systemic antibody responses. *FEMS Immunol Med Microbiol* 1999, **23**(1), 5–12.

48. Carol, H., Nieto, A., Villacres-Eriksson, M. & Morein, B. Intranasal immunization of mice with *Echinococcus granulosus* surface antigens ISCOMs evokes a strong immune response, biased towards glucidic epitopes. *Parasite Immunol* 1997, **19**(5), 197–205.

49. Thapar, M.A., Parr, E.L., Bozzola, J.J. & Parr, M.B. Secretory immune responses in the mouse vagina after parenteral or intravaginal immunization with an immunostimulating complex (ISCOM). *Vaccine* 1991, **9**(2), 129–133.

50. Hägglund, S., Hu, K.F., Larsen, L.E., Hakhverdyan, M., Valarcher, J.F., Taylor, G. et al. Bovine respiratory syncytial virus ISCOMs: protection in the presence of maternal antibodies. *Vaccine* 2004, **23**(5), 646–655.

51. Hoglund, S., Akerblom, L., Ozel, M., Villacres, M., Eriksson, M., Gelderblom, H.R. et al. Characterization of immunostimulating complexes (ISCOMs) of HIV-1. *Viral Immunol* 1990, **3**(3), 195–206.

52. Putkonen, P., Bjorling, E., Akerblom, L., Thorstensson, R., Lovgren, K., Benthin, L. et al. Long-standing protection of macaques against cell-free HIV-2 with a HIV-2 ISCOM vaccine. *J Acquir Immune Defic Syndr* 1994, **7**(6), 551–559.

53. Nordengrahn, A., Rusvai, M., Merza, M., Ekstrom, J., Morein, B. & Belak, S. Equine herpesvirus type 2 (EHV-2) as a predisposing factor for *Rhodococcus equi* pneumonia in foals: prevention of the bifactorial disease with EHV-2 immunostimulating complexes. *Vet Microbiol* 1996, **51**(1–2), 55–68.

54. Trudel, M., Boulay, G., Seguin, C., Nadon, F. & Lussier, G. Control of infectious bovine rhinotracheitis in calves with a BHV-1 subunit-ISCOM vaccine. *Vaccine* 1988, **6**(6), 525–529.

55. Merza, M., Tibor, S., Kucsera, L., Bognar, G. & Morein, B. ISCOM of BHV-1 envelope glycoproteins protected calves against both disease and infection. *Zentralbl Veterinarmed B* 1991, **38**(4), 306–314.

56. Carlsson, U., Alenius, S. & Sundquist, B. Protective effect of an ISCOM bovine virus diarrhoea virus (BVDV) vaccine against an experimental BVDV infection in vaccinated and non-vaccinated pregnant ewes. *Vaccine* 1991, **9**(8), 577–580.

57. Trudel, M., Nadon, F., Seguin, C., Simard, C. & Lussier, G. Experimental polyvalent ISCOMs subunit vaccine induces antibodies that neutralize human and bovine respiratory syncytial virus. *Vaccine* 1989, **7**(1), 12–16.

58. Trudel, M., Nadon, F., Seguin, C., Brault, S., Lusignan, Y. & Lemieux, S. Initiation of cytotoxic T-cell response and protection of BALB/c mice by vaccination with an experimental ISCOMs respiratory syncytial virus subunit vaccine. *Vaccine* 1992, **10**(2), 107–112.

59. Fekadu, M., Shaddock, J.H., Ekstrom, J., Osterhaus, A., Sanderlin, D.W., Sundquist, B. et al. An immune stimulating complex (ISCOM) subunit rabies vaccine protects dogs and mice against street rabies challenge. *Vaccine* 1992, **10**(3), 192–197.

60. Osterhaus, A., Weijer, K., UytdeHaag, F., Knell, P., Jarrett, O., Akerblom, L. et al. Serological responses in cats vaccinated with FeLV ISCOM and an inactivated FeLV vaccine. *Vaccine* 1989, **7**(2), 137–141.

61. Visser, I.K., Vedder, E.J., van de Bildt, M.W., Orvell, C., Barrett, T. & Osterhaus, A.D. Canine distemper virus ISCOMs induce protection in harbour seals (*Phoca vitulina*) against phocid distemper but still allow subsequent infection with phocid distemper virus-1. *Vaccine* 1992, **10**(7), 435–438.

62. Sundquist, B.G., Czifra, G. & Stipkovits, L. Protective immunity induced in chicken by a single immunization with *Mycoplasma gallisepticum* immunostimulating complexes (ISCOMs). *Vaccine* 1996, **14**(9), 892–897.

63. Rimmelzwaan, G.F., Claas, E.C., van Amerongen, G., de Jong, J.C. & Osterhaus, A.D. ISCOM vaccine induced protection against a lethal challenge with a human H5N1 influenza virus. *Vaccine* 1999, **17**(11–12), 1355–1358.

64. Osterhaus, A., de Vries, P., Morein, B., Akerblom, L. & Heeney, J. Comparison of protection afforded by whole virus ISCOM versus MDP adjuvanted formalin-inactivated SIV vaccines from IV cell-free or cell-associated homologous challenge. *Aids Res Hum Retroviruses* 1992, **8**, 1507–1510.

65. Heeney, J.L., Teeuwsen, V.J., van Gils, M., Bogers, W.M., De Giuli Morghen, C., Radaelli, A. et al. Beta-chemokines and neutralizing antibody titers correlate with sterilizing immunity generated in HIV-1 vaccinated macaques. *Proc Natl Acad Sci USA* 1998, **95**(18), 10803–10808.

66. Verschoor, E.J., Mooij, P., Oostermeijer, H., van der Kolk, M., ten Haaft, P., Verstrepen, B. et al. Comparison of immunity generated by nucleic acid-, MF59-, and

ISCOM-formulated human immunodeficiency virus type 1 vaccines in rhesus macaques: evidence for viral clearance. *J Virol* 1999, **73**, 3292–3300.

67. Davis, D., Morein, B., Akerblom, L., Lovgren-Bengtsson, K., van Gils, M.E., Bogers, W.M. et al. A recombinant prime, peptide boost vaccination strategy can focus the immune response on to more than one epitope even though these may not be immunodominant in the complex immunogen. *Vaccine* 1997, **15**(15), 1661–1669.

68. Crouch, C.F., Daly, J., Hannant, D., Wilkins, J. & Francis, M.J. Immune responses and protective efficacy in ponies immunised with an equine influenza ISCOM vaccine containing an "American lineage" H3N8 virus. *Vaccine* 2004, **23**(3), 418–425.

69. Mumford, J.A., Jessett, D.M., Rollinson, E.A., Hannant, D. & Draper, M.E. Duration of protective efficacy of equine influenza immunostimulating complex/tetanus vaccines. *Vet Rec* 1994, **134**(7), 158–162.

70. Mumford, J.A., Jessett, D., Dunleavy, U., Wood, J., Hannant, D., Sundquist, B. et al. Antigenicity and immunogenicity of experimental equine influenza ISCOM vaccines. *Vaccine* 1994, **12**(9), 857–863.

71. Van Pinxteren, L.A., Campbell, I., Clarke, C.J., Snodgrass, D.R. & Bruce, M.G. A single oral dose of inactivated rotavirus and ISCOM matrices induces partial protection in lambs. *Biochem Soc Trans* 1997, **25**(2), 340S.

72. Pearse, M.J. & Drane, D. ISCOMATRIX adjuvant for antigen delivery. *Adv Drug Deliv Rev* 2005, **57**(3), 465–474.

73. Stewart, T.J., Drane, D., Malliaros, J., Elmer, H., Malcolm, K.M., Cox, J.C., Edwards, S.J., Frazer, I.H. & Fernando, G.J.P. ISCOMATRIX adjuvant: an adjuvant suitable for use in anticancer vaccines. *Vaccine* 2004, **22**, 3738–3743.

74. Rimmelzwaan, G.F., Baars, M., van Amerongen, G., van Beek, R. & Osterhaus, A. D. A single dose of an ISCOM influenza vaccine induces long-lasting protective immunity against homologous challenge infection but fails to protect *Cynomolgus* macaques against distant drift variants of influenza A (H3N2) viruses. *Vaccine* 2001, **20**(1–2), 158–163.

75. Slovin, S.F., Ragupathi, G., Musselli, C., Fernandez, C., Diani, M., Verbel, D. et al. Thomsen-Friedenreich (TF) antigen as a target for prostate cancer vaccine: clinical trial results with TF cluster (c)-KLH plus QS21 conjugate vaccine in patients with biochemically relapsed prostate cancer. *Cancer Immunol Immunother* 2005, **54**(7), 694–702.

76. Evans, T.G., McElrath, M.J., Matthews, T., Montefiori, D., Weinhold, K., Wolff, M. et al. QS-21 promotes an adjuvant effect allowing for reduced antigen dose during HIV-1 envelope subunit immunization in humans. *Vaccine* 2001, **19**(15–16), 2080–2091.

77. van Schaik, S.M., Welliver, R.C. & Kimpen, J.L. Novel pathways in the pathogenesis of respiratory syncytial virus disease. *Pediatr Pulmonol* 2000, **30**(2), 131–138.

78. van Schaik, S.M., Obot, N., Enhorning, G., Hintz, K., Gross, K., Hancock, G.E. et al. Role of interferon gamma in the pathogenesis of primary respiratory syncytial virus infection in BALB/c mice. *J Med Virol* 2000, **62**(2), 257–266.

79. Bengtsson, K.L. & Sjolander, A. Adjuvant activity of iscoms; effect of ratio and co-incorporation of antigen and adjuvant. *Vaccine* 1996, **14**(8), 753–760.

80. Katz, J.M., Plowden, J., Renshaw-Hoelscher, M., Lu, X., Tumpey, T.M. & Sambhara, S. Immunity to influenza: the challenges of protecting an aging population. *Immunol Res* 2004, **29**(1–3), 113–124.

81. Sambhara, S., Woods, S., Arpino, R., Kurichh, A., Tamane, A., Bengtsson, K.L. et al. Influenza (H1N1)-ISCOMs enhance immune responses and protection in aged mice. *Mech Ageing Dev* 1997, **96**(1–3), 157–169.

82. Sambhara, S., Woods, S., Arpino, R., Kurichh, A., Tamane, A., Underdown, B. et al. Heterotypic protection against influenza by immunostimulating complexes is associated with the induction of cross-reactive cytotoxic T lymphocytes. *J Infect Dis* 1998, **177**(5), 1266–1274.

83. Sambhara, S., Kurichh, A., Miranda, R., Tumpey, T., Rowe, T., Renshaw, M. et al. Heterosubtypic immunity against human influenza A viruses, including recently emerged avian H5 and H9 viruses, induced by flu-ISCOM vaccine in mice requires both cytotoxic T-lymphocyte and macrophage function. *Cell Immunol* 2001, **211**(2), 143–153.

84. Agrawal, L., Haq, W., Hanson, C.V. & Rao, D.N. Generating neutralizing antibodies, Th1 response and MHC non restricted immunogenicity of HIV-I env and gag peptides in liposomes and ISCOMs with in-built adjuvanticity. *J Immune Based Ther Vaccines* 2003, **1**(1), 5.

85. Osterhaus, A., van Amerongen, G. & van Binnendijk, R. Vaccine strategies to overcome maternal antibody mediated inhibition of measles vaccine. *Vaccine* 1998, **16**(14–15), 1479–1481.

86. de Vries, P., Heeney, J.L., Boes, J., Dings, M.E., Hulskotte, E.G., Dubbes, R. et al. Protection of rhesus macaques from SIV infection by immunization with different experimental SIV vaccines. *Vaccine* 1994, **12**(15), 1443–1452.

87. Ennis, F.A., Cruz, J., Jameson, J., Klein, M., Burt, D. & Thipphawong, J. Augmentation of human influenza A virus-specific cytotoxic T lymphocyte memory by influenza vaccine and adjuvanted carriers (ISCOMs). *Virology* 1999, **259**(2), 256–261.

88. Blomqvist, G.A., Lovgren-Bengtsson, K. & Morein, B. Influence of maternal immunity on antibody and T-cell response in mice. *Vaccine* 2003, **21**(17–18), 2022–2031.

# 10

# SURFACE-CHARGED POLY(LACTIDE-CO-GLYCOLIDE) MICROPARTICLES AS NOVEL ANTIGEN DELIVERY SYSTEMS

MANMOHAN SINGH, JAMES CHESKO, JINA KAZZAZ,
MILDRED UGOZOLLI, PADMA MALYALA, AND
DEREK T. O'HAGAN

## 10.1 INTRODUCTION

New-generation antigens for vaccines will be complex molecules and often poorly immunogenic. These new antigens will need optimal delivery systems with or without an additional immunopotentiator to generate strong immune responses. Several technologies including emulsions, micro- and nanoparticles, aluminum salt adjuvants, and immunostimulating complexes (ISCOMS) have been used as delivery systems to boost immune responses to new vaccine antigens. Enhanced immune responses are induced through a variety of mechanisms, including increased persistence of antigen at the site of injection, improved targeting to antigen-presenting cells, or enhanced protection of antigenic epitopes against degradation. Microparticles and other particulate formulations have been explored extensively for the delivery of antigens with and within these systems.

Particulate delivery systems (e.g., microparticles, emulsions, liposomes, virosomes, and viruslike particles) are more attractive because they have dimensions comparable to those of pathogens which the immune system has evolved to combat over time. Therefore, these particulate delivery systems and associated antigens are rapidly taken up by the antigen-presenting cells (APCs). This

*Vaccine Adjuvants and Delivery Systems,* Edited by Manmohan Singh
Copyright © 2007 John Wiley & Sons, Inc.

enhanced uptake by APCs is an important factor in the ability of these delivery systems to induce more robust immune responses than those of soluble formulations. Furthermore, particulate delivery systems present multiple copies of antigens to the immune system and promote trapping and retention of antigens in local lymph nodes. Immunostimulatory adjuvants may also be included in the particulate delivery systems to enhance the level of immune response or to polarize the T-cell response toward a desired pathway (Th1 versus Th2). In addition, formulating potent immunostimulatory adjuvants into delivery systems may limit their potential to induce systemic adverse events, through restricting the circulation of the adjuvant. This is major advantage for small molecules and water-soluble immunopotentiators.

The adjuvant effect observed as a consequence of the binding of antigens to synthetic particles has been known for several years [1]. The biodegradable and biocompatible polyesters, the poly(lactide-co-glycolides) (PLGs), are the primary candidates for the development of microparticles as delivery systems and adjuvants since they have been used in humans for many years as resorbable suture material and as controlled-release drug delivery systems [2,3]. The adjuvant effect achieved through the encapsulation of antigens into PLG microparticles was first demonstrated in the early 1990s by several independent groups [4–7]. In contrast to alum, PLG microparticles were shown to be effective for the induction of cytotoxic T-lymphocyte (CTL) responses in rodents [8–10].

It is assumed that the uptake of microparticles into APCs underpins the ability of the particles to perform as vaccine delivery systems or adjuvants. The uptake of microparticles ($<5\,\mu m$) both in vitro and in vivo by phagocytic cells has been demonstrated on many occasions [11–13]. In an early paper Kanke et al. [11] described the uptake of microparticles (1 to $3\,\mu m$) by macrophages, but showed that microparticles of $12\,\mu m$ were not taken up. Similarly, Tabata and Ikada showed that maximal uptake of microparticles into macrophages occurred with particles of $<2\,\mu m$ [12–13]. In addition, surface charge and hydrophobicity of the microparticles was also shown to modify uptake [13]. It has been reported that monocytes which carry microparticles to lymph nodes can mature into dendritic cells (DCs) [14]. In addition, uptake of PLG microparticles by DCs in vitro [15] and in vivo has been demonstrated [16].

## 10.2   MICROPARTICLES AS ADJUVANTS

Particulate antigen delivery systems can also enhance immune responses as observed with traditional adjuvants (e.g., alum). Therefore, these formulations are often referred to as adjuvant formulations as they engender the same response by efficient delivery of the antigen. The adjuvant effect achieved through the encapsulation of antigens into PLG microparticles was demonstrated by several groups in the early 1990s [4–7]. O'Hagan et al. [5] showed that microparticles with entrapped ovalbumin (OVA) had comparable immunogenicity to OVA in Freund's adjuvant. While Eldridge et al. [4] confirmed that antigens entrapped in microparticles had immunogenicity comparable to that of Freund's adjuvant, using staphylococcal B enterotoxoid as an antigen. Similar observations were also made by Singh et al. [6] with encapsulated diphtheria toxoid. Particle size was shown to be an important parameter affecting the immunogenicity of microparticles, since smaller particles (<10 μm) were significantly more immunogenic than larger ones [4,17]. The effect of particle size on immunogenicity is likely to be a consequence of enhanced uptake into APCs for the smaller particles. The adjuvant effect of microparticles can be further enhanced by their administration in vehicles with additional adjuvant activity [7,18], or by microencapsulation of adjuvants.

## 10.3  CHARGED MICROPARTICLES FOR ANTIBODY INDUCTION TO ADSORBED PROTEIN ANTIGENS

We have recently developed a novel approach to using PLG microparticles as vaccine delivery systems in which the antigen is adsorbed onto the surface of the particles, which have been modified through surface charge to promote adsorption [19,20]. We have previously described studies which showed that anionic PLG microparticles were effective delivery systems for adsorbed vaccine antigens, including recombinant proteins from *Neisseria meningitidis* type B (MB) and human immunodeficiency virus (HIV) [21,22]. Modification of the PLG microparticles to yield positive or negative surface charge allows higher antigen loading levels (predominantly for plasmid DNA) than can be achieved through microencapsulation. The PLG polymer used in these formulations has a long history of safe use in humans. Work with the traditional aluminum salt adjuvants has elucidated rules for predicting protein adsorption, encouraging adjuvant–antigen interactions, and highlighted the significance of surface electrostatic forces in antigen adsorption and release. Anionic PLG microparticles represent an alternative approach to aluminum salt adjuvants or the MF59 emulsion adjuvant. The loading efficiency of protein antigens on anionic microparticles is higher than that obtained with microencapsulation mainly because the adsorption step does not include any washing, centrifugation, or decantation steps, which often lead to losses. In addition, the exposure of antigens to harsh and potentially damaging conditions during the encapsulation process is avoided through using surface adsorption. Moreover, the antigen is immediately available from the particle surface to be processed following uptake of the microparticles into cells and does not require the polymer to be degraded to release entrapped antigen.

This PLG formulation was used to adsorb recombinant antigens from *Neisseria meningitidis* serotype B (MB1) and induced higher serum bactericidal activity than that of a comparable alum adjuvant and was comparable to that

**TABLE 10.1   Serum Antibody Titers and Serum Bactericidal Responses in Groups of 10 Mice Immunized with PLG and Alum Formulations of a Recombinant Antigen from *Neisseria meningitidis* Type B (MB1) ± CpG Adjuvant and Compared to Complete Freund's Adjuvant[a]**

| Formulation | Serum Antibody Titers (GMT) to MB1 | Serum Bactericidal Titers |
|---|---|---|
| PLG | 227,981 | 1,024 |
| PLG + CpG | 382,610 | 16,384 |
| Alum | 50,211 | 256 |
| Alum + CpG | 56,867 | 4,096 |
| CFA | 253,844 | 8,192 |

[a]The geometric mean titer and serum bactericidal titer for each group are represented.

**TABLE 10.2   Charged PLG Microparticles with Adsorbed Antigens Compared with Alum for Their Ability to Induce Binding (ELISA) Functional Antibodies (Toxin Neutralizing and Bactericidal Antibodies) Against a Range of Vaccine Antigens[a]**

| Antigen/Readout | Formulation | |
|---|---|---|
| | Alum | PLG |
| DT (ELISA) | 8,568 | 3,426 |
| DT neuts. (EU/mL) | 141 | 96 |
| TT (ELISA) | 31,028 | 18,052 |
| MenC (ELISA) | 11,117 | 14,875 |
| MenC (BCA) | 19,766 | 28,440 |
| HBsAg (ELISA) | 9,118 | 3,641 |
| MenB (ELISA) | 8,143 | 48,323 |
| MenB (BCA) | 512 | 8,192 |

[a]The geometric mean titer (GMT) is shown for each ELISA result.

of complete Freund's adjuvant (Table 10.1). In another study [22], anionic PLG microparticles when adsorbed with various vaccine antigens generated potent responses in comparison to alum (Table 10.2).

A related approach has also been described elsewhere [23], in which a novel charged polymer was used to prepare nanoparticles which were able to adsorb tetanus toxoid for mucosal delivery. This formulation also induced antigen-specific immune response after mucosal immunization.

To assess the strengths and limitations of anionic PLG microparticles as a delivery system for diverse antigens, we measured physicochemical properties such as charge and ionic strength that have been implicated as important to the description of binding and release of a range of proteins from polymer surfaces. In addition to vaccine-relevant protein antigens, model proteins with a wide range of isoelectric points (pIs), including ovalbumin (pI 4.6), carbonic anhydrase (pI 6.0) and lysozyme (pI 10.7), were evaluated to represent proteins with diverse electrostatic properties.

Evaluation of the physicochemical properties of anionic PLG micropartcles, including size, charge, and surface morphology, may provide information that can be used to guide formulation strategies for diverse proteins. The complex behavior of protein on solid surfaces has been studied by various techniques to suggest the formation of structures including multilayers. Using atomic force microscopy (AFM) to image the microparticles before and after protein adsorption, we visualized the surface morphology to determine if it was consistent with our hypothesis of the formation of an adsorbed protein monolayer. Furthermore, we investigated the physicochemical properties influencing protein adsorption to microparticles and demonstrated their ability to bind and release diverse protein antigens over a range of loads, pH, and ionic strengths. In the following section we describe in detail some of these physiochemical characterization.

### 10.3.1 Charcterization of Charged Microparticles with Adsorbed Protein Antigens

The size distribution of the microparticles is determined using a particle size analyzer (Master Sizer, Malvern Instruments, UK). The electrokinetic mobility of PLG microparticles is measured in phosphate-buffered saline on a Malvern ZetaSizer (Malvern Instruments, UK) and the zeta potential (roughly equivalent to particle surface charge) is determined. Model proteins with a range of isoelectric points were chosen to span a range of buffer adsorption conditions: lysozyme with pI 10.7, lactic acid dehydrogenase with pI 6.8, carbonic anhydrase with pI 6.0, and ovalbumin with pI 4.6. The PLG content of the suspension is measured by aliquoting 1 mL of the suspension into preweighed vials which were lyophilized and weighed again, and the average net weight change was used as PLG content/mL suspension.

Atomic force microscopy is employed in noncontact mode using a Digital Instruments BioScope (Digital Instruments, Santa Barbara, California). The polymer suspension is imaged before and after exposure to protein (1% weight protein to polymer, overnight rocking at 4°C, then sample and control lyophilized overnight). The powder samples are dispersed on a flat adhesive surface, with representative sampling of numerous (>10) particle populations both within the preparation exposed to protein and in the control not exposed to protein. The AFM phase image showed greater contrast and detail than the topographic profile. Morphological differences measured between the untreated and protein-treated samples are attributed to the presence and interactions of the protein on the polymer surface. The phase-contrast AFM images show notable differences in surface features, with the addition of protein, resulting in nearly contiguous islands about 50 nm wide and 5 nm in height that are attributed to clusters of proteins approximately one molecular layer thick (data not shown).

The PLG/dioctyl sodium sulfosuccinate (DSS) microparticles have a zeta potential of −55 mV in 10 mM pH 5.0 citrate buffer, while PLG/DSS particles

**TABLE 10.3   Surface Charge on PLG Microparticles Before and After Protein Adsorption**

| | $\xi$, mV, pH 5.0 (Citrate) | $\xi$, mV, pH 7.0 (PBS) | $\xi$, mV, pH 9.0 (Borate) | Max. $\Delta\xi$ | % Binding at Max. $\Delta\xi$ |
|---|---|---|---|---|---|
| PLG particles | −55 | −28 | −82 | | |
| PLG + OVA | −49 | −23 | −82 | +5 | 62 |
| PLG + CAN | −25 | −26 | −80 | +30 | 87 |
| PLG + LYS | −9 | −24 | −25 | +57 | 100 |
| PLG + gp120dV2 | −8 | −29 | −37 | +47 | 98 |

with 1% w/w adsorbed gp120dV2 protein had a zeta potential in the same buffer of −8 mV, indicating that charge neutralization at the protein–polymer interface had occurred as a result of protein adsorption (Table 10.3). The proteins studied show a general trend of increasing (less negative) zeta potential as additional protein was adsorbed, as would be predicted for a potentiometric titration of the anionic microparticles with the protein. In cases where the protein carried a net negative charge, the adsorption was generally inefficient and the Stern layer structure may involve inclusion of positive counterions and solvation molecules.

### 10.3.2   Protein Adsorption and Release

Microparticles with protein adsorbed for release studies are made by combining a suspension containing 100 mg of PLG with 1 mg of protein in 10 mL total volume of 10 mM of the appropriate buffer and left overnight on a lab rocker (aliquot mixer, Miles Laboratories) at 4°C. Particles were incubated with the proteins ovalbumin (OVA), bovine serum albumin (BSA), carbonic anhydrase (CAN), lysozyme (LYS), MB1, MB2, and gp120 with a variety of buffer conditions and target loads ranging from 0 to 10% w/w polymer. Buffers used for pH 5 and 5.5 were histidine and citrate, for pH 7, phosphate, and for pH 9, borate. Samples used for protein release studies were lyophilized after overnight adsorption (16 to 18 hours).

To determine the amount of protein adsorbed, the microparticles are separated from the incubation medium by centrifugation, before the final lyophilization step, with the pellet washed three times with distilled water, then lyophilized. The loading level of protein adsorbed to the microparticles was determined by dissolving 10 mg of the microparticles in 2 mL of 5% sodium dodecyl sulfate (SDS)–0.2 M sodium hydroxide solution at room temperature followed by the bicinchoninic acid (BCA) protein assay (Pierce, Rockford, Illinois) using BSA concentration standards. To confirm that the BSA standard was appropriate, representative samples were analyzed for protein integrity and concentration by size exclusion chromatography.

### 10.3.3  Mechanism of Protein Adsorption on Anionic PLG Microparticles

Figure 10.1 shows adsorption pseudotherms for the model proteins lysozyme, bovine serum albumin, and carbonic anhydrase, in addition to vaccine antigens, HIV-envelope protein gp120dV2, and two Meningococcus B proteins, MB1 and MB2, on the PLG microparticles. As expected for an adsorption process, binding efficiency decreases with increasing protein input and is saturated as the available adsorption sites become occupied completely. This curve profile has been reproduced consistently for the numerous proteins that we have studied, and empirically finds an asymptote between 1 and 2% w/w protein to polymer. For species such as carbonic anhydrase and the two MenB proteins, the asymptotic saturation is less distinct.

A simplistic treatment of the protein adsorption data is explained quite adequately by a linear Langmuir plot of a ratio of aqueous protein concentration divided by surface-adsorbed protein versus aqueous protein concentration, supporting the description that a monolayer of adsorbed protein is formed (data not shown) The PLG surface fully covered with a protein monolayer has a reduced affinity to adsorb additional protein as compared to the polymer–surfactant interface with sites unoccupied by protein. For proteins adsorbing close to their isoelectric point, binding of the primary monolayer appears weaker, due perhaps to reduced electrostatic forces, but subsequent layer formation is more facile as a result of reduced charge repulsion. The existence of an upper limit (maximum binding capacity) for adsorption is consistent with the idea that a monolayer (finite, defined surface area available for adsorption) may exist.

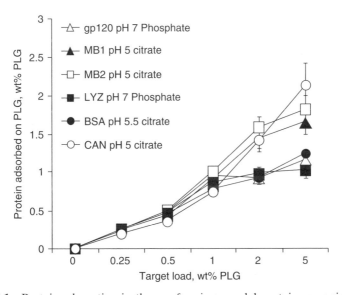

**Figure 10.1**  Protein adsorption isotherm of various model proteins on anionic PLG microparticles.

### 10.3.4    Electrostatic Interactions Are a Significant Driving Force for Adsorption to PLG Microparticles

The binding interactions of proteins to adjuvants such as aluminum salt adjuvants are often governed by a variety of forces, including electrostatic interactions between the ionic salts (aluminum phosphate and hydroxyls) and the protein. To facilitate binding to both acidic and basic proteins, the phosphate content of aluminum hydroxyphosphate must be adjusted to shift the isoelectric point to a favorable position. In contrast to aluminum salt adjuvants, PLG microparticles have surface functional groups, such as aliphatic chains and ester linkages, in addition to ionizable groups such as carboxyls and hydroxyls, that will carry a net negative charge at physiological pH, which are confirmed by the zeta potential measurements.

Proteins are heteropolymeric species, with an ensemble of acidic residues (aspartic and glutamic acid), near-neutral residues (histidine), and basic residues (lysine, arginine) that equilibrate to the buffer pH by a combination of accepting and donating protons, folding and adopting a structure whose internal microenvironments are thermodynamically favorable and kinetically accessible. It is expected that polar and charged residues will show a propensity to migrate to the solvent-accessible surface of the protein to interact strongly with surrounding water molecules.

The pI of ovalbumin is lower than the two buffer conditions tested, so the electrostatic interaction between the PLG microparticles and the ionized protein will be repulsive, due to the common, net negative charge that both species carry. Because some protein adsorption does occur, non-coloumbic forces such as van der Waals interactions are strong enough to overcome the charge repulsion. Since it is generally accepted that the solubility of proteins in aqueous solution goes through a minimum at the isoelectric point, resulting primarily from the loss of favorable interactions between water molecules and the ions they solvate, binding interactions occurring at the protein–polymer interface are expected to compete most effectively near the pI of the protein.

For the basic protein lysozyme (pI of 10.7) the charge effect is opposite, following the change in charge from positive (i.e., attractive with respect to the PLG) at middle pH values to uncharged in high-pH solutions. Since the electrostatic interaction is attractive due to unlike signs of the protein (positive) and PLG (negative), the charged lysozyme species are in excess and bind with high affinity to the microparticle surface. This is confirmed by the large positive changes measured in Table 10.3 following adsorption (PLG microparticle zeta potential). The apparent increase in binding affinity at higher pH may be a result of greater negative charge density on the PLG particles, which compensates for the slightly reduced protein charge. This effect appears to have a dramatic impact on the initial release measurement, which shows a larger fraction of protein released than for ovalbumin. We would expect this trend since the primary interaction facilitating binding is now electrostatic. Lysozyme is a relatively small (17 kD) rigid protein, with an electrostatically charged outer surface, that will have strong Coulomb interactions at the microparticle interface, with less opportunity for structural accommodation and hydrophobic interactions.

For a protein of intermediate isoelectric point such as carbonic anhydrase (pI 6.0), the contributions from electrostatic forces seem less dominant. At pH 7 the protein is primarily anionic or neutral in charge, yet 20% adsorption is observed. At lower pH (5.0) the adsorption grows to almost 90%, closely following the percentage of net cationic molecules. The amount of protein released in two hours is complementary to the fraction adsorbed. The lower adsorption efficiency of CAN at small loading levels and its gradual saturation curve suggest that it may be capable of dimerizing or self-associating with a comparable affinity to the PLG–protein interaction.

Measurement of surface charge on PLG before and after protein adsorption is a direct observation of the electrostatic attractive forces present during the binding process. The change in zeta potential following the adsorption of protein is indicative of charge neutralization (e.g., salt bridge formation) at the protein–polymer interface. We observed a direct correlation of the change in zeta potential on the microparticles to the binding efficiency of the protein, demonstrating that electrostatics drive the bound to free equilibrium strongly toward adsorption onto the microparticles (Table 10.3).

By varying buffer, pH, and ionic strength, the electrostatic forces driving protein adsorption may be altered to achieve substantial surface binding to PLG microparticles. Measurements of surface morphology by AFM and surface charge (zeta potential) support the assertion that a relatively uniform monolayer of protein is formed. The adsorption capacity of the particles can be estimated from geometric consideration of interfacial surface areas of the particles and protein molecules. Near the protein isoelectric point the strength of the electrostatic interaction greatly diminishes and other forces that we will refer to as van der Waals attraction becomes more dominant, although their contribution may also be significant at other pH values. The surface adsorption and desorption process has favorable characteristics for allowing the efficient binding, transport, and release of protein antigens while preserving immunogenic structural elements of these molecules.

## 10.4 CHARGED MICROPARTICLES FOR THE INDUCTION OF CELL-MEDIATED IMMUNITY

Several studies showed that microparticles with entrapped antigens exerted an adjuvant effect for the induction of CTL responses in rodents [8–10,24]. Moreover, microparticles induced CTL responses in mice following systemic [9] and mucosal immunization [8] with protein and peptide antigens. Microparticles also induced a delayed-type hypersensitivity (DTH) response and potent T-cell proliferative responses [8]. We showed that anionic microparticles with adsorbed p55 gag were able to induce CTL responses in rodents [19] and were more potent than microparticles with entrapped antigen (unpublished data). In addition, we also evaluated the ability of anionic microparticles with adsorbed p55 gag to induce cellular immunity in rhesus macaques. Although the anionic PLG microparticles were ineffective for CTL induction in nonhuman primates, they did induce potent antibody and T-cell proliferative responses in rhesus macaques [25].

## 10.5 CATIONIC MICROPARTICLES AS DELIVERY SYSTEMS FOR DNA VACCINES

Vaccines that result in the in situ expression of target antigens, such as attenuated live organisms, recombinant viral or bacterial vectors, and DNA vaccines, offer certain advantages over subunit proteins or inactivated organisms. These include preservation of protein native structure, particularly for viral antigens, and endogenous processing and presentation of expressed antigens by major histocompatibility complex (MHC) molecules. This latter characteristic has made these vaccine approaches highly attractive for their ability to induce cytotoxic T lymphocytes (CTLs), which are important for clearance of infections caused by viruses and intracellular bacteria. Among these types of vaccines, DNA vaccines hold great promise because of their simplicity and potential safety profile, since in their "naked" form they consist solely of highly purified plasmid DNA. As such, they do not have the same issues of potential virulence and toxicity as live organisms or vectors. In addition, studies undertaken so far have demonstrated that DNA vaccines are well tolerated in humans [26,27].

### 10.5.1 Preparation and Characterization of Cationic PLG Microparticles for DNA Delivery

Since 1993, DNA vaccines have been shown to be widely effective at inducing protective immunity in various animal models of infectious and noninfectious diseases [28–30]. These early successes and the broad potential of this technology led to high expectations for various naked DNA vaccines tested in human clinical trials. Unfortunately, with the exception of a few notable cases, naked DNA vaccines were not effective at inducing robust immune responses in humans. In 1996, Wang et al. reported that a malaria DNA vaccine could induce CTL responses in humans [31] and, more recently, Roy et al. demonstrated that a heptitis B surface antigen (HBsAg) DNA vaccine could elicit antibody and T-cell responses [32]. Otherwise, DNA vaccines have generally been disappointing. However, these data offer proof of the concept for DNA vaccines in humans and suggest that achieving efficacy will probably result through increasing vaccine potency. In this chapter we review the limitations of DNA vaccines and offer a practical technological solution to some of these limitations.

There has been great interest in DNA vaccines since they offer significant potential for the induction of potent CTL responses [33]. Nevertheless, the potency of DNA vaccines in humans has so far been disappointing, particularly in relation to their ability to induce antibody responses [31,34]. This has prompted investigators to work on adjuvants and delivery systems for DNA vaccines and also to use DNA in a prime–boost setting with alternative modalities [35–37]. An early study suggested that microparticles with entrapped DNA appeared to have potential as an adjuvant for DNA vaccines, although there was no comparison with naked DNA by the same route of immunization [38]. It soon became clear that the approach of encapsulating plasmid DNA into microparticles had several limitations. The high shear required in generating the emulsion for microencapsulation induced plasmid damage and, in addition, encapsulation efficiency was often low [39–41]. To avoid these problems, we developed a novel cationic PLG microparticle formulation that had DNA adsorbed onto the surface [42]. Importantly, the cationic microparticles induced enhanced responses in comparison to naked DNA, and this enhancement was apparent in all species evaluated, including nonhuman primates (Table 10.4) [37–39].

The preparation of cationic PLG microparticles is based on a modified solvent evaporation process that had previously been commercialized successfully. Following preparation, a detailed physicochemical characterization of the microparticles is undertaken, to include measurements of particle size and surface charge, detergent content, DNA adsorption efficiency and release rate, and DNA integrity, as described previously [42]. One particular aspect of the preparation process is its high degree of flexibility and robustness. Microparticles can be prepared with a range of cetyl trimethyl ammonium bromide (CTAB) content, which controls the adsorption efficiency and release rate for DNA [43]. High-efficiency loading (100%) of DNA is achieved routinely, with a high loading level onto the microparticles (often, >5%) [44]. In addition, although we have shown that DNA needs to be adsorbed to the microparticles to induce enhanced responses [44], we have also shown that relatively high levels of initial release of DNA can be accommodated without a reduction in

**TABLE 10.4   Levels of Enhancement of Antibody Responses Achieved with Cationic PLG–DNA Microparticles in Comparison to Naked DNA (p55 gag) Following Intramuscular Immunization in Various Animal Models at Different Dose Levels**[a]

| Species | Dose (μg) | Naked DNA | PLG–DNA |
|---|---|---|---|
| Mice | 1 | 22 | 7,664 |
| Guinea pigs | 100 | 868 | 12,882 |
| Rabbits | 250 | 644 | 8,778 |
| Rhesus macaques | 500 | 190 | 10,220 |

[a]PLG–DNA consistently shows a significantly enhanced response over naked DNA, with the fold increase ranging from 12- to 500-fold in different species.

the potency of the formulations [43]. As a consequence, microparticles can be prepared with a relatively high DNA load relative to the CTAB content, which has clear advantages from a toxicological perspective, and more than one plasmid can be loaded onto the same microparticles if required. These observations are in marked contrast to the low and inefficient loading of DNA commonly described for microencapsulation of DNA [39–41].

In addition, the cationic microparticles efficiently adsorbed DNA and could deliver several plasmids simultaneously on the same formulation at a range of loading levels [44,45]. The microparticles appeared to be effective as a consequence of efficient delivery of the adsorbed plasmids into DCs [46]. A similar approach was recently described in which DNA was adsorbed to the surface of novel cationic emulsions [47].

## 10.6 CHARGED MICROPARTICLES AS DELIVERY SYSTEMS FOR ADJUVANTS

Microparticle formulations with adsorbed antigen and adjuvant have significantly improved the response to the antigen [48]. Simultaneous delivery of antigens and adjuvants on microparticles ensures that both agents can be delivered into the same APC population. To illustrate this approach, we used cationic PLG microparticles to adsorb the polyanionic adjuvant CpG, which was coadministered with recombinant gp120 protein from HIV adsorbed onto anionic PLG microparticles [48]. This combination induced the most potent gp120-specific immune response. Similarly, we also induced enhanced serum antibody responses and protection against PA anthrax antigen and CpG when these were delivered adsorbed on PLG microparticles [49].

Microparticles can be used to deliver entrapped adjuvants to ensure long-term controlled release of these agents. Tabata and Ikada first entrapped a synthetic adjuvant, muramyl dipeptide (MDP), in microspheres of gelatin [50], and Puri and Sinko [51] showed that MDP entrapped in microspheres induced enhanced immune responses. PLG microparticles have also been used for delivery of QS21 adjuvant (Quillaria saponaria derivative) in combination with gp120 antigen [52]. It remains to be determined for different adjuvants whether it is more attractive to entrap the adjuvant in PLG, to adsorb it on the surface, or to simply coadminister them. Encapsulation or adsorption offer an opportunity to minimize adverse effects and to ensure that the adjuvant is taken up into APCs as long as the particles are small enough. In addition, both approaches offer the potential of controlled release, although the duration of release would be much longer for entrapped agents. Nevertheless, it might be desirable to have the adjuvant available early in the response, to offer optimal levels of enhancement. Simple coadministration may work well for some adjuvants, but this could be dependent on the mechanism of action of the adjuvant. If they act on the cell surface, this approach might be optimal. However, if they need to be internalized to interact with receptors in the intracellular compartment, association with the particle is preferred to promote uptake. In addition, adsorption is preferred to ensure maximal availability of adjuvant during early antigen-processing events. In addition to the use of charged microparticles to deliver adjuvants and antigens, they may also be used in conjunction with traditional adjuvants, including alum and emulsion-based adjuvants to improve their potency [53,54].

In previous studies we have evaluated CpG, either soluble or formulated with PLG microparticles, with both MenB protein (data not shown) and with HIV gp120 [55]. In these studies we found that delivering CpG, either soluble or formulated, did not make a difference in terms of the level of enhancement of immune response as long as the antigen was formulated on PLG microparticles. These findings encouraged us to continue using soluble CpG in subsequent studies. In contrast, unlike CpG, the current studies showed that encapsulation of MPL and RC529 within PLG microparticles offers a significant advantage immunologically (Table 10.5).

Additionally, in our previous studies (data not shown) vaccine formulations tested in small animals with varying early release profiles (from 30 to 60%) did not show any remarkable difference in immunogenicity. The most important aspect of these studies is the potent immunogenic effect achieved by formulating MPL and RC529 in PLG microparticles. The superior responses obtained with formulated immune potentiators over soluble formulations have great benefits. Moreover, the fact that formulating immune potentiator and antigen using the same or separate particles results in the same level of enhanced response allows great flexibility in vaccine design.

We have been successfully evaluating this approach with several Men B antigens. Encapsulating immune potentiators within microparticles and co-delivering them with formulated antigens ensures efficient delivery and

**TABLE 10.5    Adjuvant Effect of MPL and RC529 for the Induction of Antibodies Against env gp120**[a]

| Formulation | Serum Titers Two Weeks Following Third Immunization | |
| --- | --- | --- |
| | IgG | IgG2a |
| PLG/gp120 | $401 \pm 69$ | $5 \pm 0$ |
| PLG/gp120 + MPL | $1011 \pm 203$ | $92 \pm 24$ |
| PLG/gp120 + PLG/MPL | $5237 \pm 1320$ | $4432 \pm 825$ |
| PLG/gp120 + RC529 | $1084 \pm 223$ | $67 \pm 20$ |
| PLG/gp120 + PLG/RC529 | $3907 \pm 828$ | $6731 \pm 627$ |
| PLG/MPL/gp120 | $5742 \pm 595$ | $7838 \pm 2050$ |
| PLG/RC529/gp120 | $4011 \pm 479$ | $3668 \pm 3562$ |
| PLG/gp120 + CpG | $3387 \pm 785$ | $3494 \pm 494$ |

[a]Both adjuvant were evaluated either as soluble or formulated in PLG microparticles and coadministered with protein adsorbed on either the same or a separate particle and compared with adsorbed protein with and without CpG.

promotes the interaction of both antigens and immune potentiator with the key cells of the innate immune system. However, additional mechanistic studies will be undertaken to better understand and define this interaction.

## 10.7 MUCOSAL IMMUNIZATION WITH MICROPARTICLES

The selection for the use of microparticle-based formulations (<5 µm in diameter) as delivery systems for injectable vaccines is based on their ability to be taken up into a variety of phagocytic antigen-presenting cells, which has been demonstrated repeatedly both in vitro and in vivo. For example, in an early paper, Kanke et al. [11] described the uptake of microparticles (1 to 3 µm) by macrophages but showed that microparticles of 12 µm were not taken up. Hence, this early paper defined the optimal size for uptake of microparticles into APCs. Tabata and Ikada subsequently confirmed that maximal uptake of microparticles into macrophages occurred with particles of <2 µm [12,13]. They showed that surface charge and hydrophobicity of the microparticles also modified the extent of uptake [13]. More recently, it was reported that macrophages can carry microparticles to lymph nodes and then mature into DCs [14]. In addition, direct uptake of PLG microparticles into DCs in vitro [15] and in vivo has been demonstrated [16].

Microparticles were used for the oral delivery of labile compounds, including vaccine antigens, because of the observations made over many years that small microparticles (<10 µm) were taken up into the gut associated lymphoid tissues following oral administration. Detailed studies showed that microparticles appeared to be taken up preferentially by the specialized M cells, which overlie aggregates of lymphoid follicles in the gut called Peyer's patches [56,57]. It was first reported in 1993 that PLG microparticles directly bound to and were subsequently taken up by intestinal M cells [58]. Microparticles were thought to be particularly attractive as oral antigen delivery systems because they protected entrapped antigens against degradation in the gut and targeted them to the mucosal lymphoid tissue, Peyer's patches.

The ability of microparticles to perform as oral antigen delivery systems for vaccines was first described in the late 1980s [59–61]. The poly(lactide-co-glycolides) quickly established themselves as the polymers of choice for microparticle preparation for the delivery of vaccines, due to a well-established safety record and successful use in early studies [62,63]. Studies with model antigens showed that microparticles were capable of inducing potent serum and secretory antibody responses to entrapped antigens, following oral administration [64]. In addition, it was shown that antigens entrapped in microparticles were capable of inducing protective immunity following oral administration [64,65]. Subsequently, although a number of alternative polymers have been promoted for the encapsulation of antigens to protect them against degradation in the gut and promote their uptake into gut-associated lymphoid tract (GALT), no polymer has yet shown clear advantages over PLG [64].

Following from this work on oral delivery of protein antigens, some groups also evaluated the potential of microparticles for oral delivery of DNA vaccines. Jones et al. [66] were the first to describe some success following oral delivery of a DNA vaccine entrapped in PLG microparticles, using a marker gene for the induction of serum and mucosal antibody responses. Working with collaborators, this group subsequently described the induction of protective immunity against rotavirus challenge in mice following oral immunization with a DNA vaccine entrapped in PLG microparticles [67]. An alternative group described the induction of immune responses to HIV antigens and protection against a vector challenge with DNA entrapped in PLG microparticles [68]. In addition to these reports, an independent group using an alternative polymeric microparticle approach also described the transfection of Peyer's patch cells with a marker gene following oral administration [69]. An alternative polymer was also used to show that oral administration of DNA encoding an allergen could offer protection against allergy in predisposed mice [70].

Nevertheless, despite these encouraging observations in small animal models, it seems unlikely that oral delivery of DNA vaccines is likely to meet with commercial success. The studies described so far involve only small animal models and the approaches may not be easy to scale to larger animals. In general, within the field of oral vaccine delivery, observations in small animal models using relatively high doses have not scaled very well to larger species, particularly when the relative dose to body weight has needed to be significantly reduced. Most likely, the oral delivery of microencapsulated DNA vaccines will suffer a fate similar to the work on oral delivery of protein vaccines in microparticles. After encouraging observations in the early 1990s, oral delivery of protein vaccines in microparticles was advanced into several clinical trials, with mostly disappointing results [71]. Although work continues in this area, it seems likely that the extent of uptake of microparticles into the GALT is insufficient to allow the successful development of oral vaccines and that significant improvements are needed in the technology. Since the approach has shown limited efficacy for protein delivery, it seems unlikely that it will prove more successful in the more challenging arena of DNA vaccine delivery. In contrast, a novel technology involving intranasal administration of DNA vaccines adsorbed onto cationic PLG microparticles has shown some promise and outperformed naked DNA but is still a long way from commercial development [72].

Mucosal administration of vaccines offers a number of important advantages, including easier administration, reduced adverse effects, and the potential for frequent boosting. In addition, local immunization induces mucosal immunity at the sites where many pathogens initially establish infection. In general, systemic immunization has failed to induce mucosal IgA antibody responses. Oral immunization would be particularly advantageous in isolated communities, where access to health care professionals is difficult. Moreover, mucosal immunization would avoid the potential problem of infection due to the reuse of needles. Several orally administered vaccines are commercially

**TABLE 10.6 Local and Systemic T-Cell Responses in Cervical Lymph Node (CLN) and Spleen as Measured by ELISPOT Assay**[a]

| | IFNSC per 10 million MNCs | |
| Formulation | CLN | Spleen |
| --- | --- | --- |
| Naked DNA | 9 | 100 |
| PLG/DNA | 3500 | 1100 |

[a]The data represent the mean number of gag-specific IFNSC (γ-interferon secreting cells) per 10 million cells ± the standard deviation of two independent experiments.

available based on live-attenuated organisms, including vaccines against polio virus, *Vibrio cholerae*, and *Salmonella typhi*. In addition, a wide range of approaches are currently being evaluated for mucosal delivery of vaccines [73], including many approaches involving nonliving adjuvants and delivery systems, including microparticles [64,74].

Microparticles have also shown some promise for the mucosal delivery of DNA [72]. Table 10.6 shows the induction of local and systemic CD8+ T-cell responses in mice immunized intranasally with DNA encoding gag adsorbed to cationic PLG or in saline. The microparticles significantly enhanced the CD8+ CTL responses in mice and also induced enhanced antibody responses. The ability of microparticles to perform as effective adjuvants following mucosal administration is probably a consequence of their uptake into the specialized mucosal-associated lymphoid tissue (MALT).

## 10.8 MECHANISM OF ACTION OF CHARGED PLG MICROPARTICLE FORMULATIONS

Studies have been undertaken to better understand the means by which cationic PLG microparticles enhance the potency of DNA vaccines. In general, these studies have supported the original rationale of using this approach: namely, targeting APCs. In recent studies, following intramuscular administration we have evaluated PLG microparticle distribution, DNA persistence, and gene expression by microscopic tracking of fluorescence-labeled DNA (rhodamine-PNA) and PLG particles (DiI), as well as PCR analysis of both gag-specific DNA and mRNA sequences [75,76]. In this way we have mapped the location of both the injected plasmid and the cells transfected by the plasmid. We have also characterized by flow cytometry the populations of cells trafficking into the injection site. Observations at seven days' postinjection of PLG–CTAB–DNA clearly demonstrated the presence of a depot of particles at the injection site where rhodamine fluorescence persisted. The fluorescence data correlated with our demonstration by PCR that the gag sequence was

**TABLE 10.7   Cellular Influx to the Site of Injection with PLG–DNA and Naked DNA Injections**[a]

| Formulation | Cell Number ($\times 10^3$) | | |
|---|---|---|---|
| | Day 1 | Day 7 | Day 17 |
| Saline | 110 | <20 | 80 |
| Naked DNA | 110 | <20 | N.D. |
| PLG–DNA (1-μm particles) | 520 | 1600 | 150 |
| PLG–DNA (30-μm particles) | N.D. | 80 | 80 |

[a]Mice were immunized intramuscularly via the tibialis anterior with a total volume of 50 μL containing the equivalent of a 10-μg DNA dose. N.D., not done.

detectable in the muscle seven days after injection of PLG–CTAB–DNA under conditions where injection of naked DNA did not give a detectable signal. Coincident with the presence of the depot of PLG–CTAB–DNA microparticles at the periphery of the injected muscle, we have observed an influx of mononuclear cells concentrated at the muscular sheath (Table 10.7). In addition, PLG–CTAB–DNA resulted in expression of the antigen-encoded gene in the draining lymph node. This is consistent with the interpretation that microparticles enhanced transfection of antigen-presenting cells, which migrated to the draining lymph node. We have previously demonstrated transfection of mouse bone marrow–derived dendritic cells with PLG–CTAB–DNA in vitro as well as specific MHC class I–mediated presentation of a gag epitope to a gag-specific T-cell hybridoma [76]. Taken together, the data suggest that the enhanced immunogenicity obtained with the PLG–CTAB formulation could result, at least in part, from a more efficient direct priming of T cells by expression of antigen in DCs.

The relationship between the structural properties of PLG–CTAB–DNA formulations and immunological outcomes provides additional insights into the mechanism of action. It has been shown that immunization with PLG–CTAB–DNA formulated with 30 μm particles does not induce a greater antibody titer than does DNA alone. The expected depot formation has been observed with this formulation; however, neither uptake by APC nor influx of cells to the injection site was observed. In summary, it appears that uptake of PLG–CTAB–DNA microparticles of appropriate size by APCs contributes to the enhanced immunogenicity induced by the formulated DNA vaccine. Other factors that may contribute to DNA vaccine potency include (1) prolonged release of DNA from the microparticles, resulting in longer-lived transfection of muscle cells, as all active formulations have been shown to release a significant fraction of DNA rapidly within 1 to 24 hours under in vitro conditions, and (2) recruitment and activation of immune cells at the site of injection, which could facilitate presentation of antigen expressed from the DNA vaccine.

The relationship between the size of the charged microparticle formulations and the immunological outcome also provides insight into the mechanism of action. We have shown that immunization with DNA adsorbed to 30-μm PLG microparticles does not induce a greater antibody response than DNA alone; the microparticles need to be about 1 μm in size [76]. We have also demonstrated a requirement for association of the antigen with the PLG microparticles to induce enhanced responses [75,76]. We conclude that the uptake of PLG microparticles by APCs appears to be an important contributor to the enhanced immunogenicity.

## 10.9  FUTURE PROSPECTS

In the past 10 years, various naked DNA vaccines have been evaluated for safety and immunogenicity in human clinical trials. These vaccines have consisted of highly purified plasmid DNA delivered as an aqueous solution by syringe or adsorbed onto gold particles by a gene gun. In all cases so far, the DNA vaccines have proven to be very safe and well tolerated, but generally of low potency. These results have necessitated the development and application of several second-generation DNA vaccine technologies, including genetic adjuvants, DNA delivery systems, and prime–boost regimens involving other vaccine technologies used in conjunction with DNA vaccines. Many of these approaches have added complexity to the original DNA vaccines. Hence, issues of safety will need to be addressed carefully.

The prospects of safety for PLG microparticles in humans are good. First, there exists a track record of safe use of PLG in humans, as they are used in certain licensed products as controlled-release delivery systems (e.g., Lupron Depot, Zoladex, Nutropin Depot). Second, PLG-encapsulated DNA vaccines encoding human papilloma virus antigens have been tested in phase I and II human clinical trials without any reported safety issues [77,78]. Finally, PLG microparticles with adsorbed DNA vaccines encoding HIV antigens have recently entered human clinical trials in healthy volunteers. These vaccines have been fully evaluated in IND-enabling toxicology studies in animals, and we have observed no detectable integration of the plasmids into host chromosomal DNA or induction of autoimmune anti-DNA antibodies (unpublished observations). Therefore, we do not expect DNA vaccines delivered by PLG microparticles to pose significant safety issues for humans.

Importantly, the novel process we have developed, involving adsorption of DNA to preformed microparticles, eliminates many of the problems previously encountered with microencapsulation of vaccines. First, the process is highly efficient, with 100% adsorption and minimal losses of plasmid. Second, the microparticles are able to adsorb high loads of DNA, thereby allowing a significant amount of DNA to be delivered per microparticle. Third, and importantly, several plasmids can be co-formulated on the same microparticles, to allow the development of a vaccine that targets several antigens simultaneously. Finally, the process has so far shown itself to be robust, reproducible, and scalable, with many formulation parameters available for manipulation to allow the production of optimal formulations for each indication. Should this approach prove successful in the clinic, we believe the process is sufficiently robust and controllable to allow the commercial development of microparticle-adsorbed DNA vaccines.

## 10.10 SUMMARY

The most notable aspect of the current studies is the potent adjuvant effect achieved through the presentation of antigens on the surface of charged microparticles. In most studies previously described using PLG microparticles for vaccine delivery, the antigen (both protein and plasmid DNA) had been entrapped within the microparticles [4–7]. However, it has been well established that proteins and DNA are often degraded during encapsulation within microparticles [79–82]. The approach described here, involving the presentation of antigen on the microparticle surface, offers the opportunity to avoid exposing antigens to potentially denaturing conditions both during microparticle manufacture and following administration. The avoidance of these damaging conditions for protein and DNA offers enormous advantages and serves to highlight the significant potential for the development of PLG microparticles with surface-presented antigens as vaccine adjuvants.

From pharmaceutical and manufacturing perspectives, the approach of surface adsorption offers a number of important advantages. Using the anionic surface, it was shown that adsorption to microparticles was highly efficient, allowing >90% association of antigen with microparticles (data not shown). In contrast, antigen entrapment within microparticles is often an inefficient process, perhaps resulting in the loss of 50% or more of some antigens evaluated by us (data not shown). Importantly, the adsorption process is rapid, going to completion in a few hours, and is highly reproducible. This is a significant advantage over microencapsulation and makes this approach very attractive for a wide range of recombinant proteins.

In addition, the identification of specific receptors on APCs is likely to allow targeting of adjuvants for the optimal induction of potent and specific immune responses. However, further developments in novel adjuvants and their delivery will probably be driven by a better understanding of the mechanism of action of currently available adjuvants, and this is an area of research that requires additional work. Proteins can be absorbed efficiently on anionic surfaces. This mechanism for delivering protein antigens in vaccines may provide a favorable microparticle delivery system that efficiently carries recombinant proteins into antigen-presenting cells while retaining structural features that confer immunogenicity.

## ACKNOWLEDGMENTS

We would like to acknowledge the contributions of our colleagues in Novartis to the ideas contained in this chapter. We would especially like to thank all the members of the Vaccine Delivery Group for their contributions toward better understanding of these formulations.

## REFERENCES

1. O'Hagan, D.T. Microparticles as oral vaccines. In *Novel Delivery Systems for Oral Vaccines*, O'Hagan, D.T., Ed., CRC Press, Boca Raton, FL, 1994, pp. 175–205.
2. Okada, H. & Toguchi, H. Biodegradable microspheres in drug delivery. *Crit Rev Ther Drug Carrier Syst* 1995, **12**(1), 1–99.
3. Putney, S.D. & Burke, P.A. Improving protein therapeutics with sustained-release formulations. *Nat Biotechnol* 1998, **16**(2), 153–157 [published erratum appears in *Nat Biotechnol* 1998 May, **16**(5), 478].
4. Eldridge, J.H., Staas, J.K., Meulbroek, J.A., Tice, T.R. & Gilley, R.M. Biodegradable and biocompatible poly(DL-lactide-co-glycolide) microspheres as an adjuvant for staphylococcal enterotoxin B toxoid which enhances the level of toxin-neutralizing antibodies. *Infect Immun* 1991, **59**(9), 2978–2986.
5. O'Hagan, D.T., Rahman, D., McGee, J.P. et al. Biodegradable microparticles as controlled release antigen delivery systems. *Immunology* 1991, **73**(2), 239–242.
6. Singh, M., Singh, A. & Talwar, G.P. Controlled delivery of diphtheria toxoid using biodegradable poly(D,L-lactide) microcapsules. *Pharm Res* 1991, **8**(7), 958–961.
7. O'Hagan, D.T., Jeffery, H., Roberts, M.J., McGee, J.P. & Davis, S.S. Controlled release microparticles for vaccine development. *Vaccine* 1991, **9**(10), 768–771.
8. Maloy, K.J., Donachie, A.M., O'Hagan, D.T. & Mowat, A.M. Induction of mucosal and systemic immune responses by immunization with ovalbumin entrapped in poly(lactide-co-glycolide) microparticles. *Immunology* 1994, **81**(4), 661–667.
9. Moore, A., McGuirk, P., Adams, S. et al. Immunization with a soluble recombinant HIV protein entrapped in biodegradable microparticles induces HIV-specific CD8+ cytotoxic T lymphocytes and CD4+ Th1 cells. *Vaccine* 1995, **13**(18), 1741–1749.
10. Nixon, D.F., Hioe, C., Chen, P.D. et al. Synthetic peptides entrapped in microparticles can elicit cytotoxic T cell activity. *Vaccine* 1996, **14**(16), 1523–1530.
11. Kanke, M., Sniecinski, I. & DeLuca, P.P. Interaction of microspheres with blood constituents: I. Uptake of polystyrene spheres by monocytes and granulocytes and effect on immune responsiveness of lymphocytes. *J Parenter Sci Technol* 1983, **37**(6), 210–217.
12. Tabata, Y. & Ikada, Y. Macrophage phagocytosis of biodegradable microspheres composed of L-lactic acid/glycolic acid homo- and copolymers. *J Biomed Mater Res* 1988, **22**(10), 837–858.
13. Tabata, Y. & Ikada, Y. Phagocytosis of polymer microspheres by macrophages. *Adv Polym Sci* 1990, **94**, 107–141.

14. Randolph, G.J., Inaba, K., Robbiani, D.F., Steinman, R.M. & Muller, W.A. Differentiation of phagocytic monocytes into lymph node dendritic cells in vivo. *Immunity* 1999, **11**(6), 753–761.

15. Lutsiak, M.E., Robinson, D.R., Coester, C., Kwon, G.S. & Samuel, J. Analysis of poly(D,L-lactic-co-glycolic acid) nanosphere uptake by human dendritic cells and macrophages in vitro. *Pharm Res* 2002, **19**(10), 1480–1487.

16. Newman, K.D., Elamanchili, P., Kwon, G.S. & Samuel, J. Uptake of poly(D,L-lactic-co-glycolic acid) microspheres by antigen-presenting cells in vivo. *J Biomed Mater Res* 2002, **60**(3), 480–486.

17. O'Hagan, D.T., Jeffery, H. & Davis, S.S. Long-term antibody responses in mice following subcutaneous immunization with ovalbumin entrapped in biodegradable microparticles. *Vaccine* 1993, **11**(9), 965–969.

18. O'Hagan, D.T., Ugozzoli, M., Barackman, J. et al. Microparticles in MF59, a potent adjuvant combination for a recombinant protein vaccine against HIV-1. *Vaccine* 2000, **18**(17), 1793–1801.

19. Kazzaz, J., Neidleman, J., Singh, M., Ott, G. & O'Hagan, D.T. Novel anionic microparticles are a potent adjuvant for the induction of cytotoxic T lymphocytes against recombinant p55 gag from HIV-1. *J Control Release* 2000, **67**(2–3), 347–356.

20. Singh, M., Chesko, J., Kazzaz, J. et al. Adsorption of a novel recombinant glycoprotein from HIV (Env gp120dV2SF162) to anionic PLG microparticles retains the structural integrity of the protein, while encapsulation in PLG microparticles does not. *Pharm Res* 2004, **21**(12), 2148–2152.

21. Singh, M., Kazzaz, J., Chesko, J. et al. Anionic microparticles are a potent delivery system for recombinant antigens from *Neisseria meningitidis* serotype B. *J Pharm Sci* 2004, **93**(2), 273–282.

22. Singh, M., Ugozzoli, M., Kazzaz, J. et al. A preliminary evaluation of alternative adjuvants to alum using a range of established and new generation vaccine antigens. *Vaccine* 2006, **24**(10), 1680–1686.

23. Jung, T., Kamm, W., Breitenbach, A., Hungerer, K.D., Hundt, E. & Kissel, T. Tetanus toxoid loaded nanoparticles from sulfobutylated poly(vinyl alcohol)-graft-poly (lactide-co-glycolide): evaluation of antibody response after oral and nasal application in mice. *Pharm Res* 2001, **18**(3), 352–360.

24. O'Hagan, D.T., Jeffery, H. & Davis, S.S. Long-term antibody responses in mice following subcutaneous immunization with ovalbumin entrapped in biodegradable microparticles. *Vaccine* 1993, **11**(9), 965–969.

25. Otten, G.R., Doe, B., Schaefer, M. et al. Relative potency of cellular and humoral immune responses induced by dna vaccination. *Intervirology* 2000, **43**(4–6), 227–232.

26. Parker, S.E., Borellini, F., Wenk, M.L. et al. Plasmid DNA malaria vaccine: tissue distribution and safety studies in mice and rabbits. *Hum Gene Ther* 1999, **10**(5), 741–758.

27. Parker, S.E., Monteith, D., Horton, H. et al. Safety of a GM-CSF adjuvant–plasmid DNA malaria vaccine. *Gene Ther* 2001, **8**(13), 1011–1023.

28. Ulmer, J.B., Donnelly, J.J., Parker, S.E. et al. Heterologous protection against influenza by injection of DNA encoding a viral protein. *Science* 1993, **259**(5102), 1745–1749.

29. Donnelly, J.J., Ulmer, J.B., Shiver, J.W. & Liu, M.A. DNA vaccines. *Annu Rev Immunol* 1997, **15**, 617–648.

30. Gurunathan, S., Klinman, D.M. & Seder, R.A. DNA vaccines: immunology, application, and optimization. *Annu Rev Immunol* 2000, **18**, 927–974.

31. Wang, R., Doolan, D.L., Le, T.P. et al. Induction of antigen-specific cytotoxic T lymphocytes in humans by a malaria DNA vaccine. *Science* 1998, **282**(5388), 476–480.

32. Roy, M.J., Wu, M.S., Barr, L.J. et al. Induction of antigen-specific CD8+ T cells, T helper cells, and protective levels of antibody in humans by particle-mediated administration of a hepatitis B virus DNA vaccine. *Vaccine* 2000, **19**(7–8), 764–778.

33. Seder, R.A. & Gurunathan, S. DNA vaccines: designer vaccines for the 21st century. *N Engl J Med* 1999, **341**(4), 277–278.

34. Calarota, S., Bratt, G., Nordlund, S. et al. Cellular cytotoxic response induced by DNA vaccination in HIV-1-infected patients. *Lancet* 1998, **351**(9112), 1320–1325.

35. Schneider, J., Gilbert, S.C., Blanchard, T.J. et al. Enhanced immunogenicity for CD8+ T cell induction and complete protective efficacy of malaria DNA vaccination by boosting with modified vaccinia virus Ankara. *Nat Med* 1998, **4**(4), 397–402.

36. Sullivan, N.J., Sanchez, A., Rollin, P.E., Yang, Z.Y. & Nabel, G.J. Development of a preventive vaccine for Ebola virus infection in primates. *Nature* 2000, **408**(6812), 605–609.

37. Amara, R.R., Villinger, F., Altman, J.D. et al. Control of a mucosal challenge and prevention of AIDS by a multiprotein DNA/MVA vaccine. *Science* 2001, **292**(5514), 69–74.

38. Hedley, M.L., Curley, J. & Urban, R. Microspheres containing plasmid-encoded antigens elicit cytotoxic T-cell responses. *Nat Med* 1998, **4**(3), 365–368.

39. Walter, E., Moelling, K., Pavlovic, J. & Merkle, H.P. Microencapsulation of DNA using poly(DL-lactide-co-glycolide): stability issues and release characteristics. *J Control Release* 1999, **61**, 361–374.

40. Ando, S., Putnam, D., Pack, D.W. & Langer, R. PLGA microspheres containing plasmid DNA: preservation of supercoiled DNA via cryopreparation and carbohydrate stabilization. *J Pharm Sci* 1999, **88**(1), 126–130.

41. Tinsley-Bown, A.M., Fretwell, R., Dowsett, A.B., Davis, S.L. & Farrar, G.H. Formulation of poly(D,L-lactic-co-glycolic acid) microparticles for rapid plasmid DNA delivery. *J Control Release* 2000, **66**(2–3), 229–241.

42. Singh, M., Briones, M., Ott, G. & O'Hagan, D. Cationic microparticles: a potent delivery system for DNA vaccines. *Proc Natl Acad Sci USA* 2000, **97**(2), 811–816.

43. Singh, M., Ugozzoli, M., Briones, M., Kazzaz, J., Soenawan, E. & O'Hagan, D. The effect of CTAB concentration in cationic PLG microparticles on DNA adsorption and in vivo performance. *Pharm Res* 2003, **20**(2), 244–248.

44. Briones, M., Singh, M., Ugozzoli, M. et al. The preparation, characterization, and evaluation of cationic microparticles for DNA vaccine delivery. *Pharm Res* 2001, **18**(5), 709–711.

45. O'Hagan, D., Singh, M., Ugozzoli, M. et al. Induction of potent immune responses by cationic microparticles with adsorbed HIV DNA vaccines. *J Virol* 2001, **75**(19), 9037–9043.

46. Denis-Mize, K.S., Dupuis, M., MacKichan, M.L. et al. Plasmid DNA adsorbed onto cationic microparticles mediates target gene expression and antigen presentation by dendritic cells. *Gene Ther* 2000, **7**(24), 2105–2112.

47. Ott, G., Singh, M., Kazzaz, J. et al. A cationic sub-micron emulsion (MF59/DOTAP) is an effective delivery system for DNA vaccines. *J Control Release* 2002, **79**(1–3), 1–5.

48. Singh, M., Ott, G., Kazzaz, J. et al. Cationic microparticles are an effective delivery system for immune stimulatory CpG DNA. *Pharm Res* 2001, **18**(10), 1476–1479.

49. Xie, H., Gursel, I., Ivins, B.E. et al. CpG oligodeoxynucleotides adsorbed onto polylactide-co-glycolide microparticles improve the immunogenicity and protective activity of the licensed anthrax vaccine. *Infect Immun* 2005, **73**(2), 828–833.

50. Tabata, Y. & Ikada, Y. Macrophage activation through phagocytosis of muramyl dipeptide encapsulated in gelatin microspheres. *J Pharm Pharmacol* 1987, **39**(9), 698–704.

51. Puri, N. & Sinko, P.J. Adjuvancy enhancement of muramyl dipeptide by modulating its release from a physicochemically modified matrix of ovalbumin microspheres: II. In vivo investigation. *J Control Release* 2000, **69**(1), 69–80.

52. Cleland, J.L., Barron, L., Daugherty, A. et al. Development of a single-shot subunit vaccine for HIV-1: 3. Effect of adjuvant and immunization schedule on the duration of the humoral immune response to recombinant MN gp120. *J Pharm Sci* 1997, **85**(12), 1350–1357.

53. O'Hagan, D.T. HIV and mucosal immunity. *Lancet* 1991, **337**(8752), 1289.

54. O'Hagan, D.T., Singh, M., Kazzaz, J. et al. Synergistic adjuvant activity of immuno-stimulatory DNA and oil/water emulsions for immunization with HIV p55 Gag antigen. *Vaccine* 2002, **20**(27–28), 3389–3398.

55. Singh, M., Ott, G., Kazzaz, J. et al. Cationic microparticles are an effective delivery system for immune stimulatory cpG DNA. *Pharm Res* 2001, **18**(10), 1476–1479.

56. O'Hagan, D.T. Intestinal translocation of particulates: implications for drug and antigen delivery. *Adv Drug Deliv Rev* 1990, **5**, 265–285.

57. O'Hagan, D.T. The intestinal uptake of particles and the implications for drug and antigen delivery. *J Anat* 1996, **189**(Pt 3), 477–482.

58. Jepson, M.A., Simmons, N.L., O'Hagan, D.T. & Hirst, B.H. Comparison of poly(DL-lactide-co-glycolide) and polystyrene microsphere targeting to intestinal M cells. *J Drug Target* 1993, **1**(3), 245–249.

59. O'Hagan, D.T., Palin, K., Davis, S.S., Artursson, P. & Sjoholm, I. Microparticles as potentially orally active immunological adjuvants. *Vaccine* 1989, **7**(5), 421–424.

60. O'Hagan, D.T., Palin, K.J. & Davis, S.S. Poly(butyl-2-cyanoacrylate) particles as adjuvants for oral immunization. *Vaccine* 1989, **7**(3), 213–216.

61. Eldridge, J.H., Meulbroek, J.A., Staas, J.K., Tice, T.R. & Gilley, R.M. Vaccine-containing biodegradable microspheres specifically enter the gut-associated lymphoid tissue following oral administration and induce a disseminated mucosal immune response. *Adv Exp Med Biol* 1989, **251**, 191–202.

62. Challacombe, S.J., Rahman, D., Jeffery, H., Davis, S.S. & O'Hagan, D.T. Enhanced secretory IgA and systemic IgG antibody responses after oral immunization with biodegradable microparticles containing antigen. *Immunology* 1992, **76**(1), 164–168.

63. Eldridge, J.H., Hammond, C.J., Meulbroek, J.A., Staas, J.K., Gilley, R.M. & Tice, T.R. Controlled vaccine release in the gut-associated lymphoid tissues: I. Orally administered biodegradable microspheres target the Peyer's patches. *J Control Release* 1990, **11**, 205–214.

64. O'Hagan, D. Microparticles and polymers for the mucosal delivery of vaccines. *Adv Drug Deliv Rev* 1998, **34**(2–3), 305–320.

65. Jones, D.H., McBride, B.W., Thornton, C., O'Hagan, D.T., Robinson, A. & Farrar, G.H. Orally administered microencapsulated *Bordetella pertussis* fimbriae protect mice from *B. pertussis* respiratory infection. *Infect Immun* 1996, **64**(2), 489–494.

66. Jones, D.H., Corris, S., McDonald, S., Clegg, J.C. & Farrar, G.H. Poly(DL-lactide-co-glycolide)-encapsulated plasmid DNA elicits systemic and mucosal antibody responses to encoded protein after oral administration. *Vaccine* 1997, **15**(8), 814–817.

67. Chen, S.C., Jones, D.H., Fynan, E.F. et al. Protective immunity induced by oral immunization with a rotavirus DNA vaccine encapsulated in microparticles. *J Virol* 1998, **72**(7), 5757–5761.

68. Kaneko, H., Bednarek, I., Wierzbicki, A. et al. Oral DNA vaccination promotes mucosal and systemic immune responses to HIV envelope glycoprotein. *Virology* 2000, **267**(1), 8–16.

69. Mathiowitz, E., Jacob, J.S., Jong, Y.S. et al. Biologically erodable microspheres as potential oral drug delivery systems. *Nature* 1997, **386**(6623), 410–414.

70. Roy, K., Mao, H.Q., Huang, S.K. & Leong, K.W. Oral gene delivery with chitosan: DNA nanoparticles generates immunologic protection in a murine model of peanut allergy [see comments]. *Nat Med* 1999, **5**(4), 387–391.

71. O'Hagan, D. T. & Singh, M. Microparticles as vaccine adjuvants and delivery systems. *Expert Rev Vaccines* 2003, **2**(2), 269–283.

72. Singh, M., Vajdy, M., Gardner, J., Briones, M. & O'Hagan, D.T. Mucosal immunization with HIV-1 Gag DNA on cationic microparticles prolongs gene expression and enhances local and systemic immunity. *Vaccine* 2001, **20**(3–4), 594–602.

73. Levine, M.M. & Dougan, G. Optimism over vaccines administered via mucosal surfaces. *Lancet* 1998, **351**, 1375–1376.

74. Vajdy, M. & O'Hagan, D.T. Microparticles for intranasal immunization. *Adv Drug Deliv Rev* 2001, **51**(1–3), 127–141.

75. Denis-Mize, K., Dupuis, M., Singh, M. et al. Mechanisms of increased immunogenicity for DNA-based vaccines absorbed onto cationic microparticles. *Cell Immunol* 2003, **225**(1), 12–20.

76. Denis-Mize, K.S., Dupuis, M., MacKichan, M.L. et al. Plasmid DNA adsorbed onto cationic microparticles mediates target gene expression and antigen presentation by dendritic cells. *Gene Ther* 2000, **7**(24), 2105–2112.

77. Klencke, B., Matijevic, M., Urban, R.G. et al. Encapsulated plasmid DNA treatment for human papillomavirus 16-associated anal dysplasia: a phase I study of ZYC101. *Clin Cancer Res* 2002, **8**(5), 1028–1037.

78. Sheets, E.E., Urban, R.G., Crum, C.P. et al. Immunotherapy of human cervical high-grade cervical intraepithelial neoplasia with microparticle-delivered human papillomavirus 16 E7 plasmid DNA. *Am J Obstet Gynecol* 2003, **188**(4), 916–926.

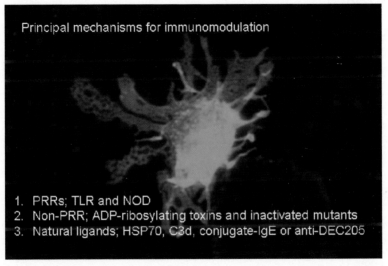

**Figure 3.1**  Summary of the major pathways for immunomodulation and adjuvant function. A dendritic cell labeled with FITC-conjugated anti-CD11c antibodies.

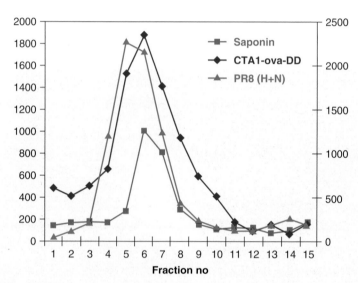

**Figure 3.2** ISCOMs as they can be studied in electron microscopy. Superimposed is the construction of ISCOMs containing two proteins: CTA1-DD adjuvant and PR8 influenza virus proteins. ISCOMs containing CTA1-DD and PR8 antigens were subjected to sucrose density gradient analysis and the fractions were analyzed for CTA1-DD/PR8 and saponin content. The peaks of incorporation of CTA1-DD, PR8 proteins, and saponin are superimposed, illustrating the successful formation of ISCOMS. The y-axis values represent arbitrary OD units at A450nm for immunodetection of HRP-labeled CTA1-DD or PR8 (left) and saponin at A214nm (right) using a spectrophotometer.

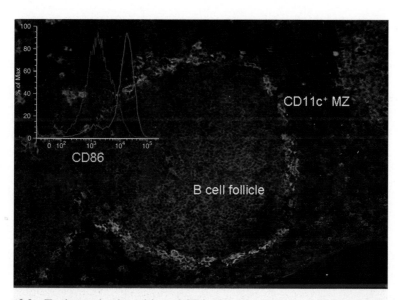

**Figure 3.3** To detect the deposition of CT in the spleen of mice following intravenous injection we conjugated CT to OVA and traced the OVA by specific labeling with an anti-OVA FITC (green) conjugated antibody. The CT-OVA localizes distinctly in the marginal zone of the spleen after 2 hours following intravenous injections. The B-cell follicle is labeled with TexasRed-conjugated anti-IgM antibody (red). Superimposed is the expression of CD86 (red) on CD11c+ cells that have been isolated from the spleen 24 hours after injection compared to cells from naive (blue) mice.

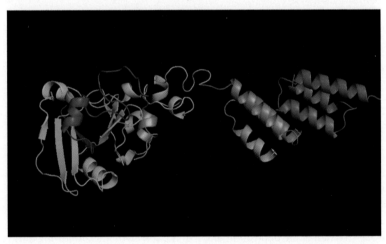

**Figure 3.4** Computer modeling of CTA1-DD adjuvant. To the left is the CTA1 and to the right is the DD-dimer.

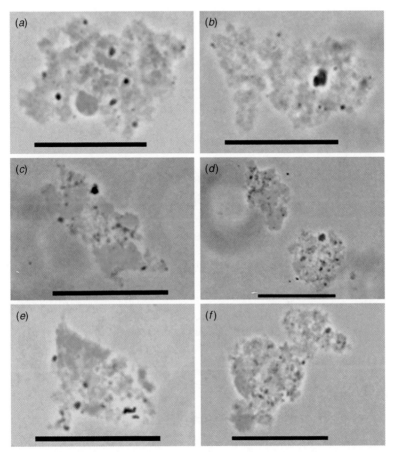

**Figure 4.14** Distribution of BODIPY FL-labeled bovine serum albumin adsorbed to aluminum hydroxide adjuvant and BODIPY TR-labeled bovine serum albumin adsorbed to aluminum hydroxide adjuvant in a combination vaccine: (*a*) BODIPY FL-labeled bovine serum albumin adsorbed to aluminum hydroxide adjuvant prior to combination; (*b*) BODIPY TR-labeled bovine serum albumin adsorbed to aluminum hydroxide adjuvant prior to combination; (*c*) 15 minutes after combination with mixing; (*d*) 30 minutes of mixing; (*e*) 45 minutes of mixing; (*f*) 60 minutes of mixing; The bars represent 5 μm. (From Ref. 64.)

**Appearance: milky white oil in water (o/w emulsion)**

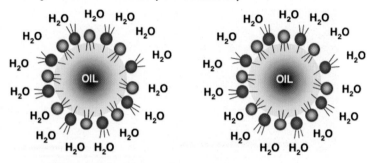

Composition:     0.5% Tween 80 - water-soluble surfactant
                    0.5% Span 85 - oil-soluble surfactant
                    4.3% Squalene oil
                    Water for injection
                    10 nM Na-citrate buffer

Density:  0.9963 g/mL                    Emulsion droplet size: ~165 nm

Viscosity:  close to water, easy to inject

**Figure 5.1**    Submicron MF59 emulsion composition.

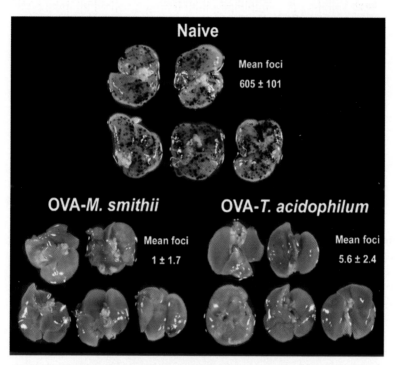

**Figure 12.11** Prophylactic vaccination against B16OVA metastatic melanoma using OVA-archaeosomes. C57BL/6 mice received injections of 15 μg of OVA entrapped in *M. smithii* or *T. acidophilum* archaeosomes on days 0 and 21. On day 35, groups of naive and OVA-archaeosome-immunized mice were challenged intravenously with B16OVA melanoma cells. On day 49, mice were euthanized, lungs harvested, and the black metastatic foci enumerated under a dissection microscope. In two of the naive control mice, because of the sheer high numbers, it was difficult to enumerate accurately the tumor foci, which were therefore assigned arbitrarily as 500. The mean ± SD of the tumor foci is indicated in the figure. OVA-archaeosome values are significantly different ($p < 0.001$) from those of naive mice as analyzed by the Mann–Whitney rank test. Data are representative of three experiments. (Reproduced after modification from Ref. 59.)

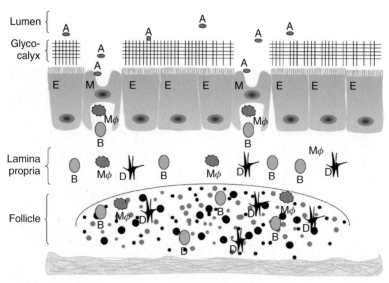

**Figure 13.3** Cellular structure of the diffuse and organized immune compartments in the gastrointestinal tract.

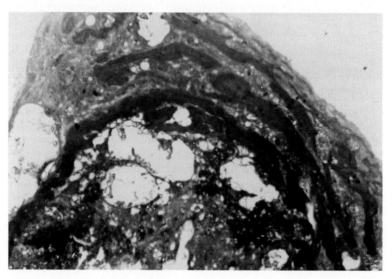

**Figure 18.1** Histological cross section of a granuloma induced by mineral oil adjuvant in BALB/c mouse. A stratified structure is seen surrounding large vacuoles where the oil droplets resided. (Preparation and photo: E. B. Lindblad and Jens Blom.)

79. Park, T.G., Lu, W. & Crotts, G. Importance of in vitro experimental conditions on protein release kinetics, stability and polymer degradation in protein encapsulated poly (D,L-lactic acid-co-glycolic acid) microspheres. *J Control Release* 1995, **33**, 211–219.

80. Shenderova, A., Burke, T.G. & Schwendeman, S.P. The acidic microclimate in poly(lactide-co-glycolide) microspheres stabilizes camptothecins. *Pharm Res* 1999, **16**(2), 241–248.

81. van de Weert, M., Hennink, W.E. & Jiskoot, W. Protein instability in poly(lactic-co-glycolic acid) microparticles. *Pharm Res* 2000, **17**(10), 1159–1167.

82. Johnson, R.E., Lanaski, L.A., Gupta, V. et al. Stability of atriopeptin III in poly(lactide-co-glycolide) microparticles. *J Control Release* 1991, **17**, 61.

# 11

# LIPOSOMAL CYTOKINES AND LIPOSOMES TARGETED TO CO-STIMULATORY MOLECULES AS ADJUVANTS FOR HIV SUBUNIT VACCINES

BULENT OZPOLAT AND LAWRENCE B. LACHMAN

## 11.1 INTRODUCTION

Liposomes are nontoxic uni- or multilamellar microvehicles consisting of a spherical phospholipid bilayer with a hydrophilic and/or aqueous inner compartment. The use of liposomes as vaccine adjuvants was first described by Allison and Gregoriadis in 1974 [1]. Their safety profiles and ability to induce an immune response makes them likely to be included in vaccine formulations.

Liposomes have been shown to be highly useful delivery systems for antigens, adjuvants, nucleic acids, and drugs. Liposomal formulations offer several major advantages. They can (1) prevent degradation of delivered antigens and adjuvants, (2) increase antigenicity of weak immunogens, (3) target antigen-presenting cells (APCs), (4) function as both a delivery system and a vaccine adjuvant at the same time, and (5) reduce the antigen and adjuvant doses required for an immune response and controlled release. Therefore, liposomes provide a safe and effective vaccine platform for subunit vaccines.

Liposomes may be prepared using phospholipids found in mammalian cell membranes and therefore nontoxic. Cholesterol is commonly added to the

*Vaccine Adjuvants and Delivery Systems*, Edited by Manmohan Singh
Copyright © 2007 John Wiley & Sons, Inc.

formulation to increase stability and bilayer rigidity. Their size may vary from several nanometers to micrometers, depending on the lipid contend and procedure. Liposomes can also be prepared with charge by incorporation of charged lipids. Addition of a small amount of anionic lipids (i.e., phosphatidylserine and phosphatidylglycerol) enhances the half-life of liposomes by reducing clearance from the circulation. Positively charged lipids (i.e., DOTAP) are used for delivery of genes and plasmids for transfection or gene vaccination purposes. Liposomes that contain neutral lipids, such as phosphatidyl cholines, form lamellar structures and increase the stability of liposomes.

Subunit antigens (i.e., peptides and synthetic small proteins) are usually weak immunogens that degrade easily in vivo and require an adjuvant and delivery system to induce protective immunity. Further targeting of liposomes to immune cells can be achieved by changing their structure, lipid content, charge, incorporating adjuvants (i.e., cytokines), and surface modifications. Immunoliposomes (liposomes coated with antibodies), virosomes (liposomes containing viral proteins), proteosomes (liposomes containing bacterial outer membrane and proteins), and archeosomes are some of the novel formulations that have been designed to target immune cells for more effective immune response [2–5]. In general, liposomes provide a safe and efficient option for the subunit vaccines.

The use of adjuvants is critical for the development of successful subunit vaccines because most peptide antigens are either nonimmunogenic or weakly immunogenic [6,7]. Adjuvants provide a mechanism for antigen persistence at the injection site and enhance the immune response to immunogens by prolonging their release and time of interaction with APCs such as macrophages, dendritic cells, and B cells [6]. Encapsulation of peptide antigens in liposomes protects them from degradation and enhances delivery of the amount of antigens to the APCs. Liposomes fall into the category of adjuvants that are considered as particulates. The function of particulate adjuvants is to increase antigen uptake by APCs and potentiate immunogenicity of associated antigen and the immune response.

APCs are highly specialized cells that can process antigens and present peptide fragments to T cells through major histocompatibility complex (MHC) class I or II molecules. MHC molecules recognize T-cell receptors (TCRs) on T cells. Identification of B7-1 (CD80) and B7-2 (CD86) as the prototypic costimulatory molecules has had a great impact on the understanding of T-cell and APC interactions [8,9] (Figure 11.1).

Cytotoxic CD8$^+$ T cells recognize antigenic peptides from proteins that have been produced in the cytoplasm and processed and presented via MHC class I. CD4$^+$ T cells, on the other hand, recognize peptides that have been uptaken by the APCs and processed through endolysosomal degradation and presented via MHC class II to TCR. Full T-cell activation requires signals from antigen-specific TCRs and the antigen-independent CD28 receptor. T-cell activation leads to the secretion of pro-inflammatory cytokines and effector T-cell

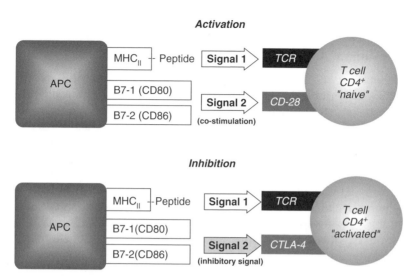

**Figure 11.1**   T-cell activation or inhibition/tolerance by antigen presenting cells. Signal 1 represents primary stimulatory signal transmitted by APCs to T-cell receptors (TCRs) through MHC molecules. Signal 2 represents co-stimulatory signal which is required to fully activate T cells via CD28–B7-1 or B7-2 (CD80 or CD86) interactions. Signal 2 transmits inhibitory signal from APCs to T cells through CTLA4–B7 molecules.

functions. CTLA4, which is expressed on T cells, binds to B7-1 and B7-2 with an affinity greater than for CD28 [10,11]. Cross-linking of CTLA4 with antibodies inhibits T-cell proliferation and IL-2 release, whereas simply blocking with CTLA4 Fab increases the proliferation of T cells [12–14]. These findings suggested that manipulation of the TCR/co-stimulatory receptor network offers therapeutic opportunities for control of hyperesponsiveness, such as autoimmune diseases, and hyporesponsiveness as exhibited by a poor response to subunit vaccine antigens and tumor antigens. Therefore, recent insights into the nature of antigen uptake by APCs and induction of immunoreactive B and T cells have led to novel approaches for vaccination of different diseases. Elucidation of novel mechanisms of T-cell activation and expression of co-stimulatory molecules on the immunological cells has led to new strategies for immunotherapy and vaccination [2].

Cytokines have significant potential as adjuvants since most adjuvants function by inducing production of cytokines from cells of the immune system [15,16]. Cytokines such as IL-1, IL-2, IL-6, IL-12, and IFN-γ have been used as adjuvants [6]. In addition to their carrier-adjuvant function, liposomes can be used for their depot effect to release antigens and other incorporated adjuvants, to increase antigen uptake by targeting antigens, and to generate strong immune responses [17,18].

## 11.2  MATERIAL AND METHODS

### 11.2.1  Antigen, Antibodies, and Cytokines

Recombinant gp120 from HIV-1 (MN strain, cholesterol-derived) and recombinant murine IFN-$\gamma$ (specific activity, $1.0 \times 10^7$ U/mg) was provided by Genentech, Inc. (South San Francisco, California). Biotinylated monoclonal anti-mouse B7-1 (CD80), anti-mouse B7-2 (GL1), anti-mouse CTLA4, anti-mouse CD28, anti-mouse CD11a, and control rat IgG$_{2a}$ were used to create immunoliposomes for vaccination purposes. Monoclonal antibodies to murine CD3(01082A) and Fc$\gamma$II/III were used as controls to demonstrate in vitro interactions between liposomes and target cells. All antibodies were purchased from Pharmingen Inc. (San Diego, California). Some antibodies were biotinylated using a biotinylation kit from Pierce Chemical Co. (Rockford, Illinois). Monoclonal antibodies were found to contain less than 12 pg of endotoxin per immunization dose.

### 11.2.2  Preparation of Liposomes Containing gp120 and Cytokines

The types of liposomes we used (Figure 11.2) were dehydration–rehydration vesicles (DRVs) created from phosphatidylcholine (PC), cholesterol (CHO), and biotinylated phosphatidylethanolamine (PE-B) (Avanti Polar Lipids Inc., Pelham, Alabama) according to the procedure of Gregoriadis et al. [19]. We prefer DRVs to traditional liposomes prepared by adding the solute to a dried lipid film because DRVs entrap greater quantities of solute [2]. When calculating the amount of phospholipids required for an experiment, one extra dose was always included for syringe retention. Thus, we routinely prepare 0.6 mL of liposomes for a group of five mice (0.1 mL/mouse) or 1.1 mL for a group of 10 mice. All lipids were used from stock solutions prepared in chloroform

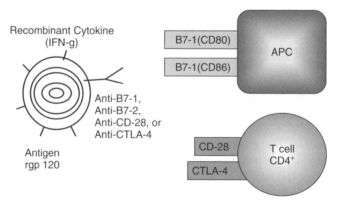

**Figure 11.2**  Immunoliposomes expressing B7, CD28, and CTLA through biotin–avidin biotin bridges at the outer surface and designed to target T and APL cells. Liposomes also carry gp120 antigen along with cytokine molecules.

**TABLE 11.1 DRV Formulation**

| Lipid Used in DRVs | One Mouse Dose (μg) |
|---|---|
| Phosphotidylcholine | 380 |
| Cholesterol | 193 |
| Phosphatidylethanolamine (biotinylated) | 10 |

($CHCl_3$). PC (500 nmol/mouse dose), CHO (500 nmol/mouse dose), and PE-B (100 nmol/mouse dose) at a 5:5:1 molar ratio are dissolved in anhydrous chloroform and placed in 25-mL sterile round-bottomed screw-capped glass centrifuge tubes (Corex, Fisher Scientific, Pittsburgh, Pennsylvania) (Table 11.1). The tubes are rotated continuously at a 45° angle while being dried by a stream of $N_2$ gas. The residual chloroform was removed by vacuum desiccation for one hour.

DRVs were prepared by creating standard liposomes from a dry lipid film and adding a small volume (50 to 100 μL) of distilled, autoclaved water sequentially until the total amount of water required has been added. The gp120 (15 μg/mouse), INF-γ (1.5 μg/mouse), and/or IL-6 (5U, R&D Systems, Minneapolis, Minnesota) were added to the water suspension of liposomes with vigorous vortex for 30 seconds. The liposome suspension was frozen in a dry ice–acetone bath (−20°C) by rotating the tubes slowly at a 45° angle to create a thin lipid film. The tubes containing the frozen lipid film were lyophilized to dryness with the screw caps loosely open. The dried liposomes were hydrated with 100-μL aliquots of water added slowly with continuous vortex. The lyophilization and hydration steps were repeated twice more to increase antigen trapping by the liposome preparation by centrifugation and washing three times with 10 mL of ice-cold phosphate-buffered saline (PBS) at 13,000 g (10,000 rpm in a Sorvall RC-5B centrifuge with an SS34 rotor) at 4°C for 30 minutes. Following the third wash the liposomes were resuspended in water (100 μL/mouse dose).

To create immunoliposomes, DRVs prepared as described above were treated with 2.5 μg/per mouse dose of avidin (Pierce, Rockford, Illinois) dissolved in water and incubated at room temperature for 30 minutes with gentle mixing of the tubes every 10 minutes [2]. After washing twice with ice-cold PBS, the liposomes were incubated with biotinylated monoclonal antibodies. In our experiments we use either monoclonal anti-mouse B7-1, anti-mouse B7-2, anti-mouse CTLA4, anti-mouse CD28 (2 μg/mouse), anti-mouse CD11a (all at 2 μg/mouse dose), or rat $IgG_{2a}$ (2 μg/mouse) as a control (Figure 11.3). Following incubation for one hour at 37°C immunoliposomes were washed twice with 10 mL of PBS at 10,000 rpm at 4°C for 30 minutes to remove unbound antibodies. Immunoliposomes were resuspended in PBS (100 μL/mouse) and used immediately for immunization of animals. All glassware was treated at 180°C for four hours to inactivate endotoxin. Sterile pyrogen-free water and PBS were used to prepare all solutions, and liposomes were prepared in a sterile tissue culture cabinet.

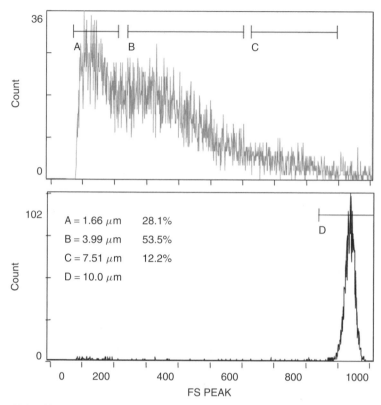

**Figure 11.3** Size of the liposomes detected by flow cytometry. Most of the liposomes range between 1.5 and 7.5 μm.

## 11.3 RESULTS

### 11.3.1 Entrapment Capacity of DRVs

Earlier, we performed a series of experiments to determine the percentage of antigen and cytokines trapped by DRV-type liposomes using radiolabeled gp120 and cytokines [20]. The liposomes were prepared in the standard manner from PC and phosphatidylserine (PS) (PC/PS, 1:1) or PC and cholesterol (PC/CHO, 7:3). As shown in Table 11.2, PC/CHO traps larger amounts of cytokines, although both types of liposomes entrap about the same amount of gp120. The trap ratios probably differ because of differences in charge and the ratio of hydrophobic amino acids in the proteins. We also measured the size of the liposomes and found that more than 90% of liposomes have a size ranged between 1.5 and 7.5 μm (Figure 11.3).

TABLE 11.2　Entrapment of Cytokines and gp120 in DRVs

| Protein | Percent Entrapment[a] | |
|---|---|---|
| | PC/CHO | PC/PS |
| IL-1α | 13.5 ± 0.4 | 5.2 ± 1.1 |
| IL-1β | 24.7 ± 2.6 | 20.2 ± 0.2 |
| IL-6 | 34.5 ± 3.3 | 17.5 ± 2.1 |
| TNFα | 24.5 ± 2.8 | 7.2 ± 1.1 |
| INF-δ | 22.7 ± 3.8 | 14.5 ± 0.3 |
| gp120 | 16.9 ± 0.2 | 16.3 ± 1.6 |

[a]Trapping ratio is the presence of added radiolabeled recombinant cytokine or gp120 associated with the DRVs following three washes in PBS.

### 11.3.2　Slow Release of Trapped Proteins by DRVs

Liposomes containing radiolabeled cytokines and gp120 were kept in RPMI 1640 medium containing 5% FCS at 37°C for up to three days to determine the level of retention of cytokines and antigen in DRVs. The percentage of radiolabeled cytokine associated with the DRVs was in the range 87 to 94% in PC/CHO liposomes after three days incubation in the medium [20]. The percentage of radiolabeled gp120 associated with DRVs was about 80% for the same period of time, suggesting that liposomes were able to release the proteins slowly to the environment.

### 11.3.3　In Vitro Binding Capability of Immunoliposomes to Target Cells

To demonstrate that immunoliposomes were able to bind target cells expressing co-stimulatory molecules, we perform analysis by flow cytometry. Fluorescent immunoliposomes were prepared by incorporating 0.1% of total PC content with NBD (7-nitro-2-1,3-benzoxadiazol-4-yl)-PC. Immunoliposomes contained monoclonal antibodies (MAb) to B7-1 were able to bind K-1735 cells transfected with B7-1 but did not bind to parental K-1735 cells after two hours of incubation at 4°C. Immunoliposomes containing either MAb to CD3 or no MAb do not bind the cells, suggested that the binding of liposomes to the cells was mediated by specific MAb to B7-1. Similarly, immunoliposomes containing MAb to B7-2 were incubated with the murine macrophage cell line, J774, which express B7-2. As shown in Figure 11.4, immunoliposomes containing MAb to B7-2-bound J744 cells, whereas immunoliposomes containing MAb to CD3 or no MAb did not bind. The presence of cell surface expression of B7-1 and B7-2 molecules was confirmed using soluble MAb to both molecules.

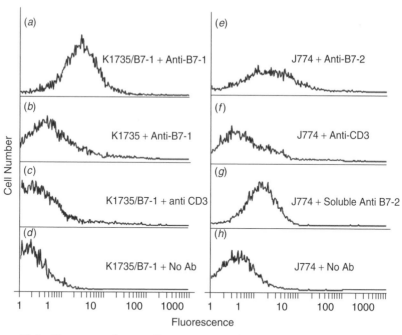

**Figure 11.4** Fluorescent immunoliposomes containing mAb to B7-1 bound to K1735 cells transfected with B7-1 and immunoliposomes containing mAb to B7-2 bound to J774 cells. Flow cytometric analysis of NBD-containing immunoliposomes prepared with anti-B7-1 demonstrated tight binding to K1735 cells transfected with B7-1 (*a*) but not to *neo*-transfected control K1735 cells (*b*). Immunoliposomes containing anti-CD3 (*c*) or lacking mAb (*d*) did not bind to B7-1-transfected K1735 cells. Immunoliposomes containing mAb to B7-2 bound to J774 cells (*e*), whereas immunoliposomes containing mAb to CD3 did not. Soluble anti-B7-2 demonstrated the presence of B7-2 on the surface of the J774 cells (*g*) and immunoliposomes lacking a mAb did not bind to J774 cells (*h*).

### 11.3.4 Immunization of Mice with Soluble gp120 and Immunoliposomes Containing Cytokines and gp120

Groups of five mice were immunized subcutaneously three times at 14-day intervals. The immunoliposomes were injected in 0.1 mL of PBS. Each injection contains 380 μg of PC, 193 μg of CH, 10 μg of gp120 of PE-B, 2.0 μg of MAb, and 10 μg of gp120, as determined previously from trapping experiments. Blood samples were collected two weeks after the last vaccine treatment on day 42. Blood samples taken from the tail vein were kept at 25°C for 30 minutes and centrifuged for 10 minutes at 5000 rpm using an Eppendorf microcentrifuge. The supernatant sera was separated into aliquots and frozen for future assays. Sera were tested for gp120-specific IgG, $IgG_1$, and $IgG_{2a}$ by ELISA. For ELISA flat-bottomed 96-well plates were coated with 20 μg of recombinant gp120. Plates were blocked with 100 μL of 1% bovine serum albumin (BSA) plus 1% polyvinylpyrrolidone (PVP-40, Sigma, St. Louis, Missouri) and were incubated at 25°C for 30 minutes. After three washes with PBS-Tween (6 mM sodium phosphate, pH 7.2, 0.15 M NaCl, 0.05% Tween 20, Sigma), serum samples were added to the plates. The blocking buffer was used as a diluent for antibodies and conjugates for all subsequent steps. To measure gp120-specific IgG response the serum was tested at a 1:50 and 1:100 dilution. For IgG, diluted samples were loaded in wells as duplicates and incubated for two hours at room temperature. After washing with buffer, 100 μL of a 1:2000 dilution of anti-mouse IgG conjugated with horseradish peroxidase (HRP) (Sigma), 1:1000 dilution of anti-mouse $IgG_1$ (Serotech, Oxford, England), or $IgG_{2a}$ (Pharmingen, San Diego, California) conjugated with HRP were added to wells and incubated for 30 minutes. Plates were washed five times with wash buffer, once with only PBS, and the color was developed by adding 100 μL of the substrate solution o-phenylenediamine dihydrochloride (Sigma) in phosphate citrate buffer. The reaction was stopped at 30 minutes by adding 3 M sulfuric acid. Optical density was measured by an ELISA plate reader (Dynatech MR5000) at 450 nm.

### 11.3.5  Delayed-Type Hypersensitivity and Humoral Immune Response in Mice After Vaccination with Liposomes and Immunoliposomes Containing Cytokines and gp120

To determine whether immunoliposomes induce an immune response to gp120 compared with liposomes lacking MAb, delayed-type hypersensitivity (DTH) responses were determined two weeks after the last immunization. DTH was measured as footpad swelling 24 hours after injection of 1 μg of rgp120 into both hind feet of mice. As shown in Figure 11.5, mice vaccinated with immunoliposomes containing MAb to B7-1, B7-2, CD28, and CTLA4 had significantly increased footpad swelling 24 hours after injection of the antigen. However, immunoliposomes containing control rat $Ig_{2a}$ antibodies or liposomes lacking MAb did not result in significant footpad swelling. Similarly, immunoliposomes that do not contain gp120 were also not able to produce significant footpad swelling. Interestingly, we did not find any additional effect in the increase of footpad swelling when IFN-γ was included in the liposomes. Our findings clearly demonstrate that the strategy of immunization by using liposomes targeted to co-stimulatory molecules induces a DTH response, and immunization with immunoliposomes does not induce statistically significant $IgG_1$ and $IgG_{2a}$ titers in the mice (Figure 11.6).

### 11.3.6  Cytokine Production by Lymph Node Cells from Mice Vaccinated with Immunoliposomes Containing Cytokines and gp120

To determine if a Th1 versus Th2 response to antigen resulted from vaccination with immunoliposomes, lymph nodes from mice vaccinated six to eight weeks earlier were isolated and incubated with liposomes containing only

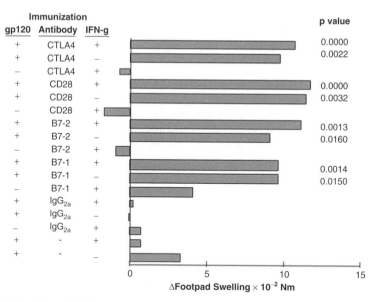

**Figure 11.5** Mean DTH responses of groups immunized three times with immunoliposomes containing rgp120 and mAbs to either CTLA4, CD28, B7-2, or B7-1 and in the presence or absence of IFN-γ. Fourteen days following the final immunization, the hind footpads were injected with 1 μg of soluble rgp120 and the amount of swelling was measured 24 hours later. Control mice were immunized with immunoliposomes containing rat IgG$_{2a}$ or no mAb. The *p* values were determined by comparing the experimental group with the control group containing a mAb but lacking rgp120. Experimental groups containing IFN-γ were compared with the control group containing IFN-γ but lacking a mAb.

gp120 for 48 hours [2]. IL-4 and IFN-γ levels in the culture medium were determined by an ELISA procedure (Pharmingen, San Diego, California). IL-4 was undetectable in all samples and IFN-γ was detected only in groups immunized with immunoliposomes containing gp120 + anti-B7-2 + IFN-γ (293 pg/mL), gp120 + anti-CTLA4 (334 pg/mL), and gp120 + anti-CTLA4 + IFN-γ (415 pg/mL), demonstrating that these immunoliposomes induce a Th1 immune response.

In conclusion, the purpose of the experiments described above was to determine if attachment of MAb to co-stimulatory molecules made liposomes more effective for the induction of a primary immune response to antigen. Presumably, the MAb to B7-1 and B7-2 would direct the liposomes to APCs and the MAb to CD28 and CTLA4 toward T lymphocytes. Our assumption that immunoliposomes containing MAb to CD28 and CTLA4 could bind to T lymphocytes must be considered in light of the fact that liposomes were cleared by phagocytic cells, some of which were APCs. Our findings supported our hypoth-

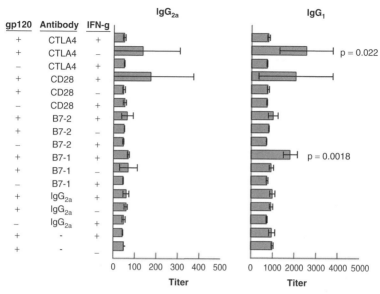

**Figure 11.6**   Measurement of mean serum IgG$_{2a}$ and IgG$_1$ Ab levels to gp120 in the serum of immunized mice. The endpoint titer for each animal was determined by ELISA as described in Section 11.2.

esis that immunoliposomes did, in fact, induce a greater immune response to gp120 than that of liposomes not containing the MAb (Figure 11.7).

In addition to IFN-g, we have previously used other cytokines, including IL-6, TNFα, IL-1α, IL-1β, and a combination of cytokines to determine antibody and cell-mediated immunity. These experiments suggested that the strategy of using liposomes containing cytokines along with gp120 antigen can significantly potentiate the immune response to the antigen.

## 11.4   CONCLUDING REMARKS

Liposomes have proven to provide safe and excellent vaccine platforms, especially for subunit vaccines. The flexibility of the formulating liposomes provides a chance for incorporation of additional immunostimulators or adjuvants (e.g., cytokines) along with antigen. Liposomes may also be targeted to cellular receptors via the inclusion of antibodies, receptor, or ligand proteins as well as mannose moiety on the surface to increase uptake by APCs to potentiate activation of the immune cells and protective immune response. Currently, several liposomal formulations for parenteral and intranasal immunization have generated promising results using antigens (influenza, hepatitis A, diphtheria, and tetanus toxoid).

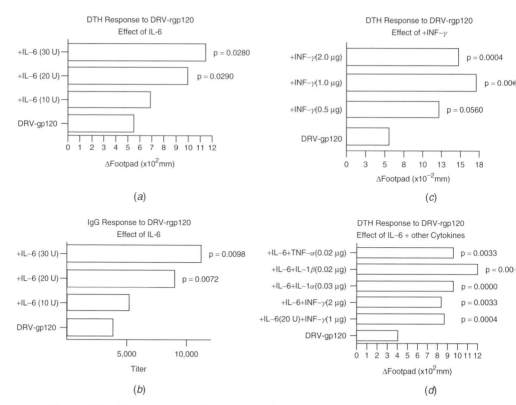

**Figure 11.7** Measurement of foot pad swelling to examine DTH and mean serum IgG antibody levels to gp120 in the serum of immunized mice. Mice were immunized with DRV liposomes containing gp120 and cytokines including IL-6 (*a* and *b*), IFN-γ (*b*), TNFα, IL-1α, and IL-1β, and a combination of cytokines (*c*) to determine antibody and cell-mediated immunity to gp120 antigen.

## ACKNOWLEDGMENTS

The authors thank Michael F. Powell, formerly of Genentech, Inc., for providing the gp120 used in this study. This research was supported by an Advanced Technology Program grant (ATP-15-100) from the Texas Higher Education Coordinating Board and Cancer Center Core grant CA16672 from the National Cancer Institute.

## REFERENCES

1. Allison, A.G. & Gregoriadis, G. Liposomes as immunological adjuvants. *Nature* 1974, **252**(5480), 252.

2. Ozpolat, B., Rao, X.M., Powell, M.F. & Lachman, L.B. Immunoliposomes containing antibodies to costimulatory molecules as adjuvants for HIV subunit vaccines. *AIDS Res Hum Retroviruses* 1998, **14**(5), 409–417.

3. Moser, C., Metcalfe, I.C. & Viret, J.F. Virosomal adjuvanted antigen delivery systems. *Expert Rev Vaccines* 2003, **2**(2), 189–196.

4. Fries, L.F., Montemarano, A.D., Mallett, C.P., Taylor, D.N., Hale, T.L. & Lowell, G. H. Safety and immunogenicity of a proteosome–*Shigella flexneri* 2a lipopolysaccharide vaccine administered intranasally to healthy adults. *Infect Immun* 2001, **69**(7), 4545–4553.

5. Patel, G.B., Zhou, H., KuoLee, R. & Chen, W. Archaeosomes as adjuvants for combination vaccines. *J Liposome Res* 2004, **14**(3–4), 191–202.

6. Allison, A.C. & Byars, N.E. Immunological adjuvants and their mode of action. *Biotechnology* 1992, **20**, 431–449.

7. Bhardwaj, N. Processing and presentation of antigens by dendritic cells: implications for vaccines. *Trends Mol Med* 2001, **7**(9), 388–394.

8. Mosmann, T.R. T lymphocyte subsets, cytokines, and effector functions. *Ann NY Acad Sci* 1992, **664**, 89–92.

9. Sprent, J. Antigen-presenting cells: professionals and amateurs. *Curr Biol* 1995, **5**(10), 1095–1097.

10. Bluestone, J.A. New perspectives of CD28-B7-mediated T cell costimulation. *Immunity* 1995, **2**(6), 555–559.

11. Chambers, M.A., Vordermeier, H., Whelan, A. et al. Vaccination of mice and cattle with plasmid DNA encoding the *Mycobacterium bovis* antigen MPB83. *Clin Infect Dis* 2000, **30**(Suppl 3), S283–287.

12. Linsley, P.S., Greene, J.L., Tan, P. et al. Coexpression and functional cooperation of CTLA-4 and CD28 on activated T lymphocytes. *J Exp Med* 1992, **176**(6), 1595–1604.

13. Krummel, M.F. & Allison, J.P. CD28 and CTLA-4 have opposing effects on the response of T cells to stimulation. *J Exp Med* 1995, **182**(2), 459–465.

14. Magott-Procelewska, M. Costimulatory pathways as a basic mechanisms of activating a tolerance signal in T cells. *Ann Transplant* 2004, **9**(3), 13–18.

15. Heath, A.W. & Playfair, J.H. The potential of cytokines as adjuvants. *AIDS Res Hum Retroviruses* 1992, **8**(8), 1401–1403.

16. Lachman, L.B., Ozpolat, B. & Rao, X.M. Cytokine-containing liposomes as vaccine adjuvants. *Eur Cytokine Netw* 1996, **7**(4), 693–698.

17. Gregoriadis, G. The immunological adjuvant and vaccine carrier properties of liposomes. *J Drug Target* 1994, **2**(5), 351–356.

18. Phillips, N.C. & Dahman, J. Immunogenicity of immunoliposomes: reactivity against species-specific IgG and liposomal phospholipids. *Immunol Lett* 1995, **45**(3), 149–152.

19. Gregoriadis, G., Davis, D. & Davies, A. Liposomes as immunological adjuvants: antigen incorporation studies. *Vaccine* 1987, **5**(2), 145–151.

20. Lachman, L.B., Shih, L.C., Rao, X.M. et al. Cytokine-containing liposomes as adjuvants for HIV subunit vaccines. *AIDS Res Hum Retroviruses* 1995, **11**(8), 921–932.

# 12

# ARCHAEOSOME VACCINE ADJUVANTS FOR CROSS-PRIMING CD8+ T-CELL IMMUNITY

LAKSHMI KRISHNAN AND G. DENNIS SPROTT

## 12.1 INTRODUCTION

Vaccination is the most cost-effective solution for control of disease. Although some infections, such as smallpox, have been fully eradicated through the intelligent use of vaccines, many other infections, such as tuberculosis, malaria, influenza, and HIV, currently lack effective prophylactic interventions. Vaccines are also desirable for cancer immunotherapy, and when used in conjunction with radio- and chemotherapy could provide long-term control of metastasis [1]. The future challenge for vaccinology is to devise efficacious nonreplicating acellular vaccines to combat chronic and/or intracellular infections and cancers.

The two main challenges in vaccine development include identification of specific antigenic targets and the ability to evoke a strong, appropriate, and long-lasting immune response. Over the past few decades, molecular and genomic approaches have contributed substantially to the identification, purification, and/or synthesis of key antigenic determinants of pathogens and tumors. However, relatively poor immunogenicity may be expected from such highly purified proteins and or peptides, limiting their ability to induce a strong protective immune response. Taking cues from the natural host response to infection, it is clear that two signals need to be coordinated for evoking a response to a purified pathogen- or tumor-specific signal: *antigen delivery* for appropriate processing and presentation of the antigen, and *danger signal*

*Vaccine Adjuvants and Delivery Systems*, Edited by Manmohan Singh
Copyright © 2007 John Wiley & Sons, Inc.

*perception* for mimicking the host response to infection. The newfound knowledge that interaction of pathogen-derived molecular patterns (PAMPs) with mammalian antigen-presenting cells (APCs) via Toll-like receptors (TLRs) and other innate receptors can signal an effective immunity cascade is driving the quest for novel molecular adjuvants [2].

The term adjuvant is often arbitrarily and interchangeably used to describe many vaccine formulation approaches with varied functions. For the purpose of discussion relevant to this article, we have broadly categorized vaccine adjuvant strategies as approaches for either antigen delivery or immunostimulation (Figure 12.1). In the category of antigen delivery systems, nonreplicating particulate carriers such as iron-coated beads are unsuitable for vaccines [3]. In contrast, particulate polymeric microspheres and liposomes can provide efficient antigen depots [4]. However, antigen delivery approaches often require additional immunostimulating components to evoke protective immunity [5,6]. Another approach has been the use of replicating systems; viral vectors direct responses to $CD8^+$ T-cell immunity, sometimes with limited $CD4^+$ T-cell help, whereas bacterial vectors bias toward $CD4^+$ T cells [7,8]. Although attenuated, replicating vectors also come with a risk of reversion to virulence. A promising approach is the use of recombinant viruslike particles that are nonreplicating but mimic viral antigen delivery pathways [9]. In contrast to delivery systems, some adjuvants are immunostimulants that recruit various innate immune cells for directing responses against the co-formulated antigen. PAMPs that interact with TLRs constitute many of the new-generation adjuvants [2,10]. In this category are included molecules such as CpG DNA [11], monophosphoryl lipid A [12], and imidizoquinolines [2]. While these immunostimulants can be efficient at triggering the innate cascade, toxicity and short half-life can be issues limiting their use in vaccines. An efficient strategy to counter this problem has been to deliver immunostimulants in

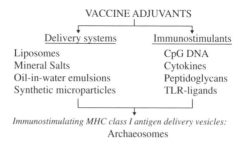

VACCINE ADJUVANTS

Delivery systems                    Immunostimulants

Liposomes                           CpG DNA
Mineral Salts                       Cytokines
Oil-in-water emulsions              Peptidoglycans
Synthetic microparticles            TLR-ligands

*Immunostimulating MHC class I antigen delivery vesicles:*
Archaeosomes

**Figure 12.1** Classification of adjuvants. Vaccine antigens are formulated in antigen delivery systems for targeting an appropriate immune effector arm, and immunostimulants are required to promote potent responses. Particulate delivery vesicles such as liposomes and microparticles, and oil and water emulsions constitute antigen delivery systems. Immunostimulants include a variety of molecular substances that can act directly on immune system cells to activate innate and consequent adaptive immunity. Archaeosomes offer both targeted antigen delivery and activation of DCs.

particulate antigen carriers [13]. This approach has the dual advantage of targeting antigen delivery and communicating danger signal perception to the same APCs for efficient priming and maintenance of immune response.

From a regulatory, cost-effective, and ease-of-compliance perspective, vaccine formulations need to be simple in their composition (including the minimal molecular components), easy to produce on a large scale, preferably not require refrigeration for storage, and most important, be capable of directing a holistic immune response (innate, B-cell, $CD4^+$ and $CD8^+$ T-cell responses). One extremely important consideration for vaccine formulations is also the need to have potent memory responses with a minimum number of immunizations.

The archaeosome technology represents a novel self-adjuvanting delivery system that combines the ability to target antigen to the MHC class I processing pathway with sufficient inherent immunostimulation for induction of potent and long-lasting immunity to coadministered subunit antigens (Figure 12.1). Archaeosomes are lipid vesicles derived from the unique ether lipids of archaea. The term *archaeosomes*, which describes this vesicle delivery system, brings together the concepts of archaea and liposomes (Figure 12.2*a*).

Archaeosomes are very akin to liposomal vesicles, with a hydrophilic core and a hydrophobic outer membrane, and are constituted as 50 to 250-nm unilamellar vesicles. The antigen may be entrapped within the hydrophilic core, anchored in the membrane, or coupled to the surface (Figure 12.2) using

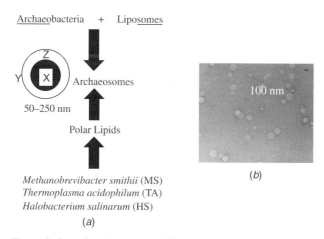

**Figure 12.2** Formulation of archaeosomes. The term *archaeosomes* was coined to bring together the concepts of Archaea + liposomes. Antigen may be entrapped either within the hydrophilic core of the archaeosomes, anchored in the membrane, or linked to the surface, as represented by X, Y, or Z antigens. Archaeosomes are 50 to 250 nm in size, unilamellar, and are constituted with the TPL unique to each archaeon. The three most tested archaeosome types in murine models include those derived from the TPL of either MS, TA, or HS (*a*). Electron micrograph of an archaeosome preparation (*b*).

standard entrapment procedures developed to formulate liposomal vesicles. The unique characteristics of archaeosomes are attributable to the novel ether glycerolipids of archaea (Figure 12.3).

In this chapter we detail studies carried out thus far on archaeosomes, their characteristics, and their ability to augment antibody and cell-mediated immunity, provide potent memory, and provide applications as infectious diseases and cancer vaccines. Although we focus strongly on the adjuvant capabilities of archaeosomes, for a better understanding of the differences between archaeosomes and conventional liposomes (constituted with synthetic ester lipids such as phosphatidylglycerol and cholesterol), we begin with a description of archaeal lipid structures and stability.

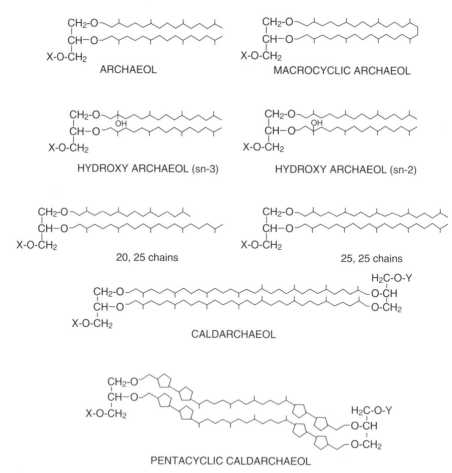

**Figure 12.3**  Polar lipids found in the various archaea. X and Y represent the divergent polar head groups found in the various strains. Core lipid structures result when X and Y are replaced by protons.

## 12.2 ARCHAEAL LIPID STRUCTURES

The domain Archaea comprises prokaryotes distinct from the domain Bacteria that occupy primarily extreme environmental habitats. Archaea includes organisms that are hyperthermophiles, thermoacidophiles, obligate anaerobes such as methanogenic archaea, extreme halophiles, and psychrophiles. Indeed, archaea have been estimated to represent 30% of the submicrometer population of the Antarctic and Alaskan waters [14]. One of the unifying features of Archaea that facilitate their survival in harsh environmental niches is the structure and composition of their cytoplasmic membrane lipids [15,16].

The total lipids of Archaea obtained by solvent extraction account for about 5% of cell dry weight and can be separated into neutral (acetone soluble) and polar (acetone-insoluble) lipid fractions [16]. The discovery by Kates and co-workers [17] of novel nonsaponifiable lipids from *Halobacterium salinarum* and their characterization as ether glycerolipids predates the discovery of a third domain of life. It is now known that the ether polar lipid structure is a distinguishing feature of Archaea [18]. The general architecture of these glycerolipids includes a core isopranoid chain of constant length with novel stereochemistry wherein the isopranoid chains are attached via ether bonds to the *sn*-2,3-glycerol carbons (Figure 12.3). In contrast, conventional phospholipids found in Bacteria and Eucarya have fatty acyl chains of variable length which may be unsaturated and attached via ester bonds to the *sn*-1,2 carbons of the glycerol. The carbon–carbon bonds of the alkyl chains from many Archaea are fully saturated, but exceptions occur, especially for the cold-adapted species [19]. The diphytanyl glycerol archaeol lipid core is ubiquitous to Archaea. However, some species synthesize additional variations, including hydroxyarchaeols, macrocyclic archaeols, archaeols with carbon-25 chains, or caldarchaeols sometimes with varying number of cyclopentane rings (Figure 12.3). Changes in the proportions of the core lipids synthesized provide an efficient way to control membrane fluidity in response to environmental changes [20]. Archaeols have one polar head group attached (monopolar), and caldarchaeols are bipolar. The polar groups encountered in Archaea are often similar to those encountered in ester glycerolipids of Eukarya and Bacteria. These include a variety of phospholipids, glycolipids, sulfated glycolipids, amino- and phosphoglycolipids, phosphopolyollipids, and phosphoaminolipids. However, an archaeal phosphatidylcholine is rarely found, with the exception of *Methanopyrus kandleri* [21] and a *N,N,N*-trimethylaminopentanetetrol variation found in *Methanospirillum hungatei* [22].

Despite their apparent novelty, archaeal lipid structures are not as esoteric as might be imagined. Isprenoid and isopranoid lipids are found in mammals as fat-soluble vitamins that have been reported to stimulate the immune system [23–26], whereas ether plasmanyl and plasmenyl lipids are typically found in mammalian leukocytes and macrophages [27]. Indeed, the mevalonic acid pathway of isopranoid biosynthesis is generally the same in eukaryotes compared to either methanogenic Archaea [28] or extreme halophiles [29].

## 12.3  ARCHAEOSOME CHARACTERISTICS

### 12.3.1  Formulation

Archaeosomes are defined as lipid vesicles made with one or more of the ether polar lipids that are unique to Archaea. The lipid membrane of archaeosomes may be entirely in the form of a bilayer (if made exclusively from monopolar archaeol lipids) or a monolayer (if made exclusively from bipolar caldarchaeol lipids). In most studies archaeosomes are formulated as unilamellar vesicles (Figure 12.2b); nevertheless, they may also be formulated as multilamellar vesicles by standard procedures used for liposome construction. In this chapter we refer to conventional liposomes as those constituted with synthetic ester-linked fatty acyl lipids of the composition, L-α-dimyristoylphosphatidylcholine/L-α-dimyristoylphosphatidylglycerol/cholesterol (1.8:0.2:1.5 molar ratio).

A historical perspective on formation of archaeal lipid vesicles has been presented in an earlier review [30] and is not repeated here. Methods to prepare archaeosomes are similar to methods used to prepare liposomes from nonarchaeal lipids [31] and have been described in detail previously [32]. To avoid the necessity of purifying lipids for archaeosome preparations, most studies have used the total polar lipids (TPLs) from various archaea. Using TPLs also has the advantage of arriving at a mixture of polar lipids that are conducive to archaeosome formation. The TPL source constituting the archaeosomes is often referred to as in the example: *Methanobrevibacter smithii* archaeosomes.

In general, the archaeon of interest is grown in 75 to 250-L fermenters, the lipids are extracted using methanol/chloroform/water (2:1:0.8v/v), and the TPL fraction is collected by cold acetone precipitation. Polar lipids are dissolved in chloroform/methanol (2:1 v/v), precipitated again with acetone, and stored in chloroform/methanol at 4°C. TPL obtained from individual archaea are dried down and hydrated in water or phosphate-buffered saline containing the antigen. After 18 to 24 hours of hydration, the average diameter of the vesicles formed is reduced in size by sonication or pressure extrusion. Vesicle size is in the range 50 to 250nm, as measured in a submicrometer particle sizer. Removal of nonentrapped antigen is achieved by high-speed centrifugation and washing. The final steps are filtration through a sterile 0.45-μm filter, and quantification of entrapped antigen. Entrapped antigen is quantified by modified SDS (sodium dodecyl sulfate) Lowry after lipid removal [33] or by SDS gel separation and densitometry. High-performance liquid chromatography may also be used for more accurate quantification, particularly for entrapped peptides. As with any vesicular system, entrapped material may include proteins, peptides, lipidated peptides, carbohydrates, and DNA. Entrapment efficiency can vary with vesicular size and characteristics of the material to be entrapped, from 30% for a soluble protein such as ovalbumin (OVA) to ≥90% for highly hydrophobic lipoproteins or lipidated peptides.

Archaeosomes may be formulated with relative ease from lipids of all the classes of Archaea, and many different types have been evaluated [30,34]. From among the various archaeosome types tested, three, those constituted of TPLs from *M. smithii*, *H. salinarum*, and *Thermoplasma acidophilum*, were narrowed down as the most interesting for vaccine development studies, based on their ability to augment strong primary and/or sustained cytotoxic T-lymphocyte (CTL) responses. Further, lipid structural differences among these Archaea represented a good sampling of lipid cores and head groups: *M. smithii* and *T. acidophilum* rich in caldarchaeols, with the former abundant in phosphatidylserine and the latter rich in β-L-gulose [35] and mannose head groups [36]; and *H. salinarum*, lacking caldarchaeols but abundant in diacidic phosphoglycerol methylphosphate head groups on an archaeol lipid core.

## 12.3.2 Stability and Tissue Distribution

One key consideration for any liposomal delivery system is physical and chemical stability. The former relates to freedom from fusion or loss of integrity, which can cause loss of encapsulated material and aggregation that can alter pharmacokinetics. Chemical stability relates to oxidation effects and/or hydrolysis on storage. Liposomes prepared from conventional ester lipids that have unsaturated fatty acyl chains may be subject to autooxidation, causing yellowing of the preparation over time. Unsaturation may also result in the formation of potentially toxic peroxides and aldehydes [37]. Thus, antioxidants often have to be included in the preparation, and storage may require inert nitrogen atmospheres. Further, ester-bonded lipids make conventional liposomes susceptible to chemical and enzymatic hydrolysis, which can lead to the formation of lysolecithins and fatty acids that affect membrane permeability, resulting in faster leakage of entrapped material [38,39].

Several unique characteristics of the archaeal lipids predicted increased physical and chemical stability. First, the archaeal saturated phytanyl lipids (as opposed to fatty acyl chains) are not oxidized in air, allowing easy storage at room temperature. Further, the *sn*-2,3 steriochemistry, as opposed to the *sn*-1,2 steriochemistry for attachment of the hydrophobic chains to the glycerol backbone, also confers inherent membrane stability and resistance to membrane permeability. Second, the stability of archaeosomes can be enhanced with the proportion of caldarchaeol lipid in the preparation. Whereas conventional ester lipids form bilayer arrangements, bipolar caldarchaeol lipids form a monolayer structure that spans the membrane with a polar group facing each side [40]. Inherently, such structures are more rigid and stable. The frequency of intramembrane fractures along the hydrophobic plane of the membrane of both archaeal cells and archaeosomes could be correlated to the amount of caldarchaeols present, indicating a membrane-spanning property of caldarchaeol lipids [41]. Notably, intramembrane fractures were essentially absent at a content of greater than 50% caldarchaeols. Although Archaea such as

*H. salinarum* are incapable of caldarchaeol synthesis, many others generate a mixture of archaeols and caldarchaeols such that the archaeosome membrane structure would be expected to change depending on the source of TPLs. The membrane arrangements of TPLs from various archaea would include a bilayer typified by the 100% archaeol content of *H. salinarum* as the lipid source, a unilayer typified by the 90% content of caldarchaeol lipids from *T. acidophilum*, or a combination of bilayer and unilayer arrangements as is the case for the TPLs from *M. smithii*.

Archaeosomes consisting largely of caldarchaeol lipids (also called tetraether or bipolar tetraether lipids) are unusually stable to thermally induced leakage of solutes and exhibit low ion permeability [42]. Leakage of entrapped carboxyfluorescein (CFSE) from archaeosomes following autoclaving declined as the caldarchaeol content increased and stabilized at a caldarchaeol content above 50% [43]. This plateau around 50% caldarchaeol content is similar to the amount that prevented intramembrane cleavage upon freeze fracturing, and clearly supports a role for caldarchaeol membrane-spanning lipids in thermostability. Archaeosomes also show increased stability and low leakage of entrapped material in physiological concentrations of bile and low pH [44]. Further, the permeability of archaeosomes composed of several TPL compositions was compared to liposomes prepared from *Escherichia coli* lipids and from commercial 1,2-di-*O*-phytanyl-*sn*-glycero-3-phosphocholine (ester linkages to phytanyl chains). From these studies it was concluded that both the ether bonds (versus ester), and the caldarchaeol lipids decreased the permeability of archaeosomes to protons [Mathai, J.C., Sprott, G.D., Zeidel, M.L. Molecular mechanisms of water and solute transport across archaebacterial lipid membranes. *J Biol Chem* 2001, **276**, 27266–27271].

One additional advantage to the use of archaeal lipids is that they lack phase transitions over a broad range of temperatures. This means that lipid vesicles can be made over a wide range of temperatures, adding further to the ease of preparation and stability. In the case of other nonarchaeal lipids, liposome formation must be performed above the phase transition temperature, wherein the heat sensitivity of the active ingredient being entrapped could be an issue. Once formed, archaeosomes remain suspended indefinitely and resist fusion or aggregation over long storage periods (more than two years) without loss of entrapped material.

## 12.4   ADJUVANT EFFECTS OF ARCHAEOSOMES

### 12.4.1   Interaction with Antigen-Presenting Cells

Two events are critical for ensuring effective T-cell immunity to a vaccine antigen. First, the antigen has to be processed and presented in the context of MHC by the APCs to allow T-cell docking. Second, the APCs have to be sufficiently activated to ensure clonal expansion of the docked T cell [45]. Indeed, it is potent APCs such as dendritic cells (DCs) that are well suited to carry out this process. Nevertheless, they need to undergo differentiation and maturation to accomplish their task [46]. The distinguishing feature of archaeosomes is that they affect both processes: antigen delivery and DC maturation.

Liposomes have long been considered as carriers for antigen delivery to APCs [47,48], as they can either undergo lipid membrane fusion with the plasma membrane to deliver their contents, or as particulate material be phagocytosed. Further, it had been reported that the uptake of small anionic conventional liposomes by phagocytic cells is greater than that of neutral, cationic, and larger anionic liposomes [49]. Interestingly, when comparing similar small (<200 nm) conventional liposomes to anionic archaeosomes, the latter was phagocytosed to a substantially greater extent than the former [50]. Further, this effect is not seen with nonphagocytic cells such as HeLa cells. Although differences exist among various archaeosomes as to their ability to be taken up by macrophages, which can be attributed to the variability in lipid structures that alter membrane characteristics, all archaeosome types are phagocytosed to a greater extent than are several liposome compositions. However, it was not clear if these vesicles would be too stable for vaccine applications by nonrelease of their antigen cargo. On further testing it was evident that once inside the cell, the structural integrity of the archaeosome membrane was lost, suggesting release of its contents. The uptake of archaeosomes by APCs was inhibited by cytochalasin D, which alters the microfilaments, ruling out fusion at the plasma membrane as a way of archaeosome cell entry [50,51].

Many of the later studies on tracking the pathway of antigen processing by archaeosomes were prompted by observations of induction of cell-mediated immunity: in particular, strong CD8[+] T-cell responses to antigen by archaeosomes. While many archaeosome types evoke strong immune responses, *M. smithii* archaeosomes were considered particularly attractive. It was reasoned that the presence of 40 wt% caldarchaeols could confer high stability to vesicles formed with these dimeric lipids [52]. Further at 40 wt% caldarchaeol content, hydration and vesicle formulation could be achieved with relative ease. In addition, *M. smithii* is characterized by 30 wt% of archaetidylserine head groups and forms archaeosomes with surface-exposed phosphoserine [52]. Thus, *M. smithii* archaeosomes interact uniquely with phosphatidylserine (PS) receptors on the surface of APCs [34,51] for receptor-mediated endocytosis, and PS liposomes can outcompete *M. smithii* archaeosomes for uptake (Figure 12.4).

PS is often everted to the outer membrane leaflet on mammalian cells that are apoptotic, and the interaction of cell-exposed PS with a PS receptor is utilized by phagocytic cells for clearance of apoptotic cells [53]. The exposed archaetidylserine head groups of *M. smithii* archaeosomes exploit this mechanism for cell entry. Once in the phagosomes, the escape of antigens from archaeosomes into the cytosol appears to occur by a mechanism that is dependent on the acidic pH of phagolysosomes [51]. This suggests either acidic pH–dependent fusion of the archaeal membrane with the phagosomal

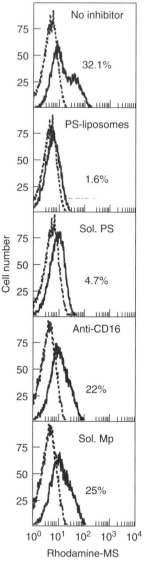

**Figure 12.4** PS-specific inhibition of phagocytosis by *M. smithii* (MS) archaeosomes. DCs were incubated with 25 μg of rhodamine-MS archaeosomes for 45 minutes in the presence or absence of various competitors, and uptake of fluorescent archaeosomes was analyzed by flow cytometry after acquisition of 20,000 events. The solid line indicates positive staining for rhodamine, and the dotted line indicates autofluorescence of DCs. The percentage within each panel denotes positive fluorescence for each condition, which is indicative of uptake of archaeosomes. Inhibitors used included PS-liposomes (4:1 ratio; 100 μg of lipid) relative to rhodamine-MS archaeosomes, *O*-phospho-L-serine (Sol. PS) at 400 μM, anti-CD16, and mannopentaose (Sol. MP; 100 μM). (Reproduced with permission from Ref. 51. Copyright © 2004 The American Association of Immunologists, Inc.)

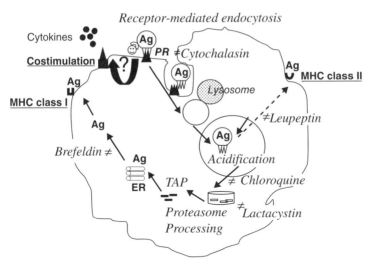

**Figure 12.5** Scheme depicting processing of antigen (Ag) encapsulated in *M. smithii* archaeosomes by an APC. Sites of inhibitor action are shown by ≠. Archaeosomes effect antigen processing to MHC classes I and II and activation of APCs by augmenting co-stimulation. Uptake is mediated by phosphatidylserine receptor (PR)-mediated endo-cytosis. The MHC class I–processing pathway is indicated by solid arrows, and the MHC class II–processing pathway is indicated by the dotted arrow. Class I Ag processing occurs in the cytosol, after escape of antigen from acidified phagolysosomal vacuole and is TAP dependent. (Based on data from Ref. 51.)

membrane, or another mechanism of antigen trafficking to the cytosol under the altered ionic conditions found in phagolysosomes. Nevertheless, the result is proteasome and peptide transporter (TAP)-dependent processing [51] for presentation of the antigen to MHC class I via the classical pathway [54] (Figure 12.5).

One important point to be noted is that although the presence of archaeti-dylserine on *M. smithii* promotes a defined receptor interaction, other archaeo-some types lacking PS are phagocytosed efficiently and evoke strong CD8+ T-cell immunity [34,54]. Thus, diverse archaeosomes may be expected to utilize other receptors and/or interactions to gain phagocytic entry into the cell.

The discovery that archaeal lipids constitute self-adjuvants capable of evoking DC maturation and differentiation was largely serendipitous. Archaea are not associated with humans except in anaerobic niches. *M. smithii* and *Methanosphaera stadtmanae* are common to the human intestinal tract, where they generate methane as part of the normal microflora [55]. However, there is no documented evidence of pathogenic archaea, despite a single recent rather speculative report suggesting that such may be the case [56]. Indeed, archaea lack endotoxin and do not synthesize any structures that resemble endotoxin. Much of the recent interest on adjuvants is focused on the ability of microbial products to interact with TLRs designed to recognize the presence of pathogen and initiate appropriate immune responses [2]. Thus, our observation that the polar lipids of nonpathogenic archaea interact with mammalian cells was unexpected.

We first documented that antigen-free archaeosomes composed of *M. smithii* can recruit macrophages and DCs in vivo after intraperitoneal injection [57]. Further, *M. smithii* archaeosomes also promoted maturation of APCs by up-regulating key co-stimulatory markers on APCs: B7-1, B7-2, and CD40, and MHC class II expression. This correlated with the increased ability of archaeosome-primed DCs to stimulate in vitro allo responses [58]. As mentioned above, two features distinguish *M. smithii* lipids: the presence of 40 wt% caldarchaeols, and a high content of archaetidylserine that may be recognized by the PS receptor. However, some studies suggest that PS receptor interaction can lead to the down-modulation of inflammation [53]. One distinguishing feature of the adjuvant effect of archaeosomes appears to be a dichotomous ability to activate dendritic cells (DCs) while overall maintaining a lower level of inflammatory cytokine production than that in other adjuvants that activate TLRs. Our preliminary studies also indicate nonengagement of TLR-2 and TLR-4 by archaeosomes (Krishnan and Sprott, unpublished). Also, *M. smithii* archaeosomes can activate immune responses even in IL-12-deficient mice [59]. Overall, while *M. smithii* utilizes the PS receptor interaction for cell entry, other core lipid (e.g., high caldarchaeol content) and/or head group features may facilitate DC activation. We have also observed that other archaea, *H. salinarum* and *T. acidophilum* which lack archaetidylserine, can activate DCs to up-regulate co-stimulatory markers (Figure 12.6).

Whereas caldarchaeol is abundant in *T. acidophilum*, *H. salinarum* is comprised only of the archaeol; thus, the presence of caldarchaeol cannot itself be essential for DC activation. Indeed, various archaea contain a combination of archaeol and caldarchaeol structures and diverse head group presentations, and thus interaction with multiple receptors such as DC-SIGN, mannose, and lecithin-binding receptors may be possible.

### 12.4.2 Induction of Antibody and T-helper Responses

Many archaeosome types can evoke strong antibody responses to diverse entrapped antigens. Nevertheless, archaeosomes composed of TPLs from *M. smithii*, *T. acidophilum*, or *H. salinarum* exhibit some differences as humoral adjuvants. The highest humoral adjuvant activity was noted for *H. salinarum* archaeosomes, often equaling that obtained with Freund's adjuvant and generally surpassing that of alum [58,60]. *M. smithii* and *T. acidophilum* prime slightly lower antibody titers, nevertheless often surpassing that of alum, and evoke titers stronger than those of conventional liposomes [58,60]. Interestingly, these three archaeosome types exhibit diversity in the abundance of caldarchaeol core, from 0 to 90%, respectively, indicating that other common lipid structural features may be sufficient for priming of immune responses.

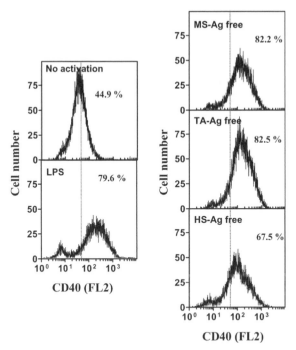

**Figure 12.6** Activation of DCs by archaeosomes. CD40 expression on DCs was determined after 24 hours in vitro culture with no activation, or treatment with 25 μg of antigen-free *M. smithii* (MS), *T. acidophilum* (TA), or *H. salinarum* (HS) archaeosomes, or 10 μg of LPS. Data are derived from 20,000 events acquired for each sample, and the percentage of CD40[high] cells beyond the vertical gate is indicated. (Data reproduced in modified form with permission from Ref. 51. Copyright © 2004 The American Association of Immunologists, Inc.)

Further, these archaeosome types localize antigen diversely, with more than 80% antigen internalized by *T. acidophilum* and only 60% internalized by *H. salinarum* [58]. Nevertheless, strong priming of antibody responses suggests that both internalized and surface-localized antigen may be processed after phagocytosis of archaeosomes, owing to intracellular vesicle degradation. Importantly, antigen needs to be physically associated with archaeosomes, as adjuvant effect is lost on admixing of antigen with preformed vesicles [58]. This suggests that entrapment ensures delivery of antigen and archaeosome to the same APCs, facilitating both antigen processing and APC activation in the same cell. Two features distinguish the adjuvant effect of archaeosomes compared to that of conventional liposomes: The response is quantitatively superior, and it is longer-lasting [34,58]. Indeed, elevated antibody titers are obtained with conventional liposomes only after co-entrapment of adjuvants such as monophosphoryl lipid A (MPL) and/or lipopolysaccharide (LPS) [61,62]. Another important observation, in keeping with the lack of endotoxin

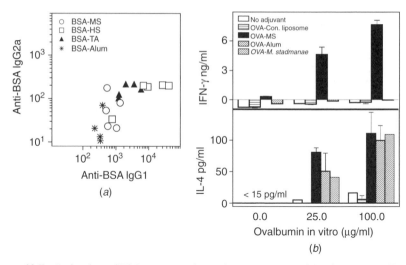

**Figure 12.7** Induction of Th1 responses by archaeosomes. Anti-bovine serum albumin (BSA) IgG2a and IgG1 antibody isotypes induced after immunization of BALB/c mice with 10 μg of BSA entrapped in *M. smithii* (MS), *H. salinarum* (HS), *T. acidophilum* (TA) archaeosomes, or formulated in alum (*a*). Values represent isotype-specific anti-BSA antibody titers of individual mice in each group. Antigen-archaeosome immunization promotes IgG2a antibody isotype indicative of Th1 immunity (*a*). Induction of both Th1 and Th2 cytokines by *M. smithii* (MS) archaeosomes (*b*). BALB/c mice were immunized (subcutaneously) with 15 μg of OVA in archaeosomes on days 0 and 21. Appropriate antigen alone (no adjuvant), antigen-conventional liposomes (Con. liposome), and antigen-alum controls were included. Spleens were harvested on day 28, and antigen-specific stimulation was set up for cells from individual mice. Cytokine levels in 72-hour supernatants were determined by ELISA. Values represent means ± standard errors of the means of cytokine production by spleen cells of individual mice in each group (*n* = 3). (Reproduced in modified form with permission from Ref. 58.)

presence in archaea is the ability of archaeosomes to evoke strong humoral responses in LPS-nonresponsive C3H/HeJ mice [58].

Strong humoral response and induction of IL-4 (Th2 immunity) switches antibody production to mainly IgG1 isotype, whereas IFN-γ (Th1 immunity) is required for skewing toward the complement-fixing IgG2a isotype [63]. One of the primary concerns of the universally approved adjuvant alum is its inability to evoke strong IgG2a and cell-mediated immunity, in particular that of Th1 cells [64,65]. Advantageously, archaeosomes strongly evoke both IgG1 and IgG2a. Further, antigen-specific activation of both Th1 and Th2 cell types is seen with archaeosomes [58] (Figure 12.7).

### 12.4.3  Induction of CD8+ T-Cell Responses

CD8+ T cells are critical for protection against many of the emerging and global diseases caused by intracellular pathogens (e.g., tuberculosis, malaria, influenza, HIV/AIDS) and cancer [66]. Naive CD8$^+$ T cells are stimulated when peptides from endogenously derived antigens are processed and presented on MHC class I. Conventionally, this process is most efficient for peptides derived from intracellular proteins being assembled in the cytosol [67]. In contrast, protein antigens from extracellular fluids that are taken up by APCs by pinocytosis and phagocytosis are processed within the endosomes and are presented on the surface of MHC class II to stimulate CD4$^+$ T cells. This dichotomy in antigen processing dictates that only the antigens of intracellular bacteria and viruses replicating within the host cell are presented in the context of MHC class I. However, professional APCs such as DCs can often break this rule and present some exogenous antigens on MHC class I even when they fail to enter the cytosol, by a process of cross-presentation/cross-priming [68,69]. However, cross-priming of soluble antigen is often weak, requires high antigen doses often not achievable physiologically in vivo, and can be short-lived. Thus, there is an urgent need for identification of new strategies to evoke a strong and sustained CD8$^+$ T-cell immunity.

One of the advantages of the archaeosome delivery system is its ability to target soluble antigen to the classical MHC class I processing pathway, resulting in the induction of a potent CD8$^+$ T-cell response. Immunization of mice with OVA entrapped in archaeosomes composed of many different TPLs, results in a strong CD8$^+$ T-cell response [34,54]. We have characterized this response using various assays [51,54,59]: expansion of antigen-specific CD8$^+$ T cells based on OVA$_{257-264}$ tetramer staining, potent CTL killing activity after restimulation of effectors in vitro, and high numbers of CD8$^+$ T cells producing IFN-$\gamma$ in response to stimulation with OVA$_{257-264}$ by Elispot assay (Figure 12.8).

More recently, we have also tracked the in vivo CD8$^+$ T-cell cytotoxicity and enumerated the circulating CD8$^+$ OVA$_{257-264}$-tetramer-specific T cells in the blood of immunized mice (Krishnan et al., unpublished). CD8$^+$ T-cell response against antigens other than OVA entrapped in archaeosomes can also be generated (i.e., listeriolysin from *Listeria monocytogenes* [70]) and cancer antigens such as tyrosine-related protein-2 (TRP-2) [71]. The CD8$^+$ T-cell response is primarily perforin dependent and can be evoked in CD4$^+$ T-cell-deficient as well as IL-12-deficient mice [54,59]. Conventional liposomes composed of dimyristoylphosphatidyl choline (DMPC)–phosphatidylglycerol (PG)–cholesterol were unable to evoke a significant CTL response to entrapped OVA (Figure 12.8) in the absence of co-adjuvanting. This correlates to the inability of conventional liposomes of the said composition to process OVA for MHC class I presentation, even in vitro [54].

### 12.4.4 Induction of Immune Memory

Infection results in profound antigenic stimulation in the context of pathogen-specific danger signals and promotes a programmed, yet massive clonal expansion phase of host-adaptive immune cells. This usually results in curtailment of infection, followed by apoptosis of the majority of effector B and T cells. A

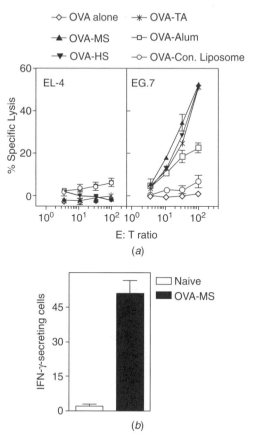

**Figure 12.8**  Induction of CD8$^+$ T-cell responses by archaeosomes. (*a*) C57BL/6 mice were immunized (subcutaneously) with 15 µg of OVA either in PBS (OVA alone), entrapped in archaeosomes composed of lipids from *M. smithii* (MS), *T. acidophilum* (TA), *H. salinarum* (HS), conventional liposomes (Con. Liposome), or adsorbed onto alum. On day 14, spleens were harvested, pooled (*n* = 3/group), and stimulated with irradiated EG.7 (OVA expressing) cells for five days, and then four-hour CTL activity on $^{51}$Cr-labeled targets (EL-4 and EG.7) was assessed. Percent specific lysis ± SD of triplicate cultures is indicated at the various effector (E)/target (T) ratios. (*b*) Enumeration of IFN-γ-secreting cells. Various number of spleen cells from immunized mice (15 µg of OVA on days 0 and 21, spleens removed on day 35) were stimulated with OVA$_{257-264}$ (10 µg/mL) for 24 hours in plates precoated with capture anti-IFN-γ-antibody. The numbers of IFN-γ-secreting cells were then evaluated by ELISPOT assay (*b*). The number of spots per 10$^6$ spleen cells is indicated. [(*a*) Reproduced after modification with permission from Ref. 54. Copyright © 2000 The American Association of Immunologists, Inc. (*b*) Reproduced with permission from Ref. 59.]

small portion of B and T cells then go on to survive as memory cells, which are a ready line of defense for a second infection onslaught [66]. The aim of vaccination is to mimic this natural host response. One key difference, however, is that vaccination, in particular with the nonreplicating adjuvants occurs in the absence of massive inflammation evoked by the pathogen danger signals. Yet vaccination is of no benefit should infection occur if it lacks strong immune memory.

Thus, following the promise of a potent adjuvant system with archaeosomes, much of our focus has been on demonstrating the ability of this adjuvant system to evoke a prolonged memory response. Our initial studies demonstrated that archaeosomes could afford a long-lasting antibody response to the co-delivered antigen [58] after just two initial injections on days 0 and 21 (Figure 12.9a).

Interestingly, adaptive immune response kicked in as early as seven days after immunization and could last well past a year after immunization. *H. salinarum* archaeosomes promoted long-lasting antibody titers, whereas *M. smithii* and *T. acidophilum* archaeosomes appear to exhibit relatively rapid decline in antibody titers beyond eight to 10 weeks (Figure 12.9a).

Compared to humoral response evoked after vaccination, it has been more challenging for researchers to demonstrate prolonged cell-mediated immunity, in particular $CD8^+$ T-cell response. Consequently, $CD8^+$ T-cell memory was evaluated and demonstrated to occur for more than one year in mice following vaccination with archaeosomes [70,71]. The ability of *M. smithii* archaeosomes to evoke a profound CTL memory to entrapped OVA is illustrated in Figure 12.9b. Interestingly, two injections of antigen–archaeosome on days 0 and 21 appear to be critical, as response after a year weakens if the animal is given only one injection (Krishnan and Sprott, unpublished). There appears to be a correlation between lipid structure and caldarchaeol content for the maintenance of immune memory, although further studies are required for a full understanding. Among nine archaeal strains tested, all evoked strong primary response, but only two, *M. smithii* and *T. acidophilum*, both constituted with caldarchaeols, evoked very strong memory recall response at more than 50 weeks [71].

The potent memory responses with *M. smithii* archaeosomes may be correlated to its potent in vivo antigen presentation. We have developed a model for tracking in vivo antigen presentation in which $CD8^+$ OVA-specific transgenic T cells from transgenic OT1 mice are labeled with a fluorescent tag (CFSE) and adoptively transferred into recipient mice that are immunized with OVA-archaeosomes. As OVA is processed and presented by endogenous APCs, the transferred transgenic OT1 cells begin to proliferate. This results in reduction of the CFSE label, which can then be tracked by flow cytometry. In the initial priming phase after OVA-archaeosome immunization, >95% of the $CD8^+$ T cells cycle in response to in vivo antigen presentation [51] (Figure 12.10). Such profound presentation usually occurs after intravenous injection with a replicating and acute viral and bacterial vectors [8,72].

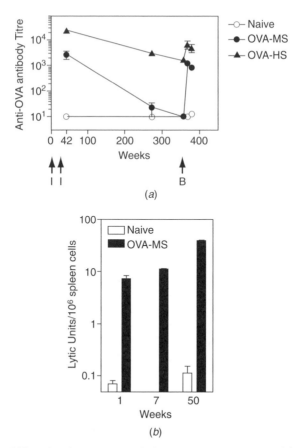

**Figure 12.9** Ability of archaeosomes to induce a memory response. (*a*) C57BL/6 mice were immunized (subcutaneously) on days 0 and 21 (indicated by I) with 15 μg of OVA entrapped in *M. smithii* (MS) or *H. salinarum* (HS) archaeosomes. A memory boost (indicated by B) of 25 μg of OVA (without adjuvant) was carried out on day 347. Mice were bled at regular intervals, and anti-OVA antibody titers were determined. Data represent mean ± standard errors of the means of mice in groups of four to six. Antibody titers are represented as endpoint dilutions exhibiting an optical density of 0.3 U above background. (*b*) Representative mice (*n* = 3) from the naive and OVA-MS groups were euthanized at 1, 7, and 50 weeks, and splenic CTL activity was assessed after five days of in vitro restimulation with EG.7 cells by standard $^{51}$Cr release assay. Splenic CTL activity is expressed as lytic units defined as the number of effector cells per $10^6$ spleen cells that yield 20% specific lysis of a population of $2.5 \times 10^4$ targets.

Various cell surface markers have been used to identify memory T cells. CD44 is by far the most reliable marker and is expressed at high levels on all memory cells. Correlating to the potent memory evoked by *M. smithii* archaeosomes, we observed that above 180 days' postimmunization, antigen recall resulted in the expansion of CD44$^+$ memory CD8$^+$ T cells [54]. Memory cells

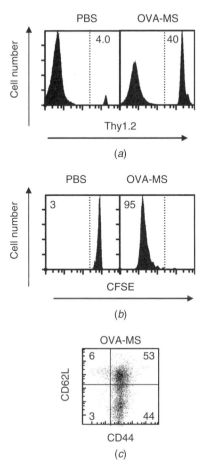

**Figure 12.10** Intensity of in vivo MHC class I presentation and stimulation of CD8+ T cells by OVA-archaeosomes. Thy 1.1+ B6.PL mice were immunized subcutaneously with 25 μg of OVA entrapped in *M. smithii* archaeosomes (OVA-MS). On day 3, naive and OVA-MS-immunized Thy 1.1+ recipients ($n = 3$) were injected with CFSE-labeled OT1 (Thy 1.2+) cells. Four days later, the number of CD8+ T cells of donor origin (Thy 1.2+) was evaluated (*a*). Furthermore, the reduction in CFSE of gated CD8+ T cells of donor origin was determined as a measure of cycling and proliferation of transferred cells (*b*). The expression of CD62L versus CD44 on gated CD8+ donor cells was evaluated (*c*) as a measure of T-cell differentiation. (Reproduced with permission from Ref. 51. Copyright © 2004 The American Association of Immunologists, Inc.)

can be further characterized into subsets: CD44$^{hi}$CD62L$^{low}$ effectors that exhibit rapid function, and a CD44$^{hi}$CD62$^{hi}$ central memory phenotype that has the capability of rapid expansion [66]. It can then be envisaged that vaccination that evokes a balanced induction of these subsets will be advantageous. We have observed that *M. smithii* archaeosomes provides such a distribution [51] (Figure 12.10).

## 12.5  VACCINE APPLICATIONS

### 12.5.1  Tumor Protection

Cell-free vaccination with the use of appropriate soluble tumor-associated antigens has long been desired for cancer immunotherapy. While considerable progress has been made in the identification of tumor-associated antigens, the availability of strong, yet safe adjuvants that can evoke cell-mediated immunity has been limiting. Currently, many peptide-based cancer vaccine trials rely on the use of incomplete Freund's adjuvant (IFA), and considerable success for immunotherapy against melanoma has been forthcoming [73]. However, IFA can evoke undesirable side effects such as erythema and induration at the injection site, and tolerance rather than immunity has also been reported in some cases [74]. As archaeosomes posed key advantages in being able to direct antigen for both MHC class I and II presentation, and evoked strong and long-term CD8+ T-cell immunity, tumor-protective responses with archaeosomes in murine model systems have been evaluated. Immunization of mice with OVA entrapped in archaeosomes composed of TPL from *M. smithii*, *T. acidophilum*, or *H. salinarum*, confers both prophylactic and therapeutic protection against the growth of EG.7 (OVA expressing) solid tumors [59,71]. Based on a number of experiments ($n > 60$), it is clear that after prophylactic vaccination, all mice remain tumor-free for the first three weeks, and more than 50% remain tumor-free after 60 days. Prophylactic tumor vaccination is effective even in IL-12-deficient mice, whereas in the absence of INF-γ, all that occurs is a significant delay in tumor growth [59]. Reducing the dose of OVA to 3 μg per injection also results in about a three-week delay in the onset of tumors [71]. CD8+ T cells are critical for this protection, as depletion of this subset completely abrogates the effect. CD4+ T cells also appear to aid CD8+ T-cell function, as depleting CD4+ T cells results in partial loss of the tumor-protective effect [59]. In the more aggressive B16 melanoma tumor model, prophylactic protection after OVA–archaeosome immunization can be afforded against both B16OVA solid and metastatic lung tumors (Figure 12.11).

When therapeutic cancer immunotherapy is evaluated with OVA–archaeosomes, with a two-injection schedule of one and seven days following tumor challenge, we routinely observe at least a three-week tumor-free period against both EG7 [59,71] and B16OVA solid tumors. Most interestingly, significant therapeutic protection is also observed with antigen-free archaeosomes of the *M. smithii* and *T. acidophilum* type in the EG7 solid tumor model [59,71]. This effect is dose dependent, saturating maximally at about 200 to 500 μg of lipid per injection. Further therapeutic protection correlates with the increased recruitment of innate immune cells, particularly NK (natural killer) cells at the tumor site [59]. Archaeosomes appear to have the potential to break tumor tolerance. Protective immunity in the mouse melanoma model can be evoked after vaccination with TRP-2 entrapped in archaeosomes [71]. Mice immu-

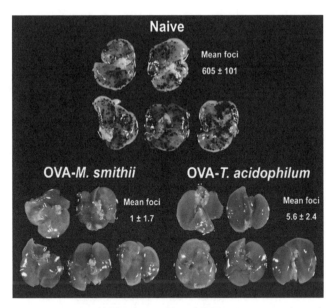

**Figure 12.11** Prophylactic vaccination against B16OVA metastatic melanoma using OVA-archaeosomes. C57BL/6 mice received injections of 15 μg of OVA entrapped in *M. smithii* or *T. acidophilum* archaeosomes on days 0 and 21. On day 35, groups of naive and OVA-archaeosome-immunized mice were challenged intravenously with B16OVA melanoma cells. On day 49, mice were euthanized, lungs harvested, and the black metastatic foci enumerated under a dissection microscope. In two of the naive control mice, because of the sheer high numbers, it was difficult to enumerate accurately the tumor foci, which were therefore assigned arbitrarily as 500. The mean ± SD of the tumor foci is indicated in the figure. OVA-archaeosome values are significantly different ($p < 0.001$) from those of naive mice as analyzed by the Mann–Whitney rank test. Data are representative of three experiments. (Reproduced after modification from Ref. 59.) (*See insert for color representation of figure.*)

nized with TRP-2 antigen entrapped in archaeosomes results in strong anti-TRP-2 CTL immunity and consequent reduction in metastatic tumor burden (Figure 12.12).

### 12.5.2   Infectious Disease Vaccine

CD8$^+$ T-cell immunity is also critical for protection against intracellular infections. Indeed, globally widespread infections currently lacking effective vaccines, such as tuberculosis, malaria, influenza, and HIV/AIDS, are caused by intracellular organisms and would benefit from development of vaccines that can evoke a CD8$^+$ T-cell response. The facultative intracellular pathogen *Listeria monocytogenes* (LM) is a typical intracellular bacterium that can

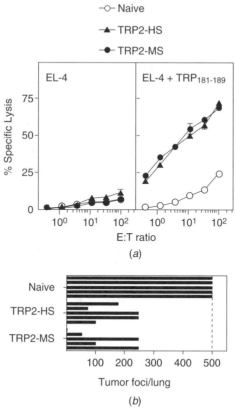

**Figure 12.12** Induction of CTL response to TRP-2$_{181-189}$ entrapped in archaeosomes and consequent protection against B16OVA metastasis. C57BL/6 mice were immunized (subcutaneously) on days 0 and 21 with 15 μg of TRP-2$_{181-189}$ peptide encapsulated in *M. smithii* (MS) or *H. salinarum* (HS) archaeosomes. (*a*) Spleens from two immunized and age-matched naive mice were obtained on day 49, and pooled spleen cells were stimulated with 0.01 μg/mL TRP-2 peptide for five days prior to assessing the 4-hour CTL activity against $^{51}$Cr-labeled targets. Targets included EL-4 (nonspecific target) and EL-4 cells preincubated for one hour with 10 μg/mL TRP-2 peptide (specific target). CTL data represent the percent specific lysis of triplicate cultures ± SD at various E/T ratios. Groups of immunized and naive control mice were also challenged with B16OVA melanoma cells on day 35. On day 49, mice were euthanized, lungs harvested, and black metastatic foci enumerated. The number of tumor foci in the lungs of individual mice in each group is indicated. It was difficult to enumerate under a dissection microscope (*b*). In naive mice, due to the large number of tumor foci, they were scored arbitrarily as >500. Tumor foci in TRP-2-archaeosome-immunized mice are significantly lower (*p* < 0.001) than the control group by the Mann–Whitney test. (Reproduced with permission from Ref. 71.)

parasitize host phagocytes. Systemically initiated infection of mice with LM has long served as a model for studying adaptive immunity to intracellular pathogens in general, as most mice strains effectively control infection within seven days, with the induction of a strong CD8+ T-cell immunity [75,76]. A major goal of vaccinology is to emulate the efficacy of such protection evoked by live infection. Interestingly, listeriolysin is the immunodominant antigen against which over 90% of the protective T-cell immunity is directed in mice that recover from a first sublethal infection of LM. The H-2K$^d$-restricted immunodominant epitope of listeriolysin is defined [77]. It has been formulated with Quil A in conventional liposomes [78], expressed with anthrax as a fusion protein or encoded in plasmid DNA [79], and expressed in recombinant vaccine virus or recombinant *Salmonella typhimurium* [80]. All these vaccines have afforded varying degrees of protection against subsequent exposure to LM.

A 13- to 20-mer lipopeptide extension of the listeriolysin CTL epitope when entrapped in archaeosomes evokes strong antigen-specific splenic CTL response (Figure 12.13) and production of IFN-γ in response to stimulation with the listeriolysin peptide in vitro [70]. This correlated to an eight- to 38 fold decrease in bacterial burden in the liver and 380-2042 fold decrease in the spleen (often, sterile spleens) after challenge with a high dose of LM compared to nonvaccinated controls, or mice receiving antigen-free archaeosomes. Mice vaccinated with the archaeosome–*Listeria* vaccine were found to be no more resistant than control nonvaccinated mice to *S. typhimurium*, indicating lack of any prolonged nonspecific innate antibacterial immunity. Kinetic studies showed that anti-*Listeria* immunity was evoked by one day of infection in the spleen and by two days in the liver [70]. Most dramatically, the anti-*Listeria* immunity was seen at 3, 5, and 10 months postvaccination (Figure 12.13). In the studies described above, lipidation of the peptide antigen was chosen to allow efficient entrapment of the peptide by insertion of the palmitoyl groups within the hydrophobic membrane domain of archaeosomes. Further, lipidation may itself enhance immunogenicity of the peptide [81,82]. Anti-*Listeria* protective immunity after vaccination with listeriolysin–archaeosomes was similar for vesicles derived from *M. smithii*, *T. acidophilum*, and *H. salinarum*. In these studies, conventional liposomes delivering the same lipopeptide were less effective. Further, to our knowledge there has been no prior demonstration of vaccination with a nonreplicating delivery system eliciting such profound protection against LM in the absence of additional immunostimulants [70].

## 12.6 BIOCOMPATIBILITY

Adjuvants and vaccine delivery systems intended for human use need to be safe and be devoid of side effects. Their biocompatibility and degradation into metabolites in vivo can also be an important consideration for practical

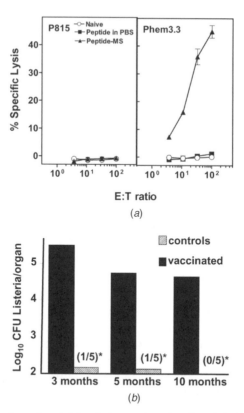

**Figure 12.13** Induction of anti-*Listeria* immunity after listeriolysin peptide–archaeo-some vaccination. BALB/c mice were vaccinated twice 29 days apart with a 13-mer dipalmitoylated listeriolysin peptide containing the nonamer H-2K$^d$ epitope (GYK-DGNEY) entrapped in *M. smithii* (MS) archaeosomes. (*a*) After 21 days of the second vaccination, spleen cells from representative mice were assessed for anti-*Listeria* CTL activity after restimulation in vitro for five days with the listeriolysin epitope, by $^{51}$Cr release assay. Killing was measured on nonspecific P815 and Phem 3.3 specific (express-ing listeriolysin) targets at various E/T ratios. (*b*) After 3, 5, and 10 months postvaccina-tion, groups of mice (*n* = 5) were challenged with 2 × 10$^4$ CFU/mouse. The bacterial burden enumerated in the spleen is indicated. In case of the archaeosome vaccine-immunized group, * indicates that the bacterial burden was detectable in only one of five mice. [(*a*) Reprinted with permission from Ref. 70. Copyright © 2001 Elsevier.]

application in humans. Archaea are a grouping of nonpathogenic, environmen-tal organisms not associated with endotoxin or other toxic metabolites, and as such, it may be anticipated that archaeosomes would be safe. In a first toxicity study of archaeal lipids in mammals, the main polar lipid of *T. acidophilum* [35] was administered to mice by intraperitoneal and oral routes and found to be free of toxicity [83]. Further, a series of TPL archaeosome compositions were tested to assess the possibility of toxicity related to their use in mammals.

Short-term repeated-dose toxicity studies based on a standard series of physical and biochemical tests revealed that intravenous (maximum 140 mg/kg per day for five consecutive days) and oral (maximally 550 mg/kg per day for 10 consecutive days) administration of unilamellar archaeosomes were very well tolerated by mice [84]. To further evaluate the safety of *M. smithii*, *T. acidophilum*, and *H. salinarum* archaeosomes as vaccine adjuvants, a study was conducted in which mice were immunized with OVA entrapped in 1.25 mg of these vesicles, administered on days 0 and 21, and then extensive blood biochemical analyses were conducted at 1, 2, 22, and 39 days. Further at day 39, animals were sacrificed and the major organs were subject to thorough biochemical analyses. These studies clearly indicated a lack of toxicity after vaccination. Further, in mice, no significant anti-archaeosome antibodies were detected following multiple injections [85].

Although the metabolic fate of the archaeal polar ether lipids in vivo is still not known, ether-linked lipids are present in plants and animals and in many mammalian tissues (e.g., nervous tissue, heart muscle, testes, kidney, erythrocytes, bone marrow skeletal tissue, macrophages) as well as in avian marine, molluskan, and protozoan species [27]. Humans can have a high dietary intake of ether-linked lipids from meats and seafood. Further, the isopranoid chains characteristic of the archaeal polar lipids are also seen in the fat-soluble vitamins and coenzyme $Q_{10}$ so essential in mammalian physiology [23,24]. Thus, the human body is constantly exposed to archaeal-like lipids, suggesting that archaeosomes will prove to be a safe adjuvant suitable for humans.

## 12.7  SUMMARY AND FUTURE CHALLENGES

The last decade has been phenomenal in revealing the tenets of initiation of immune responses following recognition of PAMPs [86,87]. Naturally, only DCs have the superior ability to recognize PAMPs efficiently and bridge the three components of host immunity: to respond innately, to direct adaptive immunity, and to promote profound memory [88]. Moreover, the mammalian host often tailors these three components of immunity advantageously to evoke a qualitatively distinct immune response with the capability of selectively and specifically combating the infectious challenge. Thus, although the research and development of novel adjuvants has increased exponentially over the last decade, it is also clear that because of the need to evoke a qualitatively specific immune response to a particular application, no single adjuvant system will be suitable for all applications.

In this context, archaeosomes represent a unique class of particulate antigen delivery vesicles with the following characteristics:

1. High stability, owing to the ether archaeal polar lipid composition conferring to the vesicle freedom from fusion, aggregation, and consequent antigen leakage

2. Efficient inherent immunostimulating ability for activating DCs

3. Ability to direct to the entrapped antigen a mixed immune response characterized by both antibody and cell-mediated immunity

4. Ability to direct efficient MHC class I presentation of the soluble antigen and consequent induction of $CD8^+$ T-cell response

5. Ability to evoke a profound memory response

While fully capable of initiating all arms of adaptive immunity, archaeosomes, stand out in their ability to target soluble antigen to the MHC class I–presenting pathway. For archaeosomes composed of *M. smithii* lipids, this ability correlates with the efficient interaction of archaeal PS with PS receptors on APCs. Nevertheless, many other archaeosome types also demonstrate the ability to cross-prime $CD8^+$ T cells, suggesting that archaeal lipid head groups may interact with many other receptors on APCs. One of the key hurdles of many new adjuvant technologies, particularly those capable of augmenting potent immunostimulation, is their apparent increased toxicity, particularly at high doses. Archaeosomes lack toxicity in rodent models and appear to be able to activate immunity in the absence of overt inflammation [51]. This may prove to be an advantage for this technology during further translation to human applications. Interestingly, while the quest for adjuvants is focused on interaction with TLRs, thus far there is no evidence for such interaction in the context of archaeal lipids (Krishnan and Sprott, unpublished). Indeed, interaction with PS receptors often signals down-regulation of immunity [89]. Nevertheless, the counterintuitive ability of *M. smithii* archaeosomes to direct potent adaptive immunity suggests that interaction of archaeal lipids with APCs may be more complex than anticipated.

With the increased desire to move away from traditional whole cell or virus vaccines to subunit vaccines, there is a need for technologies that can serve as efficient carriers for small molecules. Archaeosome adjuvant can be envisaged as having the utmost advantage for entrapment of such entities, in particular vaccination against intracellular infections and tumors that require a specific $CD8^+$ CTL response for efficacious protection. However, as archaeal phospholipids are negatively charged and archaeosomes are particulate vesicles, entrapment procedures often have to be tailored based on the physical properties of the molecule to be entrapped. Indeed, simple admixing of antigen and archaeosome vesicles does not provide the desired adjuvant effect [58]. Further an accurate quantification of the entrapped material will be required, in particular for regulatory approval. Techniques such as high-performance liquid chromatography can be used to refine antigen quantification, but again will require some standardization for each antigen.

Thus far, archaeosomes have been constituted with the TPL mixture unique to each archaeon. Such an approach has the advantage that the natural lipid mix is amenable to membrane formation. However, a complete understanding of archaeal lipid structure in relation to induction of immunity requires further study. This may aid tailoring vaccine formulation with specific synthetic lipids.

Although there is growing acceptance by vaccine producers and regulatory agencies that improved vaccine adjuvants are needed to meet the vaccine challenges of the future, new technologies carry the risk of long developmental time from bench to clinic to market. The balance of cost-effectiveness, rapidity of development, and efficacy for a desired application can dictate the commercial success of a technology. The efficacy of archaeosomes has been proven in rodent models of infection and cancer. The advantage of using archaeosomes is that they are both antigen delivery depots as well as immunostimulants (Table 12.1). Further, as a particulate delivery vehicle, archaeosomes are amenable to incorporation of other small molecular immunostimulants and/or multiple antigens. The main challenge now lies in driving this technology from the bench to the clinic and proving biocompatibility and efficacy in humans. Archaeosomes appear to have the greatest potential for applications requiring strong $CD8^+$ T-cell immunity. Archaeosomes will, however, have to compete and prove their suitability for particular applications along with the other promising technologies described in this book.

## REFERENCES

1. Yu, Z. & Restifo, N.P. Cancer vaccines: progress reveals new complexities. *J Clin Invest* 2002, **110**(3), 289–294.

2. Pashine, A., Valiante, N.M. & Ulmer, J.B. Targeting the innate immune response with improved vaccine adjuvants. *Nat Med* 2005, **11**(4 Suppl), S63–S68.

3. Kovacsovics-Bankowski, M. & Rock, K.L. A phagosome-to-cytosol pathway for exogenous antigens presented on MHC class I molecules. *Science* 1995, **267**(5195), 243–246.

4. Singh, M. & O'Hagan, D.T. Recent advances in vaccine adjuvants. *Pharm Res* 2002, **19**(6), 715–728.

5. Lingnau, K., Egyed, A., Schellack, C., Mattner, F., Buschle, M. & Schmidt, W. Poly-L-arginine synergizes with oligodeoxynucleotides containing CpG-motifs (CpG-ODN) for enhanced and prolonged immune responses and prevents the CpG-ODN-induced systemic release of pro-inflammatory cytokines. *Vaccine* 2002, **20**(29–30), 3498–3508.

6. Richards, R.L., Rao, M., Wassef, N.M., Glenn, G.M., Rothwell, S.W. & Alving, C.R. Liposomes containing lipid A serve as an adjuvant for induction of antibody and cytotoxic T-cell responses against RTS,S malaria antigen. *Infect Immun* 1998, **66**(6), 2859–2865.

7. Medina, E. & Guzman, C.A. Use of live bacterial vaccine vectors for antigen delivery: potential and limitations. *Vaccine* 2001, **19**(13–14), 1573–1580.

8. Murali-Krishna, K., Altman, J.D., Suresh, M., Sourdive, D.J., Zajac, A.J., Miller, J.D. et al. Counting antigen-specific CD8 T cells: a reevaluation of bystander activation during viral infection. *Immunity* 1998, **8**(2), 177–187.

9. Moron, V.G., Rueda, P., Sedlik, C. & Leclerc, C. In vivo, dendritic cells can cross-present virus-like particles using an endosome-to-cytosol pathway. *J Immunol* 2003, **171**(5), 2242–2250.

10. Kaisho, T. & Akira, S. Toll-like receptors as adjuvant receptors. *Biochim Biophys Acta* 2002, **1589**(1), 1–13.

11. Krieg, A.M. The role of CpG motifs in innate immunity. *Curr Opin Immunol* 2000, **12**(1), 35–43.

12. Martin, M., Michalek, S.M. & Katz, J. Role of innate immune factors in the adjuvant activity of monophosphoryl lipid A. *Infect Immun* 2003, **71**(5), 2498–2507.

13. Gursel, I., Gursel, M., Ishii, K.J. & Klinman, D.M. Sterically stabilized cationic liposomes improve the uptake and immunostimulatory activity of CpG oligonucleotides. *J Immunol* 2001, **167**(6), 3324–3328.

14. Olsen, G.J. Microbial ecology: Archaea, Archaea, everywhere. *Nature* 1994, **371**(6499), 657–658.

15. Kates, M. Archaebacterial lipids: structure, biosynthesis and function. *Biochem Soc Symp* 1992, **58**, 51–72.

16. Sprott, G.D. Structures of archaebacterial membrane lipids. *J Bioenerg Biomembr* 1992, **24**(6), 555–566.

17. Kates, M., Kushwaha, S.C. & Sprott, G.D. Lipids of purple membrane from extreme halophiles and methanogenic bacteria. *Methods Enzymol* 1982, **88**, 98–111.

18. Kates, M. Structure, physical properties, and function of archaebacterial lipids. *Prog Clin Biol Res* 1988, **282**, 357–384.

19. Nichols, D.S., Miller, M.R., Davies, N.W., Goodchild, A., Raftery, M. & Cavicchioli, R. Cold adaptation in the Antarctic Archaeon *Methanococcoides burtonii* involves membrane lipid unsaturation. *J Bacteriol* 2004, **186**(24), 8508–8515.

20. Sprott, G.D., Meloche, M. & Richards, J.C. Proportions of diether, macrocyclic diether, and tetraether lipids in *Methanococcus jannaschii* grown at different temperatures. *J Bacteriol* 1991, **173**(12), 3907–3910.

21. Sprott, G.D., Agnew, B.J. & Patel, G.B. Structural features of ether lipids in the archaeobacterial thermophiles *Pyrococcus furiosus, Methanopyrus kandleri, Methanothermus fervidus*, and *Sulfolobus acidocaldarius*. *Can J Microbiol* 1997, **43**, 467–476.

22. Sprott, G.D., Ferrante, G. & Ekiel, I. Tetraether lipids of *Methanospirillum hungatei* with head groups consisting of phospho-*N,N*-dimethylaminopentanetetrol, phospho-*N,N,N*-trimethylaminopentanetetrol, and carbohydrates. *Biochim Biophys Acta* 1994, **1214**(3), 234–242.

23. Tengerdy, R.P. & Lacetera, N.G. Vitamin E adjuvant formulations in mice. *Vaccine* 1991, **9**(3), 204–206.

24. Folkers, K., Morita, M. & McRee, J., Jr. The activities of coenzyme Q10 and vitamin B6 for immune responses. *Biochem Biophys Res Commun* 1993, **193**(1), 88–92.

25. Cippitelli, M. & Santoni, A. Vitamin D3: a transcriptional modulator of the interferon-gamma gene. *Eur J Immunol* 1998, **28**(10), 3017–3030.

26. Clavreul, A., D'Hellencourt, C.L., Montero-Menei, C., Potron, G. & Couez, D. Vitamin D differentially regulates B7.1 and B7.2 expression on human peripheral blood monocytes. *Immunology* 1998, **95**(2), 272–277.

27. Lee, T.C. Biosynthesis and possible biological functions of plasmalogens. *Biochim Biophys Acta* 1998, **1394**(2–3), 129–145.

28. Ekiel, I., Smith, I.C. & Sprott, G.D. Biosynthetic pathways in *Methanospirillum hungatei* as determined by $^{13}$C nuclear magnetic resonance. *J Bacteriol* 1983, **156**(1), 316–326.

29. Ekiel, I., Sprott, G.D. & Smith, I.C. Mevalonic acid is partially synthesized from amino acids in *Halobacterium cutirubrum*: a $^{13}$C nuclear magnetic resonance study. *J Bacteriol* 1986, **166**(2), 559–564.

30. Patel, G.B. & Sprott, G.D. Archaeobacterial ether lipid liposomes (archaeosomes) as novel vaccine and drug delivery systems. *Crit Rev Biotechnol* 1999, **19**(4), 317–357.

31. New, R. *Preparation of Liposomes*, IRL Press, Oxford, pp. 33–104.

32. Sprott, G.D., Patel, G.B. & Krishnan, L. Archaeobacterial ether lipid liposomes as vaccine adjuvants. *Methods Enzymol* 2003, **373**, 155–172.

33. Wessel, D. & Flugge, U.I. A method for the quantitative recovery of protein in dilute solution in the presence of detergents and lipids. *Anal Biochem* 1984, **138**(1), 141–143.

34. Sprott, G.D., Sad, S., Fleming, L.P., Dicaire, C.J., Patel, G.B. & Krishnan, L. Archaeosomes varying in lipid composition differ in receptor-mediated endocytosis and differentially adjuvant immune responses to entrapped antigen. *Archaea* 2003, **1**(3), 151–164.

35. Swain, M., Brisson, J.R., Sprott, G.D., Cooper, F.P. & Patel, G.B. Identification of beta-L-gulose as the sugar moiety of the main polar lipid of *Thermoplasma acidophilum*. *Biochim Biophys Acta* 1997, **1345**(1), 56–64.

36. Shimada, H., Nemoto, N., Shida, Y., Oshima, T. & Yamagishi, A. Complete polar lipid composition of *Thermoplasma acidophilum* HO-62 determined by high-performance liquid chromatography with evaporative light-scattering detection. *J Bacteriol* 2002, **184**(2), 556–563.

37. Szebeni, J. & Toth, K. Lipid peroxidation in hemoglobin-containing liposomes: effects of membrane phospholipid composition and cholesterol content. *Biochim Biophys Acta* 1986, **857**(2), 139–145.

38. Crommelin, D.J. & Storm, G. Liposomes: from the bench to the bed. *J Liposome Res* 2003, **13**(1), 33–36.

39. Aramaki, Y., Tomizawa, H., Hara, T., Yachi, K., Kikuchi, H. & Tsuchiya, S. Stability of liposomes in vitro and their uptake by rat Peyer's patches following oral administration. *Pharm Res* 1993, **10**(8), 1228–1231.

40. Gliozzi, A., Relini, A. & Chong, P.L.G. Structure and permeability properties of biomimetic membranes of bolaform archaeal tetraether lipids. *J Membr Sci* 2002, **206**, 131–147.

41. Beveridge, T.J., Choquet, C.G., Patel, G.B. & Sprott, G.D. Freeze-fracture planes of methanogen membranes correlate with the content of tetraether lipids. *J Bacteriol* 1993, **175**(4), 1191–1197.

42. Choquet, C.G., Patel, G.B., Beveridge, T.J. & Sprott, G.D. Stability of pressure-extruded liposomes made from archaeobacterial ether lipids. *Appl Microbiol Biotechnol* 1994, **42**(2–3), 375–384.

43. Sprott, G.D., Dicaire, C.J., Fleming, L.P. & Patel, G.B. Stability of liposomes prepared from archaeobacterial lipids and phsphatidylcholine mixtures. *Cells Mater* 1996, **6**, 143–155.

44. Patel, G.B., Agnew, B.J., Deschatelets, L., Fleming, L.P. & Sprott, G.D. In vitro assessment of archaeosome stability for developing oral delivery systems. *Int J Pharm* 2000, **194**(1), 39–49.

45. Grakoui, A., Bromley, S.K., Sumen, C., Davis, M.M., Shaw, A.S., Allen, P.M. et al. The immunological synapse: a molecular machine controlling T cell activation. *Science* 1999, **285**(5425), 221–227.

46. Banchereau, J. & Steinman, R.M. Dendritic cells and the control of immunity. *Nature* 1998, **392**(6673), 245–252.

47. Copland, M.J., Baird, M.A., Rades, T., McKenzie, J.L., Becker, B., Reck, F. et al. Liposomal delivery of antigen to human dendritic cells. *Vaccine* 2003, **21**(9–10), 883–890.

48. Nair, S., Zhou, F., Reddy, R., Huang, L. & Rouse, B.T. Soluble proteins delivered to dendritic cells via pH-sensitive liposomes induce primary cytotoxic T lymphocyte responses in vitro. *J Exp Med* 1992, **175**(2), 609–612.

49. Makabi-Panzu, B., Lessard, C., Beauchamp, D., Desormeaux, A., Poulin, L., Tremblay, M. et al. Uptake and binding of liposomal 2′,3-dideoxycytidine by RAW 264.7 cells: a three-step process. *J Acquir Immune Defic Syndr Hum Retrovirol* 1995, **8**(3), 227–235.

50. Tolson, D.L., Latta, R.K., Patel, G.B. & Sprott, G.D. Uptake of archaeobacterial and conventional liposomes by phagocytic cells. *J Liposome Res* 1996, **6**, 755–776.

51. Gurnani, K., Kennedy, J., Sad, S., Sprott, G.D. & Krishnan, L. Phosphatidylserine receptor-mediated recognition of archaeosome adjuvant promotes endocytosis and MHC class I cross-presentation of the entrapped antigen by phagosome-to-cytosol transport and classical processing. *J Immunol* 2004, **173**(1), 566–578.

52. Sprott, G.D., Brisson, J., Dicaire, C.J., Pelletier, A.K., Deschatelets, L.A., Krishnan, L. et al. A structural comparison of the total polar lipids from the human archaea *Methanobrevibacter smithii* and *Methanosphaera stadtmanae* and its relevance to the adjuvant activities of their liposomes. *Biochim Biophys Acta* 1999, **1440**(2–3), 275–288.

53. Somersan, S. & Bhardwaj, N. Tethering and tickling: a new role for the phosphatidylserine receptor. *J Cell Biol* 2001, **155**(4), 501–504.

54. Krishnan, L., Sad, S., Patel, G.B. & Sprott, G.D. Archaeosomes induce long-term CD8+ cytotoxic T cell response to entrapped soluble protein by the exogenous cytosolic pathway, in the absence of CD4+ T cell help. *J Immunol* 2000, **165**(9), 5177–5185.

55. Miller, T.L. & Wolin, M.J. Methanogens in human and animal intestinal tracts. *Syst Appl Microbiol* 1986, **7**, 223–229.

56. Lepp, P.W., Brinig, M.M., Ouverney, C.C., Palm, K., Armitage, G.C. & Relman, D. A. Methanogenic Archaea and human periodontal disease. *Proc Natl Acad Sci USA* 2004, **101**(16), 6176–6181.

57. Krishnan, L., Sad, S., Patel, G.B. & Sprott, G.D. The potent adjuvant activity of archaeosomes correlates to the recruitment and activation of macrophages and dendritic cells in vivo. *J Immunol* 2001, **166**(3), 1885–1893.

58. Krishnan, L., Dicaire, C.J., Patel, G.B. & Sprott, G.D. Archaeosome vaccine adjuvants induce strong humoral, cell-mediated, and memory responses: comparison to conventional liposomes and alum. *Infect Immun* 2000, **68**(1), 54–63.

59. Krishnan, L., Sad, S., Patel, G.B. & Sprott, G.D. Archaeosomes induce enhanced cytotoxic T lymphocyte responses to entrapped soluble protein in the absence of interleukin 12 and protect against tumor challenge. *Cancer Res* 2003, **63**(10), 2526–2534.

60. Sprott, G.D., Tolson, D.L. & Patel, G.B. Archaeosomes as novel antigen delivery systems. *FEMS Microbiol Lett* 1997, **154**(1), 17–22.

61. Abraham, E. & Shah, S. Intranasal immunization with liposomes containing IL-2 enhances bacterial polysaccharide antigen-specific pulmonary secretory antibody response. *J Immunol* 1992, **149**(11), 3719–3726.

62. Gregoriadis, G. Liposomes as immunological adjuvants: approaches to immunopotentiation including ligand-mediated targeting to macrophages. *Res Immunol* 1992, **143**(2), 178–185.

63. Krishnan, L. & Mosmann, T.R. Functional subpopulations of CD4+ T lymphocytes. In *T Lymphocyte Subpopulations in Immunotoxicology*, Kimber, I. & Selgrade, M.K., Eds., Wiley, Chichester, West Sussey, England, pp. 7–32.

64. Brewer, J.M., Conacher, M., Hunter, C.A., Mohrs, M., Brombacher, F. & Alexander, J. Aluminium hydroxide adjuvant initiates strong antigen-specific Th2 responses in the absence of IL-4- or IL-13-mediated signaling. *J Immunol* 1999, **163**(12), 6448–6454.

65. Gupta, R.K., Rost, B.E., Reyveld, R.E. & Siber, G.S. Adjuvant properties of aluminum and calcium compounds. In *Vaccine Design*, Powell, M.F. & Newman, M.J., Eds., Plenum Press, New York, 1995, pp. 229–248.

66. Sad, S. & Krishnan, L. Maintenance and attrition of T-cell memory. *Crit Rev Immunol* 2003, **23**(1–2), 129–147.

67. Pamer, E. & Cresswell, P. Mechanisms of MHC class I–restricted antigen processing. *Annu Rev Immunol* 1998, **16**, 323–358.

68. den Haan, J.M. & Bevan, M.J. Antigen presentation to CD8+ T cells: cross-priming in infectious diseases. *Curr Opin Immunol* 2001, **13**(4), 437–441.

69. Zinkernagel, R.M. On cross-priming of MHC class I-specific CTL: rule or exception? *Eur J Immunol* 2002, **32**(9), 2385–2392.

70. Conlan, J.W., Krishnan, L., Willick, G.E., Patel, G.B. & Sprott, G.D. Immunization of mice with lipopeptide antigens encapsulated in novel liposomes prepared from the polar lipids of various archaeobacteria elicits rapid and prolonged specific protective immunity against infection with the facultative intracellular pathogen, *Listeria monocytogenes*. *Vaccine* 2001, **19**(25–26), 3509–3517.

71. Krishnan, L. & Sprott, G.D. Archaeosomes as self-adjuvanting delivery systems for cancer vaccines. *J Drug Target* 2003, **11**(8–10), 515–524.

72. Shen, H., Slifka, M.K., Matloubian, M., Jensen, E.R., Ahmed, R. & Miller, J.F. Recombinant *Listeria monocytogenes* as a live vaccine vehicle for the induction of protective anti-viral cell-mediated immunity. *Proc Natl Acad Sci USA* 1995, **92**(9), 3987–3991.

73. Rosenberg, S.A., Yang, J.C., Schwartzentruber, D.J., Hwu, P., Marincola, F.M., Topalian, S.L. et al. Immunologic and therapeutic evaluation of a synthetic peptide vaccine for the treatment of patients with metastatic melanoma. *Nat Med* 1998, **4**(3), 321–327.

74. Toes, R.E., Blom, R.J., Offringa, R., Kast, W.M. & Melief, C.J. Enhanced tumor outgrowth after peptide vaccination: functional deletion of tumor-specific CTL induced by peptide vaccination can lead to the inability to reject tumors. *J Immunol* 1996, **156**(10), 3911–3918.

75. Cheers, C., McKenzie, I.F., Pavlov, H., Waid, C. & York, J. Resistance and suscepti-
    bility of mice to bacterial infection: course of listeriosis in resistant or susceptible
    mice. *Infect Immun* 1978, **19**(3), 763–770.

76. Edelson, B.T. & Unanue, E.R. Immunity to *Listeria* infection. *Curr Opin Immunol*
    2000, **12**(4), 425–431.

77. Pamer, E.G., Harty, J.T. & Bevan, M.J. Precise prediction of a dominant class I
    MHC-restricted epitope of *Listeria monocytogenes*. *Nature* 1991, **353**(6347),
    852–855.

78. Lipford, G.B., Wagner, H. & Heeg, K. Vaccination with immunodominant peptides
    encapsulated in Quil A–containing liposomes induces peptide-specific primary
    CD8+ cytotoxic T cells. *Vaccine* 1994, **12**(1), 73–80.

79. Ballard, J.D., Collier, R.J. & Starnbach, M.N. Anthrax toxin-mediated delivery of a
    cytotoxic T-cell epitope in vivo. *Proc Natl Acad Sci USA* 1996, **93**(22), 12531–
    12534.

80. Hess, J., Gentschev, I., Miko, D., Welzel, M., Ladel, C., Goebel, W. et al. Superior
    efficacy of secreted over somatic antigen display in recombinant *Salmonella* vaccine
    induced protection against listeriosis. *Proc Natl Acad Sci USA* 1996, **93**(4), 1458–
    1463.

81. Tsunoda, I., Sette, A., Fujinami, R.S., Oseroff, C., Ruppert, J., Dahlberg, C. et al.
    Lipopeptide particles as the immunologically active component of CTL inducing
    vaccines. *Vaccine* 1999, **17**(7–8), 675–685.

82. Loing, E., Andrieu, M., Thiam, K., Schorner, D., Wiesmuller, K.H., Hosmalin, A.
    et al. Extension of HLA-A*0201-restricted minimal epitope by N epsilon-palmitoyl-
    lysine increases the life span of functional presentation to cytotoxic T cells.
    *J Immunol* 2000, **164**(2), 900–907.

83. Freisleben, H.J., Bormann, J., Litzinger, D.C., Lehr, F., Rudolph, P., Schatton, W.
    et al. Toxicity and biodistribution of liposomes of the main phospholipid from the
    archaeobacterium *Thermoplasma acidophilum*. *J Liposome Res* 1995, **5**, 215–223.

84. Omri, A., Agnew, B.J. & Patel, G.B. Short-term repeated-dose toxicity profile of
    archaeosomes administered to mice via intravenous and oral routes. *Int J Toxicol*
    2003, **22**(1), 9–23.

85. Patel, G.B., Omri, A., Deschatelets, L. & Sprott, G.D. Safety of archaeosome adju-
    vants evaluated in a mouse model. *J Liposome Res* 2002, **12**(4), 353–372.

86. Aderem, A. & Ulevitch, R.J. Toll-like receptors in the induction of the innate
    immune response. *Nature* 2000, **406**(6797), 782–787.

87. Gordon, S. Pattern recognition receptors: doubling up for the innate immune
    response. *Cell* 2002, **111**(7), 927–930.

88. Visintin, A., Mazzoni, A., Spitzer, J.H., Wyllie, D.H., Dower, S.K. & Segal, D.M.
    Regulation of Toll-like receptors in human monocytes and dendritic cells. *J Immunol*
    2001, **166**(1), 249–255.

89. Li, M.O., Sarkisian, M.R., Mehal, W.Z., Rakic, P. & Flavell, R.A. Phosphatidylserine
    receptor is required for clearance of apoptotic cells. *Science* 2003, **302**(5650),
    1560–1563.

# 13

# MUCOSAL ADJUVANTS AND DELIVERY SYSTEMS

MICHAEL VAJDY

## 13.1 INTRODUCTION AND DEFINITIONS

The word *adjuvant* is derived from the Latin root of *adjuvare*, meaning to help. As such, any protein, DNA, RNA, live bacteria or virus, or chemical structures that help to enhance an immune response can be considered an adjuvant. However, adjuvants may be divided into immunopotentiating adjuvants, with direct acute effects on the immune system, and nonreplicating or replicating delivery systems, which may or may not have an acute or long-term effect on the immune system (Figure 13.1).

To further enhance the adjuvanticity of immunopotentiating adjuvants or delivery systems for mucosal immunizations, they may be combined [e.g., poly(lactide-co-glycolide) (PLG)-formulated protein or DNA combined with CpG or *Escherichia coli* heat-labile enterotoxin (LT)-mutant adjuvants]. Furthermore, both mucosal and systemic routes of immunizations may be used to induce optimal responses at both mucosal and systemic effector sites. The mucosal inductive sites (e.g., draining lymph nodes or Peyer's patches) are sites where antigen is presented by antigen-presenting cells (APCs) to T cells, which in turn also help B-cell differentiation, and proliferation and an immune response is thus generated. Activated B and T cells then migrate from mucosal inductive sites to mucosal effector sites (i.e., the subepithelial or intraepithelial regions of the mucosa) to exert their effector functions.

The mammalian immune system is divided into innate (antigen-unspecific) and adaptive (antigen-specific) immune responses. The innate response against

*Vaccine Adjuvants and Delivery Systems,* Edited by Manmohan Singh
Copyright © 2007 John Wiley & Sons, Inc.

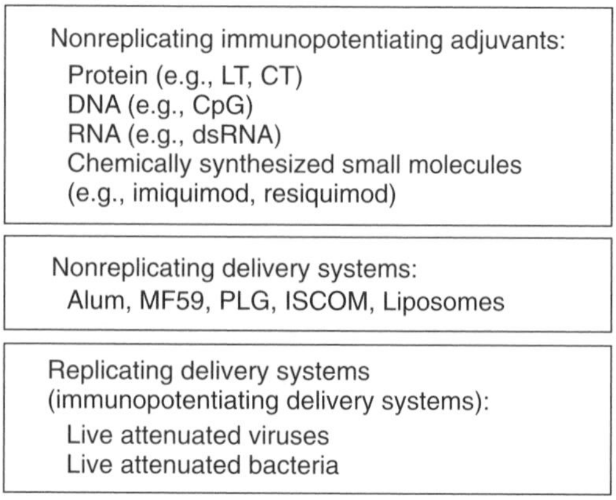

Nonreplicating immunopotentiating adjuvants:
Protein (e.g., LT, CT)
DNA (e.g., CpG)
RNA (e.g., dsRNA)
Chemically synthesized small molecules
(e.g., imiquimod, resiquimod)

Nonreplicating delivery systems:
Alum, MF59, PLG, ISCOM, Liposomes

Replicating delivery systems
(immunopotentiating delivery systems):
Live attenuated viruses
Live attenuated bacteria

**Figure 13.1**   Mucosal adjuvants and delivery systems.

previously unencountered or encountered antigens is one of the first immune responses mounted. The type and the persistence of the antigen then dictate whether the antigen can be eliminated by the innate system or whether adaptive responses, including immunological memory responses, will also be induced. The link between innate and adaptive responses has recently received much attention. The major aim of the application of mucosal immunopotentiating adjuvants and delivery systems in effective vaccines is to induce appropriate innate followed by adaptive immune responses in order to confer protection against pathogens. Throughout this chapter, examples given with regards to the application of the various immunopotentiating adjuvants and delivery systems will focus primarily on experimental HIV vaccines.

## 13.2   INNATE AND ADAPTIVE RESPONSES AT MUCOSAL INDUCTIVE AND EFFECTOR SITES

The interaction between innate and adaptive immune responses is paramount in generating an antigen-specific immune response (Figure 13.2). Most antigens are eradicated or neutralized by the innate antigen-unspecific responses. The adaptive antigen-specific immune system, apparently developed much later in evolution, follows the innate system, and as such the quality and quantity of the adaptive responses is dependent on the quality and quantity of the innate responses. Cellular components of the innate system include, but are not limited to, various antigen-presenting cells [APCs; e.g., dendritic cells (DCs), macrophages or B cells, mast cells, neutrophils, and natural killer (NK) cells]. These cells are found abundantly at mucosal inductive and effector sites. Considering the myriad of pathogenic and nonpathogenic antigens that attempt or do gain access to mucosal sites, one may argue that the main function of

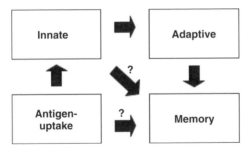

**Figure 13.2** Current understanding of the link between innate, adaptive, and memory immune responses.

the mucosal immune system, particularly in the respiratory and gastrointestinal tract, is to induce tolerance rather than an immune response. If this were not the case, chronic inflammatory responses would afflict the mucosal sites from early after birth. Because of this tolerance-inducing feature, it is most likely that induction of immune responses are more difficult following intranasal (IN) and particularly oral immunizations, as most protein antigens without an adjuvant induce tolerance rather than immunity. Thus, the function of adjuvants and delivery systems used for mucosal administrations is that of breaking tolerance and inducing an immune response. It has become increasingly clear that to induce an immune response, which in the case of mucosal vaccines has to be an adaptive memory response, a specific type of innate immune response has to be induced.

The initiation of innate immune responses begins with the interaction of pathogen-associated molecular patterns (PAMPs) on pathogens or antigens with Toll-like receptors (TLRs) on the cells involved in innate immunity (e.g., DCs). Many TLRs with specificity to recognize specific PAMPs have been discovered. Examples of this specificity include CpG interacting with TLR9, lipopolysaccharide (LPS) with TRL4, flagellin with TLR5, and imiquimod with TLR7. What follows after this interaction is the initiation of a series of intracellular signaling events followed by the production of reactive oxygen and nitrogen intermediates, up-regulation of co-stimulatory surface markers, and the production of cytokines and chemokines.

As the first step in the generation of an immune response is uptake of the vaccine by antigen-presenting cells either before (e.g., in the genitourinary tract) or after (e.g., in the gastrointestinal tract) traversing an epithelial layer, special focus has to be placed on APCs for understanding the requirements of generating innert and adaptive immune responses. It has been known for some time that DCs play a major role in initiating immune responses through activation of innate immunity [1]. More recently it has been demonstrated that DCs not only initiate immune responses but can also direct the immune responses toward T helper 1 (TH1) or TH2 responses. DCs from intestinal Peyer's patches (PPs) appear to induce TH2-type responses (IL-4 and IL-6) compared to DCs from spleen (SP), which induce TH1-type responses (IFN-γ and IL-2) [2]. The PP DCs are comprised of three different subpopulations: CD11b+/ CD8α-myeloid DCs in the subepithelial dome, the CD11b–/CD8α+ lymphoid DCs in the interfollicular regions, and the CD11b–/CD8α– DCs at both sites [3]. The latter subpopulation is predominant only in mucosal tissues. The double negative and lymphoid PP DCs produce IL-12p70 upon activation microbial, whereas the myeloid PP DCs produce IL-10 upon CD40L, *Staphylococcus aureus*, or IFN-γ activation [3]. The various populations of PP DCs migrate through the PPs based on the expression of various chemokine receptors toward various chemokine gradients, such as MIP-3α [4]. DCs have been shown to reduce the requirement for antigen concentration for cytotoxic T-lymphocyte (CTL) activity as well as acting directly on B cells to induce isotype switch differentiation [1]. As direct evidence that targeting mucosal DCs can indeed increase mucosal immune responses, it was shown that treatment of mice with Flt3L, which expanded the intestinal DCs, and coadministration of IL-1 or CT increased immune responses to co-fed antigen [5]. A recent review discusses in detail the interaction of CpG with TLR on dendritic cells [6].

## 13.3   STRUCTURE AND CELLULAR COMPONENTS OF MUCOSAL TISSUES AND ANTIGEN UPTAKE FOLLOWING VARIOUS ROUTES OF IMMUNIZATIONS

The first step in the initiation of an immune response through vaccination generally is the uptake of the vaccine at the mucosal surface by or through the epithelial layer, followed by uptake and presentation by antigen-presenting cells. However, there is a considerable difference in the anatomy of the respiratory, gastrointestinal, and genitourinary tracts with regard to their epithelial cells as well as the composition of the various antigen-presenting cells and lymphocytes in the various mucosal tracts. In the following, a brief description of the anatomy and the antigen-presenting cell and lymphocyte composition of the various mucosal tracts is given.

### 13.3.1   Nasal

The structure and localization of the nasopharyngeal lymphoid tissues and accumulating data suggest that they function as local inductive sites for the upper aerodigestive tract. The pharynx is divided into three parts: the nasopharynx (posterior to the nose and superior to the soft palate), the oropharynx (posterior to the mouth), and the laryngopharynx (posterior to the larynx) [7]. The lymphoid tissue in the pharynx forms an incomplete circular lymphoid structure called *Waldeyer's ring*. This lymphoid tissue is aggregated to form masses of lymph node (LN) called *tonsils*. The pharyngeal tonsil, known as *adenoids*, is in the mucous membrane of the roof and posterior wall of the nasopharynx [7]. The palatine tonsils are LNs at each side of the oropharynx between the palatine arches [7]. Unlike peripheral LNs, which are not directly associated with the mucosal lumen, the surface epithelium of the tonsils, similar to the mucosal-associated lymphoid tissue (MALT) of the gastrointestinal tract (e.g., Peyer's patches), is in direct contact with the lumen. The palatine tonsils and adenoids are covered with lymphoepithelium, consisting of ciliary and nonciliary epithelial cells, goblet cells, and M cells, the latter showing many invaginating lymphoid cells [8]. DCs are numerous within and underneath the epithelial layer of the tonsils and are in close contact with the neighboring B and T cells [9]. Direct uptake of antigens through the epithelial cells of the tonsils has been demonstrated and suggests that the tonsils play a major role as local inductive sites for mucosal immunity.

In addition to the lymphoid aggregates in the epithelium, local draining lymph nodes also represent important inductive sites for local and systemic immunity following application of antigens to the oral cavity. In the oral cavity, the lymphatic vessels of the parotid and submandibular glands drain to superficial and deep cervical LN [7]. Lymph from the end of the tongue drains to the superior deep cervical LN, whereas lymph from the tip of the tongue drains to the submental LN. Lymph from the sides and the middle of the tongue drain to the inferior deep cervical LN and to the submandibular LN, respectively [7].

Antigens administered to the nasal cavity are believed to be taken up by M cells overlying the follicle-associated epithelium of the nasal-associated lymphoid tissue (NALT) [10,11]. It is well established that M cells are highly efficient in the uptake of particulate antigens and microparticles and delivery to underlying antigen-presenting cells in the local lymphoid structure [10,12]. Following intratracheal (IT) delivery, it is likely that microparticles are engulfed by macrophages or dendritic cells and transported to bronchus-associated lymphoid tissue and then to local draining lymph nodes. Alternatively, vaccines may be cleared from the lung through the mucociliary elevator and swallowed. Nevertheless, delivery to the lungs is technically difficult and may be associated with potential toxicity issues, including hypersensitivity problems. Thus, in terms of vaccination strategies, the intranasal (IN) route appears more practical and feasible compared to delivery of vaccines to the lungs.

Two subtypes of IgA exist in humans, IgA1 and IgA2. The latter is most frequent in the upper aerodigestive tracts, while the former predominates in the large intestine [13]. In this regard, tonsilar IgA+ cells are predominantly of the IgA1 subtype [14,15], providing further evidence that they function as local inductive sites that seed the mucosal effector sites of the upper aerodigestive tracts. The palatine tonsils and the adenoids comprise 30 to 35% CD3$^+$ cells, 20 to 28% CD4$^+$ cells, and 5 to 6% CD8$^+$ cells, and the majority of the T cells appear to express TCRγδ. While the activation marker IL2R (CD25) is up-regulated on only 3 to 8% of the cells, CD28 is expressed on 23 to 36% of the cells, mostly in the adenoids [15]. T cells comprise about half of tonsilar intraepithelial mononuclear cells, with equal ratios of CD4$^+$ and CD8$^+$ cells. In the deeper interfollicular regions, the ratio of αβTCR$^+$ to γδTCR$^+$ cells is 10:1, whereas in the superficial areas a reduction in the number of $\alpha\beta TCR^+$ cells reduces this ratio to 2:1 [16]. Taken together, the presence of a lymphoepithelial structure, professional antigen-presenting cells [e.g., DCs, follicular dendritic cells (FDCs), and the germinal center machinery], as well as functional $CD4^+$ and $CD8^+$ T cells within the epithelial layers as well as deep in the LN, suggest that the oro/nasopharyngeal lymphoid tissues act as both immune inductive and effector sites for the upper aerodigestive tract.

### 13.3.2 Oral

Although oral immunization is the most attractive route of mucosal immunization, it is also the most difficult route for induction of immune responses. This is probably due to the structure of the intestinal mucosa, which has been developed evolutionarily to induce tolerance (unresponsiveness) rather than immunity (Figure 13.3). The small intestine is divided into duodenum (adjacent to the stomach), jejunum, and ileum. The large intestine may be divided between the colon and the rectum. The entire intestine is covered by a single layer of epithelial cells (enterocytes). Throughout the small intestine are found macroscopically visible lymph node–resembling structures, called *Peyer's patches* (PPs). Enterocytes also overlie the PPs, although interspersed between them are found special epithelial cells, called *M* (microfold) *cells*. Enterocytes and M cells covering the PP are called *follicle-associated epithelium* (FAE). The majority of intestinal pathogens as well as particulate delivery systems are taken up by M cells, whereafter they are taken up by DCs, macrophages, or

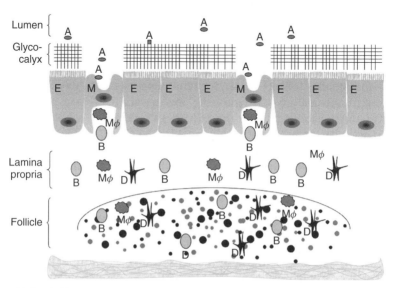

**Figure 13.3**  Cellular structure of the diffuse and organized immune compartments in the gastrointestinal tract. (*See insert for color representation of figure.*)

B cells underneath the M cells to initiate an immune response. Characteristics of the intestinal DCs are discussed in Section 2.

### 13.3.3 Rectal

The rectal mucosa of several mammalian species, including humans, contain macroscopically invisible solitary lymphoid nodules that resemble Peyer's patches of the small intestine in their cellular structure and phenotype. These structures are overlaid with microfold (M) cells that are specialized in antigen uptake [12]. Of note, both the rectal and vaginal mucosa are drained by the iliac lymph nodes, and there is indirect evidence that SIgA-secreting cells in the vaginal mucosa originate from the solitary lymphoid nodules of rectum [17–19]. In nonhuman primates [20] as well as in humans [21], the rectal and small intestinal lamina propria (LP) contain high numbers of $CD69^+$ macrophages that are concentrated under the single layer of epithelial cells (enterocytes), whereas cells with dendrites, which are far fewer in number (probably DCs) form a reticular framework throughout the LP. The rectal mucosa may serve as a vaccine delivery route, and because the vaccine does not have to go through the entire digestive tract and the intestine, lower amounts of antigen are required for intrarectal compared to oral immunizations. However, it may not be an attractive route of immunization, for socioethical reasons.

### 13.3.4 Vaginal

The vaginal mucosa is covered with multilayered squamous epithelia, while the uterus, cervix, and fallopian tubes are covered with pseudosquamous and simple columnar epithelia. Underneath the epithelial layers of the vagina, uterus, and fallopian tubes is the lamina propria compartment, comprising a large array of B cells, $CD4^+$ and $CD8^+$ T cells, and antigen-presenting cells (APCs) [22]. The presence of lymphoid aggregates in the female genital tract has also been reported, although whether these aggregates have follicle-associated epithelium, as is the case with nasal-associated lymphoid tissue (NALT) and Peyer's patches, remains to be elucidated [22,23]. DCs and $CD8^+$ T cells with cytotoxic activity are found interspersed within the squamous epithelium of the vagina [24–26]. Thus, the vaginal mucosa contains DCs as well as CTL and can mount antiviral cytotoxic T-cell responses that can be protective. The vagina is considered to be a component of the common mucosal immune system, and oral immunization in mice with microparticles has been shown to induce a vaginal antibody response [27]. In addition, IN immunization with microparticles also induced antibodies in the lower genital tract of mice [28]. Although there is no evidence to indicate the presence of lymphoid follicles or M cells in the vaginal mucosa [29], intravaginal (IVAG) immunization in humans induced local antibody responses [30]. However, IVAG immunization protocols in small animal models have not normally met with great success, despite the use of novel delivery systems and adjuvants [31–33], although a more recent report showed that vaginal or rectal but not IN or intramuscular immunizations with alpha virus–based replicon particles encoding HIV-1 gag protected against IVAG challenge with vaccinia virus encoding HIV-1 gag [34]. Moreover, the local immune response in the vagina is subject to significant hormonal regulation, with major changes in local antibodies at different stages of the menstrual cycle [35]. A study in mice showed that the IN route of immunization was more effective than the IVAG route for the induction of immune responses in the vagina [36]. In female humans the IN route of immunization may be exploited for the induction of genital tract antibody response [37]. Thus, although the vaginal mucosa contains the necessary immunological machinery to mount a local immune response, the IN immunization appears to be a more suitable route. However, it remains to be seen whether in the resting memory phase a more rapid local response is induced in the vaginal mucosa following IVAG immunization compared to IN immunization.

## 13.4 SOLUBLE SMALL IMMUNOPOTENTIATING MUCOSAL ADJUVANTS

### 13.4.1 CpG

In the last few years, an entirely new class of adjuvant actives have been identified, following the demonstration that bacterial DNA, but not vertebrate DNA, had direct immunostimulatory effects on immune cells in vitro [38,39]. The immunostimulatory effect was due to the presence of unmethylated CpG dinucleotides [40], which are underrepresented and methylated in vertebrate DNA. Unmethylated CpG in the context of selective flanking sequences are thought to be recognized by cells of the innate immune system to allow discrimination of pathogen-derived DNA from self DNA [41]. It has been shown that cellular responses to CpG DNA are dependent on the presence of TLR9 [42]. In addition, it has been reported that CpG are taken up by nonspecific endocytosis and that endosomal maturation is necessary for cell activation and the release of pro-inflammatory cytokines [43]. The TH1 adjuvant effect of CpG appears to be maximized by their conjugation to protein antigens [44]. Importantly, CpGs also appear to have potential for the modulation of preexisting immune responses, which may be useful in various clinical settings, including allergies [45]. Details and mechanisms of the interaction of CpG with TLRs on DCs have been studied extensively and continue to be a major focus of current and future studies [6].

### 13.4.2 Imiquimod and Resiquimod

Imiquimod and its related resiquimod are synthetic small molecule immunopotentiating adjuvants, which is discussed in detail in Chapter 10. Here, a brief description with emphasis on their use as mucosal adjuvants is given. There are very few studies where imiquinod or resiquimod has been used as a mucosal adjuvant. Resiquimod, also known as R848, which binds TLR7 and TLR8, was recently used as an adjuvant for mucosal and parenteral immunizations and compared with the adjuvant effects of CpG [46]. Both R848 and CpG enhanced antigen-specific responses. However, whereas CpG had to be administered by the same route and simultaneously as the antigen, this was not the case for R848 [46]. For vaccination purposes, though, it would be more practical to administer both the adjuvant and the antigen at the same time and through the same route.

## 13.5  SOLUBLE PROTEIN IMMUNOPOTENTIATING MUCOSAL ADJUVANTS

### 13.5.1  Mutants of Heat-Labile Enterotoxin from *Escherichia coli*

Genetically detoxified mutants of heat-labile enterotoxin (LT) have been shown to be potent adjuvants for inducing mucosal and systemic immune responses. To retain the adjuvanticity of these molecules but reduce their toxicity, several mutants have been generated by site-directed mutagenesis. Of these, two mutants of the enzymatic A subunit, LTK63 and LTR72, maintain a high degree of adjuvanticity. LTK63 is the result of a substitution of serine 63 with a lysine in the A subunit which renders it enzymatically inactive and nontoxic [47–51]. LTR72 is derived from a substitution of alanine 72 with an arginine in the A subunit and contains about 0.6% of the enzymatic activity of wild-type LT. LTR72 is shown to be 100,000-fold less toxic than wild-type LT in Y1 cells in vitro and 25 to 100 times less toxic than wild-type LT in the rabbit ileal loop assay [52].

In a study it was shown that an influenza vaccine given IN together with LTK63 in a novel bioadhesive microsphere delivery system induced enhanced serum IgG as well as nasal IgA responses in mice and pigs [53]. Thus, the combination of LT mutant adjuvants with a microparticle system enhanced local and systemic humoral responses. Collectively, these data show that LT mutants are effective mucosal adjuvants in small and larger animal models and can be used in combination with microparticle formulations to enhance immune responses.

The ability of LT mutants to induce CTL responses against HIV-1 p55 gag, following IN, oral, or IM immunization was described in 2000 [54]. Interestingly, in this paper evidence was provided that LTK63 and LTR72 had diverse effects when used as mucosal adjuvants for oral versus IN immunizations. LTK63 induced stronger CTL responses following IN immunization with p55 compared with LTR72. Conversely, LTR72 induced stronger CTL responses against p55 when given orally, and it also induced local CTL responses. Thus, it appears that if induction of CTL responses is the objective, some ADP-ribosyltransferase activity of the LT mutant may be required for oral but not for IN immunizations. These studies showed that IN immunization with protein vaccines and LT mutant adjuvants can be an effective means for the induction of cell-mediated immunity. However, some antigens may require optimization of the delivery systems as well as inclusion of adjuvants.

Since LT mutants are potent antigens following mucosal immunization, there is a concern that preexisting immunity to LT might affect the potency of these molecules when used as adjuvants. Therefore, the potency of LTK63 in mice and pigs with preexisting immunity to the adjuvant was evaluated in a publication. It was found that preexisting immunity to LTK63 did not affect its potency as an adjuvant, when used for IN immunization with a second vaccine, soon after the first. In addition, these studies also showed the potency of LT mutants for a protein polysaccharide conjugate vaccine (*Neisseria meningitidis* group C CRM conjugate) and extended their use into a larger animal model, the pig [55].

### 13.5.2 Mutants of CT

Another important genetically detoxified immunopotentiaing adjuvant is the CTA1-DD adjuvant derived from the genetic fusion of the enzymatically active CTA1 gene from whole cholera toxin, from *Vibrio cholera*, to a gene encoding a synthetic analog of *Staphylococcus aureus* protein A with a high affinity toward B cells. Studies have shown that this adjuvant has at least 100- to 1000-fold reduced toxicity compared to that of the whole wild-type cholera toxin and that up to 500 μg injected intraperitoneally to mice did not show any toxic effects in spleen, liver, or kidneys. Moreover, this adjuvant appeared to give protection against rotavirus comparable to LTR192 (another mutant of the heat-labile enterotoxin) and CpG (immunostimulatory dinucleotides; see below) [56].

## 13.6 SYNTHETIC, NONREPLICATING DELIVERY SYSTEMS FOR MUCOSAL IMMUNIZATIONS

### 13.6.1 Microparticles

As opposed to soluble antigens, particulates are known to induce better immune responses, probably because particulate antigens mimic in certain structural aspects the pathogens that cause disease. Particulate delivery systems are more readily taken up by M cells at mucosal inductive sites and then can better be taken up by local APCs such as DCs, macrophages, and B cells. Therefore, much emphasis has been placed on finding particulate antigen-delivery systems that can readily be taken up at mucosal surfaces. Charged microparticles as antigen delivery systems are discussed in detail in Chapter 7. Here a brief description of microparticles used as mucosal delivery systems is given. One of the most popular particulate delivery systems has been the biodegradable and biocompatible polyesters, the poly(lactide-co-glycolides)

(PLGs). The immune enhancing effect of antigens encapsulated in PLG was demonstrated over a decade ago when it was shown that such a formulation increased both B- and T-cell responses [57–63].

Alternative microencapsulation approaches other than PLGs has also been made. Strong evidence for dissemination of antigen-specific antibody-secreting cells from nasal-associated lymphoid tissue to the cervical lymph nodes and spleen following IN immunizations has been provided by Heritage et al. [64]. These local and systemic humoral responses were generated by entrapment of human serum albumin (HSA) in polymer-grafted microparticles [3-(triethoxysilyl)-propyl-terminated polydimethylsiloxane (TS-PDMS)] with a size range of 1 to 100 μm. McDermott et al. reported that polymer-grafted starch microparticles have been used as an alternative to PLG particles and were shown to deliver antigens effectively following oral or IN immunizations and elicit local and systemic humoral responses [65]. However, in comparison to PLGs, these microparticles are poorly defined and their biocompatibility has not been tested in humans.

In another study, a single IN or oral immunization with a *Schistosoma mansoni* antigen entrapped in PLG or polycaprolactone (PCL) microparticles resulted in sustained serum IgG responses [66]. However, this vaccine strategy failed to induce IgA responses in serum or BAL fluids following oral immunization, while IN immunization resulted in both serum and BAL fluid IgA responses. Interestingly, only the PLG-entrapped, not the PCL-entrapped, vaccine resulted in strong neutralizing antibody responses following either IN or oral immunization. Moreover, the humoral responses were detectable earlier following PLG versus PCL immunizations, presumably due to the physicochemical differences between the two polymers and different rates of antigen release.

Following IN immunization of anesthetized mice with hemagglutinin (HA) from influenza virus entrapped in one of four microparticle resins (sodium polystyrene sulfonate, calcium polystyrene sulfonate, polystyrene benzyltrimetylammonium chloride, or polystyrene divinylbenzene) sized to 20 to 45-μm enhanced anti-HA serum hemagglutinin-inhibiting antibodies and nasal wash IgA antibodies were induced [67]. Importantly, this study showed that this immunization strategy reduced viral burden in the lungs following IN administration of virus to the lungs of anesthetized mice. Interestingly, these resins induced enhanced serum IFN-γ levels following IT immunizations, while the levels of IL-4, IL-2, and IL-6 remained unchanged, suggesting a TH1-type of response [67]. More recently, Kim et al. demonstrated mucosal immune responses following oral immunization with rotavirus antigens encapsulated in alginate microspheres [68].

### 13.6.2 Iscoms

The immunostimulatory fractions from Quillaja saponaria (Quil A) have been incorporated into lipid particles comprising cholesterol, phospholipids, and cell membrane antigens, which are called ISCOMs [69]. In a study in macaques, an influenza ISCOM vaccine was shown to be more immunogenic than a classical subunit vaccine and induced enhanced protective efficacy [70]. A similar formulation has been evaluated in human clinical trials and was shown to induce CTL responses [71]. The principal advantage of the preparation of ISCOMs is to allow a reduction in the dose of the hemolytic Quil A adjuvant and to target the formulation directly to APCs. In addition, within the ISCOM structure, the Quil A is bound to cholesterol and is not free to interact with cell membranes. Therefore, the hemolytic activity of the saponins is significantly reduced [69,72]. However, a potential problem with ISCOMs is that inclusion of antigens into the adjuvant is often difficult and may require extensive antigen modification [73].

### 13.6.3  Liposomes

Liposomes are discussed in detail in Chapter 12. Here, they are described briefly. Liposomes are phospholipid vesicles that have been evaluated both as adjuvants and as delivery systems for antigens and adjuvants [74,75]. Liposomes have commonly been used in complex formulations, often including MPL, which makes it difficult to determine the contribution of the liposome to the overall adjuvant effect. Nevertheless, several liposomal vaccines based on viral membrane proteins (virosomes) without additional immunostimulators, have been evaluated extensively in the clinic and are approved as products in Europe for hepatitis A and influenza [76].

### 13.6.4  Bioadhesives, Chitosan, and Lectins

*Bioadhesives*, as their name implies, are used to help vaccines to adhere to a mucosal surface. As such, bioadhesives that bind to mucosal surfaces include chitosan, carbopol, hyaluronic acid, polyacrylic acid, carboxymethylcellulose, and hydroxypropylcellulose among others. Of these, the most popular for use as mucosal delivery systems is chitosan.

*Chitosan* is a cationic polysaccharide, a copolymer of *N*-acetyl-D-glucosamine (ca. 20%) and glucosamine (ca. 80%). It is derived from crustacean shells and has been used extensively as a food additive. Chitosan enhances transepithelial transport of antigen to the nasal mucosa through exerting an effect on tight junctions of the epithelial layer and decreasing mucociliary clearance [77]. In an in vitro model, chitosan microparticles induced enhanced transport across in vitro–derived M cells [78].

Chitosan has been used as a delivery system for vaccines against various pathogens in several studies for oral or IN routes of immunizations. For IN immunizations, chitosan has been used successfully to enhance antibody responses against protein antigens derived from influenza and diphtheria in clinical trials [79,80]. Chitosan has also been used as a DNA delivery system for tetanus toxoid [81]. Pulmonary or immunizations with DNA entrapped in chitosan induced T-cell responses against *Mycobacterium tuberculosis* [82] and respiratory syncytial viruses (RSV) [83,84]. Oral immunizations with chitosan–DNA promoted protection against peanut allergy [85]. The simultaneous use of chitosan and an LT-mutant adjuvant (LTK63) for IN immunizations against mengicoccal antigens induced higher antibody responses than did parenteral immunization [86,87]. A combination of hyaluronic acid with LTK63 for IN delivery of influenza antigens enhanced the local and systemic antibody responses [88].

*Lectins* may also be classified as bioadhesives. Lectins can be derived from both plants and animals, although as vaccine delivery systems, plant-derived lectins have been used more predominantly. There are several plant lectins, including wheatgerm agglutinin (*Triticum aestivum*), *Ulex europaeus* agglutinin-1 (UEA-1), *Lycopersicum esculentium* (LEA), and mistletoe lectin 1 (ML1). Of these, ML1 and UEA-1 have been shown to be particularly potent as mucosal adjuvants for IN immunizations [89–91]. Lectins have also been investigated as an oral delivery system for specifically targeting intestinal epithelial cells, including M cells [92–94]. Another type of lectin, C-type lectins, includes the important DC-expressed surface receptor, DC-specific intercellular adhesion molecule 3-grabbing nonintegrin (DC-SIGN). Because DCs play an important role in the initiation and direction of innate and adaptive responses, DC-SIGN targeting strategies have proved promising for the enhancement of immune responses [95] that have the potential to be effective for mucosal immunizations.

## 13.7 COMBINATIONS OF NONREPLICATING DELIVERY SYSTEMS AND IMMUNOPOTENTIATING ADJUVANTS FOR MUCOSAL IMMUNIZATION

It has been demonstrated that immunopotentiating adjuvants and various delivery systems can be combined synergistically to induce higher immune responses. As induction of immune responses following IN, and in particular oral, immunizations is difficult, such combinations may be required in many instances in order to achieve immunity. The potency of microparticle formulations with adsorbed antigens has been shown to improve significantly by their coadministration with adsorbed adjuvants [96]. Simultaneous delivery of antigens and adjuvants on microparticles ensures that both agents can be delivered into the same APC population. To illustrate this approach, cationic PLG microparticles were adsorbed with the immunopotentiaing adjuvant, CpG and the HIV-p55 gag protein [96]. In addition, enhanced serum antibody responses against HIV-1 gp120 protein were generated when administered in combination with PLG/CpG microparticles in mice. Microparticles can also be used to deliver entrapped adjuvants, to ensure long-term controlled release of these agents. Tabata and Ikada first entrapped a synthetic adjuvant, muramyl dipeptide (MDP), in microspheres of gelatin [97], while Puri and Sinko [98] showed that MDP entrapped in microspheres induced enhanced immune responses. Importantly, Tabata and Ikada [99] had also shown that the pyrogenicity of MDP was reduced by microencapsulation, establishing that microparticles can improve potency and also reduce the reactogenicity of adjuvants. PLG microparticles have also been used for delivery of QS21 adjuvant in combination with gp120 antigen [100]. However, although studies with an implantable osmotic pump showed that higher titers were obtained with a discontinuous rather than a continuous release profile for the adjuvant [100], the studies also showed that while the adjuvant was critical for the induction of high initial titers, it was not required for the secondary response. Therefore, the easiest approach was to suspend the microparticles containing the antigen in a solution containing the adjuvant so that it was immediately available to enhance titers. In addition to the use of microparticles to deliver adjuvants, microparticles may also be used in conjunction with traditional adjuvants, including alum [101,102]. In addition, microparticles may also be combined with emulsion-based adjuvants to improve their potency [103,104].

## 13.8 INACTIVATED BACTERIA-, INACTIVATED VIRUS-, BACTERIAL GHOST VECTORS, AND VLP-BASED VACCINES

The idea of attenuating many pathogenic viruses is impractical due to obvious issues with reverting to the wild type and causing disease. Also, inactivating the many highly pathogenic viruses would pose a risk of a few viruses escaping inactivation and causing disease. For both live attenuated and inactivated highly pathogenic viruses, large-scale production will require a very high-biosafety-level facility, which again would be impractical. To use the HIV model as an example, inactivated HIV- or SHIV-capturing nanoparticles have been produced and used to induce vaginal antibody responses in mice and rhesus macaques following IN immunizations [105,106]. Moreover, inactivated HIV-1 plus CpG adjuvant induced genital CTL and antibody responses in mice that were subsequently protected against vaginal challenge with recombinant vaccinia virus [107,108]. There have also been examples of IN immunizations with recombinant bacillus Calmette–Guérin (BCG) bacteria encoding an HIV-1 antigen [109] or heat-inactivated bacteria conjugated to HIV-env antigen [110]. As stated above, live attenuated or inactivated viruses that cause high pathogenicity will probably not serve as a vaccine candidate for human use due to serious safety concerns.

Bacterial ghost vectors (BGVs) are nonreplicating gram-negative bacterial cell envelopes that do not contain cytoplasmic granules while maintaining cell surface antigenic structures [111]. These vectors can be used as delivery systems for both protein and DNA. Similar to viruslike particles (VLPs), they can be freeze-dried, significantly increasing shelf life and stability. Bacteria used for producing BGVs include *Escherichia coli, Salmonella typhimurium, Vibrio cholera, Helicobacter pylori,* and *Pseudomonas putida,* among others [111]. In general, BGV vaccine candidates from each bacterium has been used directly to induce immune responses against the parental bacterium (i.e., oral immunizations with BGVs from *H. pylori* has induced protective immunity against *H. pylori* challenge) [112].

VLPs, which have been used as candidate and licensed vaccines, are produced by transfecting eukaryotic cells, yeast, insect, or mammalian cells with DNA encoding a gene of interest, usually delivered by baculovirus or vaccinia virus. The VLPs are then formed, without a genome, in the cell and are either secreted or remain in the cell to be purified. For example, VLPs made of HIV-gag alone or with HIV-env expressed on the surface of gag-VLP have been developed and used for IN immunizations in animal models [113,114]. As a different approach, IN immunizations of mice with chimeric influenza HA/SHIV VLP induced mucosal and systemic humoral and cellular responses [115]. An inherent problem with most studies using baculovirus-based VLP is that no or insufficient efforts were made to purify VLPs in the absence of baculovirus. Baculoviruses induce innate antiviral effects and can clearly enter mammalian cells, and although they do not replicate, the DNA enters the nucleus, thus raising concerns about host cell chromosomal integration [116].

## 13.9   RNA-BASED DELIVERY SYSTEMS FOR MUCOSAL IMMUNIZATIONS

Anyone who has purified RNA can attest to its labile nature, as there are abundant RNases ready to denature RNA everywhere. Thus, the idea of RNA-based vaccines may intuitively seem problematic. However, several RNA delivery systems have been invented which result in the infection of target cells and delivery of RNA encoding the gene of interest. In general, then, such an approach has a clear advantage over the more popular plasmid DNA immunization, as RNA, unlike DNA, does not require access to the nucleus and thus minimizes the possibility of chromosomal integration [117].

A number of RNA delivery systems as potential vaccine candidates have been described, including purified cDNA-transcribed RNA, neuraminidase-deficient influenza A virus, tickborne encephalitis virus, *Listeria monocytogenes,* nontransmissible Sendai virus, liposome-entrapped mRNA, and alphaviruses [117–125]. Of these, perhaps the most popular RNA delivery systems for an anti-HIV vaccine have been alphaviruses, either as replicating or nonreplicating replicon particles.

Alphaviruses, including Sindbis virus (SIN), Semliki forest virus, and Venezuelan equine encephalitis virus (VEE), are enveloped RNA viruses that have been developed into replication-defective "suicide" vectors [126,127]. Alphavirus replicon RNA vectors maintain the nonstructural protein gene and *cis* replication sequences required to drive abundant expression of heterologous antigens from the viral subgenomic 26S promoter but are devoid of any alphaviral structural protein genes required for propagation and spread. These vectors also offer the prospect of natural adjuvanticity and stimulation of the innate immune response, in addition to the antigen-specific adaptive response arising from the cytoplasmic amplification of these vectors through double-stranded RNA intermediates [128]. Replicon vectors have been widely evaluated as vaccine immunogens, both as plasmid DNA replicon vaccines and as viruslike replicon particles [129].

Replicating VEE expressing HIV-1 matrix/capsid were used to inject mice subcutaneously, which induced IgA as well as CTL responses [130]. Cynomolgus macaques immunized parenterally with SFV RNA vectors expressing HIV-IIIB gp160 and challenged parenterally with SHIV-4 were not protected against high viral loads [131], even though a mouse study suggested that compared to a plasmid DNA encoding HIV-env, or HIV-env protein, SFV expressing HIV-env induced the highest serum anti-env antibodies [132].

## 13.10 REPLICATING VIRUS-BASED DELIVERY SYSTEMS FOR MUCOSAL IMMUNIZATIONS

Live attenuated (L.A.) viruses as antigen-delivery systems in general offer relatively high potency. However, most have the problem of inducing high antivirus vector immunity, thus making their multiple or even subsequent use obsolete. In this regard, a report using a mouse model indicated that mucosal vaccination (which in this case was intrarectal) overcomes the barrier (i.e., immunity against the vaccinia vector) to recombinant vaccinia virus (VV) immunization caused by preexisting poxvirus immunity [133]. If the L.A. virus is the infectious virus with attenuations to eliminate or reduce disease, there have been clear examples of reverting to the wild-type phenotype or genotype and causing severe disease [134–137]. The level of attenuation appears to correlate inversely with potency since the more attenuated strains are less immunogenic [138,139]. Below, some examples of L.A. viruses are provided for IN or other routes of immunization for induction of anti-HIV immune responses.

VV-based L.A. vaccines have been used extensively to induce anti-HIV responses through IN immunizations alone or as priming or boosting immunizations. IN priming with HIV-env-expressing influenza virus and IN boosting with HIV-env-expressing VV in mice, induced systemic cellular responses in spleen and local responses in the genitorectal draining lymph nodes [140]. IN priming with DNA and IN boosting with VV expressing HIV-env induced mucosal and systemic humoral and cellular responses [141]. In a nonhuman primate model, IN, intramuscular and intrarectal immunizations with a live attenuated poxvirus (NYVAC) expressing an immunodominant CD8+ CTL gag-epitope induced CTL responses in the small intestine [142]. An IL-2-augmented DNA IN prime/VV IN boost induced mucosal and systemic humoral and cellular responses, and protected from disease (i.e., all individuals became infected but maintained CD4 cell counts and did not develop AIDS) [143].

A popular approach for using vaccinia-based anti-HIV vaccines has been DNA priming followed by vaccinia boosting, in effect avoiding strong antivaccinia immune responses. However, except for the papers listed above, the other reports have used systemic or nonnasal mucosal routes for priming or boosting immunizations. A DNA prime–vaccinia boost immunization containing multiple HIV genes controlled viral loads following intrarectal challenge [144]. A tat/rev/net-based DNA prime–tat/rev/nef vaccinia boost regimen induced rectal CTL and controlled the acute phase, but not long-term SIVmac239 viral loads [145]. This lack of long-term protection may have been due to induction of acute but not memory T-cell responses [146]. A SHIV DNA/vaccinia virus rectal vaccination of rhesus macaques induced inconsistent cellular, but mucosal and systemic humoral responses and delayed progression to AIDS [147]. To enhance the immunogenicity of the DNA prime–vaccinia boost model further, plasmid chemokines and colony-stimulating factors have been shown to be effective [148].

Several problems have been reported with regards to the use of vacciniabased anti-HIV vaccines. Gender differences have been found in mice for induction of HIV-specific CD8 responses in the reproductive tract and colon following IN peptide priming and vaccinia virus boosting immunizations, in that such a regimen induced strong responses in female but not male mice, and both priming and boosting with the vaccinia virus–based vaccine was required to induce optimal responses in the male mice [149]. It has also been reported that smallpox vaccine does not protect macaques with AIDS from a lethal monkey pox virus challenge, making the use of a therapeutic vacciniabased vaccine improbable [150]. It has also been shown that the *nef* gene of HIV and SIV, which is important for pathogenicity and maintenance of high virus loads, is a negative attenuating and cell-to-cell infection factor for vaccinia virus, hence prohibiting the use of a *nef* gene–expressing vaccinia virus as an anti-HIV vaccine [151]. It was recently shown in humans that a *nef* gene–expressing vaccinia virus was used for three vaccinations on AIDS patients on HAART therapy, after which the therapy was interrupted. However, although strong anti-vaccinia virus and some anti-*nef* responses were detected, the viral loads rebounded in all patients, but as the CD4 counts have remained above the pre-HAART levels, six of 14 patients have remained off HAART, offering some hope for this approach [152]. Recent reviews discuss in general the use of vaccinia viruses for HIV vaccines through all routes of immunizations [153–155].

Another L.A. virus used for IN immunization against HIV is vesicular stomatitis virus (rVSV). Combination of parenteral and mucosal immunizations with L.A. rVSV was shown to prevent AIDS in rhesus macaques [156]. A comparison of IN and IM routes of immunization with rVSV in rhesus macaques demonstrated that the IN route induced higher cellular as well as nasal and saliva responses, but both routes of immunization conferred protection against vaginal SHIV challenge [157]. The cellular responses were not measured in the vaginal mucosa, and the humoral responses at this site were low after IN immunizations, making deciphering the correlates of protection unallowable. However, it was shown that IN immunization was as good or perhaps better than IM immunizations for protection against AIDS. However, in another study, intraperitoneal (another systemic route of immunization) versus IN immunizations with rVSV, expressing HIV-gag and HIV-env, induced far higher CD8+ CTL responses in spleens [158]. Although these studies indicate that rVSV may serve as a vaccine delivery system for a potential anti-HIV vaccine, serious potential hazards exist with regard to this virus. VSV belongs to the family of Rhabdoviridae and causes severe vesiculation and/or ulceration of the tongue, oral tissues, feet, and teats of horses, pigs, and cattle (symptoms that are indistinguishable from foot-and-mouth disease) and as an important zoonotic pathogen can cause disease in humans [159,160]. Therefore, the use of this vector as an ultimate human vaccine delivery system against HIV seems unlikely.

Adenoviruses (Ads), proposed to be used as a vaccine delivery system for anti-HIV vaccines, are icosahedral, nonenveloped, double-stranded DNA viruses belonging to the family Adenoviridae and infect all mammalian species [161]. Most human and nonhuman primates have preexisting antibodies against most Ads. Ads are responsible for approximately 5% of upper respiratory tract infections and 8% of childhood pneumonia incidences [162]. Adenoviruses have caused major outbreaks in the past [162] and emergent strains are on the rise in North American populations [163,164]. There are six subgroups (A to F) and several serotypes within each subgroup (e.g., Ad1, Ad2, Ad5, and A6 are in subgroup C). Each serotype is known to cause a different disease or a specific severity of disease [165]. Other safety concerns stem from the fact that most Ads bind specifically to liver parenchaymal cells [166] and localize to the central nervous system following IN immunizations [167]. Indeed, administration of the popular Ad5 resulted in the death of a patient [168]. Nonetheless, optimism that the strains used as vaccine delivery vehicles are not disease causing or cause only minor symptoms, as well as the use of replication-defective vectors, have kept adenovirus-based vaccines as possible vaccine candidates.

Ad-based vaccines have been considered as a vaccine delivery system for anti-HIV vaccines for over a decade [169–173]. In early reports, Ads expressing HIV-env induced serum HIV-neutralizing antibodies following intratracheal immunizations of dogs [169]. This study was soon followed by two chimpanzee studies using IN immunizations as a stand-alone or boosting modality with Ads expressing HIV-gag and HIV-env inducing variable mucosal and systemic humoral responses [170,171]. To avoid preexisting immunity, it has been suggested that DNA priming followed by Ad boosting would enhance the responses [174,175]. Also, priming with Ads and boosting with HIV-env have been reported to reduce acute viremia following challenge [176,177]. Several reports have also demonstrated immunogenicity of replication-incompetent

Ad vectors as anti-HIV vaccine vectors [173,177,178]. Thus, in general, although replication-competent Ad vectors pose some safety concerns, they can induce mucosal and systemic humoral responses following IN immunizations as a stand-alone modality, and they can induce strong CTL responses following parenteral immunizations as a stand-alone or a boosting modality. The use of replication-incompetent Ad vectors might circumvent the safety issues but induce significantly reduced immune responses which may nonetheless be sufficient to reduce viremia.

Influenza virus binds epithelial cells of the upper and lower respiratory tract and induces a relatively strong immune response against itself. Therefore, using an influenza virus as a mucosal vaccine delivery vector may be feasible, although again safety concerns do prevail, and thus far the mucosal use of such vectors has not been reported in primate models. Recombinant influenza A virus containing an HIV-*nef* insert, which did not replicate in the respiratory tract, was used for a single IN immunization of mice and induced $CD8^+$ T-cell responses in spleen and in lymph nodes draining the respiratory tract and urogenital tract [179]. Moreover, a chimeric influenza virus that did replicate in the respiratory tract of mice, containing an HIV-env neutralizing epitope, induced humoral responses in spleen, lungs, and urogenital tracts [180,181]. Interestingly, IN immunization with inactivated influenza virus enhanced immune responses to coadministered SHIV VLP, suggesting that influenza virus viability may not be required for the adjuvant action of influenza viruses [182]. Other L.A. virus approaches for mucosal immunization against HIV include the use of a rhabdovirus (rabies virus), polio virus, and a plant virus (cowpea mosaic virus) which have induced mucosal and systemic humoral and cellular responses [183–185].

## 13.11 REPLICATING BACTERIA-BASED DELIVERY SYSTEMS FOR MUCOSAL IMMUNIZATIONS

Replicating bacteria-based delivery systems for mucosal immunizations have been used to deliver the gene of interest into the host and then produce the antigen of interest in vivo. Immunization with invasive bacterial systems such as *Shigella*, *Salmonella*, and *Listeria* as a vector system for effective gene transfer into the cytoplasm of infected cells has been explored by several groups in the last few years [186]. The use of a live attenuated bacterial delivery system as an effective modality for generating strong responses to antigens after intranasal immunization has also been explored, along with other approaches described above.

The following examples of mucosal immunizations with replicating bacterial delivery systems use HIV as a model. The first documented report of HIV protective immunity with a live *Shigella* DNA vector was published by Shata and Hone [187]. A single intranasal dose of *Shigella*/HIV-1 gp120 vaccine vector in mice induced strong CD8 T-cell response comparable to systemic immunization with a vaccinia-env vector [187]. In addition, this single immunization was also effective in affording protection against a vaccinia-env challenge. The group also showed that this vector was effective orally in mice [188]. Vecino et al. [189] published their findings with attenuated *Shigella* and *Salmonella* delivering HIV-1 gp120 evaluated intranasally in mice. They reported that attenuated *Shigella* was much more effective than attenuated *Salmonella* in generating CD8 T-cell responses after a single intranasal immunization. The intranasal route was more effective than the intramuscular route for generating higher IgA titers in the vaginal washes. Xu et al. [186] reported that a single intranasal dose of *Shigella fleneri* 2a mutant encoding for HIV-1 SF2 gag was effective in inducing gag-specific T-cell responses in both spleen and lung. Live bacterial delivery as an approach is currently being explored more through the oral route than the intranasal route, for ease of administration.

## REFERENCES

1. Steinman, R.M. & Pope, M. Exploiting dendritic cells to improve vaccine efficacy. *J Clin Invest* 2002, **109**(12), 1519–1526.
2. Everson, M.P., Lemak, D.G., McDuffie, D.S., Koopman, W.J., McGhee, J.R. & Beagley, K.W. Dendritic cells from Peyer's patch and spleen induce different T helper cell responses. *J Interferon Cytokine Res* 1998, **18**(2), 103–115.
3. Iwasaki, A. & Kelsall, B.L. Unique functions of CD11b+, CD8 alpha+, and double-negative Peyer's patch dendritic cells. *J Immunol* 2001, **166**(8), 4884–4890.
4. Iwasaki, A. & Kelsall, B.L. Localization of distinct Peyer's patch dendritic cell subsets and their recruitment by chemokines macrophage inflammatory protein (MIP)-3alpha, MIP-3beta, and secondary lymphoid organ chemokine. *J Exp Med* 2000, **191**(8), 1381–1394.
5. Williamson, E., Westrich, G.M. & Viney, J.L. Modulating dendritic cells to optimize mucosal immunization protocols. *J Immunol* 1999, **163**(7), 3668–3675.
6. Pulendran, B. Modulating vaccine responses with dendritic cells and Toll-like receptors. *Immunol Rev* 2004, **199**, 227–250.
7. Moore, K.L. & Dalley, A.F. *Clinically Oriented Anatomy.* Lippincott Williams & Wilkins, Baltimore, MD, 1999, pp. 935–935.
8. Fujimura, Y. Evidence of M cells as portals of entry for antigens in the nasopharyngeal lymphoid tissue of humans. *Virchows Arch* 2000, **436**(6), 560–566.
9. Pope, M. Mucosal dendritic cells and immunodeficiency viruses. *J Infect Dis* 1999, **179**(Suppl 3), S427–S430.
10. Van der Ven, I. & Sminia, T. The development and structure of mouse nasal-associated lymphoid tissue: an immuno- and enzyme-histochemical study. *Reg Immunol* 1993, **5**(2), 69–75.
11. Giannasca, P.J., Boden, J.A. & Monath, T.P. Targeted delivery of antigen to hamster nasal lymphoid tissue with M-cell-directed lectins. *Infect Immun* 1997, **65**(10), 4288–4298.
12. Neutra, M.R., Frey, A. & Kraehenbuhl, J.P. Epithelial M cells: gateways for mucosal infection and immunization. *Cell* 1996, **86**(3), 345–348.

13. Brandtzaeg, P. Humoral immune response patterns of human mucosae: induction and relation to bacterial respiratory tract infections. *J Infect Dis* 1992, **165**(Suppl 1), S167–S176.

14. Brandtzaeg, P., Baekkevold, E.S., Farstad, I.N. et al. Regional specialization in the mucosal immune system: what happens in the microcompartments? *Immunol Today* 1999, **20**(3), 141–151.

15. Boyaka, P.N., Wright, P.F., Marinaro, M. et al. Human nasopharyngeal-associated lymphoreticular tissues: functional analysis of subepithelial and intraepithelial B and T cells from adenoids and tonsils. *Am J Pathol* 2000, **157**(6), 2023–2035.

16. Graeme-Cook, F., Bhan, A.K. & Harris, N.L. Immunohistochemical characterization of intraepithelial and subepithelial mononuclear cells of the upper airways. *Am J Pathol* 1993, **143**(5), 1416–1422.

17. Kutteh, W.H., Moldoveanu, Z. & Mestecky, J. Mucosal immunity in the female reproductive tract: correlation of immunoglobulins, cytokines, and reproductive hormones in human cervical mucus around the time of ovulation. *AIDS Res Hum Retroviruses* 1998, **14**(Suppl 1), S51–S55.

18. Crowley-Nowick, P.A., Bell, M.C., Brockwell, R. et al. Rectal immunization for induction of specific antibody in the genital tract of women. *J Clin Immunol* 1997, **17**(5), 370–379.

19. Mestecky, J. & Fultz, P.N. Mucosal immune system of the human genital tract. *J Infect Dis* 1999, **179**(Suppl 3), S470–S474.

20. Vajdy, M., Veazey, R.S., Knight, H.K., Lackner, A.A. & Neutra, M.R. Differential effects of simian immunodeficiency virus infection on immune inductive and effector sites in the rectal mucosa of rhesus macaques. *Am J Pathol* 2000, **157**(2), 485–495.

21. Pavli, P., Maxwell, L., van de Pol, E. & Doe, W.F. Distribution of human colonic dendritic cells and macrophages: functional implications. *Adv Exp Med Biol* 1995, **378**, 121–123.

22. Yeaman, G.R., White, H.D., Howell, A., Prabhala, R. & Wira, C.R. The mucosal immune system in the human female reproductive tract: potential insights into the heterosexual transmission of HIV. *Aids Res Hum Retroviruses* 1998, **14**(Suppl 1), S57–S62.

23. Johansson, E.L., Rudin, A., Wassen, L. & Holmgren, J. Distribution of lymphocytes and adhesion molecules in human cervix and vagina. *Immunology* 1999, **96**(2), 272–277.

24. Hu, J., Gardner, M.B. & Miller, C.J. Simian immunodeficiency virus rapidly penetrates the cervicovaginal mucosa after intravaginal inoculation and infects intraepithelial dendritic cells. *J Virol* 2000, **74**(13), 6087–6095.

25. Lohman, B.L., Miller, C.J. & McChesney, M.B. Antiviral cytotoxic T lymphocytes in vaginal mucosa of simian immunodeficiency virus-infected rhesus macaques. *J Immunol* 1995, **155**(12), 5855–5860.

26. McChesney, M.B., Collins, J.R. & Miller, C.J. Mucosal phenotype of antiviral cytotoxic T lymphocytes in the vaginal mucosa of SIV-infected rhesus macaques. *AIDS Res Hum Retroviruses* 1998, **14**(Suppl 1), S63–S66.

27. Challacombe, S.J., Rahman, D. & O'Hagan, D.T. Salivary, gut, vaginal and nasal antibody responses after oral immunization with biodegradable microparticles. *Vaccine* 1997, **15**(2), 169–175.

28. Ugozzoli, M., O'Hagan, D.T. & Ott, G.S. Intranasal immunization of mice with herpes simplex virus type 2 recombinant gD2: the effect of adjuvants on mucosal and serum antibody responses. *Immunology* 1998, **93**(4), 563–571.

29. Parr, M.B. & Parr, E.L. Immunohistochemical localization of immunoglobulins A, G and M in the mouse female genital tract. *J Reprod Fertil* 1985, **74**(2), 361–370.

30. Johansson, E.L., Rask, C., Fredriksson, M., Eriksson, K., Czerkinsky, C. & Holmgren, J. Antibodies and antibody-secreting cells in the female genital tract after vaginal or intranasal immunization with cholera toxin B subunit or conjugates. *Infect Immun* 1998, **66**(2), 514–520.

31. Thaparr, M.A., Parr, E.L., Bozzola, J.J. & Parr, M.B. Secretory immune responses in the mouse vagina after parenteral or intravaginal immunization with an immunostimulating complex (ISCOM). *Vaccine* 1991, **9**(2), 129–133.

32. O'Hagan, D.T., Rafferty, D., McKeating, J.A. & Illum, L. Vaginal immunization of rats with a synthetic peptide from human immunodeficiency virus envelope glycoprotein. *J Gen Virol* 1992, **73**(Pt 8), 2141–2145.

33. O'Hagan, D.T., Rafferty, D., Wharton, S. & Illum, L. Intravaginal immunization in sheep using a bioadhesive microsphere antigen delivery system. *Vaccine* 1993, **11**(6), 660–664.

34. Vajdy, M., Gardner, J., Neidleman, J. et al. Human immunodeficiency virus type 1 Gag-specific vaginal immunity and protection after local immunizations with sindbis virus-based replicon particles. *J Infect Dis* 2001, **184**(12), 1613–1616.

35. Wira, C.R., Richardson, J. & Prabhala, R. Endocrine regulation of mucosal immunity: effect of sex hormones and cytokines on the afferent and efferent arms of the immune system in the female reproductive tract. In *Handbook of Mucosal Immunology*, Ogra, P.L., Lamm, M.E., Mcghee, J.R., Mestecky, J., Strober, W. & Bienenstock, J., Eds., Academic Press, San Diego, CA, 1994, pp. 705–718.

36. Di Tommaso, A., Saletti, G., Pizza, M. et al. Induction of antigen-specific antibodies in vaginal secretions by using a nontoxic mutant of heat-labile enterotoxin as a mucosal adjuvant. *Infect Immun* 1996, **64**(3), 974–979.

37. Bergquist, C., Johansson, E.L., Lagergard, T., Holmgren, J. & Rudin, A. Intranasal vaccination of humans with recombinant cholera toxin B subunit induces systemic and local antibody responses in the upper respiratory tract and the vagina. *Infect Immun* 1997, **65**(7), 2676–2684.

38. Messina, J.P., Gilkeson, G.S. & Pisetsky, D.S. Stimulation of in vitro murine lymphocyte proliferation by bacterial DNA. *J Immunol* 1991, **147**(6), 1759–1764.

39. Tokunaga, T., Yamamoto, H., Shimada, S. et al. Antitumor activity of deoxyribonucleic acid fraction from *Mycobacterium bovis* BCG: I. Isolation, physicochemical characterization, and antitumor activity. *J Natl Cancer Inst* 1984, **72**(4), 955–962.

40. Krieg, A.M., Yi, A.K., Matson, S. et al. CpG motifs in bacterial DNA trigger direct B-cell activation. *Nature* 1995, **374**(6522), 6546–6549.

41. Bird, A.P. CpG islands as gene markers in the vertebrate nucleus. *Trends Genet* 1987, **3**, 342–347.

42. Hemmi, H., Takeuchi, O., Kawai, T. et al. A Toll-like receptor recognizes bacterial DNA. *Nature* 2000, **408**(6813), 740–745.

43. Sparwasser, T., Koch, E.S., Vabulas, R.M. et al. Bacterial DNA and immunostimulatory CpG oligonucleotides trigger maturation and activation of murine dendritic cells. *Eur J Immunol* 1998, **28**(6), 2045–2054.

44. Klinman, D.M., Barnhart, K.M. & Conover, J. CpG motifs as immune adjuvants. *Vaccine* 1999, **17**(1), 19–25.

45. Broide, D., Schwarze, J., Tighe, H. et al. Immunostimulatory DNA sequences inhibit IL-5, eosinophilic inflammation, and airway hyperresponsiveness in mice. *J Immunol* 1998, **161**(12), 7054–7062.

46. Vasilakos, J.P., Smith, R.M., Gibson, S.J. et al. Adjuvant activities of immune response modifier R-848: comparison with CpG ODN. *Cell Immunol* 2000, **204**(1), 64–74.

47. Clements, J.D. & Finkelstein, R.A. Isolation and characterization of homogeneous heat-labile enterotoxins with high specific activity from *Escherichia coli* cultures. *Infect Immun* 1979, **24**(3), 760–769.

48. Clements, J.D., Yancey, R.J. & Finkelstein, R.A. Properties of homogeneous heat-labile enterotoxin from *Escherichia coli*. *Infect Immun* 1980, **29**(1), 91–97.

49. Pizza, M., Domenighini, M., Hol, W. et al. Probing the structure-activity relationship of *Escherichia coli* LT-A by site-directed mutagenesis. *Mol Microbiol* 1994, **14**(1), 51–60.

50. Pizza, M., Fontana, M.R., Giuliani, M.M. et al. A genetically detoxified derivative of heat-labile *Escherichia coli* enterotoxin induces neutralizing antibodies against the A subunit. *J Exp Med* 1994, **180**(6), 2147–2153.

51. Partidos, C.D., Pizza, M., Rappuoli, R. & Steward, M.W. The adjuvant effect of a non-toxic mutant of heat-labile enterotoxin of *Escherichia coli* for the induction of measles virus-specific CTL responses after intranasal co-immunization with a synthetic peptide. *Immunology* 1996, **89**(4), 483–487.

52. Giuliani, M.M., Del Giudice, G., Giannelli, V. et al. Mucosal adjuvanticity and immunogenicity of LTR72, a novel mutant of *Escherichia coli* heat-labile enterotoxin with partial knockout of ADP-ribosyltransferase activity. *J Exp Med* 1998, **187**, 1123–1132.

53. Singh, M., Briones, M. & O'Hagan, D. Hyaluronic acid biopolymers for mucosal delivery of vaccines. In *New Frontiers in Medical Sciences: Redefining Hyaluronan*, Abatangelo, G. & Weigel, P.H., Eds., Elsevier Science, Amsterdam, 2000, pp. 163–170.

54. Neidleman, J.A., Vajdy, M., Ugozzoli, M., Ott, G. & O'Hagan, D. Potent non-toxic mutant heat-labile enterotoxins as adjuvants for induction of HIV-1$_{SF2}$ p55 gag-specific cytotoxic responses through mucosal immunization. *Immunology* 2000, (101), 154–160.

55. Ugozzoli, M., Santos, G., Donnelly, J. & O'Hagan, D.T. Potency of a genetically detoxified mucosal adjuvant derived from the heat-labile enterotoxin of *Escherichia coli* (LTK63) is not adversely affected by the presence of preexisting immunity to the adjuvant. *J Infect Dis* 2001, **183**(2), 351–354.

56. Lycke, N. From toxin to adjuvant: the rational design of a vaccine adjuvant vector, CTA1-DD/ISCOM. *Cell Microbiol* 2004, **6**(1), 23–32.

57. Maloy, K.J., Donachie, A.M., O'Hagan, D.T. & Mowat, A.M. Induction of mucosal and systemic immune responses by immunization with ovalbumin entrapped in poly(lactide-co-glycolide) microparticles. *Immunology* 1994, **81**(4), 661–667.

58. Moore, A., McGuirk, P., Adams, S. et al. Immunization with a soluble recombinant HIV protein entrapped in biodegradable microparticles induces HIV-specific

CD8+ cytotoxic T lymphocytes and CD4+ Th1 cells. *Vaccine* 1995, **13**(18), 1741–1749.

59. Nixon, D.F., Hioe, C., Chen, P.D. et al. Synthetic peptides entrapped in microparticles can elicit cytotoxic T cell activity. *Vaccine* 1996, **14**(16), 1523–1530.

60. Eldridge, J.H., Staas, J.K., Meulbroek, J.A., Tice, T.R. & Gilley, R.M. Biodegradable and biocompatible poly(DL-lactide-co-glycolide) microspheres as an adjuvant for staphylococcal enterotoxin B toxoid which enhances the level of toxin-neutralizing antibodies. *Infect Immun* 1991, **59**(9), 2978–2986.

61. O'Hagan, D.T., Rahman, D., McGee, J.P. et al. Biodegradable microparticles as controlled release antigen delivery systems. *Immunology* 1991, **73**(2), 239–242.

62. O'Hagan, D.T., Jeffery, H., Roberts, M.J., McGee, J.P. & Davis, S.S. Controlled release microparticles for vaccine development. *Vaccine* 1991, **9**(10), 768–771.

63. O'Hagan, D.T., Jeffery, H. & Davis, S.S. Long-term antibody responses in mice following subcutaneous immunization with ovalbumin entrapped in biodegradable microparticles. *Vaccine* 1993, **11**(9), 965–969.

64. Heritage, P.L., Brook, M.A., Underdown, B.J. & McDermott, M.R. Intranasal immunization with polymer-grafted microparticles activates the nasal-associated lymphoid tissue and draining lymph nodes. *Immunology* 1998, **93**(2), 249–256.

65. McDermott, M.R., Heritage, P.L., Bartzoka, V. & Brook, M.A. Polymer-grafted starch microparticles for oral and nasal immunization. *Immunol Cell Biol* 1998, **76**(3), 256–262.

66. Baras, B., Benoit, M.A., Dupre, L. et al. Single-dose mucosal immunization with biodegradable microparticles containing a *Schistosoma mansoni* antigen. *Infect Immun* 1999, **67**(5), 2643–2648.

67. Higaki, M., Takase, T., Igarashi, R. et al. Enhancement of immune response to intranasal influenza HA vaccine by microparticle resin. *Vaccine* 1998, **16**(7), 741–745.

68. Kim, B., Bowersock, T., Griebel, P. et al. Mucosal immune responses following oral immunization with rotavirus antigens encapsulated in alginate microspheres. *J Control Release* 2002, **85**(1–3), 191–202.

69. Barr, I.G., Sjolander, A. & Cox, J.C. ISCOMs and other saponin based adjuvants. *Adv Drug Deliv Rev* 1998, **32**, 247–271.

70. Rimmelzwaan, G.F., Baars, M., van Beek, R. et al. Induction of protective immunity against influenza virus in a macaque model: comparison of conventional and ISCOM vaccines. *J Gen Virol* 1997, **78**(Pt 4), 757–765.

71. Ennis, F.A., Cruz, J., Jameson, J., Klein, M., Burt, D. & Thipphawong, J. Augmentation of human influenza A virus-specific cytotoxic T lymphocyte memory by influenza vaccine and adjuvanted carriers (ISCOMs). *Virology* 1999, **259**(2), 256–261.

72. Soltysik, S., Wu, J.Y., Recchia, J. et al. Structure/function studies of QS-21 adjuvant: assessment of triterpene aldehyde and glucuronic acid roles in adjuvant function. *Vaccine* 1995, **13**(15), 1403–1410.

73. Lovgren-Bengtsson, K. & Morein, B. The ISCOM™ technology. In *Vaccine Adjuvants: Preparation Methods and Research Protocols*, Vol. 42, O'Hagan, D., Ed., Humana Press, Totowa, NJ, 2000, pp. 239–258.

74. Alving, C.R. Immunologic aspects of liposomes: presentation and processing of liposomal protein and phospholipid antigens. *Biochim Biophys Acta* 1992, **1113**(3–4), 307–322.

75. Gregoriadis, G. Immunological adjuvants: a role for liposomes. *Immunol Today* 1990, **11**(3), 89–97.

76. Ambrosch, F., Wiedermann, G., Jonas, S. et al. Immunogenicity and protectivity of a new liposomal hepatitis A vaccine. *Vaccine* 1997, **15**(11), 1209–1213.

77. Aspden, T.J., Mason, J.D., Jones, N.S., Lowe, J., Skaugrud, O. & Illum, L. Chitosan as a nasal delivery system: the effect of chitosan solutions on in vitro and in vivo mucociliary transport rates in human turbinates and volunteers. *J Pharm Sci* 1997, **86**(4), 509–513.

78. van der Lubben, I.M., van Opdorp, F.A., Hengeveld, M.R. et al. Transport of chitosan microparticles for mucosal vaccine delivery in a human intestinal M-cell model. *J Drug Target* 2002, **10**(6), 449–456.

79. Read, R.C., Naylor, S.C., Potter, C.W. et al. Effective nasal influenza vaccine delivery using chitosan. *Vaccine* 2005, **23**(35), 4367–4374.

80. Mills, K.H., Cosgrove, C., McNeela, E.A. et al. Protective levels of diphtheria-neutralizing antibody induced in healthy volunteers by unilateral priming-boosting intranasal immunization associated with restricted ipsilateral mucosal secretory immunoglobulin a. *Infect Immun* 2003, **71**(2), 726–732.

81. Alpar, H.O., Somavarapu, S., Atuah, K.N. & Bramwell, V.W. Biodegradable mucoadhesive particulates for nasal and pulmonary antigen and DNA delivery. *Adv Drug Deliv Rev* 2005, **57**(3), 411–430.

82. Bivas-Benita, M., van Meijgaarden, K.E., Franken, K.L. et al. Pulmonary delivery of chitosan-DNA nanoparticles enhances the immunogenicity of a DNA vaccine encoding HLA-A*0201-restricted T-cell epitopes of *Mycobacterium* tuberculosis. *Vaccine* 2004, **22**(13–14), 1609–1615.

83. Mohapatra, S.S. Mucosal gene expression vaccine: a novel vaccine strategy for respiratory syncytial virus. *Pediatr Infect Dis J* 2003, **22**(2 Suppl), S100–S103; discussion, S103–S104.

84. Iqbal, M., Lin, W., Jabbal-Gill, I., Davis, S.S., Steward, M.W. & Illum, L. Nasal delivery of chitosan-DNA plasmid expressing epitopes of respiratory syncytial virus (RSV) induces protective CTL responses in BALB/c mice. *Vaccine* 2003, **21**(13–14), 1478–1485.

85. Roy, K., Mao, H.Q., Huang, S.K. & Leong, K.W. Oral gene delivery with chitosan–DNA nanoparticles generates immunologic protection in a murine model of peanut allergy [see comments]. *Nat Med* 1999, **5**(4), 387–391.

86. Baudner, B.C., Giuliani, M.M., Verhoef, J.C., Rappuoli, R., Junginger, H.E. & Giudice, G.D. The concomitant use of the LTK63 mucosal adjuvant and of chitosan-based delivery system enhances the immunogenicity and efficacy of intranasally administered vaccines. *Vaccine* 2003, **21**(25–26), 3837–3844.

87. Baudner, B.C., Morandi, M., Giuliani, M.M. et al. Modulation of immune response to group C meningococcal conjugate vaccine given intranasally to mice together with the LTK63 mucosal adjuvant and the trimethyl chitosan delivery system. *J Infect Dis* 2004, **189**(5), 828–832 (Epub 2004 Feb 2018).

88. Singh, M., Briones, M. & O'Hagan, D.T. A novel bioadhesive intranasal delivery system for inactivated influenza vaccines. *J Control Release* 2001, **70**(3), 267–276.

89. Lavelle, E.C., Grant, G., Pusztai, A. et al. Mistletoe lectins enhance immune responses to intranasally co-administered herpes simplex virus glycoprotein D2. *Immunology* 2002, **107**(2), 268–274.

90. Manocha, M., Pal, P.C., Chitralekha, K.T. et al. Enhanced mucosal and systemic immune response with intranasal immunization of mice with HIV peptides entrapped in PLG microparticles in combination with Ulex Europaeus-I lectin as M cell target. *Vaccine* 2005, **23**(48–49), 5599–5617.

91. Lavelle, E.C., Grant, G., Pfuller, U. & O'Hagan, D.T. Immunological implications of the use of plant lectins for drug and vaccine targeting to the gastrointestinal tract. *J Drug Target* 2004, **12**(2), 89–95.

92. Giannasca, P.J., Giannasca, K.T., Falk, P., Gordon, J.I. & Neutra, M.R. Regional differences in glycoconjugates of intestinal M cells in mice: potential targets for mucosal vaccines. *Am J Physiol* 1994, **267**(6 Pt 1), G1108–G1121.

93. Mantis, N.J., Frey, A. & Neutra, M.R. Accessibility of glycolipid and oligosaccharide epitopes on rabbit villus and follicle-associated epithelium. *Am J Physiol Gastrointest Liver Physiol* 2000, **278**(6), G915–G923.

94. Jepson, M.A., Clark, M.A. & Hirst, B.H. M cell targeting by lectins: a strategy for mucosal vaccination and drug delivery. *Adv Drug Deliv Rev* 2004, **56**(4), 511–525.

95. Tacken, P.J., de Vries, I.J., Gijzen, K. et al. Effective induction of naive and recall T-cell responses by targeting antigen to human dendritic cells via a humanized anti-DC-SIGN antibody. *Blood* 2005, **106**(4), 1278–1285.

96. Singh, M., Ott, G., Kazzaz, J. et al. Cationic microparticles are an effective delivery system for immune stimulatory CpG DNA. *Pharm Res* 2001, **18**(10), 1476–1479.

97. Tabata, Y. & Ikada, Y. Macrophage activation through phagocytosis of muramyl dipeptide encapsulated in gelatin microspheres. *J Pharm Pharmacol* 1987, **39**(9), 698–704.

98. Puri, N. & Sinko, P.J. Adjuvancy enhancement of muramyl dipeptide by modulating its release from a physicochemically modified matrix of ovalbumin microspheres: II. In vivo investigation. *J Control Release* 2000, **69**(1), 69–80.

99. Tabata, Y. & Ikada, Y. Macrophage phagocytosis of biodegradable microspheres composed of L-lactic acid/glycolic acid homo- and copolymers. *J Biomed Mater Res* 1988, **22**(10), 837–858.

100. Cleland, J.L., Barron, L., Daugherty, A. et al. Development of a single-shot subunit vaccine for HIV-1: 3. Effect of adjuvant and immunization schedule on the duration of the humoral immune response to recombinant MN gp120. *J Pharm Sci* 1997, **85**(12), 1350–1357.

101. Singh, M., Li, X.M., Wang, H. et al. Immunogenicity and protection in small-animal models with controlled-release tetanus toxoid microparticles as a single-dose vaccine. *Infect Immun* 1997, **65**(5), 1716–1721.

102. Singh, M., Carlson, J.R., Briones, M. et al. A comparison of biodegradable microparticles and MF59 as systemic adjuvants for recombinant gD from HSV-2. *Vaccine* 1998, **16**(19), 1822–1827.

103. O'Hagan, D.T. HIV and mucosal immunity. *Lancet* 1991, **337**(8752), 1289.

104. O'Hagan, D.T. & Lavelle, E. Novel adjuvants and delivery systems for HIV vaccines. *Aids* 2002, **16**(Suppl 4), S115–S124.

105. Akagi, T., Kawamura, M., Ueno, M. et al. Mucosal immunization with inactivated HIV-1-capturing nanospheres induces a significant HIV-1-specific vaginal antibody response in mice. *J Med Virol* 2003, **69**(2), 163–172.

106. Miyake, A., Akagi, T., Enose, Y. et al. Induction of HIV-specific antibody response and protection against vaginal SHIV transmission by intranasal immunization with inactivated SHIV-capturing nanospheres in macaques. *J Med Virol* 2004, **73**(3), 368–377.

107. Dumais, N., Patrick, A., Moss, R.B., Davis, H.L. & Rosenthal, K.L. Mucosal immunization with inactivated human immunodeficiency virus plus CpG oligodeoxynucleotides induces genital immune responses and protection against intravaginal challenge. *J Infect Dis* 2002, **186**(8), 1098–1105 (Epub 2002 Sep 1030).

108. Jiang, J.Q., Patrick, A., Moss, R.B. & Rosenthal, K.L. CD8+ T-cell-mediated cross-clade protection in the genital tract following intranasal immunization with inactivated human immunodeficiency virus antigen plus CpG oligodeoxynucleotides. *J Virol* 2005, **79**(1), 393–400.

109. Kawahara, M., Matsuo, K., Nakasone, T. et al. Combined intrarectal/intradermal inoculation of recombinant *Mycobacterium bovis* bacillus Calmette–Guerin (BCG) induces enhanced immune responses against the inserted HIV-1 V3 antigen. *Vaccine* 2002, **21**(3–4), 158–166.

110. Golding, B., Eller, N., Levy, L. et al. Mucosal immunity in mice immunized with HIV-1 peptide conjugated to *Brucella abortus*. *Vaccine* 2002, **20**(9–10), 1445–1450.

111. Walcher, P., Mayr, U.B., Azimpour-Tabrizi, C. et al. Antigen discovery and delivery of subunit vaccines by nonliving bacterial ghost vectors. *Expert Rev Vaccines* 2004, **3**(6), 681–691.

112. Panthel, K., Jechlinger, W., Matis, A. et al. Generation of *Helicobacter pylori* ghosts by PhiX protein E-mediated inactivation and their evaluation as vaccine candidates. *Infect Immun* 2003, **71**(1), 109–116.

113. Deml, L., Speth, C., Dierich, M.P., Wolf, H. & Wagner, R. Recombinant HIV-1 Pr55gag virus-like particles: potent stimulators of innate and acquired immune responses. *Mol Immunol* 2005, **42**(2), 259–277.

114. Yao, Q., Bu, Z., Vzorov, A., Yang, C. & Compans, R.W. Virus-like particle and DNA-based candidate AIDS vaccines. *Vaccine* 2003, **21**(7–8), 638–643.

115. Yao, Q. Enhancement of mucosal immune responses by chimeric influenza HA/SHIV virus-like particles. *Res Initiat Treat Action* 2003, **8**(2), 20–21.

116. Gronowski, A.M., Hilbert, D.M., Sheehan, K.C., Garotta, G. & Schreiber, R.D. Baculovirus stimulates antiviral effects in mammalian cells. *J Virol* 1999, **73**(12), 9944–9951.

117. Cannon, G. & Weissman, D. RNA based vaccines. *DNA Cell Biol* 2002, **21**(12), 953–961.

118. Martinon, F., Krishnan, S., Lenzen, G. et al. Induction of virus-specific cytotoxic T lymphocytes in vivo by liposome-entrapped mRNA. *Eur J Immunol* 1993, **23**(7), 1719–1722.

119. Polo, J.M., Gardner, J.P., Ji, Y. et al. Alphavirus DNA and particle replicons for vaccines and gene therapy. *Dev Biol (Basel)* 2000, **104**, 181–185.

120. Li, H.O., Zhu, Y.F., Asakawa, M. et al. A cytoplasmic RNA vector derived from nontransmissible Sendai virus with efficient gene transfer and expression. *J Virol* 2000, **74**(14), 6564–6569.

121. Schoen, C., Kolb-Maurer, A., Geginat, G. et al. Bacterial delivery of functional messenger RNA to mammalian cells. *Cell Microbiol* 2005, **7**(5), 709–724.

122. Michiels, A., Tuyaerts, S., Bonehill, A. et al. Electroporation of immature and mature dendritic cells: implications for dendritic cell-based vaccines. *Gene Ther* 2005, **12**(9), 772–782.

123. Yoshii, K., Hayasaka, D., Goto, A. et al. Packaging the replicon RNA of the Far-Eastern subtype of tick-borne encephalitis virus into single-round infectious particles: development of a heterologous gene delivery system. *Vaccine* 2005, **23**(30), 3946–3956 (Epub 2005 Mar 3925).

124. Shinya, K., Fujii, Y., Ito, H. et al. Characterization of a neuraminidase-deficient influenza a virus as a potential gene delivery vector and a live vaccine. *J Virol* 2004, **78**(6), 3083–3088.

125. Ahlquist, P., Schwartz, M., Chen, J., Kushner, D., Hao, L. & Dye, B.T. Viral and host determinants of RNA virus vector replication and expression. *Vaccine* 2005, **23**(15), 1784–1787.

126. Rayner, J.O., Dryga, S.A. & Kamrud, K.I. Alphavirus vectors and vaccination. *Rev Med Virol* 2002, **12**(5), 279–296.

127. Schlesinger, S. Alphavirus vectors: development and potential therapeutic applications. *Expert Opin Biol Ther* 2001, **1**(2), 177–191.

128. Schlesinger, S. & Dubensky, T.W. Alphavirus vectors for gene expression and vaccines. *Curr Opin Biotechnol* 1999, **10**(5), 434–439.

129. Leitner, W.W., Hwang, L.N., deVeer, M.J. et al. Alphavirus-based DNA vaccine breaks immunological tolerance by activating innate antiviral pathways. *Nat Med* 2003, **9**(1), 33–39.

130. Caley, I.J., Betts, M.R., Irlbeck, D.M. et al. Humoral, mucosal, and cellular immunity in response to a human immunodeficiency virus type 1 immunogen expressed by a Venezuelan equine encephalitis virus vaccine vector. *J Virol* 1997, **71**(4), 3031–3038.

131. Berglund, P., Quesada-Rolander, M., Putkonen, P., Biberfeld, G., Thorstensson, R. & Liljestrom, P. Outcome of immunization of cynomolgus monkeys with recombinant Semliki forest virus encoding human immunodeficiency virus type 1 envelope protein and challenge with a high dose of SHIV-4 virus. *AIDS Res Hum Retroviruses* 1997, **13**(17), 1487–1495.

132. Brand, D., Lemiale, F., Turbica, I., Buzelay, L., Brunet, S. & Barin, F. Comparative analysis of humoral immune responses to HIV type 1 envelope glycoproteins in mice immunized with a DNA vaccine, recombinant Semliki forest virus RNA, or recombinant Semliki forest virus particles. *AIDS Res Hum Retroviruses* 1998, **14**(15), 1369–1377.

133. Belyakov, I.M., Moss, B., Strober, W. & Berzofsky, J.A. Mucosal vaccination overcomes the barrier to recombinant vaccinia immunization caused by preexisting poxvirus immunity. *Proc Natl Acad Sci USA* 1999, **96**(8), 4512–4517.

134. Hofmann-Lehmann, R., Vlasak, J., Williams, A.L. et al. Live attenuated, nef-deleted SIV is pathogenic in most adult macaques after prolonged observation. *AIDS* 2003, **17**(2), 157–166.

135. Ruprecht, R.M. Live attenuated AIDS viruses as vaccines: Promise or peril? *Immunol Rev* 1999, **170**, 135–149.

136. Baba, T.W., Liska, V., Khimani, A.H. et al. Live attenuated, multiply deleted simian immunodeficiency virus causes AIDS in infant and adult macaques. *Nat Med* 1999, **5**(2), 194–203.

137. Baba, T.W., Jeong, Y.S., Pennick, D., Bronson, R., Greene, M.F. & Ruprecht, R.M. Pathogenicity of live, attenuated SIV after mucosal infection of neonatal macaques. *Science* 1995, **267**(5205), 1820–1825.

138. Wyand, M.S., Manson, K., Montefiori, D.C., Lifson, J.D., Johnson, R.P. & Desrosiers, R.C. Protection by live, attenuated simian immunodeficiency virus against heterologous challenge. *J Virol* 1999, **73**(10), 8356–8363.

139. Johnson, R.P., Lifson, J.D., Czajak, S.C. et al. Highly attenuated vaccine strains of simian immunodeficiency virus protect against vaginal challenge: inverse relationship of degree of protection with level of attenuation. *J Virol* 1999, **73**(6), 4952–4961.

140. Gherardi, M.M., Najera, J.L., Perez-Jimenez, E., Guerra, S., Garcia-Sastre, A. & Esteban, M. Prime–boost immunization schedules based on influenza virus and vaccinia virus vectors potentiate cellular immune responses against human immunodeficiency virus env protein systemically and in the genitorectal draining lymph nodes. *J Virol* 2003, **77**(12), 7048–7057.

141. Gherardi, M.M., Perez-Jimenez, E., Najera, J.L. & Esteban, M. Induction of HIV immunity in the genital tract after intranasal delivery of a MVA vector: enhanced immunogenicity after DNA prime-modified vaccinia virus Ankara boost immunization schedule. *J Immunol* 2004, **172**(10), 6209–6220.

142. Stevceva, L., Alvarez, X., Lackner, A.A. et al. Both mucosal and systemic routes of immunization with the live, attenuated NYVAC/simian immunodeficiency virus SIV(gpe) recombinant vaccine result in gag-specific CD8(+) T-cell responses in mucosal tissues of macaques. *J Virol* 2002, **76**(22), 11659–11676.

143. Bertley, F.M., Kozlowski, P.A., Wang, S.W. et al. Control of simian/human immunodeficiency virus viremia and disease progression after IL-2-augmented DNA-modified vaccinia virus Ankara nasal vaccination in nonhuman primates. *J Immunol* 2004, **172**(6), 3745–3757.

144. Amara, R.R., Villinger, F., Altman, J.D. et al. Control of a mucosal challenge and prevention of AIDS by a multiprotein DNA/MVA vaccine. *Science* 2001, **292**(5514), 69–74.

145. Vogel, T.U., Reynolds, M.R., Fuller, D.H. et al. Multispecific vaccine-induced mucosal cytotoxic T lymphocytes reduce acute-phase viral replication but fail in long-term control of simian immunodeficiency virus SIVmac239. *J Virol* 2003, **77**(24), 13348–13360.

146. Santra, S., Barouch, D.H., Korioth-Schmitz, B. et al. Recombinant poxvirus boosting of DNA-primed rhesus monkeys augments peak but not memory T lymphocyte responses. *Proc Natl Acad Sci USA* 2004, **101**(30), 11088–11093 (Epub 2004 Jul 11016).

147. Wang, S.W., Bertley, F.M., Kozlowski, P.A. et al. An SHIV DNA/MVA rectal vaccination in macaques provides systemic and mucosal virus-specific responses and protection against AIDS. *AIDS Res Hum Retroviruses* 2004, **20**(8), 846–859.

148. Barouch, D.H., McKay, P.F., Sumida, S.M. et al. Plasmid chemokines and colony-stimulating factors enhance the immunogenicity of DNA priming–viral vector boosting human immunodeficiency virus type 1 vaccines. *J Virol* 2003, **77**(16), 8729–8735.

149. Peacock, J.W., Nordone, S.K., Jackson, S.S. et al. Gender differences in human immunodeficiency virus type 1-specific CD8 responses in the reproductive tract and colon following nasal peptide priming and modified vaccinia virus Ankara boosting. *J Virol* 2004, **78**(23), 13163–13172.

150. Edghill-Smith, Y., Bray, M., Whitehouse, C.A. et al. Smallpox vaccine does not protect macaques with AIDS from a lethal monkeypox virus challenge. *J Infect Dis* 2005, **191**(3), 372–381 (Epub 2005 Jan 2004).

151. Chan, K.S., Verardi, P.H., Legrand, F.A. & Yilma, T.D. Nef from pathogenic simian immunodeficiency virus is a negative factor for vaccinia virus. *Proc Natl Acad Sci U S A* 2005, **102**(24), 8734–8739 (Epub 2005 Jun 8731).

152. Harrer, E., Bauerle, M., Ferstl, B. et al. Therapeutic vaccination of HIV-1-infected patients on HAART with a recombinant HIV-1 *nef*-expressing MVA: safety, immunogenicity and influence on viral load during treatment interruption. *Antivir Ther* 2005, **10**(2), 285–300.

153. Franchini, G., Gurunathan, S., Baglyos, L., Plotkin, S. & Tartaglia, J. Poxvirus-based vaccine candidates for HIV: two decades of experience with special emphasis on canarypox vectors. *Expert Rev Vaccines* 2004, **3**(4 Suppl), S75–S88.

154. Im, E.J. & Hanke, T. MVA as a vector for vaccines against HIV-1. *Expert Rev Vaccines* 2004, **3**(4 Suppl), S89–S97.

155. Amara, R.R. & Robinson, H.L. A new generation of HIV vaccines. *Trends Mol Med* 2002, **8**(10), 489–495.

156. Rose, N.F., Marx, P.A., Luckay, A. et al. An effective AIDS vaccine based on live attenuated vesicular stomatitis virus recombinants. *Cell* 2001, **106**(5), 539–549.

157. Egan, M.A., Chong, S.Y., Rose, N.F. et al. Immunogenicity of attenuated vesicular stomatitis virus vectors expressing HIV type 1 env and SIV gag proteins: comparison of intranasal and intramuscular vaccination routes. *AIDS Res Hum Retroviruses* 2004, **20**(9), 989–1004.

158. Haglund, K., Leiner, I., Kerksiek, K., Buonocore, L., Pamer, E. & Rose, J.K. High-level primary CD8(+) T-cell response to human immunodeficiency virus type 1 gag and env generated by vaccination with recombinant vesicular stomatitis viruses. *J Virol* 2002, **76**(6), 2730–2738.

159. Schmitt, B. Vesicular stomatitis. *Vet Clin North Am Food Anim Pract* 2002, **18**(3), 453–459, vii–viii.

160. Letchworth, G.J., Rodriguez, L.L. & Delcbarrera, J. Vesicular stomatitis. *Vet J* 1999, **157**(3), 239–260.

161. Kovacs, G.M., Harrach, B., Zakhartchouk, A.N. & Davison, A.J. Complete genome sequence of simian adenovirus 1: an Old World monkey adenovirus with two fiber genes. *J Gen Virol* 2005, **86**(Pt 6), 1681–1686.

162. Choi, E.H., Kim, H.S., Eun, B.W. et al. Adenovirus type 7 peptide diversity during outbreak, Korea, 1995–2000. *Emerg Infect Dis* 2005, **11**(5), 649–654.

163. Gray, G.C., Setterquist, S.F., Jirsa, S.J., DesJardin, L.E. & Erdman, D.D. Emergent strain of human adenovirus endemic in Iowa. *Emerg Infect Dis* 2005, **11**(1), 127–128.

164. Chen, H.L., Chiou, S.S., Hsiao, H.P. et al. Respiratory adenoviral infections in children: a study of hospitalized cases in southern Taiwan in 2001–2002. *J Trop Pediatr* 2004, **50**(5), 279–284.

165. Lichtenstein, D.L. & Wold, W.S. Experimental infections of humans with wild-type adenoviruses and with replication-competent adenovirus vectors: replication, safety, and transmission. *Cancer Gene Ther* 2004, **11**(12), 819–829.

166. Roelvink, P.W., Lizonova, A., Lee, J.G. et al. The coxsackievirus–adenovirus receptor protein can function as a cellular attachment protein for adenovirus serotypes from subgroups A, C, D, E, and F. *J Virol* 1998, **72**(10), 7909–7915.

167. Lemiale, F., Kong, W.P., Akyurek, L.M. et al. Enhanced mucosal immunoglobulin A response of intranasal adenoviral vector human immunodeficiency virus vaccine and localization in the central nervous system. *J Virol* 2003, **77**(18), 10078–10087.

168. Thomas, C.E., Ehrhardt, A. & Kay, M.A. Progress and problems with the use of viral vectors for gene therapy. *Nat Rev Genet* 2003, **4**(5), 346–358.

169. Natuk, R.J., Chanda, P.K., Lubeck, M.D. et al. Adenovirus–human immunodeficiency virus (HIV) envelope recombinant vaccines elicit high-titered HIV-neutralizing antibodies in the dog model. *Proc Natl Acad Sci USA* 1992, **89**(16), 7777–7781.

170. Natuk, R.J., Lubeck, M.D., Chanda, P.K. et al. Immunogenicity of recombinant human adenovirus—human immunodeficiency virus vaccines in chimpanzees. *AIDS Res Hum Retroviruses* 1993, **9**(5), 395–404.

171. Lubeck, M.D., Natuk, R.J., Chengalvala, M. et al. Immunogenicity of recombinant adenovirus–human immunodeficiency virus vaccines in chimpanzees following intranasal administration. *AIDS Res Hum Retroviruses* 1994, **10**(11), 1443–1449.

172. Barouch, D.H. & Nabel, G.J. Adenovirus vector-based vaccines for human immunodeficiency virus type 1. *Hum Gene Ther* 2005, **16**(2), 149–156.

173. Shiver, J.W. & Emini, E.A. Recent advances in the development of HIV-1 vaccines using replication-incompetent adenovirus vectors. *Annu Rev Med* 2004, **55**, 355–372.

174. Santra, S., Seaman, M.S., Xu, L. et al. Replication-defective adenovirus serotype 5 vectors elicit durable cellular and humoral immune responses in nonhuman primates. *J Virol* 2005, **79**(10), 6516–6522.

175. Mascola, J.R., Sambor, A., Beaudry, K. et al. Neutralizing antibodies elicited by immunization of monkeys with DNA plasmids and recombinant adenoviral vectors expressing human immunodeficiency virus type 1 proteins. *J Virol* 2005, **79**(2), 771–779.

176. Gomez-Roman, V.R., Patterson, L.J., Venzon, D. et al. Vaccine-elicited antibodies mediate antibody-dependent cellular cytotoxicity correlated with significantly reduced acute viremia in rhesus macaques challenged with SIVmac251. *J Immunol* 2005, **174**(4), 2185–2189.

177. Casimiro, D.R., Bett, A.J., Fu, T.M. et al. Heterologous human immunodeficiency virus type 1 priming–boosting immunization strategies involving replication-defective adenovirus and poxvirus vaccine vectors. *J Virol* 2004, **78**(20), 11434–11438.

178. Gomez-Roman, V.R. & Robert-Guroff, M. Adenoviruses as vectors for HIV vaccines. *AIDS Rev* 2003, **5**(3), 178–185.

179. Ferko, B., Stasakova, J., Sereinig, S. et al. Hyperattenuated recombinant influenza A virus nonstructural-protein-encoding vectors induce human immunodeficiency

virus type 1 *nef*-specific systemic and mucosal immune responses in mice. *J Virol* 2001, **75**(19), 8899–8908.

180. Ferko, B., Katinger, D., Grassauer, A. et al. Chimeric influenza virus replicating predominantly in the murine upper respiratory tract induces local immune responses against human immunodeficiency virus type 1 in the genital tract. *J Infect Dis* 1998, **178**(5), 1359–1368.

181. Muster, T., Ferko, B., Klima, A. et al. Mucosal model of immunization against human immunodeficiency virus type 1 with a chimeric influenza virus. *J Virol* 1995, **69**(11), 6678–6686.

182. Kang, S.M., Guo, L., Yao, Q., Skountzou, I. & Compans, R.W. Intranasal immunization with inactivated influenza virus enhances immune responses to coadministered simian–human immunodeficiency virus-like particle antigens. *J Virol* 2004, **78**(18), 9624–9632.

183. Crotty, S. & Andino, R. Poliovirus vaccine strains as mucosal vaccine vectors and their potential use to develop an AIDS vaccine. *Adv Drug Deliv Rev* 2004, **56**(6), 835–852.

184. Tan, G.S., McKenna, P.M., Koser, M.L. et al. Strong cellular and humoral anti-HIV Env immune responses induced by a heterologous rhabdoviral prime–boost approach. *Virology* 2005, **331**(1), 82–93.

185. Durrani, Z., McInerney, T.L., McLain, L. et al. Intranasal immunization with a plant virus expressing a peptide from HIV-1 gp41 stimulates better mucosal and systemic HIV-1-specific IgA and IgG than oral immunization. *J Immunol Methods* 1998, **220**(1–2), 93–103.

186. Xu, F., Hong, M. & Ulmer, J.B. Immunogenicity of an HIV-1 gag DNA vaccine carried by attenuated *Shigella*. *Vaccine* 2003, **21**(7–8), 644–648.

187. Shata, M.T. & Hone, D.M. Vaccination with a *Shigella* DNA vaccine vector induces antigen-specific CD8(+) T cells and antiviral protective immunity. *J Virol* 2001, **75**(20), 9665–9670.

188. Devico, A.L., Fouts, T.R., Shata, M.T., Kamin-Lewis, R., Lewis, G.K. & Hone, D. M. Development of an oral prime-boost strategy to elicit broadly neutralizing antibodies against HIV-1. *Vaccine* 2002, **20**(15), 1968–1974.

189. Vecino, W.H., Morin, P.M., Agha, R., Jacobs, W.R., Jr. & Fennelly, G.J. Mucosal DNA vaccination with highly attenuated *Shigella* is superior to attenuated *Salmonella* and comparable to intramuscular DNA vaccination for T cells against HIV. *Immunol Lett* 2002, **82**(3), 197–204.

# 14

# CYTOKINES AS VACCINE ADJUVANTS

Nejat K. Egilmez

## 14.1 VACCINES AND VACCINE ADJUVANTS

The traditional whole organism vaccines mimic true infections and have been highly successful in inducing protective immunity to a number of deadly diseases. However, for organisms that cause persistent or latent infections such as human immunodeficiency virus (HIV), hepatitis C virus (HCV), herpesviruses, mycobacteria, and parasites, the use of whole organism–based vaccines is either not effective or too risky [1–3]. With the advent of recombinant DNA technology, vaccines based on purified recombinant proteins (subunit vaccines) have been investigated as safer alternatives. Although these more elegant vaccines eliminate the risk of attenuated organisms regaining pathogenic properties through mutation, they are weak immune stimulators [1,4]. In addition, whereas most subunit vaccines induce strong antibody responses, they generally promote weak T-cell activity, which is critical to the eradication of viruses and intracellular bacteria. The use of immunological adjuvants, which function primarily by enhancing the inflammatory mechanisms during the priming process, can enhance the potency of subunit vaccines [4]. Currently used vaccine adjuvants include agents such as mineral salts (aluminum hydroxide and aluminum phosphate), tensoactive adjuvants (QS-21), bacteria-derived adjuvants (cell wall peptidoglycan or lipopolysaccharide of gram-negative bacteria, CG-rich oligonucleotides), adjuvant emulsions (oil-in-water emulsions such as Freund's incomplete adjuvant, Montanide, Adjuvant 65, and Lipovant), liposomes, polymeric microspheres, carbohydrate adjuvants, and

*Vaccine Adjuvants and Delivery Systems*, Edited by Manmohan Singh
Copyright © 2007 John Wiley & Sons, Inc.

cytokines [4]. In this chapter we focus on the use of cytokines as vaccine adjuvants.

## 14.2 CYTOKINES AND THEIR POTENTIAL AS VACCINE ADJUVANTS

Cytokines are small proteins (about 25 kD on average) that are released by many different cell types in response to immunological stimuli and function in regulating immune activity and homeostasis [2,5]. Over 50 different cytokines with unique as well as redundant functions have been identified in the past 25 years [6]. The great majority of cytokines have short in vivo half-lives (on the order of minutes) and act primarily in an autocrine or paracrine manner at low (nano- to picomolar range) physiological concentrations, via specific receptors that are expressed on leukocyte subsets. Thus, cytokines potentially represent a more biologically relevant alternative to nonspecific adjuvants as they mediate their immunological effects in a more specific and controlled manner. Adjuvant use of cytokines in infectious disease and cancer vaccination is discussed in the following sections, summarizing the past experience, the recent developments, and future directions.

## 14.3 BIOLOGICAL PROPERTIES OF CYTOKINES THAT HAVE BEEN EMPLOYED AS VACCINE ADJUVANTS

The biological properties of the cytokines that have commonly been evaluated as immunological adjuvants are discussed briefly below.

1. *Granulocyte–macrophage colony-stimulating factor* (GM-CSF). GM-CSF was originally characterized as a 14.6-kD protein that is produced by monocytes or macrophages and activated T cells with the ability to promote the development of granulocyte and macrophage colonies from precursor cells in mouse bone marrow [7,8]. Recent evidence supports an immunologically more essential role that includes the activation of mature granulocytes (particularly, eosinophils) and macrophages, and more important, functioning as a growth factor that specifically promotes the generation and recruitment of dendritic cells in response to external inflammatory stimuli [8,9]. GM-CSF also enhances the up-regulation of MHC classes I and II, as well as co-stimulatory molecule expression on antigen-presenting cells (APCs) [8]. Given its central role in antigen presentation, GM-CSF represents an attractive adjuvant candidate, and not surprisingly has been the primary choice in studies evaluating the potential of cytokines as vaccine adjuvants [8,9].

2. *Interleukin-12* (IL-12). IL-12 is a heterodimeric 70-kD pro-inflammatory cytokine that induces the production of IFN-γ and promotes the development of T helper 1 (Th1) cells, forming a link between innate resistance and adaptive

immunity [10]. IL-12 is produced primarily by phagocytes (monocytes/macrophages and neutrophils) and dendritic cells in response to inflammatory stimuli. The targets of IL-12 include NK (natural killer) cells, NKT cells, and T cells, for which IL-12 induces proliferation and augments cytotoxic activity and the production of secondary cytokine mediators such as IFN-γ, TNFα, IL-8, and GM-CSF. In addition, IL-12 also induces the production of Th1-type antibody production. Due to its broad immune-stimulatory properties, IL-12 has been evaluated extensively for its therapeutic potential in cancer and infectious disease [11]. These studies demonstrate that while systemic administration of IL-12 as a nonspecific immune stimulant is associated with severe toxicity, it is highly effective when administered locally as a vaccine adjuvant [11].

3. *Interferons* (IFN-α, β, and γ). Interferons can be divided broadly in two groups: type I (interferon α, β, and other minor subtypes, ~19 kD) and type II (interferon γ, 17 kD) based on their source, receptors, and functional properties [12,13]. Type I interferons are secreted primarily by myeloid leukocytes (plasmacytoid dendritic cells) and many other types of cells, although at lower levels. Their potent antiviral function is mediated through their ability to induce MHC class I up-regulation on a variety of cells, immature dendritic cell activation, potentiation of effector cell (NK/T) cytotoxicity, and B-cell activation [12]. Type II interferon (IFN-γ), which functions through a distinct pathway that includes the activation of over 200 immunologically relevant genes, has potent pro-inflammatory activity [13]. IFN-γ is produced primarily by activated NK, NK/T, and TH1 CD4+ T cells. The major immunological activities of IFN-γ include macrophage, neutrophil, and NK-cell activation, up-regulation of MHC class I and II expression, induction of other cytotoxic cytokines (e.g., TNFα), and induction of T- and B-cell differentiation [13]. The broad as well as specific immune-activatory characteristics of interferons have resulted in their use as vaccine adjuvants. The primary use for type I interferons has been in the systemic setting to nonspecifically stimulate vaccine efficacy, whereas due to its functional characteristics and severe toxicity, IFN-γ has been used as a local adjuvant.

4. *Interleukin-2* (IL-2). IL-2 is a 15.4-kD protein, which is produced primarily by CD4+ T cells, although production by CD8+ T cells, dendritic cells, and thymic cells has also been reported [14]. IL-2 acts through a multisubunit receptor which is primarily up-regulated on activated T and NK cells [14]. Binding of IL-2 stimulates both the proliferation and activation T and NK cells [14]. Along with type I interferons, IL-2 was among the first cytokines to be available in recombinant form and has been evaluated extensively in cancer patients for its antitumor activity, resulting in FDA approval for late-stage cancers [15]. Although IL-2 has primarily been used as a T-cell growth factor in preclinical and clinical studies, the recent finding that loss of IL-2 or its receptor in knockout mice results in the development of lethal immunoproliferative disease has caused a paradigm shift [14]. It is now thought that the primary function of IL-2 might actually involve immune regulation through

its ability to enhance the survival of suppressor T cells [14]. Therefore, the potential use of IL-2 in the adjuvant setting might need to be reevaluated.

5. *Interleukin-4* (IL-4). IL-4 is a 15-kD pleiotropic cytokine that regulates diverse T- and B-cell responses, including cell proliferation, survival, and gene expression [16]. Produced by mast cells, T cells, and bone marrow cells, IL-4 is antagonistic to IFN-$\gamma$ and regulates the differentiation of naive CD4$^+$ T cells into helper Th2 cells that produce IL-4, IL-5, IL-6, IL-10, and IL-13, which favor a humoral immune response. An additional important function of IL-4 is the regulation of immunoglobulin class switching to the IgG1 and IgE isotypes.

6. *Interleukin-18* (IL-18). IL-18 is a pro-inflammatory cytokine structurally related to IL-1 which targets primarily T cells. Initially known as IFN-$\gamma$-inducing factor, it induces IFN-$\gamma$ production primarily by Th1 CD4$^+$ T cells and NK cells [17]. IL-18 is produced primarily by mononuclear cells (monocytes and macrophages) and also by keratinocytes constitutively, and is up-regulated in response to macrophage stimulators. It has been demonstrated that the IFN-$\gamma$-inducing properties of IL-18 are largely dependent on synergistic co-stimulation with IL-12 [17]. In addition to IFN-$\gamma$, IL-18 induces T-cell production of IL-2, GM-CSF, and TNF$\alpha$. Other properties of IL-18 include the chemokine induction in monocytes and NK cells and the inhibition of IgE production by B cells [17].

7. *Interleukin-15* (IL-15). IL-15 is a pleiotropic 15-kD cytokine with stimulatory effects on both innate and adaptive immunity. It is produced primarily by macrophages and monocytes and targets NK, NKT, and CD8$^+$ T cells [18]. IL-15 binds to a multimeric receptor that consists of the IL-15 R$\alpha$, IL-2R$\beta$, and IL-2R$\gamma$ subunits and shares similar signaling and activation pathways with IL-2 [18]. IL-15 is a key regulator of NK- and NKT-cell development, homeostasis, and activity and is also involved in the survival and expansion of naive and memory CD8$^+$ T cells [18].

8. *Interleukin-7* (IL-7). IL-7 is a 25-kD protein that has effects on both T- and B-cell biology [19]. IL-7 is produced primarily by fetal liver, thymic, and bone marrow stromal cells and has been shown to be essential to both T- and B-cell development. The biological effects of IL-7 is mediated through the IL-7 receptor, which is expressed on T cells, B cells, NK cells, dendritic cells, monocytes, intestinal epithelial cells, and keratinocytes. In addition to its role in lymphocyte development, IL-7 has been shown to be critical to lymphocyte homeostasis, primarily that of naive and memory T cells [19].

9. *Interleukin-1$\alpha$, $\beta$* (IL-1$\alpha$, $\beta$). IL-1$\alpha$ and $\beta$ are 17.5-kD pro-inflammatory cytokines produced in a variety of cells, including monocytes, tissue macrophages, keratinocytes, and other epithelial cells [20]. They both bind to the same receptor and have similar biological properties. Their ranges of activities include stimulation of thymocyte proliferation, B-cell maturation and proliferation, and mitogenic fibroblast growth factor-like activity. In addition, IL-1 induces the expression of over 90 genes, including cytokines (TNF, IL-2, IL-12,

GM-CSF, and interferons, among others), as well as chemokines, adhesion molecules, pro-inflammatory mediators, growth factors, and extracellular matrix enzymes [20]. The major difference between $\alpha$ and $\beta$ forms is that whereas IL-1$\beta$ is a secreted cytokine, IL-1$\alpha$ is predominantly cell associated.

10. *Fetal liver tyrosine kinase 3 ligand* (Flt3L). Flt3L is a ~17-kD growth factor that regulates the proliferation of early hematopoietic cells [21]. It is expressed by various hematopietic cell lineages and binds to primitive hematopoietic progenitor cells expressing the tyrosine kinase receptor Flt3. Flt3L synergizes with other colony-stimulating factors and interleukins (IL-3, IL-6, IL-11, IL-12, KIT-ligand, and GM-CSF) to induce the growth and differentiation of myeloid cells, primarily that of myeloid dendritic cells [21]. In vivo administration of Flt3L can induce an increase of up to 40-fold in the number of secondary lymphoid organ dendritic cells.

11. *Tumor necrosis factor $\alpha$* (TNF$\alpha$). TNF$\alpha$ is a 17-kD pro-inflammatory cytokine that is produced primarily by activated macrophages, NK cells, and T cells, although some production in fibroblasts, astrocytes, Kuppfer cells, and keratinocytes has also been demonstrated [22]. TNF functions through its receptors, TNFR-1 expressed on all cells and TNFR-2 expressed on immune and endothelial cells. The primary receptor is TNFR-1, which upon binding of TNF$\alpha$, mediates a pro-apoptotic/necrotic cascade that results in the death of the target cell [22]. Tumor cells appear to be especially sensitive to the cytotoxic effects of TNF$\alpha$, and thus TNF$\alpha$ has been used as an anticancer agent alone or in combination with other antineoplastic agents [22]. However, systemic delivery of TNF$\alpha$ results in severe toxicity before clinically effective doses can be reached. As a result, loco-regional administration has been explored with superior results.

## 14.4 CYTOKINES AS ADJUVANTS IN INFECTIOUS DISEASE VACCINATION

The potential of cytokines as vaccine adjuvants has been evaluated in numerous bacterial, viral, or parasitic infection models in the past two decades. Between 1984 and 1995, fewer than 20 reports were published investigating the potential of IL-1, IL-2, and the interferons as vaccine adjuvants. These studies have been summarized in several reviews [23,24]. In contrast, a brief survey of the literature in the past decade identified over 100 reports in which numerous cytokines were utilized as adjuvants in studies involving peptide, protein, DNA, and whole organism vaccines. A cytokine-centric review of this literature is presented below, with individual discussions of cytokines ordered in their relative popularity.

1. *GM-CSF.* The adjuvant potential of GM-CSF has been evaluated extensively with numerous viral, bacterial, and parasitic vaccines in murine, rabbit, bovine, simian, and human models. The most extensive testing has been

performed with antiviral vaccines including hepatitis A and B [25–29], HIV/ SIV [30–34], influenza [29,35], papilloma virus [36], herpes [37], and bovine viral diarrhea virus [38]. In addition, GM-CSF has been tested as a component of antibacterial vaccines, including tuberculosis [39,40], tetanus [29], diphteria [29], chlamydia [41], and rickettsia [42], as well as in antiparasitic vaccines for schistosoma [43–45] and leishmania [46]. The use of GM-CSF as adjuvant resulted in the augmentation of both antibody and cellular responses in the overwhelming majority (22 of 25) of these studies. More specifically, GM-CSF, either as soluble protein or as a DNA vaccine component, promoted significantly higher antibody titers (primarily IgG1 in systemic vaccination and IgA in mucosal vaccination) to either bacterial, viral, or parasitic subunit proteins [25,26,28,30,38,40,42,44–47], to peptide antigens [32,36,48,49], or to inactivated whole organisms [41]. In addition, adjuvanting with GM-CSF resulted in enhanced CTL and helper T-cell responses in a number of these studies [28,30,32,34,36–38,40–42,48,49]. These findings (i.e., coaugmentation of both humoral and cellular responses) are consistent with the well-established role of GM-CSF as a potent stimulator of antigen presentation. In a number of these studies, GM-CSF-mediated potentiation of humoral and cellular immunity resulted in superior protection from challenge in viral [26,32,33,36,37], bacterial [40,41], and parasitic [44,45] infection models. Others showed that GM-CSF was synergistic when combined with other cytokines, most notably with IL-12 and TNFα [2,49]. In several studies, adjuvanting vaccines with GM-CSF did not enhance immunity beyond that obtained with antigen alone. Although this was observed in only a fraction of the studies (4/25), they involved either macaque monkeys [31] or human subjects [27,29,35], with possible important clinical implications. In one macaque study, coadministration of IL-12- and GM-CSF-expressing DNA plasmids with a simian immunodeficiency virus (SIV) library vaccine [31] induced superior immunity (compared to SIV vaccine library alone) initially; however, long-term viral load or survival were inferior in adjuvanted mice. In this study the dose of cytokine DNA delivered was very low (25 µg per treatment). Ordinarily, much larger quantities of DNA (1 to 10 mg) are required per inoculation in primates and clinical trials, and it is possible that the low dose of cytokine DNA administered in this study resulted in an abortive response, thus worsening protection. For example, in two other SIV vaccine/macaque monkey studies, where GM-CSF provided significant enhancement of both humoral and cellular anti-SIV responses, 0.24 and 1 mg of DNA was used per vaccination [30,33]. In the human studies, recombinant GM-CSF in saline was delivered near the vaccine site separately from antigen, either prior to or shortly after the vaccination. As physical colocalization of adjuvant with antigen appears to be critical to achieving a potent immune response [34,45,47], the lack of efficacy observed in the human studies could be associated, at least in part, with the delivery method. These few exceptions notwithstanding, the studies outlined here demonstrate that GM-CSF is a highly effective and promising adjuvant that can promote both humoral and cellular responses in different vaccine settings.

2. *IL-12.* The only cytokine to equal the popularity of GM-CSF as vaccine adjuvant, IL-12 has been evaluated extensively in viral, bacterial, and parasitic vaccine studies in murine, rat, feline, and rhesus macaque models. Evaluation of IL-12 as vaccine adjuvant in humans has lagged, possibly due to toxicity concerns based on previous experience with its systemic use in cancer patients, even though it is delivered locally in the vaccine setting. Due to its cytotoxic T-lymphocyte (CTL)-enhancing properties, IL-12 has been evaluated extensively as a viral vaccine adjuvant in SIV/HIV [32,33,49–54], hepatitis B [55], measles [56], HSV-2 [57], and feline leukemia virus (FLV) [58] models. The ability of IL-12 to induce the Th1 cytokines IFN-γ, TNFα, and IL-2 provided the rationale for its evaluation in bacterial vaccines, including tetanus [59], tuberculosis [60,61], pneumonia [62], tularemia [63], and meningitis [64], as well as in parasite vaccines, including schistosoma [45,65,66], toxoplasma [67], leishmania [68], trypanosome infection [69], and malaria [70]. In all of these studies, IL-12 consistently induced a switch from a Th2 to a Th1 type of response with substantial IFN-γ production, promoted the development of antigen-specific CTL, and induced IgG2a (in systemic vaccination studies) and IgA (in mucosal vaccination studies) isotype antibodies, independent of formulation (soluble protein or DNA), delivery route (subcutaneous, intramuscular, intranasal), dosage, or schedule of administration. In addition, IL-12 provided superior protection from challenge to that of no cytokine controls in the overwhelming majority of the studies (95%). An additional observation was the enhancement of long-term memory T-cell activity [45]. The clear conclusion from the studies noted above is that IL-12 represents a highly promising adjuvant for the induction of Th1-type cellular and humoral immunity against pathogens in animal models. Whether this will hold true in human trials remains to be shown.

3. *Interferon-α, β and γ.* Interferons were among the first cytokines to be tested as vaccine adjuvants, due to their early availability in recombinant form [71,72]. Type I interferons have been used to enhance responses to viral (hepatitis B [71,73], hepatitis C [74], and influenza [75]) and parasitic (malaria [72,76]) vaccines. Type I interferons were shown to enhance both antibody (primarily, IgG1) and CTL responses in these studies. In one study, intranasal administration of IFNαβ adjuvanted influenza vaccine provided complete protection from challenge [75]. The efficacy of soluble systemic interferon was dose dependent [74]. IFN-γ, a pro-inflammatory Th1 cytokine has been tested more extensively than IFN-α or β, primarily as a viral vaccine adjuvant for HIV [30,49,51,77–80], HSV-1 [81], hepatitis B [28], and influenza [82]. There also are several examples of its use as a bacterial vaccine adjuvant for *Yersinia pestis* [83], *Mycobacterium tuberculosis* [84], and coccidiosis in chickens [85]. Use of IFN-γ as adjuvant promotes a switch from IgG1 to IgG2a isotype antibodies, enhances Th1 T-cell activity resulting in further IFN-γ production, and promotes CTL development in murine and macaque monkey models [28,30,49,79,80,82,84,86]. In several studies, adjuvanting with IFN-γ provided

protection from viral [82] or bacterial challenge [84,85]. In contrast, use of IFN-γ either failed to enhance [31,81] or reduced [83] protection in SIV, HSV-1, or plague models in comparison to parental vaccine, respectively. The failure to protect in the case of SIV occurred in the same macaque monkey study discussed above [31]. The failure to enhance the efficacy in the HSV-1 model occurred despite development of strong CTL responses and IgG2a levels [81]. In the same study, vaccination with an IL-4 expressing plasmid was more effective, suggesting that a Th2 response may be more appropriate in this ocular challenge model. In the case of the *Yersinia* study, the inability of IFN-γ to provide protection is not surprising, as the efficacy of V-antigen-based vaccination in *Yersinia* correlates with IgG1 responses [83]. Another group reported the use of a novel approach involving the use of a DNA vaccine expressing a fusion protein between HIV gp120 antigen and IFN-γ [79,80]. The authors demonstrated that the plasmid expressing the fusion protein was superior to a mixture of gp120 and IFN-γ plasmids in stimulating both antibody and T-cell responses. Other observations include the finding that delivery of IFN-γ in emulsion [84] or as a liposomal formulation [82] was superior to soluble IFN-γ. Finally, in the case of recombinant cytokine, dose appeared to be critical to efficacy [82,84].

4. *IL-2*. The earliest report on the use of IL-2 as a vaccine adjuvant involved co-delivery of recombinant IL-2 with sperm whale myoglobin in incomplete Freund's adjuvant (IFA), which demonstrated enhanced antibody responses [87]. Based on its characteristics as a stimulator of T-cell growth and activity, subsequent studies focused on the use of IL-2 as a systemic stimulator of vaccine responses in guinea pig–HSV, bovine herpes virus, and mouse-rabies vaccine models [88–92]. The overall conclusions from these studies were that systemic use of IL-2 augmented protection significantly, which correlated with enhanced cellular responses rather than antibody-mediated protection. A number of studies investigated the potency of IL-2 as a local adjuvant starting with a report where liposomal IL-2 was used to adjuvant *Pseudomonas aeruginosa* antigens delivered intranasally [93]. This mucosal vaccination strategy enhanced a drastic increase in mucosal IgA levels compared to antigen-alone controls and significantly reduced challenge-associated mortality. In a separate study a fusion protein between HSV-1 glycoprotein D and IL-2 was shown to enhance both antibody and DTH responses, providing complete protection from HSV-1 challenge in mice [94]. In one study where IL-2 was delivered locally with an HIV peptide in adjuvant in a mouse model, the vaccine failed to induce a significant CTL response [49]. In contrast, in numerous studies involving DNA-based vaccination where IL-2 plasmid DNA was employed as adjuvant, IgG2a antibody, Th1 and in some cases CTL responses were obtained in models of HIV [86], hepatitis B [28], influenza [95], papilloma virus [96], foot-and-mouth disease [97], BVDV [35], HSV-1 [88], and PRV [98]. Several novel approaches, including liposomal antigen + IL-2 [95] and viral particle-

encapsidated IL-2 plasmid [96], were used in these studies. In a number of these studies, the timing of the administration of IL-2 plasmid was critical to enhancing immunity [86]. The most impressive results with regard to protection from challenge were obtained in a macaque monkey/SIV trial, where the vaccination of mice with plasmid DNA expressing viral antigens plus exogenous IL-2 or IL-2/Ig (either as soluble protein or DNA) resulted in 100% protection from clinical disease upon viral challenge [99]. In contrast, in a recent report, use of low-dose systemic IL-2 in chronically infected macaque monkeys vaccinated with a canarypox recombinant SIV vaccine resulted in the expansion of both CD8+ T cells and CD4+ CD25+ T-suppressor cells, and loss of virus-specific CD4+ T cells, demonstrating the complex nature of IL-2-based therapy [100].

5. *IL-4.* IL-4 has been tested primarily as an adjuvant for viral subunit vaccines, including HIV/SIV [49,77,78,101], hepatitis B [28], and HSV-1 [81]. In addition, two recent studies evaluated the ability of IL-4 to augment antiparasitic vaccines [43,44]. Analysis of postvaccine immunity demonstrated enhanced Th2 type antibody responses, primarily of the IgG1 and IgG3 isotype, against the specific antigen. In studies where CTL activity could be evaluated, use of IL-4 suppressed CTL activity [28,49,77], with one exception [81]. In challenge studies, IL-4 did not provide any protection in immunized animals in SIV/macaque monkey studies [101] or from schistosoma [43,44]. The only exception was a HSV-1 study, where vaccination of mice with IL-4-expressing virus provided effective suppression of viral titers upon ocular challenge in mice [81]. Collectively, the studies above demonstrate that IL-4 does not represent an effective vaccine adjuvant in vaccines against pathogens that require a Th1 response.

6. *IL-18.* Systemic administration of IL-18 in mice infected with either HSV-1 or leishmania major demonstrated the ability of IL-18 to enhance preexisting immune responses and suppress disease progression [102,103]. In these experiments, the antipathogenic activity of IL-18 was shown to be mediated by innate mechanisms and IFN-γ. Later studies, where IL-18 was used as a local adjuvant in vaccination against HIV [50,104], feline leukemia virus (FLV) [58], HSV-1 [105], Newcastle disease virus [106], and tuberculosis [40], established that IL-18 enhances CTL development, production of IgG2a and IgA antibodies (when administered mucosally), and Th1 cell proliferation. In the FLV study, kittens vaccinated with a combination of IL-12 and IL-18 adjuvant were protected from a challenge [58], demonstrating the clinical potential of IL-18 as vaccine adjuvant.

7. *Other cytokines.* Numerous other cytokines have also been evaluated for their adjuvant capacity in sporadic studies. Adjuvant potential of IL-15 was first demonstrated when IL-15 and a model antigen (tetanus toxoid) were coadministered in liposomal formulations, resulting in up to 10-fold increases in antitoxoid IgG responses in comparison to toxoid alone [107]. In another

study, vaccination of IL-15 transgenic mice with *Mycobacterium tuberculosis* BCG resulted in the development of a drastic improvement in the antibacterial CD8$^+$ T-cell responses and effective resistance to challenge versus nontransgenic mice [108]. Similarly, in a recent study, systemic administration of recombinant IL-15 to mice that are immunized against the male antigen HY-enhanced effector T-cell populations and CD8$^+$ T cells to subdominant antigens [109]. Finally, immunization of mice with plasmids expressing IL-15 and HIV antigens promoted a significant enhancement of antigen-specific CD8$^+$ T-cell proliferation, IFN-$\gamma$ secretion, and long-term CD8$^+$ memory T-cell activity in comparison to antigen plasmid alone [110]. In the same study, vaccination of mice with influenza A hemagglutinin-expressing DNA and IL-15 plasmid promoted similar CD8$^+$ T-cell activity, resulting in the complete protection of mice from a lethal mucosal challenge with influenza virus [110]. IL-7, which is critical to both T- and B-cell homeostasis, promoted a Th2 response and suppressed CTL activity when delivered locally in an HIV vaccine study in mice [49]. In contrast, systemic delivery of IL-7 during vaccination in the HY-antigen immunization study dramatically increased the antigen-specific CD8$^+$ T-cell response [109]. Interleukin-1$\beta$ and its noninflammatory synthetic peptide derivative [111] have been tested for adjuvant activity. Use of the IL-1$\beta$ peptide enhanced antibody responses to a hepatitis B vaccine in two of three mouse strains [112]. Systemic administration of recombinant IL-1$\beta$ to calves augmented neutralizing antibody titers and cytotoxic responses induced by bovine herpesvirus-1 vaccine and reduced viral loads [113]. Local administration of IL-1$\beta$ with antigen failed to enhance CTL or helper T-cell responses in an HIV peptide antigen vaccine study [49]. Similarly, vaccination of mice with an HIV peptide-IL-1$\beta$ fusion construct failed to induce CTL or antibody responses in a separate study [114]. In this study a DTH response was induced by the fusion protein; however, an equally strong DTH response could be induced by a fusion protein of antigen–KLH (keyhole limpet hemocyanin). Nasal immunization of mice with an HIV peptide immunogen and IL-1$\alpha$ induced high levels of anti-HIV antibodies [50]. T-cell responses were not evaluated. Other cytokines tested include: Flt3L which in combination with either GM-CSF [42] or IL-1$\beta$ peptide-antigen fusion protein [114], promoted dendritic cell recruitment, increased antigen-specific T-cells and IFN-$\gamma$ production, and TNF$\alpha$, which by itself did not enhance the potency of an HIV peptide vaccine but was strongly synergistic with IL-12 [49].

## 14.5 CYTOKINES IN CANCER THERAPY

Somewhat surprisingly, cytokines have been evaluated more extensively in tumor therapy than in infectious disease as immunological adjuvants. A brief survey of the literature in the past two decades identified well over 200 publications in this area. In early studies, systemic administration of recombinant interferon $\alpha$ and IL-2 resulted in the suppression of murine tumors [115–118], and later in cancer patients [15,119]. As a result, IFN-α and IL-2 are now standard therapy for hematological malignancies and advanced melanoma/renal carcinoma, respectively. Subsequently, many other cytokines, including IL-3, IL-4, IL-6, IL-10, IL-12, M-CSF, GM-CSF, IFN-γ, TNFα, and Flt3L, were tested in phase I to III clinical trials [15,119]. The general conclusion from these studies is that although systemic cytokine therapy can have beneficial effects in a small percentage of cancer patients, toxic side effects limit the clinical utility of this strategy [15,120].

## 14.6  CYTOKINES AS ADJUVANTS FOR CELLULAR TUMOR VACCINES

The potential utility of cytokines as local adjuvants was first shown by Forni and colleagues in studies where peritumoral injection of IL-2 enhanced immune-mediated tumor rejection and long-term immunity in mice [121]. This finding was followed by the development of gene-modified tumor cell vaccines (GMTVs), where tumor cells are genetically modified to secrete cytokines and then administered to patients as whole cell vaccines [122,123]. The list of cytokines tested in this setting is extensive, with the finding that most cytokines, in the context of GMTV, provided protection from tumor challenge in mice [124–126]. Based on these results, numerous clinical trials were undertaken in patients with IL-2, IL-4, IL-6, IL-7, IL-12, IFN-γ, GM-CSF, and lymphotactin GMTV [15,125]. In contrast to the preclinical studies, results from the completed trials demonstrated that GMTV, although effective in inducing antitumor immunity, achieved limited clinical responses [15,126]. The major difference between the murine studies and the clinical trials was that the preclinical vaccines were tested primarily in the preventive setting, whereas in the clinic they were administered to patients with advanced disease as a therapeutic vaccine. In a small number of studies, administration of cytokine-adjuvanted tumor vaccines achieved superior clinical responses when performed in patients with minimal residual disease following standard therapy [127].

Of the numerous cytokines evaluated as GMTV adjuvants, GM-CSF and IL-12 stand out as being most effective [119,124,126]. Of the two, GM-CSF has been evaluated more extensively [8,9,15]. In a comparative study, Dranoff and colleagues demonstrated that in the context of GMTV, GM-CSF stimulated a more potent and long-lasting antitumor immunity than 28 other cytokines and chemokines in mice [9,15]. In subsequent studies the authors demonstrated that GM-CSF mediates its potent antitumor activity via the stimulation of local infiltration of antigen-presenting cell (APC) populations (dendritic cells, macrophages) and granulocytes [9,15]. Based on these studies, a number of phase I studies were undertaken with melanoma patients using autologous GMTV [8,9,15]. These trials demonstrated the induction of inflammatory responses at the vaccine site as well as at distant metastases. The inflammatory infiltrates consisted of dendritic cells and eosinophils at the injection site and plasma cells and T cells in distant metastatic lesions, resulting in extensive tumor destruction [8,15]. In this initial study, one partial clinical response was achieved (out of 14 patients) and 14% of the patients remained disease-free postsurgery at last follow-up [8]. Based on these results, similar studies were carried out in metastatic renal, prostate, lung, and colon carcinoma patients with similar results [8]. Recent modifications to this therapy (i.e., administration of CTLA-4-blocking antibodies to patients who have received GM-CSF GMTV therapy previously) appears to improve tumor destruction significantly [15]. IL-12-based GMTV was also shown to induce effective tumor eradication and protection from challenge in preclinical studies, especially in the established disease setting [15,124,126]. To this end, several clinical trials evaluated the potential of IL-12-GMTV in patients [128,129]. These studies demonstrated the development of antitumor T-cell activity and DTH in most patients with transient tumor regressions in some. To date, the most successful use of IL-12 in the adjuvant setting has been in the treatment of cutaneous T-cell lymphoma, where soluble IL-12 was injected locally at or near the tumor site [130]. In this limited study, clinical responses were observed in seven of 11 evaluated patients (two complete responses), and biopsies of regressing lesions demonstrated $CD8^+$ T-cell infiltrates.

## 14.7 CYTOKINES AS ADJUVANTS FOR PROTEIN- AND PEPTIDE-BASED TUMOR VACCINES

Cytokines have also been used to adjuvant protein- and peptide-based tumor vaccines. Since the cloning of the first melanoma-associated antigen, MAGE-1 [131], many tumor-associated antigens have been identified [132]. With the availability of this information and the ability to identify both MHC classes I and II T-cell epitopes, protein- and/or peptide-based tumor vaccines targeting T-cell activation have been evaluated in a number of clinical trials, a significant number of which included the use of cytokines as adjuvant [125,133].

Several studies investigated the use of recombinant tumor antigens with cytokine as adjuvant [134–138]. In one study, administration of IFN-α systemically concurrent with ganglioside GM-2-KLH conjugate vaccination in melanoma patients, enhanced postsurgical survival significantly [133,134]. In B-cell lymphoma patients, local administration of GM-CSF with antigen increased the efficacy of a KLH-conjugated idiotype vaccine dramatically, resulting in the complete molecular remission of disease in 85% of patients [136]. The presence of GM-CSF was critical to the efficacy of the vaccine in these studies. Based on these results, a number of phase I and II clinical trials in B-cell lymphoma and multiple myeloma patients have been completed with phase III trials currently under way [136]. More recently, efficacy of GM-CSF as a vaccine adjuvant with recombinant Ep-Cam-KLH conjugate was demonstrated in colorectal cancer patients [137].

The literature on peptide-based cancer vaccines is considerable, probably due to the ease of producing large quantities of peptides synthetically [119,125,133,139]. A recent summary of peptide-based tumor vaccine clinical trials lists 32 trials, 22 of which utilized cytokine adjuvants [125]. Of these, 16 involved the use of GM-CSF, four of which combined GM-CSF with IL-2. In the remaining studies, IL-2, Flt3L, and G-CSF were tested. In some studies the efficacy of cytokine-adjuvanted vaccines could be compared to controls where peptides were administered either alone or with a noncytokine adjuvant. Enhanced induction of CD8+ peptide-specific T-cells and objective tumor responses were reported in an early study following the addition of GM-CSF to a multipeptide vaccine [139]. In two consecutive small phase I trials in melanoma patients, however, addition of GM-CSF did not enhance peptide-specific T cell responses [140]. In other studies, vaccination of resected stage III melanoma patients with peptides emulsified in IFA with or without GM-CSF. GM-CSF increased the frequency of tumor-specific T cells and resulted in enhanced disease-free survival [141,142]. The efficacy of peptide immunization was evaluated with three adjuvants – IFA, QS21, and GM-CSF – in a separate melanoma trial where both GM-CSF and QS21 induced peptide-specific T-cell responses, whereas IFA was not effective [143]. Others have demonstrated similar potentiating effects of GM-CSF with her-2/neu peptides in breast cancer [144]. Finally, in a recent trial combining GM-CSF adjuvanted peptide vaccine with imatinib (a tyrosine kinase inhibitor), complete cytogenetic remission in chronic myelogenous leukemia (CML) patients was enhanced significantly compared to imatinib-only historical controls [145]. In other studies performed at the National Cancer Institute, systemic administration of IL-2 or IL-12 to melanoma patients receiving peptide vaccines resulted in a decrease in the precursor frequency of peptide-specific T cells, which was associated with enhanced tumor regression in these patients compared to patients receiving peptide alone [146,147]. In two other studies, adjuvant use of IL-12 promoted peptide-specific T-cell responses in melanoma patients [148,149]. In the studies above, IL-2 and IL-12 were administered systemically. In a recent study, coadministration of peptides with IL-12 in a slow-release matrix enhanced antigen-specific T-cell responses in mice [150].

## 14.8   CYTOKINES IN DNA-BASED TUMOR VACCINES

In contrast to infectious disease, reports on DNA-based vaccines with cytokine adjuvant are relatively few and preliminary in cancer. In preclinical studies, vaccination of mice with plasmids coexpressing her2/neu with GM-CSF or IL-12, or melanoma antigen MAGE-1 and GM-CSF, enhanced antitumor immunity and protection from challenge [151,152]. In a clinical trial involving the vaccination of B-cell lymphoma patients with plasmids expressing GM-CSF and the idiotype improved humoral and cellular responses against the idiotype as compared to vaccination with plasmid expressing the idiotype alone [153].

## 14.9   CYTOKINE ADJUVANT FORMULATIONS

An important finding that emerges from this review is that in addition to the specific biological properties of individual cytokines, the method of delivery is also critical to their adjuvanting efficacy. Studies with numerous cytokines, including GM-CSF [34,45,47,151,152,154], IL-12 [55,57,150,155], IFNγ [79,80], IL-15 [107], and IL-18 [152], demonstrate that delivery approaches which favor local and sustained release of the cytokine at the vaccination site, and in close physical contact with the antigen, provide superior immunity. This is not surprising, as under physiological conditions cytokines function locally within the inflammatory microenvironment where antigen release occurs. Thus, development of adjuvant formulations that would approximate these physiological conditions would be expected to provide the most effective vaccine strategy. To this end, a brief analysis of the literature from the perspective of different delivery approaches is presented below.

1. *Soluble cytokine with or without traditional adjuvants.* The simplest method for delivery of cytokine adjuvants is direct injection of soluble recombinant protein. However, the short in vivo half-lives of most cytokines necessitate frequent injections of high doses of cytokine, resulting in toxicity and reduced efficacy. Therefore, in many studies, recombinant cytokine is administered in emulsion with traditional adjuvants to provide local and sustained release at the vaccine site [45,47,49,55,57,70,84,155,156]. These studies demonstrate that cytokines in emulsion promote superior immunity compared to injection of soluble cytokine [47,57,70,84,155,156]. The adjuvants utilized include alum [45,47,55,57,70] and oil-in-water [49,156] and lipid-based reagents [84,157]. Importantly, in some of the studies, coadsorption of antigen and cytokine to alum promoted immunity superior to that of mixtures of antigen + alum and cytokine + alum, demonstrating the importance of close physical linkage of antigen and cytokine [45,47]. Similarly, fusion proteins of cytokine and antigen have been found to be more effective immune adjuvants than mixtures of antigen and cytokine [79,80,94]. Although the use of cytokine and/or antigen in emulsion with adjuvant is a simple and effective approach, the manipulations that need to be performed prior to vaccination, plus potential limitations associated with stability and/or storage, may pose significant drawbacks for such formulations in the clinical setting.

2. *DNA vectors.* Genetic immunization represents an attractive alternative to the whole-organism or peptide/protein subunit-based vaccines [3,158–160]. The major advantage of genetic vaccines is the ability to deliver antigens to the intracellular antigen-processing pathway for presentation by MHC class I and the induction of CD8+ cytotoxic T-cell (CTL) responses. This approach has been used to deliver antigen with cytokine in different combinations, including antigen DNA with recombinant cytokine [43,52,161], mixtures of antigen DNA and cytokine DNA [28,30,31,33,37,38,40,54,77,78], or use of DNA vectors that coexpress antigen and cytokine [34,56,151,152]. Similar to the alum coadsorption reports discussed above, a number of these studies

found that coexpression of antigen and cytokine on the same DNA vector promoted superior immunity compared to mixtures of DNA vectors expressing either the antigen or the cytokine confirming the importance of close physical linkage [31,151,152]. The major limitation of DNA-based vaccines is the fact that transfection efficiencies can be low, resulting in the production of limited amounts of cytokine and antigen at the immunization site. This drawback is especially apparent in primate and human studies, where large amounts of plasmid DNA need to be administered repeatedly to obtain significant immune activity. Use of viral vectors can provide more effective transfection; however, this strategy is associated with potential health risks, vector immunogenicity, and the possibility of skewing the immune response to viral proteins.

3. *Particulate-sustained delivery formulations.* This category includes vehicles such as liposomes and biodegradable polymer particles which were originally designed for delivery of small molecule drugs [162]. These formulations are designed to enhance the bioavailability of encapsulated agents by increasing their in vivo half-lives, and in the case of biodegradable particles, provide local and sustained release of drugs, characteristics that are ideally suited to vaccine adjuvants [163]. In addition, the particulate nature of these adjuvants promotes enhanced uptake by antigen-presenting cells [164]. Both lipid-and polymer-based systems have been evaluated extensively as antigen delivery systems, with promising results [164–166]. Liposomal formulations of cytokines have been tested successfully in the treatment of both infectious disease and cancer, primarily as systemic therapeutics [167]. Several studies tested the potential of liposomal cytokine and antigen formulations in vaccination [95,107,151,168–171]. In all cases, liposomal formulations induced superior immunity compared to soluble antigen and/or cytokine. In one study, the authors demonstrated that association of antigen and cytokine on the same liposomal vesicle was superior to mixtures of liposomes containing antigen or cytokine separately [107]. Biodegradable polymer microspheres have also been evaluated for their adjuvant capacity [163–166]. Polymer particles provide the advantage of local and sustained release (for days to weeks) of encapsulated antigens and can be stored in lyophilized form without refrigeration. Whereas the potential of microparticles for antigen delivery, either in DNA or protein form, has been evaluated in many studies [164–166], their use for in vivo delivery of biologically active macromolecules such as cytokines has been limited. In an early study, coadministration of GM-CSF-encapsulated microspheres with irradiated tumor cells was shown to be as effective as a GM-CSF-secreting tumor cell vaccine in protecting mice from tumor challenge [172]. Subsequent studies in our laboratory demonstrated that intratumoral injection of IL-12-encapsulated microspheres induced the complete regression of established tumors in a T- and NK-cell-dependent manner and promoted the development of long-term protective immunity in mice [173]. In other studies, intratumoral injection of IL-12 + GM-CSF microspheres was found to be more effective than either cytokine alone in inducing the

suppression of advanced tumors and achieved complete eradication of disseminated disease in a metastatic murine lung tumor model [174]. Characterization of these formulations established that up to 90% of the cytokine activity could be preserved after encapsulation, and physiologically relevant levels of cytokines were released for up to one month [175]. These studies established the adjuvant potential of sustained-release particulate cytokine formulations in cancer therapy. Their adjuvant potential in subunit vaccines remains to be determined.

## 14.10  FUTURE DIRECTIONS

The studies summarized above establish that cytokines represent specific effective adjuvants for vaccination. Although the proof of principle for their efficacy has been established, a number of important issues concerning the translation of these results to the clinic lay ahead. One issue is the choice of the most promising cytokine(s) for evaluation in human studies. The current literature suggests that GM-CSF, IL-12, and IL-2 represent initial promising choices; however, further evaluation of the more recently discovered cytokines, such as IL-15 and IL-18, is certainly warranted. Additionally, as a number of studies demonstrate, different cytokines can be highly synergistic, and novel synergies between cytokines that reflect their natural interactions within the inflammatory microenvironment should be investigated further. Other critical parameters appear to be the dose, timing, and the mode of administration of cytokines. As individual cytokines possess unique biophysical/functional characteristics, careful optimization of both the dose and the timing of their administration should be performed for each cytokine prior to their clinical evaluation. The mode of delivery is another crucial element, and as physical co-localization of cytokine and antigen appears to be critical to efficacy, strategies involving the co-adsorption of antigen with recombinant cytokine on traditional adjuvants, such as alum, or the use of DNA vectors that coexpress both antigen and cytokine, should be pursued. In addition, delivery methods that mimic the natural physiology and kinetics of antigen and cytokine release within the infection microenvironment should be targeted. To this end, particulate slow-release systems such as liposomes and polymeric biodegradable particles represent especially attractive future technologies, as they have the potential to provide controlled co-delivery of cytokine and antigen in a local and sustained manner.

## REFERENCES

1. Berzofsky, J.A., Ahlers, J.D., Janik, J. et al. Progress on new vaccine strategies against chronic viral infections. *J Clin Invest* 2004, **114**(4), 450–462.

2. Ahlers, J.D., Belyakov, I.M. & Berzofsky, J.A. Cytokine, chemokine, and costimulatory molecule modulation to enhance efficacy of HIV vaccines. *Curr Mol Med* 2003, **3**(3), 285–301.

3. Eriksson, K. & Holmgren, J. Recent advances in mucosal vaccines and adjuvants. *Curr Opin Immunol* 2002, **14**(5), 666–672.

4. Petrovsky, N. & Aguilar, J.C. Vaccine adjuvants: current state and future trends. *Immunol Cell Biol* 2004, **82**(5), 488–496.

5. Villinger, F. Cytokines as clinical adjuvants: How far are we? *Expert Rev Vaccines* 2003, **2**(2), 317–326.

6. Janeway, C.A., Travers, P., Walport, M. & Shlomchik, M.J. *Immunobiology: The Immune System in Health and Disease*, Garland Science Publishing, New York, 2005.

7. Hamilton, J.A. & Anderson, G.P. GM-CSF biology. *Growth Factors* 2004, **22**(4), 225–231.

8. Chang, D.Z., Lomazow, W., Joy Somberg, C., Stan, R. & Perales, M.A. Granulocyte–macrophage colony stimulating factor: an adjuvant for cancer vaccines. *Hematology* 2004, **9**(3), 207–215.

9. Dranoff, G. GM-CSF-based cancer vaccines. *Immunol Rev* 2002, **188**, 147–154.

10. Trinchieri, G. Interleukin-12 and the regulation of innate resistance and adaptive immunity. *Nat Rev Immunol* 2003, **3**(2), 133–146.

11. Portielje, J.E., Gratama, J.W., van Ojik, H.H., Stoter, G. & Kruit, W.H. IL-12: a promising adjuvant for cancer vaccination. *Cancer Immunol Immunother* 2003, **52**(3), 133–144.

12. Theofilopoulos, A.N., Baccala, R., Beutler, B. & Kono, D.H. Type I Interferons (A/B) in immunity and autoimmunity. *Annu Rev Immunol* 2005, **23**, 307–336.

13. Boehm, U., Klamp, T., Groot, M. & Howard, J.C. Cellular responses to interferon-gamma. *Annu Rev Immunol* 1997, **15**, 749–795.

14. Nelson, B.H. IL-2, regulatory T cells, and tolerance. *J Immunol* 2004, **172**(7), 3983–3988.

15. Dranoff, G. Cytokines in cancer pathogenesis and cancer therapy. *Nat Rev Immunol* 2004, **4**, 11–22.

16. Paludan, S.R. Interleukin-4 and interferon-gamma: the quintessence of a mutual antagonistic relationship. *Scand J Immunol* 1998, **48**(5), 459–468.

17. Dinarello, C.A. Interleukin-18. *Methods* 1999, **19**(1), 121–132.

18. Alpdogan, O. & van den Brink, M.R. IL-7 and IL-15: therapeutic cytokines for immunodeficiency. *Trends Immunol* 2005, **26**(1), 56–64.

19. Fry, T.J. & Mackall, C.L. The many faces of IL-7: from lymphopoiesis to peripheral T cell maintenance. *J Immunol* 2005, **174**(11), 6571–6576.

20. Dinarello, C.A. Interleukin-1. *Cytokine Growth Factor Rev* 1997, **8**(4), 253–265.

21. Shurin, M.R., Esche, C. & Lotze, M.T. FLT3: receptor and ligand: biology and potential clinical application. *Cytokine Growth Factor Rev* 1998, **9**(1), 37–48.

22. Mocellin, S., Rossi, C.R., Pilati, P. & Nitti, D. Tumor necrosis factor, cancer and anticancer therapy. *Cytokine Growth Factor Rev* 2005, **16**(1), 35–53.

23. Taylor, C.E. Cytokines as adjuvants for vaccines: antigen-specific responses differ from polyclonal responses. *Infect Immun* 1995, **63**(9), 3241–3244.

24. Lin, R., Tarr, P.E. & Jones, T.C. Present status of the use of cytokines as adjuvants with vaccines to protect against infectious diseases. *Clin Infect Dis* 1995, **21**(6), 1439–1449.

25. Verkade, M.A., van de Wetering, J., Klepper, M., Vaessen, L.M., Weimar, W. & Betjes, M.G. Peripheral blood dendritic cells and GM-CSF as an adjuvant for hepatitis B vaccination in hemodialysis patients. *Kidney Int* 2004, **66**(2), 614–621.

26. Sasaki, M.G., Foccacia, R. & de Messias-Reason, I.J. Efficacy of granulocyte–macrophage colony-stimulating factor (GM-CSF) as a vaccine adjuvant for hepatitis B virus in patients with HIV infection. *Vaccine* 2003, **21**(31), 4545–4549.

27. Hasan, M.S., Agosti, J.M., Reynolds, K.K., Tanzman, E., Treanor, J.J. & Evans, T.G. Granulocyte macrophage colony-stimulating factor as an adjuvant for hepatitis B vaccination of healthy adults. *J Infect Dis* 1999, **180**(6), 2023–2026.

28. Chow, Y.H., Chiang, B.L., Lee, Y.L. et al. Development of Th1 and Th2 populations and the nature of immune responses to hepatitis B virus DNA vaccines can be modulated by codelivery of various cytokine genes. *J Immunol* 1998, **160**(3), 1320–1329.

29. Somani, J., Lonial, S., Rosenthal, H., Resnick, S., Kakhniashvili, I. & Waller, E.K. A randomized, placebo-controlled trial of subcutaneous administration of GM-CSF as a vaccine adjuvant: effect on cellular and humoral immune responses. *Vaccine* 2002, **21**(3–4), 221–230.

30. Lena, P., Villinger, F., Giavedoni, L., Miller, C.J., Rhodes, G. & Luciw, P. Co-immunization of rhesus macaques with plasmid vectors expressing IFN-gamma, GM-CSF, and SIV antigens enhances anti-viral humoral immunity but does not affect viremia after challenge with highly pathogenic virus. *Vaccine* 2002, **20** (Suppl 4), A69–A79.

31. Sykes, K.F., Lewis, M.G., Squires, B. & Johnston, S.A. Evaluation of SIV library vaccines with genetic cytokines in a macaque challenge. *Vaccine* 2002, **20**(17–18), 2382–2395.

32. Belyakov, I.M., Ahlers, J.D., Clements, J.D., Strober, W. & Berzofsky, J.A. Interplay of cytokines and adjuvants in the regulation of mucosal and systemic HIV-specific CTL. *J Immunol* 2000, **165**(11), 1320–1326.

33. O'Neill, E., Martinez, I., Villinger, F. et al. Protection by SIV VLP DNA prime/protein boost following mucosal SIV challenge is markedly enhanced by IL-12/GM-CSF co-administration. *J Med Primatol* 2002, **31**(4–5), 217–227.

34. Barouch, D.H., Santra, S., Tenner-Racz, K. et al. Potent CD4+ T cell responses elicited by a bicistronic HIV-1 DNA vaccine expressing gp120 and GM-CSF. *J Immunol* 2002, **168**(2), 562–568.

35. Pauksen, K., Linde, A., Hammarstrom, V. et al. Granulocyte–macrophage colony-stimulating factor as immunomodulating factor together with influenza vaccination in stem cell transplant patients. *Clin Infect Dis* 2000, **30**(2), 342–348.

36. Hu, J., Cladel, N.M., Wang, Z., Han, R., Pickel, M.D. & Christensen, N.D. GM-CSF enhances protective immunity to cottontail rabbit papillomavirus E8 genetic vaccination in rabbits. *Vaccine* 2004, **22**(9–10), 1124–1130.

37. Sin, J.I., Kim, J.J., Ugen, K.E., Ciccarelli, R.B., Higgins, T.J. & Weiner, D.B. Enhancement of protective humoral (Th2) and cell-mediated (Th1) immune responses against herpes simplex virus-2 through co-delivery of granulocyte–macrophage

colony-stimulating factor expression cassettes. *Eur J Immunol* 1998, **28**(11), 3530–3540.

38. Nobiron, I., Thompson, I., Brownlie, J. & Collins, M.E. Cytokine adjuvancy of BVDV DNA vaccine enhances both humoral and cellular immune responses in mice. *Vaccine* 2001, **19**(30), 4226–4235.

39. Wang, J., Zganiacz, A. & Xing, Z. Enhanced immunogenicity of BCG vaccine by using a viral-based GM-CSF transgene adjuvant formulation. *Vaccine* 2002, **20**(23–24), 2887–2898.

40. Kyoung, M.B., Ko, S.-Y., Lee, M., Lee, J.-S., Kim, J.-O. & Ko, H.-J. Comparative analysis of effects of cytokine gene adjuvants in DNA vaccination against *Mycobacterium* tuberculosis heat shock proteins 65. *Vaccine* 2003, **21**, 3684–3689.

41. Lu, H., Xing, Z. & Brunham, R.C. GM-CSF transgene-based adjuvant allows the establishment of protective mucosal immunity following vaccination with inactivated *Chlamydia trachomatis*. *J Immunol* 2002, **169**(11), 6324–6331.

42. Mwangi, W., Brown, W.C., Lewin, H.A. et al. DNA-encoded fetal liver tyrosine kinase 3 ligand and granulocyte macrophage-colony-stimulating factor increase dendritic cell recruitment to the inoculation site and enhance antigen-specific CD4+ T cell responses induced by DNA vaccination of outbred animals. *J Immunol* 2002, **169**(7), 3837–3846.

43. Siddiqui, A.A., Pinkston, J.R., Quinlin, M.L. et al. Characterization of protective immunity induced against *Schistosoma mansoni* via DNA priming with the large subunit of calpain (Sm-p80) in the presence of genetic adjuvants. *Parasite* 2005, **12**(1), 3–8.

44. Siddiqui, A.A., Phillips, T., Charest, H. et al. Induction of protective immunity against *Schistosoma mansoni* via DNA priming and boosting with the large subunit of calpain (Sm-p80): adjuvant effects of granulocyte–macrophage colony-stimulating factor and interleukin-4. *Infect Immun* 2003, **71**(7), 3844–3851.

45. Argiro, L., Henri, S., Dessein, H., Dessein, A.J. & Bourgois, A. Induction of a protective immunity against *Schistosoma mansoni* with ovalbumin-coupled Sm37-5 coadsorbed with granulocyte–macrophage colony stimulating factor (GM-CSF) or IL-12 on alum. *Vaccine* 1999, **17**(1), 13–18.

46. Follador, I., Araujo, C., Orge, G. et al. Immune responses to an inactive vaccine against American cutaneous leishmaniasis together with granulocyte–macrophage colony-stimulating factor. *Vaccine* 2002, **20**(9–10), 1365–1368.

47. Argiro, L., Henri, S., Dessein, H., Kouriba, B., Dessein, A.J. & Bourgois, A. Induction of a protection against *S. mansoni* with a MAP containing epitopes of Sm37-GAPDH and Sm10-DLC: effect of coadsorption with GM-CSF on alum. *Vaccine* 2000, **18**(19), 2033–2038.

48. Iwasaki, A., Stiernholm, B.J., Chan, A.K., Berinstein, N.L. & Barber, B.H. Enhanced CTL responses mediated by plasmid DNA immunogens encoding costimulatory molecules and cytokines. *J Immunol* 1997, **158**(10), 4591–4601.

49. Ahlers, J.D., Dunlop, N., Alling, D.W., Nara, P.L. & Berzofsky, J.A. Cytokine-in-adjuvant steering of the immune response phenotype to HIV-1 vaccine constructs: granulocyte–macrophage colony-stimulating factor and TNF-alpha synergize with IL-12 to enhance induction of cytotoxic T lymphocytes. *J Immunol* 1997, **158**(8), 3947–3958.

50. Bradney, C.P., Sempowski, G.D., Liao, H.X., Haynes, B.F. & Staats, H.F. Cytokines as adjuvants for the induction of anti-human immunodeficiency virus peptide immunoglobulin G (IgG) and IgA antibodies in serum and mucosal secretions after nasal immunization. *J Virol* 2002, **76**(2), 517–524.

51. Kim, J.J., Nottingham, L.K., Tsai, A. et al. Antigen-specific humoral and cellular immune responses can be modulated in rhesus macaques through the use of IFN-gamma, IL-12, or IL-18 gene adjuvants. *J Med Primatol* 1999, **28**(4–5), 214–223.

52. Gherardi, M.M., Ramirez, J.C. & Esteban, M. Towards a new generation of vaccines: the cytokine IL-12 as an adjuvant to enhance cellular immune responses to pathogens during prime–booster vaccination regimens. *Histol Histopathol* 2001, **16**(2), 655–667.

53. Albu, D.I., Jones-Trower, A., Woron, A.M. et al. Intranasal vaccination using interleukin-12 and cholera toxin subunit B as adjuvants to enhance mucosal and systemic immunity to human immunodeficiency virus type 1 glycoproteins. *J Virol* 2003, **77**(10), 5589–5597.

54. Chattergoon, M.A., Saulino, V., Shames, J.P., Stein, J., Montaner, L.J. & Weiner, D.B. Co-immunization with plasmid IL-12 generates a strong T-cell memory response in mice. *Vaccine* 2004, **22**(13–14), 1744–1750.

55. Wang, S., Liu, X. & Caulfield, M.J. Adjuvant synergy in the response to hepatitis B vaccines. *Vaccine* 2003, **21**(27–30), 4297–4306.

56. Hoffman, S.J., Polack, F.P., Hauer, D.A. et al. Vaccination of rhesus macaques with a recombinant measles virus expressing interleukin-12 alters humoral and cellular immune responses. *J Infect Dis* 2003, **188**(10), 1553–1561.

57. Cooper, D., Pride, M.W., Guo, M. et al. Interleukin-12 redirects murine immune responses to soluble or aluminum phosphate adsorbed HSV-2 glycoprotein D towards Th1 and CD4+ CTL responses. *Vaccine* 2004, **23**(2), 236–246.

58. Hanlon, L., Argyle, D., Bain, D. et al. Feline leukemia virus DNA vaccine efficacy is enhanced by coadministration with interleukin-12 (IL-12) and IL-18 expression vectors. *J Virol* 2001, **75**(18), 8424–8433.

59. Boyaka, P.N., Marinaro, M., Jackson, R.J. et al. IL-12 is an effective adjuvant for induction of mucosal immunity. *J Immunol* 1999, **162**(1), 122–128.

60. Martin, E., Kamath, A.T., Briscoe, H. & Britton, W.J. The combination of plasmid interleukin-12 with a single DNA vaccine is more effective than *Mycobacterium bovis* (Bacille Calmette–Guerin) in protecting against systemic *Mycobacterim avium* infection. *Immunology* 2003, **109**(2), 308–314.

61. Uzonna, J.E., Chilton, P., Whitlock, R.H., Habecker, P.L., Scott, P. & Sweeney, R.W. Efficacy of commercial and field-strain *Mycobacterium* paratuberculosis vaccinations with recombinant IL-12 in a bovine experimental infection model. *Vaccine* 2003, **21**(23), 3101–3109.

62. Lynch, J.M., Briles, D.E. & Metzger, D.W. Increased protection against pneumococcal disease by mucosal administration of conjugate vaccine plus interleukin-12. *Infect Immun* 2003, **71**(8), 4780–4788.

63. Duckett, N.S., Olmos, S., Durrant, D.M. & Metzger, D.W. Intranasal interleukin-12 treatment for protection against respiratory infection with the *Francisella tularensis* live vaccine strain. *Infect Immun* 2005, **73**(4), 2306–2311.

64. Buchanan, R.M., Arulanandam, B.P. & Metzger, D.W. IL-12 enhances antibody responses to T-independent polysaccharide vaccines in the absence of T and NK cells. *J Immunol* 1998, **161**(10), 5525–5533.

65. Fonseca, C.T., Brito, C.F., Alves, J.B. & Oliveira, S.C. IL-12 enhances protective immunity in mice engendered by immunization with recombinant 14 kDa *Schistosoma mansoni* fatty acid-binding protein through an IFN-gamma and TNF-alpha dependent pathway. *Vaccine* 2004, **22**(3–4), 503–510.

66. Bungiro, R.D., Jr., Goldberg, M., Suri, P.K. & Knopf, P.M. Interleukin-12 as an adjuvant for an antischistosome vaccine consisting of adult worm antigens: protection of rats from cercarial challenge. *Infect Immun* 1999, **67**(5), 2340–2348.

67. Letscher-Bru, V., Villard, O., Risse, B., Zauke, M., Klein, J.P. & Kien, T.T. Protective effect of vaccination with a combination of recombinant surface antigen 1 and interleukin-12 against toxoplasmosis in mice. *Infect Immun* 1998, **66**(9), 4503–4506.

68. Santos, W.R., de Lima, V.M., de Souza, E.P., Bernardo, R.R., Palatnik, M. & Palatnik de Sousa, C.B. Saponins, IL12 and BCG adjuvant in the FML-vaccine formulation against murine visceral leishmaniasis. *Vaccine* 2002, **21**(1–2), 30–43.

69. Katae, M., Miyahira, Y., Takeda, K. et al. Coadministration of an interleukin-12 gene and a *Trypanosoma cruzi* gene improves vaccine efficacy. *Infect Immun* 2002, **70**(9), 4833–4840.

70. Su, Z., Tam, M.F., Jankovic, D. & Stevenson, M.M. Vaccination with novel immunostimulatory adjuvants against blood-stage malaria in mice. *Infect Immun* 2003, **71**(9), 5178–5187.

71. Grob, P.J., Joller-Jemelka, H.I., Binswanger, U., Zaruba, K., Descoeudres, C. & Fernex, M. Interferon as an adjuvant for hepatitis B vaccination in non- and low-responder populations. *Eur J Clin Microbiol* 1984, **3**(3), 195–198.

72. Playfair, J.H. & de Souza, J.B. Recombinant gamma interferon is a potent adjuvant for a malaria vaccine in mice. *Clin Exp Immunol* 1987, **67**(1), 5–10.

73. Goldwater, P.N. Randomized comparative trial of interferon-alpha versus placebo in hepatitis B vaccine non-responders and hyporesponders. *Vaccine* 1994, **12**(5), 410–414.

74. Gehring, S., Gregory, S.H., Kuzushita, N. & Wands, J.R. Type 1 interferon augments DNA-based vaccination against hepatitis C virus core protein. *J Med Virol* 2005, **75**(2), 249–257.

75. Bracci, L., Canini, I., Puzelli, S. et al. Type I IFN is a powerful mucosal adjuvant for a selective intranasal vaccination against influenza virus in mice and affects antigen capture at mucosal level. *Vaccine* 2005, **23**(23), 2994–3004.

76. Sturchler, D., Berger, R., Etlinger, H. et al. Effects of interferons on immune response to a synthetic peptide malaria sporozoite vaccine in non-immune adults. *Vaccine* 1989, **7**(5), 457–461.

77. Kim, J.J., Yang, J.S., Manson, K.H. & Weiner, D.B. Modulation of antigen-specific cellular immune responses to DNA vaccination in rhesus macaques through the use of IL-2, IFN-gamma, or IL-4 gene adjuvants. *Vaccine* 2001, **19**(17–19), 2496–2505.

78. Kim, J.J., Yang, J.S., VanCott, T.C. et al. Modulation of antigen-specific humoral responses in rhesus macaques by using cytokine cDNAs as DNA vaccine adjuvants. *J Virol* 2000, **74**(7), 3427–3429.

79. McCormick, A.L., Thomas, M.S. & Heath, A.W. Immunization with an interferon-gamma-gp120 fusion protein induces enhanced immune responses to human immunodeficiency virus gp120. *J Infect Dis* 2001, **184**(11), 1423–1430.

80. Nimal, S., McCormick, A.L., Thomas, M.S. & Heath, A.W. An interferon gamma-gp120 fusion delivered as a DNA vaccine induces enhanced priming. *Vaccine* 2005, **23**(30), 3984–3990.

81. Osorio, Y. & Ghiasi, H. Comparison of adjuvant efficacy of herpes simplex virus type 1 recombinant viruses expressing Th1 and Th2 cytokine genes. *J Virol* 2003, **77**(10), 5774–5783.

82. van Slooten, M.L., Hayon, I., Babai, I. et al. Immunoadjuvant activity of interferon-gamma-liposomes co-administered with influenza vaccines. *Biochim Biophys Acta* 2001, **1531**(1–2), 99–110.

83. Griffin, K.F., Eyles, J.E., Spiers, I.D., Alpar, H.O. & Williamson, E.D. Protection against plague following immunisation with microencapsulated V antigen is reduced by co-encapsulation with IFN-gamma or IL-4, but not IL-6. *Vaccine* 2002, **20**(31–32), 3650–3657.

84. Hovav, A.H., Fishman, Y. & Bercovier, H. Gamma interferon and monophos-phoryl lipid A-trehalose dicorynomycolate are efficient adjuvants for *Myco-bacterium* tuberculosis multivalent acellular vaccine. *Infect Immun* 2005, **73**(1), 250–257.

85. Ding, X., Lillehoj, H.S., Quiroz, M.A., Bevensee, E. & Lillehoj, E.P. Protective immunity against *Eimeria acervulina* following in ovo immunization with a recom-binant subunit vaccine and cytokine genes. *Infect Immun* 2004, **72**(12), 6939–6944.

86. Calarota, S.A. & Weiner, D.B. Enhancement of human immunodeficiency virus type 1-DNA vaccine potency through incorporation of T-helper 1 molecular adju-vants. *Immunol Rev* 2004, **199**, 84–99.

87. Kawamura, H., Rosenberg, S.A. & Berzofsky, J.A. Immunization with antigen and interleukin 2 in vivo overcomes Ir gene low responsiveness. *J Exp Med* 1985, **162**(1), 381–386.

88. Weinberg, A. & Merigan, T.C. Recombinant interleukin 2 as an adjuvant for vaccine-induced protection: immunization of guinea pigs with herpes simplex virus subunit vaccines. *J Immunol* 1988, **140**(1), 294–299.

89. Reddy, P.G., Blecha, F., Minocha, H.C. et al. Bovine recombinant interleukin-2 augments immunity and resistance to bovine herpesvirus infection. *Vet Immunol Immunopathol* 1989, **23**(1–2), 61–74.

90. Nunberg, J.H., Doyle, M.V., York, S.M. & York, C.J. Interleukin 2 acts as an adju-vant to increase the potency of inactivated rabies virus vaccine. *Proc Natl Acad Sci USA* 1989, **86**(11), 4240–4243.

91. Hughes, H.P., Campos, M., Godson, D.L. et al. Immunopotentiation of bovine herpes virus subunit vaccination by interleukin-2. *Immunology* 1991, **74**(3), 461–466.

92. Hughes, H.P., Campos, M., van Drunen Littel-van den Hurk, S. et al. Multiple administration with interleukin-2 potentiates antigen-specific responses to subunit vaccination with bovine herpesvirus-1 glycoprotein IV. *Vaccine* 1992, **10**(4), 226–230.

93. Abraham, E. & Shah, S. Intranasal immunization with liposomes containing IL-2 enhances bacterial polysaccharide antigen-specific pulmonary secretory antibody response. *J Immunol* 1992, **149**(11), 3719–3726.

94. Hazama, M., Mayumi-Aono, A., Asakawa, N., Kuroda, S., Hinuma, S. & Fujisawa, Y. Adjuvant-independent enhanced immune responses to recombinant herpes

simplex virus type 1 glycoprotein D by fusion with biologically active interleukin-2. *Vaccine* 1993, **11**(6), 629–636.

95. Ben-Yehuda, A., Joseph, A., Barenholz, Y. et al. Immunogenicity and safety of a novel IL-2-supplemented liposomal influenza vaccine (INFLUSOME-VAC) in nursing-home residents. *Vaccine* 2003, **21**(23), 3169–3178.

96. Oh, Y.K., Sohn, T., Park, J.S. et al. Enhanced mucosal and systemic immunogenicity of human papillomavirus-like particles encapsidating interleukin-2 gene adjuvant. *Virology* 2004, **328**(2), 266–273.

97. Yadav, S., Sharma, R. & Chhabra, R. Interleukin-2 potentiates foot-and-mouth disease vaccinal immune responses in mice. *Vaccine* 2005, **23**(23), 3005–3009.

98. Lin, Y., Qigai, H., Xiaolan, Y., Weicheng, B. & Huanchun, C. The co-administrating of recombinant porcine IL-2 could enhance protective immune responses to PRV inactivated vaccine in pigs. *Vaccine* 2005, **23**(35), 4436–4441.

99. Barouch, D.H., Santra, S., Schmitz, J.E. et al. Control of viremia and prevention of clinical AIDS in rhesus monkeys by cytokine-augmented DNA vaccination. *Science* 2000, **290**(5491), 486–492.

100. Nacsa, J., Edghill-Smith, Y., Tsai, W.P. et al. Contrasting effects of low-dose IL-2 on vaccine-boosted simian immunodeficiency virus (SIV)–specific CD4+ and CD8+ T cells in macaques chronically infected with SIVmac251. *J Immunol* 2005, **174**(4), 1913–1921.

101. Boyer, J.D., Nath, B., Schumann, K. et al. IL-4 increases simian immunodeficiency virus replication despite enhanced SIV immune responses in infected rhesus macaques. *Int J Parasitol* 2002, **32**(5), 543–550.

102. Fujioka, N., Akazawa, R., Ohashi, K., Fujii, M., Ikeda, M. & Kurimoto, M. Interleukin-18 protects mice against acute herpes simplex virus type 1 infection. *J Virol* 1999, **73**(3), 2401–2409.

103. Ohkusu, K., Yoshimoto, T., Takeda, K. et al. Potentiality of interleukin-18 as a useful reagent for treatment and prevention of *Leishmania major* infection. *Infect Immun* 2000, **68**(5), 2449–2456.

104. Billaut-Mulot, O., Idziorek, T., Loyens, M., Capron, A. & Bahr, G.M. Modulation of cellular and humoral immune responses to a multiepitopic HIV-1 DNA vaccine by interleukin-18 DNA immunization/viral protein boost. *Vaccine* 2001, **19**(20–22), 2803–2811.

105. Zhu, M., Xu, X., Liu, H. et al. Enhancement of DNA vaccine potency against herpes simplex virus 1 by co-administration of an interleukin-18 expression plasmid as a genetic adjuvant. *J Med Microbiol* 2003, **52**(Pt 3), 223–228.

106. Degen, W.G., van Zuilekom, H.I., Scholtes, N.C., van Daal, N. & Schijns, V.E. Potentiation of humoral immune responses to vaccine antigens by recombinant chicken IL-18 (rChIL-18). *Vaccine* 2005, **23**(33), 4212–4218.

107. Gursel, M. & Gregoriadis, G. Interleukin-15 acts as an immunological co-adjuvant for liposomal antigen in vivo. *Immunol Lett* 1997, **55**(3), 161–165.

108. Umemura, M., Nishimura, H., Saito, K. et al. Interleukin-15 as an immune adjuvant to increase the efficacy of *Mycobacterium bovis* bacillus Calmette–Guerin vaccination. *Infect Immun* 2003, **71**(10), 6045–6048.

109. Melchionda, F., Fry, T.J., Milliron, M.J., McKirdy, M.A., Tagaya, Y. & Mackall, C.L. Adjuvant IL-7 or IL-15 overcomes immunodominance and improves survival of the CD8+ memory cell pool. *J Clin Invest* 2005, **115**(5), 1177–1187.

110. Kutzler, M.A., Robinson, T.M., Chattergoon, M.A. et al. Coimmunization with an optimized IL-15 plasmid results in enhanced function and longevity of CD8 T cells that are partially independent of CD4 T cell help. *J Immunol* 2005, **175**(1), 112–123.

111. Antoni, G., Presentini, R., Perin, F. et al. Interleukin 1 and its synthetic peptides as adjuvants for poorly immunogenic vaccines. *Adv Exp Med Biol* 1989, **251**, 153–160.

112. Manivel, V. & Rao, K.V. Interleukin-1 derived synthetic peptide as an added co-adjuvant in vaccine formulations. *Vaccine* 1991, **9**(6), 395–397.

113. Reddy, D.N., Reddy, P.G., Minocha, H.C. et al. Adjuvanticity of recombinant bovine interleukin-1 beta: influence on immunity, infection, and latency in a bovine herpesvirus-1 infection. *Lymphokine Res* 1990, **9**(3), 295–307.

114. Pisarev, V.M., Parajuli, P., Mosley, R.L. et al. Flt3 ligand and conjugation to IL-1beta peptide as adjuvants for a type 1, T-cell response to an HIV p17 gag vaccine. *Vaccine* 2002, **20**(17–18), 2358–2368.

115. Gresser, I., Bourali, C., Levy, J.P., Fontaine-Brouty-Boye, D. & Thomas, M.T. Increased survival in mice inoculated with tumor cells and treated with interferon preparations. *Proc Natl Acad Sci USA* 1969, **63**(1), 51–57.

116. Cheever, M.A., Greenberg, P.D., Fefer, A. & Gillis, S. Augmentation of the anti-tumor therapeutic efficacy of long-term cultured T lymphocytes by in vivo administration of purified interleukin 2. *J Exp Med* 1982, **155**(4), 968–980.

117. Bubenik, J., Perlmann, P., Indrova, M., Simova, J., Jandlova, T. & Neuwirt, J. Growth inhibition of an MC-induced mouse sarcoma by TCGF (IL 2)-containing preparations: preliminary report. *Cancer Immunol Immunother* 1983, **14**(3), 205–206.

118. Donohue, J.H., Rosenstein, M., Chang, A.E., Lotze, M.T., Robb, R.J. & Rosenberg, S.A. The systemic administration of purified interleukin 2 enhances the ability of sensitized murine lymphocytes to cure a disseminated syngeneic lymphoma. *J Immunol* 1984, **132**(4), 2123–2128.

119. Smyth, M.J., Cretney, E., Kershaw, M.H. & Hayakawa, Y. Cytokines in cancer immunity and immunotherapy. *Immunol Rev* 2004, **202**, 275–293.

120. Antonia, S., Mule, J.J. & Weber, J.S. Current developments of immunotherapy in the clinic. *Curr Opin Immunol* 2004, **16**(2), 130–136.

121. Forni, G., Cavallo, G.P., Giovarelli, M. et al. Tumor immunotherapy by local injection of interleukin 2 and non-reactive lymphocytes: experimental and clinical results. *Prog Exp Tumor Res* 1988, **32**, 187–212.

122. Bubenik, J., Voitenok, N.N., Kieler, J. et al. Local administration of cells containing an inserted IL-2 gene and producing IL-2 inhibits growth of human tumours in nu/nu mice. *Immunol Lett* 1988, **19**(4), 279–282.

123. Tepper, R.I., Pattengale, P.K. & Leder, P. Murine interleukin-4 displays potent anti-tumor activity in vivo. *Cell* 1989, **57**(3), 503–508.

124. Parmiani, G., Rodolfo, M. & Melani, C. Immunological gene therapy with ex vivo gene-modified tumor cells: a critique and a reappraisal. *Hum Gene Ther* 2000, **11**(9), 1269–1275.

125. Ribas, A., Butterfield, L.H., Glaspy, J.A. & Economou, J.S. Current developments in cancer vaccines and cellular immunotherapy. *J Clin Oncol* 2003, **21**(12), 2415–2432.

126. Mach, N. & Dranoff, G. Cytokine-secreting tumor cell vaccines. *Curr Opin Immunol* 2000, **12**(5), 571–575.

127. Bendandi, M., Gocke, C.D., Kobrin, C.B. et al. Complete molecular remissions induced by patient-specific vaccination plus granulocyte–monocyte colony-stimulating factor against lymphoma. *Nat Med* 1999, **5**(10), 1171–1177.

128. Sun, Y., Jurgovsky, K., Moller, P. et al. Vaccination with IL-12 gene-modified autologous melanoma cells: preclinical results and a first clinical phase I study. *Gene Ther* 1998, **5**(4), 481–490.

129. Kang, W.K., Park, C., Yoon, H.L. et al. Interleukin 12 gene therapy of cancer by peritumoral injection of transduced autologous fibroblasts: outcome of a phase I study. *Hum Gene Ther* 2001, **12**(6), 671–684.

130. Rook, A.H., Wood, G.S., Yoo, E.K. et al. Interleukin-12 therapy of cutaneous T-cell lymphoma induces lesion regression and cytotoxic T-cell responses. *Blood* 1999, **94**(3), 902–908.

131. Jager, E., Jager, D. & Knuth, A. Antigen-specific immunotherapy and cancer vaccines. *Int J Cancer* 2003, **106**(6), 817–820.

132. Novellino, L., Castelli, C. & Parmiani, G. A listing of human tumor antigens recognized by T cells: March 2004 update. *Cancer Immunol Immunother* 2005, **54**(3), 187–207.

133. Scheibenbogen, C., Letsch, A., Schmittel, A., Asemissen, A.M., Thiel, E. & Keilholz, U. Rational peptide-based tumour vaccine development and T cell monitoring. *Semin Cancer Biol* 2003, **13**(6), 423–429.

134. Kirkwood, J.M., Ibrahim, J., Lawson, D.H. et al. High-dose interferon alfa-2b does not diminish antibody response to GM2 vaccination in patients with resected melanoma: results of the Multicenter Eastern Cooperative Oncology Group Phase II Trial E2696. *J Clin Oncol* 2001, **19**(5), 1430–1436.

135. Kirkwood, J.M., Ibrahim, J.G., Sosman, J.A. et al. High-dose interferon alfa-2b significantly prolongs relapse-free and overall survival compared with the GM2-KLH/QS-21 vaccine in patients with resected stage IIB–III melanoma: results of intergroup trial E1694/S9512/C509801. *J Clin Oncol* 2001, **19**(9), 2370–2380.

136. Coscia, M. & Kwak, L.W. Therapeutic idiotype vaccines in B lymphoproliferative diseases. *Expert Opin Biol Ther* 2004, **4**(6), 959–963.

137. Neidhart, J., Allen, K.O., Barlow, D.L. et al. Immunization of colorectal cancer patients with recombinant baculovirus-derived KSA (Ep-Cam) formulated with monophosphoryl lipid A in liposomal emulsion, with and without granulocyte–macrophage colony-stimulating factor. *Vaccine* 2004, **22**(5–6), 773–780.

138. Parmiani, G., Pilla, L., Castelli, C. & Rivoltini, L. Vaccination of patients with solid tumours. *Ann Oncol* 2003, **14**(6), 817–824.

139. Jager, E., Ringhoffer, M., Dienes, H.P. et al. Granulocyte-macrophage-colony-stimulating factor enhances immune responses to melanoma-associated peptides in vivo. *Int J Cancer* 1996, **67**(1), 54–62.

140. Scheibenbogen, C., Schadendorf, D., Bechrakis, N.E. et al. Effects of granulocyte–macrophage colony-stimulating factor and foreign helper protein as immunologic adjuvants on the T-cell response to vaccination with tyrosinase peptides. *Int J Cancer* 2003, **104**(2), 3012–3024.

141. Weber, J., Sondak, V.K., Scotland, R. et al. Granulocyte–macrophage-colony-stimulating factor added to a multipeptide vaccine for resected stage II melanoma. *Cancer* 2003, **97**(1), 186–200.

142. Letsch, A., Keilholz, U., Fluck, M. et al. Peptide vaccination after repeated resection of metastases can induce a prolonged relapse-free interval in melanoma patients. *Int J Cancer* 2005, **114**(6), 936–941.

143. Schaed, S.G., Klimek, V.M., Panageas, K.S. et al. T-cell responses against tyrosinase 368–376(370D) peptide in HLA*A0201+ melanoma patients: randomized trial comparing incomplete Freund's adjuvant, granulocyte macrophage colony-stimulating factor, and QS-21 as immunological adjuvants. *Clin Cancer Res* 2002, **8**(5), 967–972.

144. Disis, M.L., Grabstein, K.H., Sleath, P.R. & Cheever, M.A. Generation of immunity to the HER-2/neu oncogenic protein in patients with breast and ovarian cancer using a peptide-based vaccine. *Clin Cancer Res* 1999, **5**(6), 1289–1297.

145. Bocchia, M., Gentili, S., Abruzzese, E. et al. Effect of a p210 multipeptide vaccine associated with imatinib or interferon in patients with chronic myeloid leukaemia and persistent residual disease: a multicentre observational trial. *Lancet* 2005, **365**(9460), 657–662.

146. Rosenberg, S.A., Yang, J.C., Schwartzentruber, D.J. et al. Immunologic and therapeutic evaluation of a synthetic peptide vaccine for the treatment of patients with metastatic melanoma. *Nat Med* 1998, **4**(3), 321–327.

147. Rosenberg, S.A., Yang, J.C., Schwartzentruber, D.J. et al. Impact of cytokine administration on the generation of antitumor reactivity in patients with metastatic melanoma receiving a peptide vaccine. *J Immunol* 1999, **163**(3), 1690–1695.

148. Lee, P., Wang, F., Kuniyoshi, J. et al. Effects of interleukin-12 on the immune response to a multipeptide vaccine for resected metastatic melanoma. *J Clin Oncol* 2001, **19**(18), 3836–3847.

149. Cebon, J., Jager, E., Shackleton, M.J. et al. Two phase I studies of low dose recombinant human IL-12 with Melan-A and influenza peptides in subjects with advanced malignant melanoma. *Cancer Immun* 2003, **3**, 7.

150. Salem, M.L., Kadima, A.N., Zhou, Y. et al. Paracrine release of IL-12 stimulates IFN-gamma production and dramatically enhances the antigen-specific T cell response after vaccination with a novel peptide-based cancer vaccine. *J Immunol* 2004, **172**(9), 5159–5167.

151. Sun, X., Hodge, L.M., Jones, H.P., Tabor, L. & Simecka, J.W. Co-expression of granulocyte–macrophage colony-stimulating factor with antigen enhances humoral and tumor immunity after DNA vaccination. *Vaccine* 2002, **20**(9–10), 1466–1474.

152. Chang, S.Y., Lee, K.C., Ko, S.Y., Ko, H.J. & Kang, C.Y. Enhanced efficacy of DNA vaccination against Her-2/neu tumor antigen by genetic adjuvants. *Int J Cancer* 2004, **111**(1), 86–95.

153. Timmerman, J.M., Singh, G., Hermanson, G. et al. Immunogenicity of a plasmid DNA vaccine encoding chimeric idiotype in patients with B-cell lymphoma. *Cancer Res* 2002, **62**(20), 5845–5852.

154. Kass, E., Panicali, D.L., Mazzara, G., Schlom, J. & Greiner, J.W. Granulocyte/macrophage-colony stimulating factor produced by recombinant avian poxviruses enriches the regional lymph nodes with antigen-presenting cells and acts as an immunoadjuvant. *Cancer Res* 2001, **61**(1), 206–214.

155. Hancock, G.E., Smith, J.D. & Heers, K.M. The immunogenicity of subunit vaccines for respiratory syncytial virus after co-formulation with aluminum hydroxide adjuvant and recombinant interleukin-12. *Viral Immunol* 2000, **13**(1), 57–72.

156. Chianese-Bullock, K.A., Pressley, J., Garbee, C. et al. MAGE-A1-, MAGE-A10-, and gp100-derived peptides are immunogenic when combined with granulocyte–macrophage colony-stimulating factor and montanide ISA-51 adjuvant and administered as part of a multipeptide vaccine for melanoma. *J Immunol* 2005, **174**(5), 3080–3086.

157. Liao, H.X., Cianciolo, G.J., Staats, H.F. et al. Increased immunogenicity of HIV envelope subunit complexed with alpha2-macroglobulin when combined with monophosphoryl lipid A and GM-CSF. *Vaccine* 2002, **20**(17–18), 2396–2403.

158. Toka, F.N., Pack, C.D. & Rouse, B.T. Molecular adjuvants for mucosal immunity. *Immunol Rev* 2004, **199**, 100–112.

159. Murphy, A., Westwood, J.A., Teng, M.W., Moeller, M., Darcy, P.K. & Kershaw, M.H. Gene modification strategies to induce tumor immunity. *Immunity* 2005, **22**(4), 403–414.

160. Barouch, D.H., Letvin, N.L. & Seder, R.A. The role of cytokine DNAs as vaccine adjuvants for optimizing cellular immune responses. *Immunol Rev* 2004, **202**, 266–274.

161. Disis, M.L., Shiota, F.M., McNeel, D.G. & Knutson, K.L. Soluble cytokines can act as effective adjuvants in plasmid DNA vaccines targeting self tumor antigens. *Immunobiology* 2003, **207**(3), 179–186.

162. Moses, M.A., Brem, H. & Langer, R. Advancing the field of drug delivery: taking aim at cancer. *Cancer Cell* 2003, **4**(5), 337–341.

163. Lofthouse, S. Immunological aspects of controlled antigen delivery. *Adv Drug Deliv Rev* 2002, **54**(6), 863–870.

164. O'Hagan, D.T., Singh, M. & Ulmer, J.B. Microparticles for the delivery of DNA vaccines. *Immunol Rev* 2004, **199**, 191–200.

165. O'Hagan, D.T., MacKichan, M.L. & Singh, M. Recent developments in adjuvants for vaccines against infectious diseases. *Biomol Eng* 2001, **18**(3), 69–85.

166. O'Hagan, D.T. & Valiante, N.M. Recent advances in the discovery and delivery of vaccine adjuvants. *Nat Rev Drug Discov* 2003, **2**(9), 727–735.

167. Ten Hagen, T.L. Liposomal cytokines in the treatment of infectious diseases and cancer. *Methods Enzymol* 2005, **391**, 125–145.

168. Krup, O.C., Kroll, I., Bose, G. & Falkenberg, F.W. Cytokine depot formulations as adjuvants for tumor vaccines: I. Liposome-encapsulated IL-2 as a depot formulation. *J Immunother* 1999, **22**(6), 525–538.

169. Babai, I., Samira, S., Barenholz, Y., Zakay-Rones, Z. & Kedar, E. A novel influenza subunit vaccine composed of liposome-encapsulated haemagglutinin/neuraminidase and IL-2 or GM-CSF: I. Vaccine characterization and efficacy studies in mice. *Vaccine* 1999, **17**(9–10), 1223–1238.

170. Babai, I., Barenholz, Y., Zakay-Rones, Z. et al. A novel liposomal influenza vaccine (INFLUSOME-VAC) containing hemagglutinin-neuraminidase and IL-2 or GM-CSF induces protective anti-neuraminidase antibodies cross-reacting with a wide spectrum of influenza A viral strains. *Vaccine* 2001, **20**(3–4), 505–515.

171. Wales, J.R., Baird, M.A., Davies, N.M. & Buchan, G.S. Fusing subunit antigens to interleukin-2 and encapsulating them in liposomes improves their antigenicity but not their protective efficacy. *Vaccine* 2005, **23**(17–18), 2339–2341.

172. Golumbek, P.T., Azhari, R., Jaffee, E.M. et al. Controlled release, biodegradable cytokine depots: a new approach in cancer vaccine design. *Cancer Res* 1993, **53**(24), 5841–5844.

173. Egilmez, N.K., Jong, Y.S., Sabel, M.S., Jacob, J.S., Mathiowitz, E. & Bankert, R.B. In situ tumor vaccination with interleukin-12-encapsulated biodegradable microspheres: induction of tumor regression and potent antitumor immunity. *Cancer Res* 2000, **60**(14), 3832–3837.

174. Hill, H.C., Conway, T.F., Jr., Sabel, M.S. et al. Cancer immunotherapy with interleukin 12 and granulocyte-macrophage colony-stimulating factor-encapsulated microspheres: coinduction of innate and adaptive antitumor immunity and cure of disseminated disease. *Cancer Res* 2002, **62**(24), 7254–7263.

175. Sharma, A., Harper, C.M., Hammer, L., Nair, R.E., Mathiowitz, E. & Egilmez, N.K. Characterization of cytokine-encapsulated controlled-release microsphere adjuvants. *Cancer Biother Radiopharm* 2004, **19**(6), 764–769.

# 15

# POLYPHOSPHAZENES AS VACCINE ADJUVANTS

Alexander K. Andrianov

## 15.1 INTRODUCTION

Since the discovery of water-soluble polyphosphazenes as vaccine adjuvants in the early 1990s [1–3], much research has been carried out on their in vivo performance. The development of a lead compound, its advancement to clinical trials, and progress in the mechanistic studies were facilitated with the emergence of new synthetic approaches and the establishment of highly controlled production processes [4–7]. Dozens of scientific papers and patents have been dedicated to the field. It became also apparent that evolution of the technology through the discovery of new, more potent derivatives [8,9] and development of aqueous microencapsulation methods on the basis of such systems marks the advent of a new versatile technology platform. A comprehensive review of our current knowledge on polyphosphazene adjuvants, including discussion of the potential and challenges of this remarkable system, is now long overdue.

Polyphosphazenes are large synthetic macromolecules with an inorganic backbone consisting of alternating phosphorus and nitrogen atoms, which is "decorated" with covalently attached organic side groups (Scheme 15.1). The physicochemical and functional properties of such macromolecules vary dramatically depending on the nature of pendant groups [10]. Fluorinated elastomers, solid-state polyelectrolytes, hydrophobic biodegradable materials, water-soluble polyelectrolytes, liquid-crystalline polymers, and hydrogels are just few examples of polyphosphazene materials that can be constructed

*Vaccine Adjuvants and Delivery Systems*, Edited by Manmohan Singh
Copyright © 2007 John Wiley & Sons, Inc.

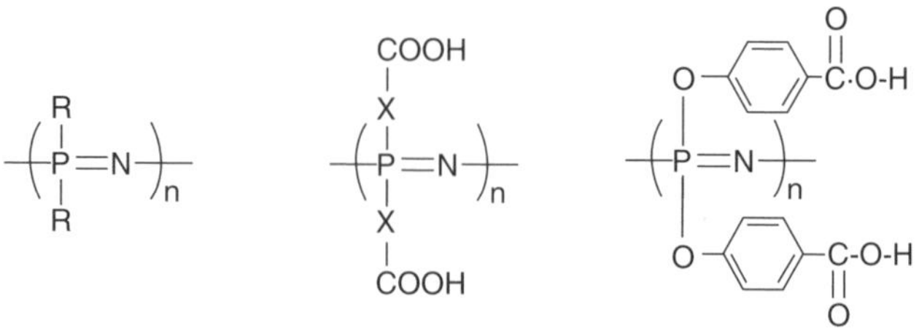

Scheme 15.1  Polyphosphazene structures.

depending on the choice of the side group. However, some of the unique properties, which are especially valuable for life sciences applications, stem from the characteristics of the phosphorus–nitrogen backbone itself. The ability of some phosphazenes to undergo hydrolytical breakdown, accompanied by the cleavage of side groups and release of phosphate and ammonium ions, is one of the most biologically important features. The rate of degradation can be modulated through changes in the structure of the pendant group. Phosphazene backbone is also one of the most flexible main chains currently known, which is a good starting point for the design of polymers for biological interactions and self-assembly. Once again, the flexibility of macromolecule can be regulated through the selection of bulkier or smaller side groups. High functional density, such as ionic density, can be achieved for polyphosphazenes since two side groups are connected to one monomeric unit. Finally, the unique macromolecular substitution route utilized in the construction of polyphosphazene structures, which is discussed below in greater detail, makes polyphosphazenes one of the largest and most structurally diverse classes of synthetic polymers. This versatility of the platform, combined with the advancement of high-throughput synthetic methods, opens unprecedented opportunities for the efficient discovery and rapid optimization of vaccine adjuvants and delivery systems.

All polyphosphazene immunoadjuvants synthesized to date are water-soluble polymers containing acidic groups (Scheme 15.1). Preparation of such polymers is carried out through multistep chemical reactions, and control of their molecular characteristics requires high degree of sophistication in synthetic and characterization methods. The lead polyphosphazene immunoadjuvant, poly[di(carboxylatophenoxy)phosphazene] (PCPP), is the most investigated representative of this class of phosphazene polymers.

In this chapter we summarize data on the in vivo performance of polyphosphazene adjuvants in the context of current knowledge of their chemistry, macromolecular architecture, and formulation behavior.

## 15.2 SYNTHESIS OF POLYPHOSPHAZENE ADJUVANTS

### 15.2.1 Synthetic Pathway and Recent Developments in Polyphosphazene Chemistry

In contrast to traditional polymer chemistry methods, synthesis of polyorganophosphazenes does not depend on the polymerization of structurally diverse monomers but relies on the chemical derivatization of a macromolecular precursor, polydichlorophosphazene (PDCP) [10]. This extremely reactive inorganic intermediate (Scheme 15.2) is substantially a "naked" phosphorus–nitrogen backbone trimmed with chlorine atoms. This is also the only compound that has to be synthesized by polymerization (Scheme 2). Once the intermediate is prepared, the diverse arsenal of organic chemistry can be mobilized to replace chlorine atoms with the desired organic side groups (Scheme 2). It is the high reactivity of P–Cl bonds that leads to the

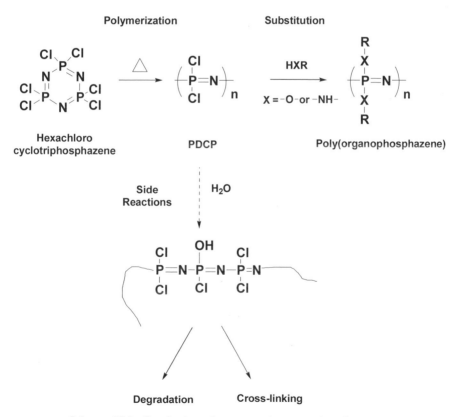

**Scheme 15.2** Synthetic pathway to polyorganophosphazenes.

unprecedented structural diversity of polyphosphazenes. However, this very advantage of polyphosphazene chemistry also constitutes its biggest challenge. Hydrolytic reactions occurring with chlorine atoms of the polymer during the synthetic course result in undesirable processes of polymer degradation and cross-linking in the presence of even trace amounts of water [5]. Both of these processes can cause problems with control of the polymer's molecular weight and reproducibility. Thus, command of hydrolytic reactions of chlorine atoms of PDCP is a key to a successful development of polymers for biological applications. Advancement of the lead polyphosphazene immunoadjuvant, PCPP, became possible due to significant improvements that occurred in the area of polyphosphazene production and characterization [4–7,11–15].

From a practical standpoint, especially in terms of achieving molecular-weight reproducibility and simplifying the reaction workup, a critical breakthrough was achieved with the discovery of the stabilizing effect on PDCP of diglyme. This addressed many of the issues listed above, eliminated the need for a frequent PDCP synthesis, and allowed production and storage of PDCP in large quantities in a ready-to-use form [5,11]. The stabilization technique also resulted in a highly reliable direct analysis of PDCP using light scattering/chromatographic methods and simplified PDCP conversion in the organo-substituted polymer [5,6,16]. In our experience, solutions of PDCP in diglyme remained stable for at least five years. By comparison, PDCP gelation in an anhydrous aprotic solvent such as tetrahydrofuian (THF) can occur within approximately one week. Although the exact mechanism of the stabilizing effect is not clear, it is speculated that it is due to the coordination and "inhibition" of water and intermediates with diglyme in the hydrolysis process [5]. This hypothesis can be supported by the well-known ability of oligo(ethylene oxides) to form complexes with water molecules.

Another aspect of polyphosphazene production includes the ability to modulate the molecular weight of polyphosphazene in the synthetic process. If necessary, molecular-weight control can be introduced in the polymerization reaction, since the degree of polymerization increases with the degree of conversion [5]. Another method involves "controlled degradation" of the polymer on the substitution reaction through careful control of the nucleophile/PDCP ratio [12].

Finally, new approaches emerged in the synthesis of polyphosphazene polyacids which involve the use of noncovalent methods of protection [7]. Polyphosphazene "sulfonation" can be conducted in a single-step reaction by the direct replacement of the chlorine atoms in PDCP with a sulfonic acid–containing nucleophile: hydroxybenzene sulfonic acid. The method makes use of "noncovalent" protection of sulfonic acid functionality with hydrophobic dimethyldipalmitylammonium ion, which then can easily be removed after completion of the reaction. No noticeable degradation of polyphosphazene backbone was observed under the conditions of the synthesis.

### 15.2.2   Production Process

Polyphosphazene production includes two main stages: polymerization and macromolecular substitution (Scheme 15.2) [4,5]. Polymerization is a common step for all polyphosphazenes that are of interest as vaccine adjuvants, so there is no need to introduce any changes in this process when a new polyphosphazene derivative goes into the development and production phase.

The polymerization procedure, developed for cGMP production of PCPP, includes melting of hexachlorocyclotriphosphazene in the reactor under a nitrogen blanket, purging the reaction content with anhydrous nitrogen to remove traces of moisture or hydrochloric acid, and performing the reaction at 235°C with intensive stirring in a titanium reactor [5]. Polymerization takes typically about 10 hours to complete and is monitored by measuring changes in the electrical current drawn by the stirring motor, which at a constant stirring rate correlates to the viscosity of the reaction mixture. Thus, a kinetic curve is produced and the reaction is terminated at a desired viscosity level to achieve consistency in the degree of conversion and the molecular weight.

A macromolecular substitution stage leading to PCPP and other polyacids involves conversion of polymer with hydrolytically labile P–Cl bonds (PDCP) into polymers designed for aqueous environment. In the case of PCPP the process involves (1) substitution of chlorine atoms with nucleophile containing an ester of hydroxybenzoic acid and (2) hydrolysis of the ester function under alkaline conditions (deprotection reaction). Reaction pathways in PCPP synthesis, as well as structures of potential by-products resulting from incomplete substitution and deprotection reactions are shown in Scheme 15.3.

To improve synthetic control, a single pot–single solvent approach was developed. The method utilizes diglyme as a solvent, a forced substitution reaction, an aqueous-based deprotection reaction, commercial sodium propyl paraben as a nucleophilic reagent, short reaction times, and purification based on salt precipitation of PCPP [4]. A typical production lot generates approximately 2 kg of a purified PCPP.

### 15.2.3   Polymer Characterization. Potential Contaminants and Structural Irregularities

There are clear indications that the biological performance of polyphosphazenes can be greatly affected by their molecular weight and the potential presence of structural irregularities [4,17]. Thus, it was important to synthesize polymers of various molecular weights as well as macromolecules containing "structural defects" and to investigate the effect of such possible reaction by-products on polymer performance. Potential side reactions are shown in Scheme 15.3, and mixed substituent copolymers of PCPP containing "structural defects," propyl ester functionalities and hydroxyl groups, were produced using variations in the reaction conditions [4]. The influence of such molecules

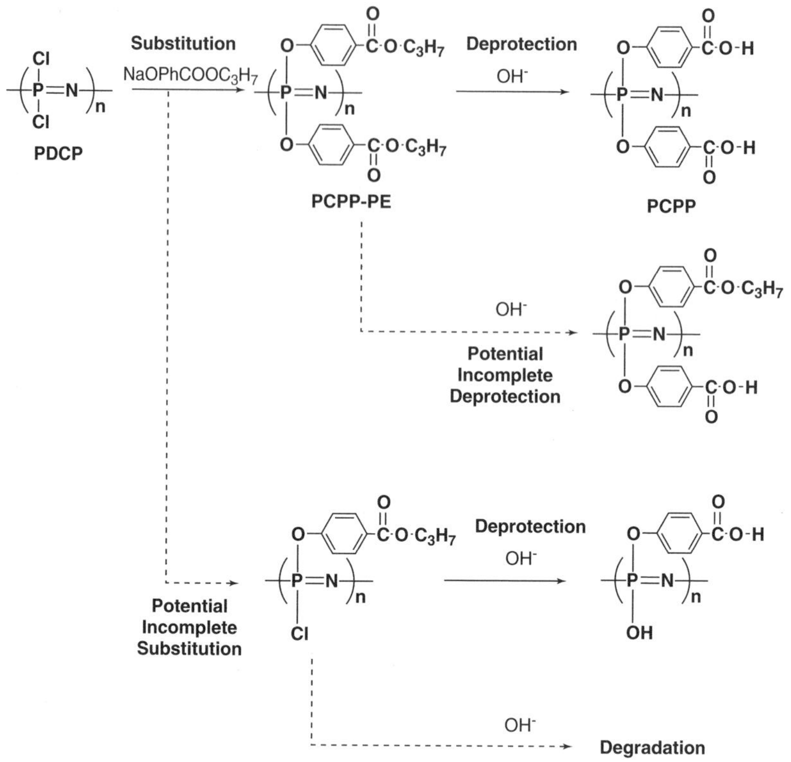

**Scheme 15.3** Synthesis of PCPP.

on polymer degradation characteristics, microencapsulating properties, and biological activity is discussed below.

Monitoring of molecular-weight changes during the reaction was performed by size-exclusion high-performance liquid chromatography (HPLC) and light-scattering analysis of polyphosphazene samples before and after macromolecular substitution [4–6,16]. Reproducibility of PCPP synthesis was confirmed for more than 100 samples with a production scale ranging from 200 mg to 2 kg. For PCPP prepared using the synthetic method described above, weight-average molecular-weight fluctuations were not in excess of 10%. Polymers with various molecular weights were synthesized and fractionated using preparative HPLC, and their in vivo performance was evaluated.

Comprehensive polymer characterization was performed. PCPP structure and purity was evaluated using $^1$H, $^{31}$P, $^{13}$C NMR, Fourier-transformed infrared spectrophotometry (FTIR), elemental analysis, multiangle laser light scattering coupled with size-exclusion chromatography, viscometry, Karl–Fisher titration, atomic absorption spectrometry, and inductively coupled plasma mass spectroscopy.

## 15.3  ADJUVANT ACTIVITY OF PCPP: A LEAD COMPOUND

### 15.3.1  Overview of PCPP Performance in Animal Studies

PCPP is recognized as the most investigated polyphosphazene immunoadjuvant. It is estimated that PCPP was tested with 10 bacterial and 13 viral antigens in 11 animal models. PCPP proved to be a potent adjuvant for multiple vaccines and antigens, such as trivalent influenza virus vaccine, hepatitis B surface antigen, herpes simplex virus, glycoprotein gD2, tetanus toxoid, polyribosylribitolphosphate from *Haemophilus influenzae* type b, toxin co-regulated pilin TcpA4 and TcpA6 synthetic peptides, inactivated rotavirus particles, formalin-inactivated HIV-1 LAI virus [3,17–25]. Taken together, these results clearly demonstrate the breadth of PCPP utility as an adjuvant.

One of the most comprehensive studies of PCPP as an adjuvant was conducted in mice for an influenza vaccine using both commercial trivalent influenza vaccine and formalin-inactivated influenza virions [17]. PCPP was found to be very potent adjuvant with the mouse effective dose of 100 μg when administered by either a subcutaneous or intramuscular route. The results of immunization with formalin-inactivated X-31 influenza particle suspension formulated with PCPP showed at least 10-fold increases in IgG and IgM titers compared to the antigen alone 3 weeks postimmunization, and these levels were maintained for the length of the experiment, 21 weeks. PCPP formulations also induced functional antibodies as assayed by hemagglutination inhibition (HAI) that were also approximately 10-fold higher than the levels detected for the antigen alone (Figure 15.1).

Very similar results were obtained for commercial vaccine containing three strains: A/Texas/36/91 (H1N1), A/Shangdong/9/93 (H3N2), and B/Panama/45/90 [17]. Addition of PCPP to the vaccine dramatically enhanced the immune response to all three influenza HA strains (Figure 15.2). Both vaccines, with and without PCPP, elicited IgG1 and IgG2a responses, the IgG1 being the dominant isotype. In summary, PCPP enhanced the IgM, IgG, and IgG1 antibody titers to influenza antigens at least 10-fold compared to the commercial vaccine alone. The IgG2a isotype titers were enhanced only about twofold. PCPP also increased the HAI antibody response 10-fold over the levels elicited by the vaccine alone.

The effect of PCPP on the immunogenicity of hepatitis B surface (HBs) antigen was studied in CD-1 mice [19]. Antibody titers for plasma-derived antigen formulated with 100 μg of PCPP were approximately 20 times higher than for the same antigen formulated on aluminum phosphate. A recombinant-expressed HBs antigen–PCPP formulation showed antibody titers that were seven times higher than that of aluminum hydroxide–formulated recombinant antigen. In both cases, maximum antibody titers for the PCPP-formulated immunogens were reached by 12 weeks and were maintained to 41 weeks.

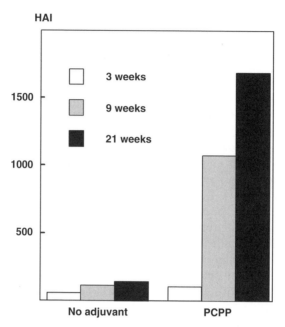

**Figure 15.1** HAI immune response kinetics after subcutaneous immunization of BALB/c mice with purified formalin inactivated X-31 influenza virus formulated with and without PCPP (five mice per group; 5 μg of antigen; 100 μl injection volume; 100 μg of PCPP; significance testing $p = 0.004$). (Reproduced with permission from Ref. 17.)

PCPP was demonstrated to be a potent adjuvant not only for particulate antigens such as influenza and HBs, but also with water-soluble antigens. The effect of PCPP on water-soluble glycoprotein was examined using recombinant expressed herpes simplex virus (HSV) type 2 glycoprotein (gD2) [19]. Addition of PCPP to this weak antigen resulted in a dramatic (up to 100-fold) increase in ELISA antibody titers, with high titers maintained for at least 15 weeks following immunization. PCPP also increased the seroconversion rate from one out of five for 2 μg of gD2 alone to five out of five for the same dose formulated with PCPP.

A comparison of the adjuvant activity of PCPP and aluminum phosphate was carried out using a capsular polysaccharide, polyribosylribitolphosphate (PRP), from *Haemophilus influenza* type B (Hib) conjugated to tetanus toxoid (Hib-T). PCPP adjuvanted Hib-T antigen elicited sixfold-higher anti-PRP antibody levels at four weeks than were seen with aluminum phosphate [19]. Peak titers for PCPP formulations were achieved at week 8, when they were 10-fold higher than those achieved with aluminum phosphate. Although the response for PCPP-adjuvanted Hib-T antigen decreased over the following weeks, at week 20 they were still twofold higher than were the maximum titers elicited by the aluminum phosphate formulation.

HAI

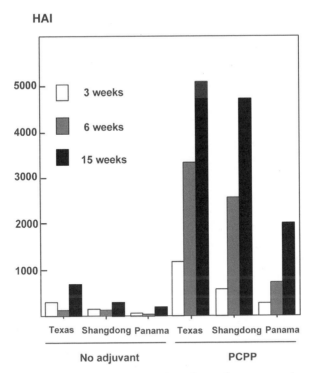

**Figure 15.2** HAI immune response kinetics after subcutaneous immunization of BALB/c mice with multivalent influenza vaccine formulated with and without PCPP [five mice per group; trivalent influenza vaccine: A/Texas/36/91 (H1N1), A Shangdong/9/93 (H3N2), and B/Panama/45/90; 15 μg of HA for each strain; 500 μl injection volume; 100 μg of PCPP; significance testing at week 15: Texas $p = 0.0001$, Shangdong $p = 0.0000001$, Panama $p = 0.00012$]. (Reproduced with permission from Ref. 17.)

PCPP was also studied as an adjuvant for synthetic peptid–toxin co-regulated pilin TcpA4 and TcpA6, which can be of interest for the development of a subunit vaccine for cholera [23]. Because of their small size, peptide-based antigens are not likely to elicit robust stimulation of the immune system and thus require an effective adjuvant. PCPP–peptide formulations were used to immunize adult female CD-1 mice subcutaneously three times at four-week intervals. Serum samples were taken four weeks after each immunization. Another polymeric adjuvant, CRL-1005, a block copolymer of ethylene oxide–propylene oxide, was also used in these studies, and two adjuvants were compared. PCPP demonstrated high immunoadjuvant activity for both peptides; the increase in anti-TcpA titers after two immunizations was approximately 100-fold compared to that of peptides alone. The differences were, however, less pronounced after the third vaccination. The response was characterized by a fast kinetic profile—TcpA4-PCPP was the only formulation capable of inducing significantly higher titers than those of peptide alone after just one immunization. PCPP–peptide formulations induced a broad spectrum of IgG subclasses (IgG1, IgG2a, IgG2b, and IgG3), with IgG1 and IgG2a being predominant, suggesting activation of both Th1- and Th2-type helper cells and lower levels of IgA. PCPP was also efficient in affording protection in infant mice born to immunized mothers against both 1 $LD_{50}$ and 10 $LD_{50}$ challenges. Interestingly, for TcpA6-PCPP formulations, the protection levels were at 100% against 1 $LD_{50}$, and 75% against 10 $LD_{50}$, whereas for CRL-1005 the values were at 75% and 33%, respectively.

The effects of PCPP and other adjuvants (QS-21, QS-7, Quil A, and Ribi adjuvant system) were compared in the studies of intramuscular immunization of mice with inactivated rotavirus particles [24]. PCPP stimulated significantly higher titers than other adjuvants for all rotavirus antibodies measured except stool IgA. Twenty-eight days following immunization with 20 mg of purified ultraviolet/psoralen-inactivated murine rotavirus (EDIM), both with and without adjuvant, BALB/c mice were orally challenged with live EDIM, and virus shedding was measured. All five adjuvants stimulated large ($p < 0.001$) increases in rotavirus antibody, but significant differences were found between adjuvants. The order of rotavirus IgG responses was the following: no adjuvant < RAS < QS-7 < Quil A < QS-21 < PCPP, and it was the same as the order of protection except that QS-21 and PCPP were reversed [24].

A small study of PCPP in primates was reported [25]. Two macaques were immunized with formalin-inactivated HIV-1 LAI virus formulated with PCPP, while the third animal received the inactivated HIV-1 LAI virus alone. After the first immunization only monkeys that received inactivated virus formulated with PCPP developed serum antibodies against HIV-1 gp120 as measured by ELISA and Western blot analysis. Secondary immunization with recombinant HIV-1 HXBc2 gp120 formulated with PCPP was conducted at week 43. Both animals that received HIV-1 LAI–PCPP in the first immunization showed titers that were significantly higher than those seen in the other monkey. All the vaccinees were then challenged intravenously with SHIV (simian/human immunodeficiency virus, chimeric virus). The animal that did not receive PCPP in the first immunization became infected. The other two monkeys remained virus isolation negative throughout 24 weeks of the experiment. Thus, vaccination of two rhesus monkeys with whole inactivated HIV-1 adjuvanted with PCPP protected the animals from becoming infected by a SHIV challenge.

### 15.3.2  Effect of the Molecular Characteristics of PCPP on the In Vivo Activity

Macromolecular characteristics of polyphosphazenes, such as their molecular weights, can affect adjuvant activity. Samples of PCPP were prepared of various molecular weights and carefully fractionated to achieve narrow polydispersity [17]. Figure 15.3 shows that the molecular weight of the polymer has a significant influence on the immune response. HAI titers to Texas strain of the tri-

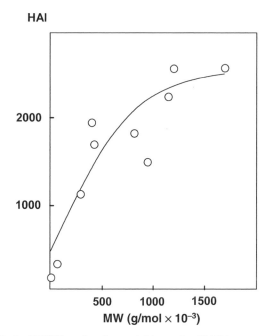

**Figure 15.3** Effect of PCPP molecular weights on the HAI immune response to multivalent influenza vaccine in mice [responses against Texas are shown; vaccine and doses are the same as in Figure 15.2; significance of testing compared to vaccine alone: 58,000 g/mol $p = 0.105$ (not significant) and all other MWs were at least $p < 0.002$]. (Reproduced with permission from Ref. 17.)

valent influenza vaccine raise as the molecular weight of PCPP increases before reaching a plateau at approximately 1,000,000 g/mol. These results demonstrate that the potency of PCPP can be maximized through modulation of the molecular-weight characteristics and emphasizes the need for careful control in the production process.

Incompletely deprotected PCPP polymers (Scheme 15.3) are potential synthetic PCPP by-products which can easily be detected even at low content using structural analysis methods such as $^1$H NMR and FTIR. They also present an interesting opportunity for the investigation of the effect of ionic density and hydrophobicity on polymer performance. The in vivo activity of such mixed substituent copolymers and PCPP was evaluated using influenza antigen in mice [4]. Groups of mice were subcutaneously immunized with antigen, detergent split X-31 influenza virus particle suspension, mixed with aqueous solutions of polymers. The immune responses (i.e., IgG titers) induced by these formulations are presented in Figure 15.4. All synthesized polymers enhanced the antibody response compared to the levels elicited by the antigen alone. It is evident from the figure that the content of carboxylic acid groups determines the activity of the polymer as an immunoadjuvant. Reduction in

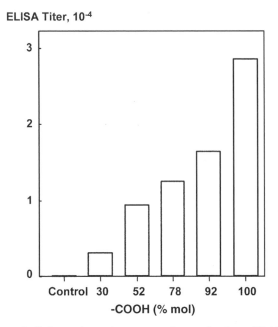

**Figure 15.4** Serum IgG titers after subcutaneous immunization of BALB/C mice with Split X-31 influenza formulated with PCPP copolymers containing variable amounts of carboxylic acid groups (for incompletely deprotected polymers, see Scheme 3; week 3 postimmunization data; antigen dose, 5 μg; control, Split X-31 influenza; all formulations are based on phosphate-buffered saline, pH 7.4). (Reproduced with permission from *Biomacromolecules*, 2004, Vol. 5, pp. 1999–2006. Copyright © 2004 American Chemical Society.)

the content of acid groups in the polymer gradually decreases the immune response.

## 15.4  PCPP IN CLINICAL TRIALS

Publications on the results of clinical trials involving PCPP remain scarce. A phase I clinical study using PCPP was carried out with influenza vaccine on both young and elderly adults (total of 96 subjects) [26]. Three doses of PCPP were tested (100, 200, and 500 μg) and nonadjuvanted vaccine was used as a control. No serious adverse events related to the vaccine were reported. The best results were obtained for A/Johannesburg/33/94 (H3N2) strain with the a 500-μg dose of PCPP found to be most efficient. For this dose, PCPP-adjuvanted vaccine produced a 14.7-fold increase in antibody titers (day 21 versus day 0 postimmunization), compared to a 3.1-fold increase for nonadjuvanted vaccine. A seroconversion rate was found to be 80% for the PCPP formulation as opposed to 51% for nonadjuvanted vaccine. PCPP was reportedly also used in clinical trials on HIV vaccine and canarypox vCP300 rgp160 (MN/LA12) [27], but no results have yet been published.

## 15.5 HYDROLIC DEGRADATION OF POLYPHOSPHAZENE POLYACIDS

Degradation of biologically active polymers is an important property defining their performance and storage conditions. PCPP is reported to undergo slow hydrolytical degradation in aqueous solutions (Figure 15.5) [4,28–30]. The degradation rate can be affected significantly by the presence of residual chlorine atoms and hydroxyl groups, irregularities produced if polyphosphazenes are incompletely substituted (Scheme 15.3). As shown in the figure, the molecular weight loss of incompletely substituted sample was significantly faster than that of the properly substituted sample and was characterized by a well-pronounced initial rapid degradation phase [4]. Careful control of the residual groups—"structural defects" or "weak links" in polyphosphazene structures—can open a pathway to the tailoring of polymer degradation characteristics.

Degradation profiles of PCPP can be modulated effectively through the introduction of hydrolytically labile side groups in the polymer structure. It has been demonstrated that mixed substituent copolymers of PCPP containing *N*-ethylpyrrolidone [29] or ethyl glycine side groups [30] degrade faster than PCPP and that the rate of hydrolysis is a function of the copolymer composition.

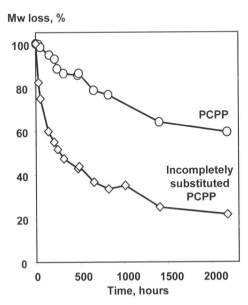

**Figure 15.5**   Molecular weight loss of PCPP and incompletely substituted PCPP versus degradation time (aqueous PBS pH 7.4; 55°C; weight-average molecular weight shown; incompletely substituted polymer (Scheme 3) was synthesized with 2% mol. deficiency of nucleophile). (Reproduced with permission from *Biomacromolecules*, 2004, Vol. 5, pp. 1999–2006. Copyright © 2004 American Chemical Society.)

## 15.6 FORMULATION ASPECTS AND NOTES ON THE MECHANISM OF ACTION

A number of observations have been reported that can be of aid in elucidating the mechanism of action of polyphosphazene immunoadjuvants. Many of them indicate the importance of interactions between the polymer and the antigen. Direct evidence of noncovalent complex formation between polyphosphazene and the antigen was reported [18] and are discussed below in more detail. The induction of highly functional antibody titers demonstrated with influenza vaccine suggests that PCPP is capable of forming very gentle interactions with a biological molecule, which preserves the antigenic integrity [17]. The relationship between the molecular weight of the polymer and adjuvant activity [17] can be explained potentially in terms of more stable complexes that larger macromolecules can form. The fact that PCPP moves out of the site of injection within 24 hours of postimmunization suggests maintenance of the water-soluble nature of the complex [17]. Long-lived IgM response can be interpreted as evidence for the persistence of the antigen [17]. The fact that the excision of the injection site has no detectable effect on the kinetics of antibody induction suggests that PCPP does not act as a depot [17]. However, the mechanism of action and role of polyphosphazene–antigen interactions in these systems largely remain open to discussion.

Previous research has shown that synthetic polyelectrolytes conjugated covalently to antigenic proteins act as vaccine adjuvants [31]. A common mechanism of action has been suggested for such systems. It proposes that the polymer–antigen conjugate targets the cell surface receptor of the lymphocyte and then adsorbs on the cell surface, where polyelectrolyte clusters on membrane proteins [31]. This stimulates intracellular ionic fluxes, activates ATPase, and eventually results in an enhanced immune response.

Multivalent receptor clustering can also be an important part of the biological activity [32]. A critical parameter in the induction of a significant in vivo effect is the existence of a strong, typically covalent linkage between the antigen and the polyelectrolyte so that the conjugate is targeted and all components are presented to B lymphocyte simultaneously (Scheme 15.4). Nonconjugated polyelectrolytes are usually very weak adjuvants.

It is clear that no covalent polymer attachment of PCPP to the antigen is required to display a potent immune response, and PCPP has been shown to be a superior adjuvant to nonpolyphosphazene polyelectrolytes such as polymethylacrylic and polyacrylic acids [19].

To understand the fundamental basis for the activity of PCPP, its interactions with a model protein antigen, bovine serum albumin (BSA), were investigated in aqueous solution using light scattering/aqueous size-exclusion HPLC [18]. Based on the decrease in BSA peak area and changes in the light-scattering profiles, the complex formation was proven. "Binding maps" showing the composition of the complexes and the efficiency of protein binding were obtained. This phenomenon was not unique for PCPP–BSA formulation and

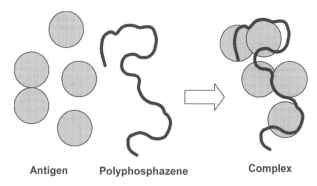

Antigen       Polyphosphazene          Complex

**Scheme 15.4**  Polyphosphazene–antigen complex formation.

similar binding maps were established for other proteins [18]. The number of protein molecules carried varied broadly in the range 1 to 140 per polyphosphazene molecule, depending on the protein, the polyphosphazene, and the conditions of the experiment. The results suggest that simple mixing of PCPP with BSA produces water-soluble protein–polyphosphazene macromolecular complexes, which can carry a significant "load" of "biologically available" protein ligands by means of noncovalent interactions.

An attempt to establish a correlation between the in vitro and in vivo behavior of such a system was carried out [18]. Three formulations with various protein/polymer ratios were investigated at a fixed polymer concentration and various protein loadings. IgG titers induced by these formulations were measured four weeks after immunization in mice. Figure 15.6 shows the results for PCPP–BSA formulations (white bars) and corresponding BSA formulations to which no PCPP was added (shaded bars). The amounts of BSA bound to the polymer and the weight-average root-mean-square radii of the complexes are also shown in the figure.

As expected, for formulations containing BSA alone, an increase in antibody titers was observed with increasing dose (Figure 15.6, shaded bars). This trend was not the case for BSA–PCPP formulations in the same protein dose range (Figure 15.6, white bars). The change in BSA dose from $2\,\mu g$ to $10\,\mu g$ in polymer formulations produced higher antibody titers. However, further increase in the dose to $50\,\mu g$ did not result in enhancement of the immune response. Interestingly, the latter increase in protein loading was accompanied by significant compaction of the complex, manifested in the reduction of its root-mean-square radius (Figure 15.6, curve 2). Such compaction caused by BSA can leave some protein molecules "trapped" in the polymer and might make them unavailable for interaction with the appropriate biological receptor. These results suggest the importance of complex conformation and its ability to present the antigenic molecule.

Thus, it appears that both the amount of bound protein (protein loading) and effective "protein presentation" play critical roles in the in vivo perfor-

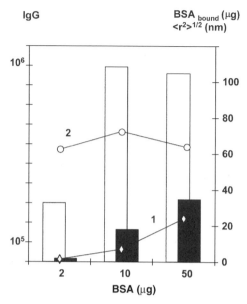

**Figure 15.6** Serum IgG titers after immunization of mice with BSA (shaded bars) and BSA formulated with PCPP (white bars) versus BSA dose. Overlaid is the amount of BSA bound to PCPP (curve 1) and root mean square radius of the complex (curve 2) versus BSA dose (5 BALB/c mice per group, 50 μg PCPP/mouse, single-dose subcutaneous injection, four-week data). (Reproduced with permission from *Biomacromolecules*, 2005, Vol. 6, pp. 1375–1379. Copyright © 2005 American Chemical Society.)

mance of the complex. The results also suggest that the remarkably flexible backbone (conformational adaptability) of the polyphosphazene carrier plays an important role in the presentation of the protein and the biological activity of this polymer.

## 15.7 NEW POLYPHOSPHAZENE IMMUNOADJUVANTS AND HIGH-THROUGHPUT METHODS FOR THEIR DISCOVERY

After the discovery of immunoadjuvant activity of PCPP, there have been efforts to synthesize polyphosphazene polyelectrolytes similar to PCPP, with the goal of understanding the effect of polymer structure on its in vivo activity. In earlier studies, mixed substituent copolymers of PCPP were synthesized containing various amounts of hydrophilic methoxyethoxyethoxy side groups [8,33,34]. The in vivo activity of such polymers, evaluated in mice using X-31 influenza antigen, surpassed that of PCPP. Serum IgG titers for the most active copolymer were increased almost 10 times and HAI titers approximately five times compared to those elicited by PCPP formulation. It has also been shown that the activity can be controlled through variations in copolymer composition. Interestingly, a polyphosphazene homopolymer containing methoxyethoxyethoxy side groups only did not display any immunostimulating activity.

Recently, a newly developed high-throughput methods of polyphosphazene synthesis were used in the construction of polyphosphazene polyelectrolytes of approximately 40 members [35,36]. New polymers included mixed substituent copolymers of PCPP as well as some new derivatives. Some representative structures are shown in Scheme 15.5.

Some synthesized polymers have already shown high potency as immunoadjuvants. Experiments in mice using HBs antigen formulated with some of the new polymers demonstrated that at least two of them dramatically surpass the activity of PCPP with this particular antigen (Figure 15.7). There is a clear

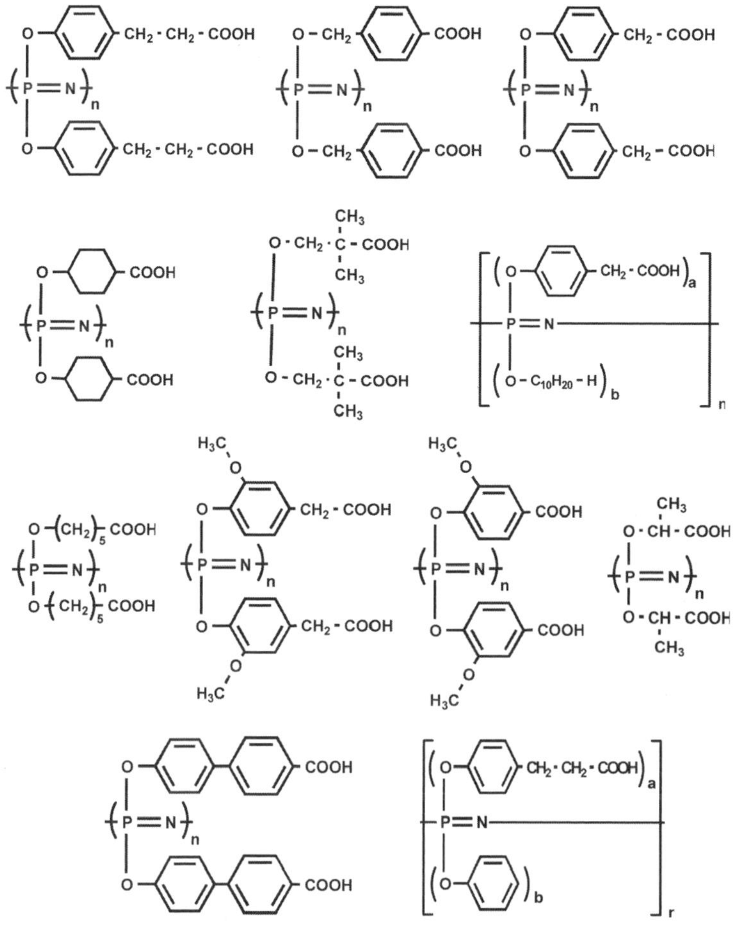

**Scheme 15.5**  New polyphosphazene polyelectrolytes synthesized for immunoadjuvant applications.

IgG Response (GMT) x10$^{-6}$

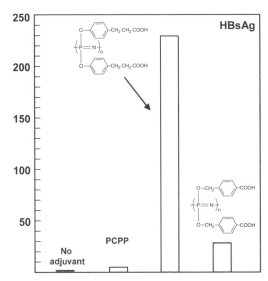

**Figure 15.7** IgG titers after immunization of mice with HBsAg formulated with PCPP, HBsAg formulated with new polymer derivatives, and HBsAg alone as a control (five BALB/c mice per group; 1 µg HBsAg/mouse; 50 µg polymers/mouse; single-dose intramuscular injection; 16-week data).

need in further investigation of new polyphosphazene derivatives. Establishment of the structure–activity relationship is one of the most important and desirable goals, which can, potentially, lead to the discovery of "superpotent" immunoadjuvants and the elucidation of their mechanism of action.

## 15.8  MICROENCAPSULATION SYSTEMS BASED ON POLYPHOSPHAZENE ADJUVANTS

Formulation of water-soluble immunoadjuvants into microspheric or nanospheric forms can extend their utility in the areas of mucosal immunization and controlled vaccine release [37–47]. The design of antigen-loaded microspheres, in which immunoadjuvant serves as the only "wall material," is also attractive because it can allow sustained release of antigen and adjuvant at the same time. Water-soluble polyphosphazene immunoadjuvants also support aqueous-based processes, which are especially appealing because of mild conditions for protein encapsulation.

Three main approaches have been developed for microsphere preparation using aqueous solutions of polyphosphazene immunoadjuvants (Scheme 15.6). All of these methods are built using the polyelectrolytic nature of the adjuvants and their ability to participate in the reactions of ionic complexation.

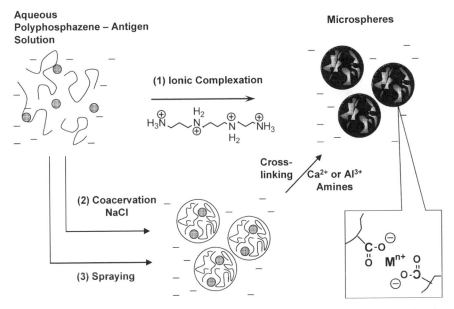

**Scheme 15.6** Microsphere preparation methods using aqueous solutions of poly-phosphazene immunoadjuvants.

Earlier microencapsulation techniques involved spraying protein–polyphosphazene mixtures using air-atomization nozzles into aqueous calcium chloride [37,38]. They required sophisticated spraying tools, containment equipment to prevent antigen's distribution in the air, and allowed only limited control over the microsphere size.

Some of these problems were addressed when a method utilizing aqueous coacervation, an alternative way to make microdroplets, was suggested (Scheme 15.6) [39–41]. In the first step of the process, microdroplets were formed through the addition of sodium chloride to a solution of polyphosphazene polyelectrolyte (coacervation), and in the second step they were cross-linked by the introduction of calcium chloride in the system. The resulting ionically cross-linked microspheres can still show some sensitivity to the salts of monovalent cations such as sodium chloride. The stability of microspheres can be improved through the treatment of microspheres with polycations, such as poly(L-lysine), but this method often leads to the formation of large aggregates or precipitation [38–42].

A single-step process involving ionic complexation of polyphosphazene adjuvants with physiologically benign organic amines such as spermine and spermidine has been developed [42–44]. This approach combines the simplicity of the coacervation approach with the stability of multivalent interactions between the organic compounds. Unlike coacervation, the spermine complexation method is applicable to a large number of polyphosphazenes, including those with carboxyl acid groups and with sulfonic acid functionalities. Recently, use of this method was reported for some water-soluble fluorinated poly-electrolytes [43,44].

Both coacervation and spermine complexation processes are simple, scal-able, and do not require complex manufacturing equipment, elevated tem-peratures, and organic solvents. The reaction time for the production of microspheres typically is in the range 2 to 60 minutes. Microsphere size can be controlled effectively through reaction parameters such as reaction time, concentration of polymer, and coacervating/cross-linking agent. Typically, microspheres can be prepared in a broad range of sizes (1 to 1000 µm), and the spermine complexation method also allows preparation of nanospheres between 200 and 1000 nm. Microencapsulation methods were used to encap-sulate model proteins, such as BSA and antigens, such as formalin-inactivated influenza virus particles with high loadings and high efficiency of encapsula-tion achieved (up to 76% for BSA and 94% for influenza) [40,42]. Typically, protein's escape from ionically cross-linked hydrogel polyphosphazene micro-spheres can be achieved through one of the following pathways: diffusion, degradation of polyphosphazene backbone, or dissolution of the polymer as a result of the cross-linker removal via ion-exchange reactions [42,45–47]. Protein release profiles were also investigated, and it was found that they can be modulated through the parameters of encapsulation, type of the cross-linker, and additional coating of microspheres with polycations.

In vivo experiments were conducted in which calcium cross-linked micro-spheres containing whole formalin-inactivated influenza virus particles were delivered by a parenteral route [22,47]. Polyphosphazene microspheres dem-onstrated antibody titers which at week 5 were approximately 30 times higher than that of the antigen alone and 15 times higher than that of antigen encap-sulated in microspheres constructed of calcium cross-linked alginate. PCPP microspheres were also the only test subject in the experiment that produced high levels of functional antibodies, as measured by HAI and neutralization assays. Thus, polyphosphazene microspheres display high immunoadjuvant activity, which dramatically exceeds the activity of microspheres based on alginic acid, a similar hydrogel-forming material.

## 15.9  CONCLUSIONS

A polyphosphazene platform offers unprecedented opportunities for the development of new vaccine adjuvants. Structural diversity of polyphospha-zenes, flexibility in tailoring of their physicochemical properties, high potency of already synthesized compounds, and the existence of high-throughput dis-covery methods differentiate this synthetic class from other compounds uti-lized in the field. The ability to conform immunoadjuvants into microsphere formulations allows modulation of the delivery properties and controlled release of both the antigen and the adjuvant.

It is apparent that one of the most important properties of polyphosphazenes in regard to their biological activity is their ability to form noncovalent water-soluble complexes with proteins. In such complexes, polyphosphazenes play an antigen-transporting role and can carry, stabilize, and protect a significant number of ligands. Structural versatility of polyphosphazenes along with the remarkable flexibility of their backbone provides an ideal interactive polymer system which allows tailoring of macromolecular complex with the desired characteristics, such as protein loading and stability. Even the most complex potential situations, involving formation of compact complex structures and water-insoluble products, can, potentially, be resolved through careful formulation development.

Development of thorough control over polymer characteristics and the production process was critical in the establishment of the technology. Although the fundamental characteristics of synthetic macromolecules, such as their polydispersity, supermolecular architecture, potential structural irregularities, and instability of the synthetic precursor, presented challenges in the development, they are now sufficiently controlled to allow simple synthesis and reproducible properties of the compounds.

One of the most desirable goals for researchers in the field is the creation of a "universal" polyphosphazene adjuvant capable of working with all antigens. How achievable this objective will be is not yet known. Based on the data obtained for the most investigated polyphosphazene adjuvant, PCPP, the breadth of its utility is significant. However, restrictions can be also foreseen from our current understanding of the physicochemical aspects of the formulation. In any case, a trade-off can be anticipated in the potency of polyphosphazene with a specific antigen and the breadth of its applications. Despite the fact that PCPP underperformed with specific antigens compared to other polymers, its utility as an adjuvant with a broad spectrum of applications can still be of great value.

Although the results of the in vivo performance of the lead compound, PCPP, and some of the newly synthesized derivatives are impressive, it is possible that the potency of the compounds yet to be discovered could be even greater. The future development of the platform can be facilitated with the establishment of a broad structure–activity relationship and in-depth mechanistic studies.

## REFERENCES

1. Andrianov, A.K., Jenkins, S.A., Payne, L.G. & Roberts, B.E. *Phosphazene Polyelectrolytes as Immunoadjuvants*, U.S. Patent 5,494,673, 1996.

2. Andrianov, A.K., Jenkins, S.A., Payne, L.G. & Roberts, B.E. *Hydrogel Microencapsulated Vaccines*, U.S. Patent 5,529,777, 1996.

3. Payne, L.G., Jenkins, S.A., Andrianov, A., Langer, R. & Roberts, B.E. Xenobiotic polymers as vaccine vehicles. *Adv Exp Med Biol* 1995, **371B**, 1475–1480.

4. Andrianov, A.K., Svirkin, Y.Y. & LeGolvan, M.P. Synthesis and biologically relevant properties of polyphosphazene polyacids. *Biomacromolecules* 2004, **5**(5), 1999–2006.

5. Andrianov, A., Chen, J. & LeGolvan, M.P. Poly(dichlorophosphazene) as a precursor for biologically active polyphosphazenes: synthesis, characterization, and stabilization. *Macromolecules* 2004, **37**(2), 414–420.

6. Andrianov, A. & LeGolvan, M.P. Characterization of poly[di(carboxylatophenoxy )phosphazene] by an aqueous gel permeation chromatography. *J Appl Polym Sci* 1996, **60**, 2289–2295.

7. Andrianov, A., Marin, A., Chen, J., Sargent, J. & Corbett, N. Novel route to sulfonated polyphosphazenes: single-step synthesis using "non-covalent protection" of sulfonic acid functionality. *Macromolecules* 2004, **37**(11), 4075–4080.

8. Andrianov, A.K., Sargent, J.R., Sule, S.S., LeGolvan, M.P., Woods, A.L., Jenkins, S.A. et al. Synthesis, physico-chemical properties and immunoadjuvant activity of water-soluble phosphazene polyacids. *J Bioact Compat Mater* 1998, **13**, 243–256.

9. Andrianov, A.K. Design and synthesis of functionalized polyphosphazens with immune modulating activity. *Am Chem Soc PMSE Prepr* 2003, **88**.

10. Allcock, H.R. *Chemistry and Applications of Polyphosphazenes*, Wiley, Hoboken, NJ, 2003.

11. Andrianov, A.K., Sargent, J.R., Sule, S.S. & LeGolvan, M.P. *Polyhalophosphazene Solutions Stable Against Gelation*, U.S. Patent 5,707,597, 1998.

12. Andrianov, A.K., LeGolvan, M.P., Svirkin, Y.Y. & Sule, S.S. *Production of Polyoranophosphazenes Having Desirable Molecular Weight*, U.S. Patent 5,869,016, 1999.

13. Andrianov, A.K., LeGolvan, M.P., Svirkin, Y.Y. & Sule, S.S. *Purification of Polyphosphazene Polyacids*, U.S. Patent 5,842,471, 1998.

14. Andrianov, A.K. & Sargent, J.R. *Production of Polyorganophosphazenes*, U.S. Patent 5,760,271, 1998.

15. Andrianov, A.K., Sargent, J.R. & Sule, S.S. *Recovery of Polyphosphazne Polyacids or Acid Salts Thereof*, U.S. Patent 5,814,704, 1998.

16. Andrianov, A.K., LeGolvan, M.P., Yu, L., Svirkin, Y.Y. & Sule, S.S. Characterization of water-soluble polyphosphazene polyelectrolye and its macromolecular precursor using multidetection gel permeation chromatography. *Polym Prepr* 1998, **39**(2), 220–221.

17. Payne, L.G., Jenkins, S.A., Woods, A.L., Grund, E.M., Geribo, W.E., Loebelenz, J. R. et al. Poly[di(carboxylatophenoxy)phosphazene] (PCPP) is a potent immunoadjuvant for an influenza vaccine. *Vaccine* 1998, **16**(1), 92–98.

18. Andrianov, A.K., Marin, A. & Roberts, B.E. Polyphosphazene polyelectrolytes: a link between the formation of noncovalent complexes with antigenic proteins and immunostimulating activity. *Biomacromolecules* 2005, **6**(3), 1375–1379.

19. Payne, L.G., Van Nest, G., Barchfeld, G.L., Siber, G.R., Gupta, R.K. & Jenkins, S.A. PCPP as a parenteral adjuvant for diverse antigens. *Dev Biol Stand* 1998, **92**, 79–87.

20. Payne, L.G., Jenkins, S.A., Andrianov, A. & Roberts, B.E. Water-soluble phosphazene polymers for parenteral and mucosal vaccine delivery. *Pharm Biotechnol* 1995, **6**, 473–493.

21. Payne, L.G., Jenkins, S.A., Andrianov, A.K. & Roberts, B.E. Water-soluble phosphazene polymers for parenteral and mucosal vaccine delivery. In *Vaccine Design*, Powell, M.F. & Newman, M.J., Eds., Plenum Press, New York, 1995, pp. 473–493.

22. Andrianov, A.K. & Payne, L.G. Polyphosphazene hydrogel microspheres for protein delivery. In *Microparticulate Systems for the Delivery of Proteins and Vaccines*, Cohen, S. & Bernstein, H., Eds., Marcel Dekker, New York, 1996, pp. 127–147.

23. Wu, J.Y., Wade, W.F. & Taylor, R.K. Evaluation of cholera vaccines formulated with toxin-coregulated pilin peptide plus polymer adjuvant in mice. *Infect Immun* 2001, **69**(12), 7695–7702.

24. McNeal, M.M., Rae, M.N. & Ward, R.L. Effects of different adjuvants on rotavirus antibody responses and protection in mice following intramuscular immunization with inactivated rotavirus. *Vaccine* 1999, **17**(11–12), 1573–1580.

25. Lu, Y., Salvato, M.S., Pauza, C.D., Li, J., Sodroski, J., Manson, K. et al. Utility of SHIV for testing HIV-1 vaccine candidates in macaques. *J Acquir Immune Defic Syndr Hum Retrovirol* 1996, **12**(2), 99–106.

26. Bouveret Le Cam, N.N., Ronco, J., Froncon, A., Blondeau, C. & Fanget, B. Adjuvants for influenza vaccine. *Res Immunol* 1998, **149**, 19–23.

27. Gilbert, P.B., Chiu, Y.L., Allen, M., Lawrence, D.N., Chapdu, C., Israel, H. et al. Long-term safety analysis of preventive HIV-1 vaccines evaluated in AIDS vaccine evaluation group NIAID-sponsored phase I and II clinical trials. *Vaccine* 2003, **21**(21–22), 2933–2947.

28. Andrianov, A.K., LeGolvan, M.P., Sule, S.S. & Payne, L.G. Degradation of poly [di(carboxylatophenoxy)phosphazene] in aqueous solution. *Am Chem Soc PMSE Prepr* 1997, **76**(1), 369–370.

29. Andrianov, A.K., Marin, A. & Peterson, P. Water-soluble biodegradable polyphosphazenes containing *N*-entylpyrrolidone groups. *Macromolecules* 2005, **38**(19), 7972–7976.

30. Andrianov, A.K., Payne, L.G., Visscher, K.B., Allcock, H.R. & Langer, R. Hydrolytic degradation of ionically cross-linked polyphosphazene microspheres. *J Appl Polym Sci* 1994, **53**, 1573–1578.

31. Kabanov, V.A. From synthetic polyelectrolytes to polymer-subunit vaccines. *Pure Appl Chem* 2004, **76**(9), 1659–1677.

32. Cairo, C.W., Gestwicki, J.E., Kanai, M. & Kiessling, L.L. Control of multivalent interactions by binding epitope density. *J Am Chem Soc* 2002, **124**(8), 1615–1619.

33. Andrianov, A.K., Payne, L.G., Sargent, J.R. & Sule, S.S. *Polyphosphazene Polyelectrolyte Immunoadjuvants*, U.S. Patent 5,855,895, 1999.

34. Andrianov, A.K., Payne, L.G., Sargent, J.R. & Sule, S.S. *Polyphosphazene Polyelectrolyte Immunoadjuvants*, U.S. Patent 6,015,563, 2000.

35. Andrianov, A.K. *Immunostimulating Polyphosphazene Compounds*, U.S. Patent Application 2006/0193820, 2006.

36. Andrianov, A.K. Water-soluble biodegradable polyphosphazenes: emerging systems for biomedical applications. *Polym Prepr* 2005, **46**(2), 715.

37. Andrianov, A.K., Langer, R., Payne, L.G., Roberts, B.E., Jenkins, S.A. & Allcock, H.R. Polyphosphazene ionotropic gel microcapsules for controlled drug delivery. In *Proceedings of the 20th International Symposium on Controlled Release of Bioactive Materials*, Washington, DC, 1993, pp. 26–27.

38. Cohen, S., Bano, M.C., Visscher, K.B., Chow, M., Allcock, H.R. & Langer, R. Ionically cross-linked polyphosphazene: a novel polymer for microencapsulation. *J Am Chem Soc* 1990, **112**, 7832–7833.

39. Andrianov, A.K. & Chen, J. Preparation of polyphophazene Microspheres by Coacervation, U.S. Patent 5,807,757, 1998.

40. Andrianov, A.K., Chen, J. & Payne, L.G. Preparation of hydrogel microspheres by coacervation of aqueous polyphosphazene solutions. *Biomaterials* 1998, **19**(1–3), 109–115.

41. Andrianov, A.K., Chen, J., Sule, S.S. & Roberts, B.E. Ionically cross-linked polyphosphazene microspheres. In *Controlled Drug Delivery: Designing Technologies for the Future*, American Chemical Society, Washington, DC, 2000, pp. 395–406.

42. Andrianov, A.K. & Chen, J. Polyphosphazene microspheres: preparation by ionic complexation of phosphazene polyacids with spermine. *J Appl Polym Sci* 2006, **101**, 414–419.

43. Andrianov, A.K., Marin, A., Peterson, P. & Chen, J. Fluorinated polyphosphazene polyelectrolytes. *J Appl Polym Sci* 2007, **103**, 53–58.

44. Andrianov, A.K., Marin, A., Peterson, P. & Chen, J. Fluorinated polyphosphazene polyelectrolytes. *Polym Prepr* 2005, **46**(2), 671–672.

45. Andrianov, A.K. & Payne, L.G. Protein release from polyphosphazene matrices. *Adv Drug Deliv Rev* 1998, **31**, 185–196.

46. Andrianov, A.K. & Payne, L.G. Polymeric carriers for oral uptake of microparticulates. *Adv Drug Deliv Rev* 1998, **34**(2–3), 155–170.

47. Andrianov, A.K., Cohen, S., Visscher, K.B., Payne, L.G., Allcock, H.R. & Langer, R. Controlled release using ionotropic polyphosphazene hydrogels. *J Control Release* 1993, **27**, 69–77.

# 16

# VACCINE DELIVERY BY TRANSDERMAL ELECTROPORATION

Amit Misra

## 16.1 INTRODUCTION

Electroporation (EP) of intact skin holds out four important promises to those of us interested in vaccine delivery. First, it can deliver antigen to immuno-competent skin, a significant portal of entry for pathogens. Second, it can do so without the use of needles. Third, it is speculated that acellular preparations meant for transdermal EP need not be sterile, since the pores created by the voltage applied would be large enough to permit the vaccine but not intact organisms. Finally, and probably most beguilingly, EP might stimulate several endogenous mechanisms of amplification of protective immune responses, leading to immunity rather than skin allergy.

EP of skin cells of living animals was first reported by Titomirov et al. [1], who described stable transformation of mouse skin fibroblasts with plasmid DNA injected intradermally. Prausnitz et al. developed the field to include needle-free transdermal drug delivery in the 1990s [2,3]. Although the use of EP to permeabilize biomembranes had been well known at the time, these results opened up fresh possibilities for the field of needle-free DNA and antigen delivery to induce immune responses.

In this chapter we examine some of the issues relating to the four promises outlined above, based on results reported by various groups using EP for administration of protein, peptide, and DNA vaccines. For convenience,

*Vaccine Adjuvants and Delivery Systems*, Edited by Manmohan Singh
Copyright © 2007 John Wiley & Sons, Inc.

non-nucleic acid vaccines (NNAVs) are distinguished from DNA and RNA delivery. A few comments on instrumentation and feasibility of adoption of EP for mass immunization are offered. EP-mediated ex vivo delivery of antigen, including nucleic acids to antigen-presenting cells (APCs) such as monocytes and dendritic cells (DCs), followed by their repopulation in the host, is outside the purview of this chapter. Cell-based vaccines are also excluded, on the premise that rather large pores would be required to deliver whole-cell vaccines through the skin, since there is equivocal evidence on delivery of particulate material using EP [4,5]. Alternative technology might, however, be more suitable for the delivery of particulate vaccines to the skin [6]. Transcutaneous adjuvanted immunization without the use of EP is also not considered, although there exists substantial literature in support of induction of protective immunity by epicutaneous exposure of adjuvanted antigen [7,8] rather than IgE-predominant Th2 responses and allergy when antigen alone is painted on skin [9,10].

The technique of skin EP involves applying electrical voltage on the order of 50 to 500 volts (V) in a series of pulses of typical duration 0.1 to 100 ms separated by interpulse intervals on the order of a few seconds across skin. The material to be administered is placed in close proximity to the electrodes that deliver the voltage pulses, and diffuses rapidly through the skin, apparently through transient aqueous pores created as a result of the EP pulses.

It continues to be difficult to describe, using first principles, how EP works. Weaver and Chizmadzhev have brilliantly reviewed the biophysics of EP and electropermeabilization [11]. These authors suggest that a voltage pulse across a bilayer membrane can provide formation energy for a hydrophilic (or rather, aqueous) pore. Pore formation is stochastic and restricted by a parabolic energy barrier. Such an energy barrier restricts the dimensions of the putative aqueous pores that can be formed on application of EP voltage of a given magnitude. Very small pores require as much energy to form as do very large ones. Also, the pore is transient in nature, lasting for at most a few seconds after a pulse that itself lasts for a few milliseconds.

Pore formation in artificial membranes has been observed by applying EP pulses across liposomes and using fast digital imaging to record the manner in which the liposomal membrane deforms, develops pores, and recovers [12]. In living membranes such as bacterial envelopes, EP-induced pores have been visualized by transmission electron microscopy after severe EP undertaken in a food-processing application [13]; by magnetic resonance imaging of skin sections from hairless rats [14], and in membranes of muscle cells by electron microscopy [15]. It is, however, sage advice that one should not be "tempt[ed] to look for the physical 'pores' predicted by electroporation (since) it is not the most useful approach" [3]. It is rather more pertinent to evaluate the effects expected as a result of the appearance of small (<10 nm), transient perturbations in the permeability of skin, as evidenced by a transitory and severalfold enhancement of transdermal diffusion or alteration of skin microstructure.

Transdermal EP as a method of drug delivery has received due attention [16,17]. It is projected that adoption of EP for drug delivery in regular clinical practice awaits just a few crucial developments in technology: electrodes, formulations, and protocols that optimize the transport of drug molecules through the skin and minimize discomfort. It has been demonstrated that the flux of drugs through the skin is enhanced severalfold when EP is employed, but macromolecules require a formulation additive such as urea or sodium thiosulfate that can enlarge the aqueous pathway of diffusion by chaotropy [18,19]. EP thus emerges as a technique worthy of the attention of vaccine delivery professionals interested in nonadjuvanted, needle-free immunization with genes and their products.

## 16.2 RATIONALE AND REALIZATION OF SKIN EP FOR VACCINE DELIVERY

### 16.2.1 Ag Presentation in the Skin

Mammalian skin is a peripheral "organ of immunity," to use the evocative phrase of J. D. Bos [20]. With reference to delivery of acellular and non-nucleic acid vaccines, professional APC types found in skin include Langerhans cell monocytes and macrophages, while keratinocytes and dendritic epidermal T cells (DETCs) have the ability to present Ag in the context of constitutively expressed major histocompatibility complex (MHC) class I or induced MHC class II molecules [21]. Skin APCs have subtly nuanced capacities to invoke "appropriate" responses to antigen challenge [22]. Although inclusion of DETCs in a list of APCs might be controversial, there is evidence to show that T cells such as the DETCs that express an invariant $\gamma\delta$ T-cell receptor (TCR) can present antigen to CD4$^+$ T cells [23], and even hint at acquiring functions pertinent to APCs, such as up-regulation of their Toll-like receptors (TLRs) under conditions of cutaneous inflammation [24]. The APC function of skin Langerhans cells (LCs) is now too well known to bear repetition, and cells of the monocyte-macrophage lineage are, of course, textbook APCs.

It is axiomatic that the LCs in the epidermis and DCs in the dermis are the preferred APCs to be targeted for skin immunization. Based on the presentation desired, it is necessary to work out the EP conditions that would deliver the vaccine for processing to either the phagosome or the cytosol. NNAV intended for presentation on MHC class I and induction of cytotoxic T-lymphocyte (CTL) responses need to reach the cytosol, especially if engineered for receptor-mediated targeting to DCs [25]. Proteins and peptides meant for presentation on MHC class II currently take their chances of pinocytosis by APCs when delivered to the extracellular milieu, since to this reviewer's knowledge, the delivery of particulate or vesicular antigen preparations by EP has not been attempted. It may be stated, nevertheless, that soluble protein [diphtheria toxoid (DT)] or a peptide derived from hepatitis B surface antigen

(HBsAg) *do* induce mature immune responses when delivered through the skin by EP. The response is characterized by in vitro recall of immunizing antigen by cells recovered from draining lymph nodes, mature Ig (IgG, IgG$_1$, IgG$_{2a}$, IgA), and a mixed Th1/Th2 phenotype [26]. Administration of ovalbumin (OVA) by EP has achieved similar results [27]. We have speculated that skin EP is likely to target LCs (and/or DCs) in the skin, since we have observed mature immune responses to antigen delivered by EP but have not been able to quantify the dose of antigen delivered under the EP conditions employed [26](and unpublished data). For example, radiolabeling OVA with $^{125}$I and attempting to recover activity from the site where antigen was applied during EP yielded results such as those shown in Figure 16.1. Since DCs can productively present miniscule amounts of antigen, they could be the APCs responsible for the effects we observed. It has also been shown that LCs are more amenable than skin DCs to EP-mediated delivery [28].

Of course, all the benefits of EP-enhanced transdermal immunization and more, such as CTL responses against papilloma virus–induced tumors [29] and mucosal immunity at sites as distant from the skin as the vaginal epithelium [30], are realized with adjuvanted transcutaneous immunization in animals [31,32] as well as humans [33]. The parallels between results obtained with and

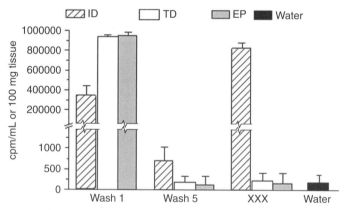

**Figure 16.1** In vitro skin permeation of ovalbumin (OVA). Excised full-thickness depilated skin samples from mice sacrificed in irrelevant experiments were used. Duplicate skin samples either received an intradermal injection (hatched bars) of 10 μg of $^{125}$I-labeled OVA in 100 μL of phosphate buffer, or were exposed to the same amount by placing 200 μL of antigen solution in latex cups stuck on the surface. In the TD group (hollow bars), no EP voltage was applied, while 10 pulses of magnitude 100 V and duration 1 ms, separated by intervals of 1 second, were applied to the EP group (shaded bar). Water gave background counts represented by the single black bar. The radioactivity per 100 mg of skin tissue was determined after washing the area of application free of surface radioactivity in five washes. Negligible amounts were detected after EP or transdermal diffusion compared to intradermal injection.

without EP strongly indicate that antigen presentation and processing would be similar in both cases. Our own experiments to directly visualize radio-labeled or fluorescein-labeled antigen after transdermal EP have not met with success, but there is direct flow-cytometry-based evidence of skin APCs taking up epicutaneously-administered antigen and migrating not only to draining lymph nodes but also to Peyer's patches of the intestine [34]. The same study also described a skin-derived migrant present in Peyer's patches that was CD11c⁻, CD3ε⁻, and CD19⁻, a surface phenotype that excludes B cells, T cells, LCs, and macrophages but may be a plasmacytoid DC or blastic natural killer (NK) cell [35]. It is therefore evident that a varied stable of APCs awaits antigen delivered to the skin, and the decision to attempt skin immunization can be guided by considerations of the types of APCs expected to be encountered by the NNAVs.

The key question that remains is to determine whether a given EP protocol would deliver a given antigen to the extracellular or intracellular milieu. Empirical evidence indicates that the permeation pathway that is created in the stratum corneum as a result of EP is rectilinear, unlike the tortuous path followed by small lipophilic molecules diffusing through the skin barrier [36,37]. It is thus easier to visualize antigen delivered by EP reaching viable epidermis situated below the stratum corneum and remaining in extracellular matrix than to imagine a further event of EP that delivers antigen to the cytosol of cells situated in the dermis.

We have attempted to examine the fate of a model antigen, OVA, delivered by different EP protocols in terms of presentation on MHC class I or II (unpublished observations). Cells harvested from the lymph nodes of mice immunized with OVA by EP were capable of presenting cognate peptides to two T-cell hybridomas, designated as B3 and 13.8, that recognize OVA-derived peptides in the context of MHC classes I and II (Figure 16.2). Current efforts are aimed at comparing various EP conditions and antigen doses that generate cleaner data, but we are predisposed to believe that some antigen does reach the proteasomal-processing pathway upon skin EP.

The requirement of delivery to the cytosol of APCs is an imperative with DNA (and RNA) vaccines. Most researchers have reported that EP-mediated immunization with naked DNA does not result in high transfection efficiency [38]. This is not to deny the *efficacy* of DNA vaccination by transdermal EP, merely to point out that only a small proportion of the dose reaches the APC cytosol. For example, Buchan et al. recently reported that vaccine dose and volume were key factors determining the efficacy of an antitumor DNA vaccine, despite inclusion of promiscuous T-helper epitopes in the construct. They also reported that EP enhanced the immune response when used for both priming and boosting [39]. These results not only highlight the contribution of EP to the magnitude of the immune response but also serve to remind us that Ag dose delivery remains the limiting step in skin immunization. Nevertheless, reports continue to pour in regarding the efficacy of antiviral [40–42] and antitumor [43,44] vaccination by transdermal EP. Antigen

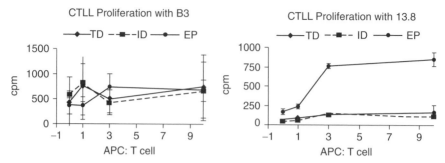

**Figure 16.2**   Ag presentation on MHC classes I and II. Four mice per group were administered 10 µg of OVA by application on a depilated area of the skin (TD), by intradermal injection (ID), or by application of 10 exponential pulses of magnitude 100 V and duration 1 ms, separated by intervals of 1 second, after applying to the skin as described in Ref. 26. Two days later their lymph nodes were harvested and the cells were co-cultured with a T-cell hybridoma specific for OVA peptides restricted by either MHC class I (B3) or class II (13.8). After 96 hours of co-culture, the supernatants were tested for IL-2 using the indicator cell line CTLL, starved of IL-2 overnight. Proliferation was recorded by [³H]thymidine incorporation.

presentation in the case of nucleic acid vaccines is also dogged by the issue of cross-presentation. Nucleic acid delivered by EP can make its way to virtually any cell, since the technique is in no way selective for any particular cell type. Under these circumstances, the outcome of foreign gene expression, translation, processing, and presentation may lead to either cross-tolerance or cross-priming. Lu and Ghazizadeh have recently constructed a model to study the process of cross-presentation. These authors used a retrovirus to transduce green fluorescent protein (GFP) DNA into primary skin cells. These were then grafted into mice that were either rendered tolerant toward GFP or treated it as a xenoantigen. The fate of cells expressing the fluorescent marker was easy to track in the two sets of mice, providing an elegant system for researchers who wish to examine the response to DNA vaccine constructs including a reporter gene such as GFP [45].

Before concluding this section, it is pertinent to mention two aspects of skin immunology that militate against the use of the skin as the portal of entry for vaccines. Nonadjuvanted antigen allowed to permeate through the skin over several days can cause the induction of an IgE-predominant Th2-driven allergic response [6], as is well known in the case of haptens. Although we did not find antigen-specific IgE in mice immunized using skin EP, this is not sufficient evidence to claim that EP somehow bypasses the events associated with skin sensitization, contact dermatitis, and so on. The immunotoxicology of skin EP remains to be investigated in full.

Second, and somewhat counterintuitively in light of the foregoing information, Janeway's highly respected group has been reporting transcutaneous immunization as a method of tolerization against autoantigens. Their results show that the delivery of autoantigenic peptides derived from myelin basic protein and relevant to induction of experimental allergic encephalomyelitis induces regulatory T cells that prevent the development of disease in a TCR transgenic mouse [46]. This occurs through a process of receptor revision involving *RAG-1* and *RAG-2* activity [47]. Bynoe and Viret have recently reviewed their experience in this area and stress that epicutaneous, but not subcutaneous antigen delivery induces CD4+/CD25+ cells with "full suppressor potential" against the immunizing antigen [48]. Although these investigators have observed that priming with antigen alone and boosting with adjuvanted antigen is essential for induction of suppressor T cells, they ascribe the phenomenon of induction of these cells to antigen presentation within the epidermis.

### 16.2.2   Getting Transdermal EP to Work

Peachman et al. have, not without justification, classified EP as an invasive method of antigen delivery to the skin. They have suggested excellent protocols and procedures for antigen delivery to the skin using seven different methods, which interested readers would do well to consult for a practical approach [49]. The predominant use of EP in vaccine delivery today is to use it for enhancement of transfection with DNA vaccines which are themselves delivered through a needle. This is, however, not being addressed here; since the objective is to stimulate interest in the objective of delivering vaccines without the use of needles. Alving's group has pioneered the development of transcutaneous adjuvanted immunization [7,21,31–34,49,50]. This approach is considered noninvasive, despite procedural steps that can affect the integrity of the stratum corneum, such as shaving, and delipidization with alcohol.

The difference between nonadjuvanted transcutaneous immunization and transdermal EP lies primarily in the *amounts* of antigen needed for evoking a sufficiently strong immune response and the *time* required for vaccine delivery. It is not unusual to see several hundred micrograms of antigen being administered to immunize a single mouse through the skin when transdermal permeation enhancement, by EP for instance, is not resorted to. In our experience, on the other hand, 1 to 20 µg of protein or peptide Ag is sufficient to induce robust responses when EP is employed to enhance permeation [26]. Similarly, the time required for transcutaneous administration of NNAV and adjuvant is on the order of days, whereas EP protocols can be completed within minutes.

The important parameters that control transdermal delivery by EP have been reviewed in detail by Denet et al. [16]. These authors point out that EP conditions need to be worked out from an evaluation of the physicochemical properties of the molecule to be delivered and the excipients used to prepare the formulation. Even a tentative list of physicochemical properties that NNAV might possess is difficult to compile, but there are certain obvious commonalities. For instance, antigen preparations are hydrophilic insofar as they are composed of biomolecules; in contrast to certain drug molecules, that may be hydrophobic in the extreme. For vaccine delivery, therefore, the experimenter need not be concerned with the diffusion of the preparation across what is thought to be an aqueous pore. The hydrophobicity of the vaccine preparation is unlikely to impose a rate-limiting step on the rate of delivery.

Molecular mass, or even physical size on the order of several micrometers, is the major constraint that NNAV preparations possess in terms of amenability to administration by transdermal EP. It is obvious that the larger the molecular weight (or physical dimension), the more difficult it is to transport biological material across the skin. The conformation of a polypeptide sequence in solution meant to be delivered by EP is an important related issue. Globular proteins are likely to diffuse relatively easily and rapidly through aqueous channels as compared to unfolded or malfolded random structures, such as those obtained on chemical denaturation. Indeed, it is possible to visualize nanosized protein molecules (or plasmids) as classical DLVO colloids, falling through EP-induced pores with Stokes velocity [51]. But for the aspect of charge, under the influence of electric fields that create transmembrane potentials on the order of 0.5 to 1 V [3,4,7,10], electrophoresis is likely to be the major motive force causing the flux of charged species through the skin. The diffusion of charged species across EP-induced pores is affected by electrode polarity [3,16,52]. Once the pulsing protocol is done with, however, diffusive flux of antigen may be expected to continue until such time as the disruption of the skin barrier allows it. Charged species other than the immunizing antigen are likely to strongly affect results. The $pK_a$ value of the Ag should serve as a guide for setting the pH of the solution, and counterions that could potentially neutralize the antigen charge need to be avoided.

Considerations of size and charge also apply to DNA (and RNA) vaccines. In vitro studies by Prausnitz's group have demonstrated that DNA vaccines meant for administration by EP should not be packaged in delivery vehicles, in the interest of higher efficiency of transfection [53]. The objective therefore, remains to keep the antigen lean, mean, and all charged up.

With antigen-related parameters taken care of, the EP conditions need to be figured out. These include the amplitude, duration, interpulse interval, and shape of the EP pulses, as well as their number. The number of pulses administered also needs to be considered in the context of their duration, as will become clearer below. The foregoing considerations apply essentially to the EP voltage-generating part of EP equipment. What the skin experiences during EP, however, is an electric field strength and an input of columbic charge. These two parameters are strongly influenced by the types of electrodes that are used to deliver EP pulses to the skin and are considered in some detail when discussing instrumentation.

The amplitude of pulses used in vivo for skin EP typically ranges from 100 to 300 V, although EP voltages of up to 1000 V have been used [3,16]. The resulting transdermal potential is likely to be in the region of 10% of this value. Although it is certainly true that pulses of larger amplitude will induce pores more efficiently (whether in terms of size, numbers, or lifetimes), increasing the EP voltage beyond 300 V is accompanied by hazards of severe damage to the skin, induction of muscle contraction, and even electric shock [26]. The enhancement of transdermal flux is also saturable: Beyond a point, there is no significant advantage offered by increasing EP voltage. Prausnitz and others ascribe rapid changes in the electrical parameters such as resistivity and capacitance to alteration in structure of skin lipids [3]. Resistivity of the skin is in the region of $10^5$ ohm·cm$^2$ under resting conditions. Within microseconds of application of an EP pulse, this value falls by three orders of magnitude, to about 100. A few milliseconds after cessation of EP, resistivity rises again and reaches values in the region of $10^4$ ohm·cm$^2$. Pretreatment values are restored within seconds to minutes.

Theoretical considerations predict that EP is restricted to a relatively small area of the skin close to the electrode. Regardless of the EP conditions employed, Weaver's group has demonstrated that maximal breakdown of the skin barrier occurs in close proximity of the electrodes [32]. As the distance from the electrode increases, there is decreasing barrier disruption by electric field. At a certain distance from the electrode, the EP energy dissipates and forms a boundary characterized by lower permitivity to diffusants [54]. These authors describe the region in which pores are created as a *local transport region* and the area where EP energy dissipates as a *local dissipative region*. It would be worthwhile to remember that the area occupied by the pores themselves is expected to be merely 1% of the local transport region [3].

The heating effect of electric current passing through tissue, and its impact on the permeability of skin, is also worth considering. Such heating is directly proportional to the amplitude of the EP pulse (as it is to the pulse duration and number of pulses) and inversely proportional to the interpulse interval. Local rise in temperature is expected to denature skin lipids, thereby disrupting the barrier properties of the stratum corneum. Pliquett et al. have recently measured temperature changes in excised skin samples subjected to EP using a liquid crystal [55]. These new results confirm that pore formation occurs readily when the electric current causes a localized temperature rise to 60 to 70°C, at which point skin sphingolipids are denatured. Although it is a side issue, the heating effect of EP needs also to be kept in mind with regard to possible effects on the antigen to be delivered: whether a few seconds of exposure to local temperatures in the region of 70°C would affect the antigen in any way.

To continue on the theme of EP conditions: the pulse duration, interpulse interval, and number of pulses deserve careful consideration. Extensive data from studies on excised skin are available to demonstrate that an optimum value for each of these lies between a minimal and largely reversible disruption of barrier properties, and a maximal injurious damage to skin tissue. The experimenter needs to work out these optimal numbers by actual trial with the antigen of choice, perhaps through a factorial experiment design. In general, high-voltage pulses of longer duration and short interval are considered more efficacious than lower-voltage pulses delivering equivalent coulombic charge, although there is evidence that pulse shape is not really important as long as the charge transferred is equal [56]. For what it is worth, our experience suggests that voltages below 100 V are inadequate, while those above 300 V cause severe discomfort (muscle contraction and itching). Pulse durations of 1 to 100 ms for 10 to 30 pulses separated by 1 second are adequate for immunizing mice against diphtheria toxoid, ovalbumin, or a 23-amino-acid myristoylated peptide.

The shape of the EP pulse has been investigated with reference to its effects on EP [38]. The most common pulse shapes used by investigators are the exponentially decaying and the square waveform. Certain advantages are realized with the use of square waves, in terms of greater control on the duration and strength of the pulse at any point in its trajectory [38,57]. A novel exponentially enhanced waveform has been reported that achieves EP with fewer pulses [58]. Other innovative pulse shapes have been considered, so that experimenters can employ any arbitrary pulse shape that suits their fancy [59].

No discussion of EP is likely to be meaningful unless the design of the electrodes delivering the electric pulse is considered. Here, again, it is intuitively obvious that designs that spread electric energy over a large surface area have to be balanced with designs that deliver all the energy at a single focus. Meandering and spiral designs of electrodes are discussed below.

Before moving on to the next section, it might be instructive to foreshadow a discussion on the consequences of EP-induced perturbations of the immune environment of the skin. This is taken up in some detail later, when the apparent adjuvanticity of skin EP is examined, but the information contained in this section augurs well for the induction of an innate stress response to the insult offered by EP.

### 16.2.3   This Is Not Going to Hurt

The electrical paraphernalia associated with EP, and at first sight, the high voltages employed, might induce reluctance in accepting the technique as a clinical procedure. Nevertheless, skin EP is remarkably innocuous in terms of long-term effects at the site of the application. It is likely to be less painful than the hypodermic needle when the technology is refined appropriately, and as discussed below is less likely to require aseptic technique during use.

Evaluation of the acceptability of an EP protocol is best conducted by determining whether the disruption induced in the skin's barrier properties is reversible. To a smaller extent, it might also be appropriate to determine whether any attendant nociceptive stimuli are perceived during EP in intact animals. Rather than going into an analysis of EP protocols used by various experimenters and the resulting observations on reversibility of the disruption induced, it might be more useful to suggest methods and outcome measures that can be used to gauge the extent of and recovery from EP-induced alterations in the skin's barrier properties.

Transepidermal insensate water loss (TEWL) is an established parameter to assess the barrier properties of the stratum corneum [60]. Measurement of TEWL is a facile, noninvasive, and reliable technique, but requires a fairly expensive and elaborate apparatus to be recorded faithfully. EP protocols regarded as safe can be characterized by a rapid return of TEWL to pretreatment values. In vivo, 10 square-wave pulses of 335 V amplitude and 5 ms duration induced an almost sixfold enhancement of TEWL, which did not return to pretreatment levels over the next 48 hours. An equal number of pulses of 1000 V and 100 μs duration, while inducing only threefold TEWL enhancement, allowed return to near-baseline values within about half an hour [42].

Reversibility of the electrical properties of skin subjected to EP has been studied in vitro for a variety of EP protocols employing excised skin from various species. Weaver's group pioneered the concept of studying electrical properties such as resistance, and Vanbever et al. have reported in detail about the key parameter indicating the electric integrity of the skin, being reflected primarily in the ohmic resistance rather than the impedance of the skin [16,17]. It has been mentioned above that the resistivity of skin displays dramatic fluctuations in response to EP protocols. The safety or otherwise of a particular protocol may therefore be assessed from the time required for recovery of skin resistivity. This parameter is measured accurately using an inductance, capacitance, and resistance (LCR) meter, and can even be recorded by a sensitive multimeter.

Resealing of the skin's barrier to diffusion of small molecules is much more difficult to measure in living animals, but would provide definitive confirmation of reversal of the effects of EP. This author is not aware of any efforts undertaken in this direction, but wonders whether it would be possible to assess the rate of resealing using a small uncharged molecule such as glucose, placing it in the receptacle that is used to contain the solution of material intended for transdermal delivery, and use a biosensor to evaluate its depletion.

Experimenters who have attempted skin EP in mice are aware of the phenomenon of muscle contraction induced when pulses are delivered. The sparse data from clinical studies on application of high-voltage pulses to the skin does indicate that pulses of up to 80 V are completely innocuous and produce no discomfort even in such a richly innervated area as the human penis [61].

## 16.3 INSTRUMENTATION

Equipment for skin EP is evolving from apparatus used for EP-mediated gene transfer to cells and tissue. It consists of a power supply that incorporates a generator and control system to modulate the output, a monitoring system to measure the characteristics of the output, and electrodes that deliver the output across the skin. No major changes in power supply or signal monitoring are required for adaptation to in vivo EP, but electrode design obviously needs to be modified for application to the skin.

Conventional power supplies are familiar to experimenters and are offered for sale by well-known firms such as Eppendorf, Bio-Rad, Invitrogen, Life Technologies, and the specialty firm BTX, all of which have significant presence on the Internet. The key feature of the power supply is the electric field strength that it generates at a given setting when adapted to the job of skin EP. For simple parallel-plate electrodes in vacuum, the electric field strength can be calculated as the ratio of the voltage to the distance of separation. When innovative electrode design and a medium of highly variable dielectric constant (the skin) intervene, heroic measures are required to calculate electric field strength. It is expedient, therefore, to work by rule of thumb and assure oneself that the power supply is capable of delivering up to 1000 V/cm in selectable steps of 100 V at the electrode. How this is done is fairly uniform across manufacturers. A capacitor is charged to the desired voltage using the electric mains supply. The stored charge is then delivered to the resistor, comprising the rest of the circuit (the apparatus, the electrodes, and the skin) through an output switch, such as the MOSFET switch used in uninterruptible power supply batteries for personal computers. Standard techniques of circuit design are used to set output characteristics such as the waveform (exponential decay or square wave and in the foreseeable future, any pulse shape describable by an equation), pulse duration, and interpulse interval, usually through timing resistors. A current-limiting resistor is often incorporated in the circuit to minimize formation of an electric arc at high voltage; an occasional, disturbing, but not overtly dangerous phenomenon involving localized thunder and lightning near the electrodes. Arcing has not been encountered by our group during skin EP, using homemade equipment and graphite electrodes at voltages up to 500 V [26].

Generators using mains power supply or even batteries tend to be bulky and heavy. A number of hand-held electroporation devices are available commercially, but typically weigh more than 500 g. Upadhyay's innovation in using piezoelectricity to generate EP voltage is therefore an extremely useful contribution [62]. The device is easily fabricated from a piezo-ceramic gas igniter used in millions of households in India and elsewhere. It is a handheld device no larger than a hypodermic syringe and weighs less than 50 g. Actuating the igniter compresses the plunger with a force of about 2000 N, generating a voltage of about 18,000 V. This voltage is delivered to the electrode within microseconds. The coulombic charge transferred to the electrodes during each actuation of the piezoelectroporation apparatus is on the order of $10^{-6}$ C, which is orders of magnitude lower than charge transferred in conventional EP.

Charge transfer, and more important, distribution of the electric field are also controlled by electrode design. Popular designs include the needle or stub, ring, plate, and "meander" or intercalating electrode (Figure 16.3). There are, of course, several other designs of electrodes, such as Ichor's Tri-Grid [37], that combine invasive transcutaneous or intramuscular delivery of antigen with EP. These are, however, invasive electrodes, and are out of the purview of this chapter.

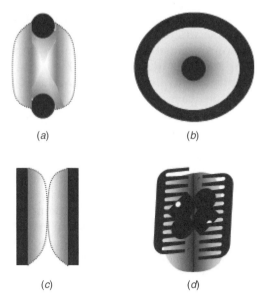

(a)    (b)

(c)    (d)

**Figure 16.3** Electrode designs. Solid areas represent the material of construction of the electrode and the shaded areas represent the distribution of the electric field: (*a*) stub or needle electrodes, (*b*) stub-in-ring, (*c*) parallel plates, and (*d*) meander electrodes, a special case of several parallel plates in series.

Sugibayashi et al. investigated the impact of electrode design on the distribution of electric field through the skin undergoing EP [63,64]. These authors report that parallel-plate electrode geometry generates nearly double the electric field strength for the same EP voltage in comparison to needle or needle-in-ring designs. The meander electrode is reported to be equally efficacious and safe for clinical applications as a plate electrode [65]. Calipers that pinch out a fold of skin have been used to accommodate electrodes and the receptacle for the solution intended for EP-enhanced transdermal delivery. However, there is a fear of undue stimulation of nerve endings and superficial burns within a skinfold held by calipers [16].

Electrodes for skin EP should not be "active" electrodes that participate in electrochemical reactions, such as the silver–silver chloride electrode. Platinum, silver, and carbon (graphite) have been used as materials of construction with equal effect. It is also possible to coat electrodes with an electrochemically stable compound such as titanium nitride, platinum, platinum iridium, or iridium oxide [66].

It has been suggested that antigen preparations intended for administration by transdermal EP need not be sterile [26]. The suggestion is premised on the perception that microorganisms would not diffuse through pores formed as a result of EP, although there is some controversy regarding the sizes of pores achievable and particle sizes of material that can cross the skin barrier after

EP. Whereas Hoffman et al. have published evocative electron micrographs of microspheres crossing the stratum corneum [4], Chen et al. have concluded that charged microparticles cannot be delivered across the skin using reasonable EP protocols [5]. Although the issue of sterility may appear trivial in its statement, it has the potential to affect the cost of vaccines. Most Ag preparations are thermolabile and cannot be subjected to sterilization with steram. Such preparations require sterile filtration on a large scale. Manufacturers would certainly save costs, effort, and time if terminal sterilization of vaccine preparations could be dispensed with. To ensure sterility at the time of vaccination, a terminal sterilization filter such as a sterile cellulose acetate membrane may be interposed between the receptacle that holds the Ag solution and the skin surface. Disinfecting the EP site before application of the Ag reservoir with the filter side down would provide additional safety.

In conclusion, it must be pointed out that if skin EP is to find a place in routine immunization, the equipment must fulfill the requirements of health professionals and vaccine recipients. It should be portable, user friendly, and should not appear intimidating. Electrodes should be designed so that they may be affixed to and removed from the skin with ease and comfort. Ag reservoirs should be easy to replenish or replace. There is much to recommend in the piezoceramic power supply design with regard to the ease of use in the field. A simple handheld device is likely to find much greater acceptance for administering a vaccine than apparatus that is plugged into an electrical supply line.

## 16.4  SKIN EP AS ADJUVANT

### 16.4.1  Adjuvanticity of Skin LC

The scope of this chapter has purposely been limited to the field of EP as a means to transport antigen across the skin, to APCs that can elicit a desired immune response. However, there is a depressingly little activity in the area of transdermal vaccine delivery by EP, as evidenced by the accessible literature. Several reasons may account for the apparently low interest in transdermal EP for antigen delivery, but one of them certainly has to do with the difficulty of assessing the dose of antigen administered using this technique. Unlike transdermal delivery of small molecules such as drugs, NNAVs are difficult to recover from the body for quantitative estimation. It is generally acknowledged, however, that microgram amounts of antigen do not make their way across the skin as a result of EP protocols. Despite this realization, our group and others profess faith in the efficacy of the method. Two main lines of speculations sustain this faith. The first has already been hinted at: that antigen delivered by EP finds its way to skin DCs, particularly LCs. Attention has been drawn earlier to the parallels between adjuvanted transcutaneous immunization [7,21,31–34,49,50] and nonadjuvanted transdermal EP for antigen delivery [26,27]. The literature addressing the former technique has identified skin DCs as the APCs responsible for mucosal immunization. The adjuvant employed in studies on transcutaneous immunization activates skin LCs whether or not they have captured antigen, and elicits bystander stimulation of those cells that are actually involved in antigen processing and presentation [67].

The adjuvant nature of DCs has been highlighted by Stienman et al. time and again (for reviews, see [68] and [69]). It is well known that DCs (including skin LCs) can take up antigen through CD205 and present them either on MHC class I or II, with phenomenally greater efficiency, evoking robust immune responses to miniscule amounts of antigen. Antigen presentation on DCs has even been considered as an absolute requirement for priming naive T cells. A virtual "who's who" of immunologists has commented and experimented on antigen presentation by DCs and its relevance to the induction of immune response. The role of LCs as a distinct subset of skin DCs in the induction of immunity or tolerance has not been worked out in sufficient detail, but methods are beginning to emerge whereby their contributions can be studied. Bennett et al. recently reported an ingenious system that selectively ablates LCs in mice for up to four weeks [70]. It would be interesting to compare the consequences of EP-enhanced transdermal immunization in mice that have a full complement of LCs with those observed in those that have had their LCs ablated. Experiments of this nature are required to establish whether or not the immune response observed on skin EP is robust because of the adjuvanticity of the APCs it targets.

The second line of thinking that supports our perception that EP is in itself possessed of adjuvanticity has much to do with skin DCs as well. It is speculated that EP protocols could activate LCs more strongly than the mere presence of antigen in the vicinity of the cells. Skin DCs might be expected to respond to the "danger" represented by the consequences of EP [71,72]. Stienman et al. have listed disruption of the integrity of the epidermis as a prime motivator of DC migration in their seminal work during the 1980s and onward. Independent lines of evidence also confirm that various forms of stress mobilize skin LCs. Damage to the skin by ultraviolet light induces migration of DCs away from the sunburn sites toward lymph nodes, as documented in several experiments, including a recent one on skin explants [73]. Mechanical injury, such as that caused by repeatedly applying and removing a strip of adhesive tape to an area of skins, has similar effects but induces IL-10, which is a key Th2-polarizing cytokine in the induction of responses such as atopic or contact dermatitis [74]. The picture of DC mobilization in chronically inflamed skin also indicates that there is a homeostatic mechanism that balances the allergenic versus protective immunologic aspects of antigen presentation by LCs. The key to eliciting protective immunity and not allergy such as that encountered during nonadjuvanted epicutaneous exposure probably lies in achieving sufficient perturbation to activate LC migration with the delivered antigen in tow, instead of inducing injury and/or cytokine production by bystander cells such as keratinocytes that could lead to establishment of allergic reaction. This is something of a double benefit: a skin immunization protocol that is highly efficacious in achieving the desired protective immune response is ipso facto likely to be safe from the point of view of immunotoxicology. An exercise to evaluate the immunobiological consequences of skin EP is necessary before any definitive claims can be made. The cytokine milieu of the EP site and surrounding skin, chemokines induced or suppressed as epidermal cells respond to EP, and the role of autacoids, prostanoids, and so on, released in the microenvironment are all likely to have their impact on LC mobilization [75,76].

There are several more reasons to expect adjuvanticity from EP protocols. Vernier et al. reported voltage pulses of even higher magnitude than EP pulses, but lasting for too short a time to induce EP induce intracellular calcium bursts [77]. Recognizing that the actual proportion of cells that are porated is likely to be quite small, it is interesting to speculate whether calcium flux in bystander cells would influence their own activation status and consequent impact on the ensuing immune response. Finally, despite the small numbers of porated cells, we may expect bidirectional flux of material from the pores. There is growing evidence that ligand-gated ion channels intervene in LC differentiation and migration. In a recent report, Granstien et al. demonstrated that a stable analog of ATPs that ligates P2 channels significantly enhanced the antigen-presentation abilities of cultured LCs [78]. These authors suggest that such nucleotide analogs of purinergic receptors represent a novel class of adjuvants. It is conceivable that ATP and nucleotides emerging from the cytosol of porated cells could contribute to the adjuvanticity of EP.

### 16.4.2  Experience with Noninvasive EP and Future Prospects

More than a decade ago, when transdermal drug delivery research was investigating the phenomenon of iontophoresis to enhance the permeation of drugs across skin, we attempted to investigate whether diphtheria toxoid (DT) could be induced to cross the stratum corneum and evoke an immune response [79]. Working with rats, we found that a barely detectable IgG response was discernible seven days after the animals were subjected to an iontophoretic current of 1 mA for 1 hour in the presence of 5 μg (3500 Lf/mL) DT. Surprisingly, we found no statistically significant difference in the DT-specific IgG when we examined sera from control rats that had received the same treatment except for iontophoretic current. It was concluded that iontophoresis did not enhance transdermal permeation, but since DT is a potent immunogen, it could elicit a response to the miniscule amounts permeating the skin.

When EP-enhanced transdermal delivery of DT without adjuvant was attempted in mice, the antibody response was lower than, but comparable to, that elicited by adjuvanted immunization and much greater than the response to nonadjuvanted intradermal injection [26]. These results prompted us to conclude that insofar as immune responses are dose dependent, EP enhances transdermal delivery of DT. The issue of adjuvanticity presented itself when we compared immune responses to a 23-amino acid peptide delivered by EP, or injected intradermally with no accompanying adjuvant. Intraperitoneal injection on Freund's complete adjuvant provided a positive control for comparison. Comparing in vitro recall of the peptide by cells harvested from lymph nodes, we were gratified to observe that lymphocytes from the EP-immunized group proliferated almost as efficiently as did those from the group administered adjuvant. However, we remained ignorant of the actual dose of the material delivered by EP to achieve this effect. And there the story stands. Preat's group more or less confirmed our findings [27], but uncharacteristically for such an active group, has not published any further results. Upadhyay demonstrated that as small a dose as 0.2 µg taken for transdermal EP with a 10-day interval between priming and boosting elicited an antibody response [51]. This study, too, did not report any estimate of the amount of antigen diffusing into skin as a result of EP. Subsequent experiments undertaken by the present author for delivering ovalbumin (Figures 16.2 and 16.3) and independently by Upadhyay and colleagues on a DNA contraceptive vaccine ([80] and unpublished observations) indicate that the responses elicited by antigen delivered by skin EP tend to be highly variable. Inbred littermates display differences of orders of magnitude in endpoint antibody titers. In vitro recall of immunizing antigen is gratifyingly prompt in some cases and frustratingly recalcitrant in others. In sum, we are still far from elucidating the EP conditions that are necessary and sufficient to invoke a robust immune response.

Despite the lack of information on dose-dependence and repeated observation on the extreme variability of the immune response, there is some data to buttress the claim of adjuvanticity inherent in skin EP. The hallmarks of adjuvant-enhanced immune response to T-dependent antigen include immunoglobulin class switching and the generation of T-cell memory. Both phenomena have been documented, along with evidence of a mixed Th1/Th2 phenotype, as discussed above [26].

Further work is necessary to establish whether or not the difficulties presented by the current scenario are surmountable. Three important areas can be suggested for the attention of readers who are interested in the possibilities of skin EP as a feasible method of immunization. Apart from the need to establish dose–response correlations, the principal issue is to determine whether or not LCs are targeted and/or activated as a consequence of EP. This author has been trying to address the issue of Ag uptake by APCs using fluorescent-labeled OVA and flow cytometry or confocal microscopic detection of epidermal migrants: once again, without much headway. The experience with noninvasive EP to deliver antigen is therefore limited, despite the large body of evidence that the technique enhances the diffusion of a wide variety of small molecules through the skin barrier.

A parting look at insulin delivery by EP could be instructive from the point of view of designing experiments in skin immunization, as well as to build morale. To the best of this reviewer's knowledge, the largest therapeutic peptide that has been demonstrated to diffuse through the skin is insulin. The first results were presented at a meeting of the Controlled Release Society in 1995 by Bommannan et al., the reference for which is now elusive. Prausnitz considers insulin delivery by skin EP as feasible [81], and Sen et al. have recently demonstrated the use of lipids in the insulin formulation to enhance permeation in vitro [82]. The molecular weight of insulin is about 6000 Da, which is well within the 10 kD limit of size diffusible through EP-induced pores, as proposed by Preat's group, albeit with reservations [56]. The size cutoff proposed here was not as sharp as is encountered in iontophoretic transdermal delivery and was determined using dextrans of different molecular weights, conjugated to FITC for fluorescence-based detection by confocal microscopy or fluorescence spectroscopy. When either a 38- or 12-kD FITC–dextran was used, the authors observed a permeation flux of ~2000 pMol/cm$^2$ per hour following EP with 10 pulses of 150 V lasting for 150 ms. More significantly, they observed localization of the 38-kD FITC–dextran in the epidermis: the target site of choice for skin immunization. The molecule was found to localize within the cytosol of keratinocytes as well as in the stratum corneum matrix. These results are extremely encouraging for vaccine delivery to the skin, for the purpose of mucosal immunization.

Forecasts of the future are inevitably tinged with the bias that prompts such forecasts, so all that can be said in conclusion is that this new field of vaccine delivery is just standing there, waiting in the wings for researchers to take an interest in its debut. The rudiments of technology are in place, the evidence for efficacy is preliminary but heartening and the scope of application is exciting. These conditions themselves warrant a bright future for research in this area.

## REFERENCES

1. Titomirov, A.V., Sukharev, S. & Kistanova, E. In vivo electroporation and stable transformation of skin cells of newborn mice by plasmid DNA. *Biochim Biophys Acta* 1991, **1088**, 131–134.

2. Prausnitz, M.R., Bose, V.G., Langer, R. & Weaver, J.C. Electroporation of mammalian skin: a mechanism to enhance transdermal drug delivery. *Proc Natl Acad Sci USA* 1993, **90**, 10504–10508.

3. Prausnitz, M.R. A practical assessment of transdermal drug delivery by skin electroporation. *Adv Drug Deliv Rev* 1999, **35**, 61–76.

4. Hofmann, G.A. & Suder, K.S. Electro-incorporation of microcarriers as a method for the transdermal delivery of large molecules. *Bioelectrochem Bioenerget* 1995, **38**, 209–222.

5. Chen, T., Langer, R. & Weaver, J.C. Charged microbeads are not transported across the human stratum corneum in vitro by short high-voltage pulses. *Bioelectrochem Bioenerget* 1999, **48**, 181–192.

6. Dean, H.J., Fuller, D. & Osorio, J.E. Powder and particle-mediated approaches for delivery of DNA and protein vaccines into the epidermis. *Comp Immunol Microbiol Infect Dis* 2003, **26**, 373–388.

7. Glenn, G.M., Rao, M., Matyas, G.R. & Alving, C.R. Skin immunization made possible by cholera toxin. *Nature* 1998, **391**, 851.

8. Partidos, C.D., Beignon, A.S., Briand, J.P. & Muller, S. Modulation of immune responses with transcutaneously deliverable adjuvants. *Vaccine* 2004, **22**, 2385–2390.

9. Wang, L.F., Lin, J.Y., Hsieh, K.H. & Lin, R.H. Epicutaneous exposure of protein antigen induces a predominant Th2-like response with high IgE production in mice. *J Immunol* 1996, **156**, 4077–4082.

10. Hsieh, K.Y., Tsai, C.C., Wu, C.H. & Lin, R.H. Epicutaneous exposure to protein antigen and food allergy. *Clin Exp Allergy* 2003, **33**, 1067–1075.

11. Weaver, J.C., Chizmadzhev, Y.A. Theory of electroporation: a review. *Bioelectrochem Bioenerget* 1996, **41**, 135–160.

12. Riske, K.A. & Dimova, R. Electro-deformation and poration of giant vesicles viewed with high temporal resolution. *Biophys J* 2005, **88**, 1143–1155.

13. Rowan, N.J., MacGregor, S.J., Anderson, J.G., Fouracre, R.A. & Farish, O. Pulsed electric field inactivation of diarrhoeagenic *Bacillus cereus* through irreversible electroporation. *Lett Appl Microbiol* 2000, **31**, 110–114.

14. Caban, J.B., Moerland, T.S., Gibbs, S.J., McFadden, L. & Locke, B.R. Transdermal water mobility in the presence of electrical fields using MR microscopy. *Magn Reson Imaging* 1999, **17**, 1183–1191.

15. Bhatt, D.L., Gaylor, D.C. & Lee, R.C. Rhabdomyolysis due to pulsed electric fields. *Plast Reconstr Surg* 1990, **86**, 1–11.

16. Denet, A.R., Vanbever, R. & Preat, V. Skin electroporation for transdermal and topical delivery. *Adv Drug Deliv Rev* 2004, **56**, 659–674.

17. Vanbever, R. & Preat, V.V. In vivo efficacy and safety of skin electroporation. *Adv Drug Deliv Rev* 1999, **35**, 77–88.

18. Ilic, L., Gowrishankar, T.R., Vaughan, T.E., Herndon, T.O. & Weaver, J.C. Spatially constrained skin electroporation with sodium thiosulfate and urea creates transdermal microconduits. *J Control Release* 1999, **61**, 185–202.

19. Zewert, T.E., Pliquett, U.F., Vanbever, R., Langer, R. & Weaver, J.C. Creation of transdermal pathways for macromolecule transport by skin electroporation and a low toxicity, pathway-enlarging molecule. *Bioelectrochem Bioenerg* 1999, **49**, 11–20.

20. Bos, J.D. The skin as an organ of immunity. *Clin Exp Immunol* 1997, **107**(Suppl 1), 3–5.

21. Glenn, G.M., Kenney, R.T., Ellingsworth, L.R., Frech, S.A., Hammond, S.A. & Zoeteweij, J.P. Transcutaneous immunization and immunostimulant strategies: capitalizing on the immunocompetence of the skin. *Expert Rev Vaccines* 2003, **2**, 253–267.

22. Partidos, C.D. & Muller, S. Decision-making at the surface of the intact or barrier disrupted skin: potential applications for vaccination or therapy. *Cell Mol Life Sci* 2005, **62**, 1418–1424.

23. Collins, R.A., Werling, D., Duggan, S.E., Bland, A.P., Parsons, K.R. & Howard, C.J. Gammadelta T cells present antigen to CD4+ alphabeta T cells. *J Leukoc Biol* 1998, **63**, 707–714.

24. Shimura, H., Nitahara, A., Ito, A., Tomiyama, K., Ito, M. & Kawai, K. Up-regulation of cell surface Toll-like receptor 4-MD2 expression on dendritic epidermal T cells after the emigration from epidermis during cutaneous inflammation. *J Dermatol Sci* 2005, **37**, 101–110.

25. Bonifaz, L., Bonnyay, D., Mahnke, K., Rivera, M., Nussenzweig, M.C. & Steinman, R.M. Efficient targeting of protein antigen to the dendritic cell receptor DEC-205 in the steady state leads to antigen presentation on major histocompatibility complex class I products and peripheral CD8+ T cell tolerance. *J Exp Med* 2002, **196**, 1627–1638.

26. Misra, A., Ganga, S. & Upadhyay, P. Needle-free, non-adjuvanted skin immunization by electroporation-enhanced transdermal delivery of diphtheria toxoid and a candidate peptide vaccine against hepatitis B virus. *Vaccine* 1999, **18**, 517–523.

27. Dujardin, N., Staes, E. & Preat, V. Needle free immunization using skin electroporation. *Proc Control Release Bioactive Mater* 2000, **27**, 7409.

28. Rozis, G., de Silva, S., Benlahrech, A. et al. Langerhans cells are more efficiently transduced than dermal dendritic cells by adenovirus vectors expressing either group C or group B fibre protein: implications for mucosal vaccines. *Eur J Immunol* 2005, **35**, 2617–2626.

29. Dell, K., Koesters, R. & Gissmann, L. Transcutaneous immunization in mice: induction of T-helper and cytotoxic T lymphocyte responses and protection against human papillomavirus-induced tumors. *Int J Cancer* 2006, **118**, 364–372.

30. Berry, L.J., Hickey, D.K., Skelding, K.A. et al. Transcutaneous immunization with combined cholera toxin and CpG adjuvant protects against *Chlamydia muridarum* genital tract infection. *Infect Immun* 2004, **72**, 1019–1028.

31. Matyas, G.R., Friedlander, A.M., Glenn, G.M., Little, S., Yu, J. & Alving, C.R.. Needle-free skin patch vaccination method for anthrax. *Infect Immun* 2004, **72**, 1181–1183.

32. Hammond, S.A., Walwender, D., Alving, C.R. & Glenn, G.M. Transcutaneous immunization: T cell responses and boosting of existing immunity. *Vaccine* 2001, **19**, 2701–2707.

33. Glenn, G.M., Taylor, D.N., Li, X., Frankel, S., Montemarano, A. & Alving, C.R. Transcutaneous immunization: a human vaccine delivery strategy using a patch. *Nat Med* 2000, **6**, 1403–1406.

34. Belyakov, I.M., Hammond, S.A., Ahlers, J.D., Glenn, G.M. & Berzofsky, J.A. Transcutaneous immunization induces mucosal CTLs and protective immunity by migration of primed skin dendritic cells. *J Clin Invest* 2004, **113**, 998–1007.

35. Maeda, T., Murata, K., Fukushima, T. et al. A novel plasmacytoid dendritic cell line, CAL-1, established from a patient with blastic natural killer cell lymphoma. *Int J Hematol* 2005, **81**, 148–154.

36. Prausnitz, M.R., Gimm, J.A., Guy, R.H., Langer, R., Weaver, J.C. & Cullander, C. Imaging regions of transport across human stratum corneum during high-voltage and low-voltage exposures. *J Pharm Sci* 1996, **85**, 1363–1370.

37. Pliquett, U.F., Zewert, T.E., Chen, T., Langer, R. & Weaver, J.C. Imaging of fluorescent molecule and small ion transport through human stratum corneum during high voltage pulsing: localized transport regions are involved. *Biophys Chem* 1996, **58**, 185–204.

38. van Drunen Littel-van den Hurk, S., Babiuk, S.L. & Babiuk, L.A. Strategies for improved formulation and delivery of DNA vaccines to veterinary target species. *Immunol Rev* 2004, **199**, 113–125.

39. Buchan, S., Gronevik, E., Mathiesen, I., King, C.A., Stevenson, F.K. & Rice, J. Electroporation as a "prime/boost" strategy for naked DNA vaccination against a tumor antigen. *J Immunol* 2005, **174**, 6292–6298.

40. Luxembourg, A., Hannaman, D., Ellefsen, B., Nakamura, G. & Bernard, R. Enhancement of immune responses to an HBV DNA vaccine by electroporation. *Vaccine* 2006, **24**, 4490–4493.

41. Chen, J., Fang, F., Li, X., Chang, H. & Chen, Z. Protection against influenza virus infection in BALB/c mice immunized with a single dose of neuraminidase-expressing DNAs by electroporation. *Vaccine* 2005, **23**, 4322–4328.

42. Medi, B.M., Hoselton, S., Marepalli, R.B. & Singh, J. Skin targeted DNA vaccine delivery using electroporation in rabbits. I. Efficacy. *Int J Pharm* 2005, **294**, 53–63.

43. Kalat, M., Kupcu, Z., Schuller, S. et al. In vivo plasmid electroporation induces tumor antigen-specific CD8+ T-cell responses and delays tumor growth in a syngeneic mouse melanoma model. *Cancer Res* 2002, **62**, 5489–5494.

44. Quaglino, E., Iezzi, M., Mastini, C. et al. Electroporated DNA vaccine clears away multifocal mammary carcinomas in her-2/neu transgenic mice. *Cancer Res* 2004, **64**, 2858–2864.

45. Lu, Z. & Ghazizadeh, S. Host immune responses in ex vivo approaches to cutaneous gene therapy targeted to keratinocytes. *Exp Dermatol* 2005, **14**, 727–735.

46. Bynoe, M.S., Evans, J.T., Viret, C. & Janeway, C.A., Jr. Epicutaneous immunization with autoantigenic peptides induces T suppressor cells that prevent experimental allergic encephalomyelitis. *Immunity* 2003, **19**, 317–328.

47. Bynoe, M.S., Viret, C., Flavell, R.A. & Janeway, C.A., Jr. T cells from epicutaneously immunized mice are prone to T cell receptor revision. *Proc Natl Acad Sci USA* 2005, **102**, 2898–2903.

48. Bynoe, M.S. & Viret, C. Antigen-induced suppressor T cells from the skin point of view: suppressor T cells induced through epicutaneous immunization. *J Neuroimmunol* 2005, **167**, 4–12.

49. Peachman, K.K., Rao, M. & Alving, C.R. Immunization with DNA through the skin. *Methods* 2003, **31**, 232–242.

50. Glenn, G.M., Scharton-Kersten, T. & Alving, C.R. Advances in vaccine delivery: transcutaneous immunisation. *Expert Opin Investig Drugs* 1999, **8**, 797–805.

51. Gapinski, J., Wilk, A., Patkowski, A. et al. Diffusion and microstructural properties of solutions of charged nanosized proteins: experiment versus theory. *J Chem Phys* 2005, **123**, 054708.

52. Prausnitz, M. Do high-voltage pulses cause changes in skin structure? *J Control Release* 1996, **40**, 321–326.

53. Coulberson, A.L., Hud, N.V., LeDoux, J.M., Vilfan, I.D. & Prausnitz, M.R. Gene packaging with lipids, peptides and viruses inhibits transfection by electroporation in vitro. *J Control Release* 2003, **86**, 361–370.

54. Pliquett, U.F. Local transport regions in human stratum corneum due to short and long high-voltage pulses. *Bioelectrochem Bioenerget* 1998, **47**, 151–161.

55. Pliquett, U., Gallo, S., Hui, S.W., Gusbeth, C. & Neumann, E. Local and transient structural changes in stratum corneum at high electric fields: contribution of Joule heating. *Bioelectrochemistry* 2005, **67**, 37–46.

56. Lombry, C., Dujardin, N. & Preat, V. Transdermal delivery of macromolecules using skin electroporation. *Pharm Res* 2000, **17**, 32–37.

57. Dujardin, N., Staes, E., Kalia, Y., Clarys, P., Guy, R. & Preat, V. In vivo assessment of skin electroporation using square wave pulses. *J Control Release* 2002, **79**, 219–227.

58. Lucas, M.L., Jaroszeski, M.J., Gilbert, R. & Heller, R. In vivo electroporation using an exponentially enhanced pulse: a new waveform. *DNA Cell Biol* 2001, **20**, 183–188.

59. Yakovenko, S.A. Electroporators based on digital formation of arbitrarily-shaped electroporation pulses. *Biomed Instrum Technol* 2004, **38**, 397–409.

60. Kalia, Y.N., Pirot, F. & Guy, R.H. Homogeneous transport in a heterogeneous membrane: water diffusion across human stratum corneum in vivo. *Biophys J* 1996, **71**, 2692–2700.

61. Zhang, L. & Rabussay, D.P. Clinical evaluation of safety and human tolerance of electrical sensation induced by electric fields with non-invasive electrodes. *Bioelectrochemistry* 2002, **56**, 233–236.

62. Upadhyay, P. Electroporation of the skin to deliver antigen by using a piezo ceramic gas igniter. *Int J Pharm* 2001, **217**, 249–253.

63. Sugibayashi, K., Yoshida, M., Mori, K., Watanabe, T. & Hasegawa, T. Electric field analysis on the improved skin concentration of benzoate by electroporation. *Int J Pharm* 2001, **219**, 107–112.

64. Mori, K., Hasegawa, T., Sato, S. & Sugibayashi, K. Effect of electric field on the enhanced skin permeation of drugs by electroporation. *J Control Release* 2003, **90**, 171–179.

65. Zhang, L., Nolan, E., Kreitschitz, S. & Rabussay, D.P. Enhanced delivery of naked DNA to the skin by non-invasive in vivo electroporation. *Biochim Biophys Acta* 2002, **1572**, 1–9.

66. Bernard, R.M., Hannaman, A.W. & Bernard, B.M. Method and apparatus for delivery of therapeutic agents. *U.S. patent* 6,912,417, 2005.

67. Guebre-Xabier, M., Hammond, S.A., Epperson, D.E., Yu, J., Ellingsworth, L. & Glenn, G.M. Immunostimulant patch containing heat-labile enterotoxin from Escherichia coli enhances immune responses to injected influenza virus vaccine through activation of skin dendritic cells. *J Virol* 2003, **77**, 5218–5225.

68. Schuler, G. & Steinman, R.M. Dendritic cells as adjuvants for immune-mediated resistance to tumors. *J Exp Med* 1997, **186**, 1183–1187.

69. Steinman, R.M. & Pope, M. Exploiting dendritic cells to improve vaccine efficacy. *J Clin Invest* 2002, **109**, 1519–1526.

70. Bennett, C.L., van Rijn, E., Jung, S. et al. Inducible ablation of mouse Langerhans cells diminishes but fails to abrogate contact hypersensitivity. *J Cell Biol* 2005, **169**, 569–576.

71. Gallucci, S., Lolkema, M. & Matzinger, P. Natural adjuvants: endogenous activators of dendritic cells. *Nat Med* 1999, **5**, 1249–1255.

72. Cumberbatch, M., Dearman, R.J., Griffiths, C.E. & Kimber, I. Langerhans cell migration. *Clin Exp Dermatol* 2000, **25**, 413–418.

73. Hofmann-Wellenhof, R., Smolle, J., Roschger, A. et al. Sunburn cell formation, dendritic cell migration, and immunomodulatory factor production after solar-simulated irradiation of sunscreen-treated human skin explants in vitro. *J Invest Dermatol* 2004, **123**, 781–787.

74. Laouini, D., Alenius, H., Bryce, P., Oettgen, H., Tsitsikov, E. & Geha, R.S. IL-10 is critical for Th2 responses in a murine model of allergic dermatitis. *J Clin Invest* 2003, **112**, 1058–1066.

75. Gangi, S. & Johansson, O. A theoretical model based upon mast cells and histamine to explain the recently proclaimed sensitivity to electric and/or magnetic fields in humans. *Med Hypotheses* 2000, **54**, 663–671.

76. Narumiya, S. Prostanoids in immunity: roles revealed by mice deficient in their receptors. *Life Sci* 2003, **74**, 391–395.

77. Vernier, P.T., Sun, Y., Marcu, L., Salemi, S., Craft, C.M. & Gundersen, M.A. Calcium bursts induced by nanosecond electric pulses. *Biochem Biophys Res Commun* 2003, **310**, 286–295.

78. Granstein, R.D., Ding, W., Huang, J. et al. Augmentation of cutaneous immune responses by ATP gamma S: purinergic agonists define a novel class of immunologic adjuvants. *J Immunol* 2005, **174**, 7725–7731.

79. Misra, A., Rafi, K. & Khar, R. Iontophoretic in vivo release of diptheria toxoid through skin. Presented at the Indian Pharmaceutical Congress, Chandigarh, India, 1994.

80. Rath, A., Batra, D., Kaur, R., Vrati, S. & Gupta, S.K. Characterization of immune response in mice to plasmid DNA encoding dog zona pellucida glycoprotein-3. *Vaccine* 2003, **21**, 1913–1923.

81. Prausnitz, M.R. Overcoming skin's barrier: the search for effective and user-friendly drug delivery. *Diabetes Technol Ther* 2001, **3**, 233–236.

82. Sen, A., Daly, M.E. & Hui, S.W. Transdermal insulin delivery using lipid enhanced electroporation. *Biochim Biophys Acta* 2002, **1564**, 5–8.

# 17

# NONCLINICAL SAFETY ASSESSMENT OF VACCINES AND THE EVALUATION OF NOVEL ADJUVANTS AND DELIVERY SYSTEMS

JOSE A. LEBRON, JAYANTHI J. WOLF, CATHERINE V. KAPLANSKI, AND BRIAN J. LEDWITH

## 17.1  INTRODUCTION

Vaccines represent the most diverse type of pharmaceutical products, including attenuated and killed pathogens (bacteria, parasites, and viruses), purified and recombinant proteins, synthetic peptides and polysaccharide preparations (free or conjugated to carriers), and recombinant DNA and viral vectors. In addition, vaccines can be formulated with various types of adjuvants [e.g., aluminum-based, oil emulsions, virosomes, immunostimulating complexes (ISCOMs), or microbial derivates] or may utilize novel delivery mechanisms (e.g., electroporation, jet injection, microneedles, or nanoparticles). Despite this vast diversity, there is a general approach to the safety assessment of vaccines, which is reviewed here.

Ensuring the safety of vaccine candidates prior to the first evaluation in humans usually involves a multicomponent approach comprising (1) evaluation of the vaccine in animal immunogenicity and efficacy studies, (2) formal testing of the vaccine in toxicology (nonclinical safety assessment) studies in animals, and (3) development of a comprehensive quality control (QC) testing

*Vaccine Adjuvants and Delivery Systems*, Edited by Manmohan Singh
Copyright © 2007 John Wiley & Sons, Inc.

program that includes product release (e.g., identity, purity, potency, mass, and sterility assays) and lot characterization assays, and tests for adventitious agents, if applicable. However, the nonclinical safety assessment program is the most important component in ensuring the safety of vaccines, as it helps minimize risks to humans by identifying potential or unknown toxicities and target organs, and eliminates potential vaccine candidates that have unacceptable toxicities prior to entry into human clinical trials. Thus, the discussion presented here will focus primarily on the strategy for designing the nonclinical safety assessment studies needed to support testing of vaccines in human clinical trials. First, regulatory expectations will be discussed, followed by a discussion of the issues to be considered when developing the nonclinical safety assessment strategy, including the choice of animal species, study design, and types of studies available to support the various clinical developmental stages of vaccines. This discussion also focuses on how the nonclinical safety assessment testing strategy can be applied to the evaluation of vaccines when they contain novel adjuvants and/or delivery mechanisms. Finally, as quality control testing programs are also an important component in ensuring the safety of vaccines, a summary of the key regulatory guidances for this type of testing is presented and a brief description of the most important components of quality control testing programs is also given.

## 17.2 CONSIDERATIONS FOR THE DESIGN OF NONCLINICAL SAFETY ASSESSMENT STUDIES

Historically, vaccines were generally considered to be inherently safe. As a result, nonclinical safety assessment studies had not been required until relatively recently, although a careful assessment of safety has long been an integral part of the evaluation of vaccines in human clinical trials. As many new technologies began to be incorporated into experimental vaccines, a need to provide compelling evidence for the safety of the vaccine prior to the initiation of human clinical trials was realized, and nonclinical safety testing is currently required for all new vaccines.

The design of the toxicity studies is based on a scientific understanding of the vaccine, including the intended pharmacological and immunological effects, method of manufacture, and intended clinical use. The latter, in particular, is used to select the route of administration, study duration, dose levels, and dosing regimen for the preclinical toxicity studies. The clinical route of administration is used whenever possible (e.g., intramuscular, subcutaneous, intradermal, oral, or intravenous), including the use of particular clinical delivery devices (e.g., microneedles, jet injection, nanoparticles, or electroporation). The administration of a full human dose equivalent to animals is recommended. If the size of the animal prevents dosing with the full human dose equivalent or requires use of a nonstandard delivery device, these deviations should be discussed with regulatory agencies prior to performing the toxicology studies. The schedule of vaccine administrations should mimic the clinical protocol, although for vaccines with wide intervals between doses in clinical studies, dosing intervals may be compressed in the preclinical studies (to approximately two to three weeks) provided that comparable immunogenicity can be demonstrated. The U.S. Food and Drug Administration (FDA) has recommended, for preventive vaccines, that animals receive one more dose than is to be administered in the clinic, often referred to as the *N + 1 rule* [1]. The number of animals in the toxicology studies must be adequate to make an appropriate evaluation, and the sample size depends on the animal model used.

## 17.3   REGULATORY EXPECTATIONS

According to existing regulatory guidelines pertaining to the safety assessment of vaccines (Table 17.1), all new vaccines should be tested in nonclinical safety assessment (toxicology) studies prior to being tested in the clinic, with some exceptions: for example, for combined vaccines containing known antigens [2]. General guidelines for nonclinical safety assessment of vaccines have been published by the European Medicines Evaluation Agency (EMEA) and the World Health Organization (WHO) [2,3]. Guidance documents for specific types of vaccine classes, such as DNA vaccines, viral vectors, combination vaccines, and recombinant protein–peptide vaccines, have also been provided by several regulatory agencies, including the FDA. Addition of a new adjuvant or novel delivery system for vaccines presents unique considerations, and toxicology studies need to be designed on a case-by-case basis to evaluate the safety profile of the adjuvant and adjuvant–vaccine combination. In 2005, the EMEA published a guideline for vaccine adjuvants which covers aspects of nonclinical and clinical testing for new adjuvants.

In addition to the guidelines listed in Table 17.1, nonclinical safety studies are performed in compliance with Good Laboratory Practices (GLP) regulations as detailed in 21 CFR Part 58 [15]. Thorough documentation is needed during all phases of a preclinical study and the GLP regulations clearly define the roles and responsibilities of the study sponsor as well as those of the study director, quality assurance unit, and management at the testing laboratory. Since laboratory animals are used in nonclinical safety assessment studies, the regulations and standards for animal welfare issued by the National Institutes of Health (NIH) and the Association for Assessment and Accreditation of Laboratory Animal Care (AAALAC) must be followed.

## 17.4   VACCINE LOTS USED IN NONCLINICAL SAFETY ASSESSMENT STUDIES

Following the regulations discussed above, the vaccine lots used in safety assessment studies must be well characterized and evaluated for identity, potency, purity, and stability. Ideally the intended clinical supplies, which are

**TABLE 17.1 Guidelines Pertaining to the Nonclinical Safety Assessment of Vaccines**

| Vaccine Class | Guideline | Ref. |
|---|---|---|
| All vaccines | EMEA Notes for Guidance of Preclinical Pharmacological and Toxicological Testing of Vaccines (1997) | 2 |
| All vaccines | WHO Guidelines on Nonclinical Evaluation of Vaccines (2003) | 3 |
| Covers vaccines (although this is more relevant for other biologics) | ICH S6: Preclinical Safety Evaluation of Biotechnology-Derived Pharmaceuticals (1997) | 4 |
| Vaccines for pregnant women and women of child-bearing potential | FDA Guidance for Industry. Considerations for Developmental Toxicity Studies for Preventive and Therapeutic Vaccines for Infectious Disease Indications (2006) | 5 |
| Adjuvanted vaccines | EMEA Guideline on Adjuvants in Vaccines for Human Use (2005) | 6 |
| Combination vaccines | EMEA Notes for Guidance on Pharmaceutical and Biological Aspects of Combined Vaccines (1998) | 7 |
| DNA vaccines | FDA Guidance for Industry (Draft). Considerations for Plasmid DNA Vaccines for Infectious Disease Indications (2005) | 8 |
| Viral vector and cell-based vaccines | FDA Guidance for Industry: Guidance for Human Somatic Cell Therapy and Gene Therapy (1998) | 9 |
| Viral vector and DNA vaccines | EMEA Note for Guidance on the Quality, Preclinical and Clinical Aspects of Gene Transfer Medicinal Products (2001) | 10 |
| Cell-based vaccines | EMEA Points to Consider on the Manufacture and Quality Control of Human Somatic Cell Therapy Medicinal Products (2001) | 11 |
| Influenza vaccines | EMEA Points to Consider on the Development of Live Attenuated Influenza Vaccines (2003) | 12 |
| Smallpox vaccines | EMEA Note for Guidance on the Development of Vaccinia Virus Based Vaccines Against Smallpox (2002) | 13 |
| Recombinant protein/peptide vaccines | FDA Points to Consider in the Production and Testing of New Drugs and Biologicals Produced by Recombinant DNA Technology (1985) | 14 |

produced according to current good manufacturing practice (cGMP), should be used in nonclinical safety assessment studies. However, in cases where cGMP materials are not available, a non-GMP vaccine lot can be used in the animal study if it is demonstrated to be analytically comparable to the clinical material by product release testing. Given that the nonclinical safety assessment studies are usually conducted over several months, vaccine stability should be monitored at the start and end of a GLP safety assessment study [15], in accordance with the GLP regulations, to ensure that the toxicity evaluation was performed using materials that have maintained their potency. However, short-term stability studies may be used to establish the stability of the vaccine under the conditions of dosing. For example, if the vaccine is administered via syringes, a short-term syringe stability study can be performed to ensure that the vaccine is stable in the syringe during the time taken for dosing the animals. The stability of adjuvants that are contained in the vaccine formulation should also be assessed using the relevant physical, chemical, and/or biological properties based on the characteristics of the adjuvant. For adjuvants that may adsorb or bind to vaccine antigens, the stability parameters may include measurements of the antigen adsorption or binding characteristics.

The formulation used during the study must be representative of the proposed clinical formulation. If formulation changes occur after pivotal toxicology studies are performed, the impact of the formulation changes on the results of the previous safety studies should be evaluated. Usually, formulation changes are made to improve the stability of the vaccine and may include changes in pH or excipient concentrations, addition of excipients, addition of an approved or a novel adjuvant, and/or changes in the concentration of active ingredients. Some of these changes may be insignificant, such as a minor change in pH or in the concentration of an excipient that is listed in the FDA *Inactive Ingredient Guide* [16], and can usually be addressed using an analytical bridging strategy. Other changes, such as the addition of an adjuvant or the use of a new delivery mechanism, may be considered significant and require additional animal safety studies to demonstrate that the safety and efficacy of the vaccine have not been compromised [17]. All pivotal safety studies should be performed in compliance with the GLPs using qualified or validated assays whenever possible.

## 17.5   SELECTION OF AN ANIMAL MODEL FOR NONCLINICAL SAFETY ASSESSMENT STUDIES

Safety assessment programs for vaccines typically use a single species (unlike the toxicology programs for new chemical entities, where two species are required); however, the chosen species must be shown to be a relevant animal model. Relevance is usually demonstrated based on the immunogenicity of the vaccine in the chosen species. In the case of live virus vaccines, the susceptibility of the animal species to infection by the wild-type organism should also be considered. Safety assessment programs for vaccines generally use rodents or rabbits; however, nonhuman primates are sometimes used as a last resort if the vaccine lacks immunogenicity in all other species, or if the organism is host restricted. In some cases, relevant animal models may not exist (e.g., use of a human cytokine adjuvant that does not have any activity in any animal species). In these cases, an animal homolog (e.g., use of the animal version of the human gene) could be used in addition to testing the vaccine candidate. Transgenic animals might also be used to address specific concerns (e.g., use of mice that express a human cell surface receptor); however, since few historical control data exist for transgenic models, it is difficult to put any unexpected findings into perspective. Typically, immunogenicity is first evaluated in an exploratory study to confirm the relevance of the animal species chosen for the definitive toxicity assessment.

## 17.6  IMMUNOGENICITY EVALUATIONS WITHIN NONCLINICAL SAFETY ASSESSMENT STUDIES

As outlined in the WHO guideline [3], it is recommended that the immunogenicity of the vaccine be measured concurrently with the evaluation of its potential for toxicity. The main objective of this measure is to confirm the expected immune response in the animal model within the toxicology studies. In addition, this evaluation might allow the correlation of any toxic effects with the degree of the immune response induced by the vaccine.

Typically, immunoassays measuring either an antigen-specific antibody response or a cellular immune response are developed and used, depending on the vaccine candidate and the arm of the immune response it targets. Most frequently, immunoassays using a standard ELISA format are used to measure an antibody-based immune response. However, multiplex assays are being used more widely for multiple antigen candidate vaccines: based on newer technologies, such as Luminex, these assays allow the simultaneous evaluation of multiple antigen-specific antibody responses in a single serum sample [18]. When the candidate vaccine targets the cellular arm of the immune response, specific assays measuring cytokine-secreting antigen-specific T lymphocytes are most appropriate, as they allow an evaluation of the most relevant antigen-specific immune response in the nonclinical safety assessment studies. Immunoassays such as γ-interferon ELIspot assays allow an efficient and sensitive assessment of the T-cell response to one or several antigens, although such assays carry the disadvantage of being quite resource intensive. Briefly, peripheral blood mononuclear cells collected from animals administered the candidate vaccine are stimulated ex vivo with the corresponding peptide antigen(s), and the number of γ-interferon secreting T lymphocytes is evaluated [19]. Finally, it should be noted that the availability of reagents may sometimes limit the development of immunoassays (particularly those measuring cell-mediated immune responses) in certain animal species used routinely for the nonclinical safety assessment studies.

## 17.7  TYPES OF NONCLINICAL SAFETY ASSESSMENT STUDIES

The GLP-compliant nonclinical safety assessment studies are designed to evaluate (1) the inherent toxicity of the vaccine; (2) the toxicity of any adjuvants, impurities, and contaminants; (3) the toxicity due to the interaction of the vaccine antigens with other components in the vaccine formulation; and (4) the toxicity linked to the immune response induced by the antigen. Studies that might be performed include single-dose toxicity, repeat-dose toxicity, local tolerance, safety pharmacology, developmental and reproductive toxicity, biodistribution and integration (for DNA or viral-vectored vaccines), genotoxicity (for some adjuvants), and virulence and neurovirulence testing. Since vaccine safety assessment programs are typically designed on a case-by-case basis, usually with input from regulatory agencies, not all of these studies need to be performed. Each of these studies, and the reason for their inclusion or exclusion, are discussed below.

The vaccine formulation used in these studies is the final formulation (e.g., for adjuvanted vaccines, the vaccine and adjuvant are tested together) and is representative of the proposed clinical formulation. The adjuvant alone is usually included as a control group in these studies, but is usually not tested by itself unless it is a new chemical-based adjuvant. In the latter case, the new adjuvant is treated as a novel chemical entity and requires single- and repeat-dose toxicity studies in two species (rodent and nonrodent), studies of pharmacokinetics and tissue distribution, assessment of genotoxicity, and determination of the potential for systemic hypersensitivity and pyrogenicity [6]. More details on considerations for new synthetic adjuvants are provided below.

### 17.7.1 Single-Dose Toxicity Studies

Single-dose toxicity studies are designed to identify the acute effects of the vaccine and usually involve the administration of a single dose of the vaccine followed by a period that allows for the development of a primary immune response (approximately 14 days) and evaluation of various parameters, such as mortality, clinical signs, body weight, and macroscopic examination. Although they are mentioned in the EMEA guideline [2], these studies are not always performed for vaccines that are used in the clinic with a repeated-dose schedule, since single-dose evaluations could be performed during a repeated-dose toxicity study. It should be noted that single-dose toxicity studies in mice and guinea pigs using the intraperitoneal route is a batch release test (general safety test) for vaccine lots that are tested in the clinic. However, this is considered to be part of the quality control of the vaccine lot rather than part of nonclinical safety assessment.

### 17.7.2 Repeated-Dose Toxicity Studies

These studies are usually the pivotal GLP toxicity studies for a vaccine product since most vaccines are administered in the clinic using a repeated-dose regimen. Repeated-dose toxicity studies incorporate an evaluation of both ante- and postmortem parameters. Antemortem parameters generally include mortality, physical signs, body weights, food consumption, ophthalmic examinations, urinalyses, hematology, serum biochemistry, coagulation, and concurrent assessment of immunogenicity. Complete necropsies are usually performed at two time points, a few days (e.g., two to three days) after the last vaccine dose (to determine the early effects after vaccine dosing) and two to four weeks after the last vaccine dose (to detect any delayed toxicity and determine whether any detected effects have resolved over time). Postmortem evaluations include gross examination of all major organs, organ weights, and a complete list of tissues (as described in the WHO vaccine guidance) are collected [3]. Histomorphological examination will depend on the vaccine and is usually focused on the immune organs (spleen, thymus, draining lymph nodes), pivotal organs (brain, kidneys, liver, reproductive organs), and the site of vaccine administration (e.g., quadriceps at the injection site and skin over the quadriceps if the vaccine is administered intramuscularly) [3]. In certain cases where there are specific theoretical safety concerns, additional parameters may be included in the study. For example, potential systemic inflammatory responses could be evaluated using a serum IL-6 measurement, and potential pathogenic autoimmune responses against a particular tissue could be evaluated by more detailed immunohistochemistry or other relevant evaluations (e.g., autoantibodies).

Treatment-related effects are often observed in repeat-dose toxicity studies, and these effects are usually indicative of local reactogenicity (e.g., redness at the injection site), enlargement and hyperplasia of lymph nodes draining the injection sites, increase in spleen weight, and hematological changes. Unless unusually severe, these are consistent with the expected and intended immunological and inflammatory responses to the vaccine and are not considered adverse effects. For vaccines, in many cases the highest dose level tested (full human dose) is the no-observed-adverse-effects level (NOAEL); however, with the introduction of novel adjuvants that might cause toxicity, it is recommended that at least two vaccine–adjuvant dose levels be tested such that a NOAEL can be determined.

### 17.7.3  Local Tolerance Studies

Local tolerance studies are designed to evaluate irritation at the injection site. A single dose of the vaccine is administered, half of the animals are subjected to an interim necropsy (a few days after the dose), and the remainder are taken to a final necropsy (approximately two weeks after the dose). The injection site is examined both macroscopically and histologically. In the interest of reducing animal use and study costs, the assessment of local tolerance can be performed within the repeated-dose toxicity study.

### 17.7.4  Safety Pharmacology Studies

Safety pharmacology studies investigate the potential for undesirable pharmacodynamic effects of a substance on physiological functions [20]. The classical safety pharmacology studies for drug products focus on the cardiovascular system, respiratory system, and central nervous system. Safety pharmacology studies are not mandated for new vaccines, but rather, are to be considered when appropriate. The EMEA guidance [2] recommends the consideration of safety pharmacology studies for vaccines, and the WHO guidance [3] states that "if data from nonclinical and/or human clinical studies suggest that the vaccine (e.g., one based on specific toxoids) may affect physiological functions (CNS, respiratory, cardiovascular, renal functions) other than the immune system, safety pharmacology studies should be incorporated into the toxicity assessment." Additionally, the FDA's Center for Biologics Evaluation and Research (CBER) has stated that safety pharmacology studies are generally not needed for vaccines, and that this should be considered on a case-by-case basis [1]. Furthermore, the International Conference on Harmonization (ICH) S7A guidance on safety pharmacology for human pharmaceuticals [20] states that "safety pharmacology studies may not be needed for locally applied agents where the pharmacology of the test substance is well characterized, and where the systemic exposure or distribution to other organs or tissues is demonstrated to be low." As the systemic exposure to vaccines is generally low, it could be argued that, in general, safety pharmacology studies may not be needed for vaccines. However, if they are needed, some of the safety pharmacology parameters (e.g., body temperature, blood pressure, and central nervous system evaluations) might be collected within a repeated-dose toxicity study in order to reduce animal use and cost.

### 17.7.5 Developmental and Reproductive Toxicity Studies

Developmental and reproductive toxicity studies provide information on potential effects of the product on fertility, fetal development, and postnatal development of the offspring. These studies are not needed for vaccines that are given during childhood, to adult men, or to women of non-child-bearing potential. However, vaccines that will be administered to women of child-bearing potential require developmental toxicity testing prior to or concurrent with large-scale (phase III) clinical studies. A guidance issued by the FDA outlines the study design for a combined embryo fetal and postnatal development study [5]. The female animals are immunized a few weeks before mating to ensure peak immune responses during the critical phases of pregnancy (e.g., organogenesis). The animals are then boosted during gestation (embryo–fetal period) and lactation (postnatal period) to evaluate potential direct embryo-toxic effects of the components of the vaccine formulation and to maintain a high level of antibody throughout the remainder of gestation. Placebo control groups should be included in the study, and if an adjuvant is included in the vaccine, an adjuvant-alone control group should be included. Sufficient animals should be assigned per group such that half of the animals (at least 20) may be taken through cesarean section for routine uterine and fetal examinations, and the other half (at least 20) would undergo natural delivery and be allowed to rear their offspring to weaning to monitor postnatal development.

Several evaluations are performed in these studies. For $F_0$ female rats, evaluations include body weight, potential signs of local toxicity, food consumption, duration of pregnancy, abortions, premature deliveries, and parturition (for maternal animals not subjected to Caesarian sectioning). For the group that undergoes Caesarean sectioning, a necropsy is performed (macroscopic examination) and maternal tissue with macroscopic findings is preserved for histological evaluations as deemed necessary by the gross findings. The number and distribution of corpora lutea, implantation sites, viable and nonviable fetuses, and early and late resorptions are recorded, and a gross evaluation of the placenta is performed. The live fetuses are examined for body weight and for gross external, visceral, and skeletal alterations. Late resorptions and dead fetuses are also examined for gross external alterations to the extent possible. All fetuses are examined internally to determine sex.

For the group that undergoes natural delivery, a calculation of fertility index, gestation index, and live birth index is performed. Animals that deliver a litter are sacrificed at the end of the preweaning period, and a gross necropsy of the thoracic, abdominal, and pelvic viscera is performed. Animals that die or are sacrificed because of moribund condition, abortion, or premature delivery should be examined for the cause of death and pregnancy status recorded. The $F_1$ generation is followed from birth to weaning to assess normal growth, body weight gain, and nursing activity as indicators for normal development. The study usually includes tests to screen for normal neurological development: for example, auditory and visual function tests. Viability and lactation indices should be determined and individual sexes should be recorded. At terminal sacrifice, a necropsy is performed and gross lesions are retained for possible histological examinations.

An assessment of the vaccine-induced antibody response is performed in this study to verify exposure of the embryo–fetus to maternal antibody. Serum samples from maternal animals prior to and following dosing are collected to assess the development of antibodies. In addition, it is recommended that cord blood samples be taken to assess placental transfer of maternal antibodies from animals in the Caesarean subgroup. Additionally, antibody levels from a representative number of pups or litters at the end of the weaning period is assessed to verify exposure of the neonates to maternal antibodies.

### 17.7.6 Biodistribution and Integration Studies

Biodistribution and integration studies are performed for nucleic acid– and viral vector-based vaccines to determine the tissue distribution following administration and the potential for the vector to integrate into the host genome. These studies may not be needed for vectors that have been tested in the clinic; therefore, regulatory agencies should be consulted for advice regarding whether these studies are needed for a particular vaccine. Biodistribution and integration studies are usually separate GLP studies, where animals are administered a single dose of the vaccine via the proposed clinical route of administration, including the use of a new delivery device (if applicable). Necropsies are then performed on cohorts of animals at different time points, usually from one day up to six months after injection to determine vector persistence in the tissues over time. Typically, a minimum of three animals per gender per group and per time point are used. DNA is extracted from the tissues and analyzed using sensitive molecular techniques (e.g., quantitative Polymerase Chain Reaction assays) to detect the presence of the vaccine in a variety of tissues, including the injection site muscle (if administered intramuscularly), skin overlying the muscle, and the draining lymphatic tissues. It is important to examine the reproductive organs to rule out any possibility of germline transmission. To determine the chance for insertional mutagenesis, a quantitative-size fractionation/gel purification assay can be performed that involves gel purification of cellular DNA and quantitative PCR for integrated vaccine vectors [21,22]. A qualitative confirmatory assay, the repeat-anchored integration-capture (RAIC) assay, can be performed to identify insertion sites if results from the gel purification assay show potential vector integration [23].

### 17.7.7 Genotoxicity Studies

According to the EMEA *Note for Guidance on Preclinical Pharmacological and Toxicological Testing of Vaccines* [2] and the *WHO Guidelines on Nonclinical Evaluation of Vaccines* [3], genotoxicity studies are not generally required for vaccines. However, genotoxicity studies should be considered if a novel chemical linker is included. For adjuvants of biological origin, genotoxicity studies might not be regarded as relevant [4]. For synthetic adjuvants, which are considered to be new chemical entities, the standard battery of tests that are used to assess the potential for gene mutation, chromosome aberrations, and primary DNA damage are needed [24] and any deviations from this panel of tests should be justified scientifically.

### 17.7.8 Carcinogenicity Studies

According to the EMEA *Note for Guidance on Preclinical Pharmacological and Toxicological Testing of Vaccines* [2] and the *WHO Guidelines on Nonclinical Evaluation of Vaccines* [3], carcinogenicity studies are not generally required for vaccines. Carcinogenicity studies are also not needed for adjuvants. According to the EMEA *Guideline on Adjuvants in Vaccines for Human Use* [6], adjuvants are intended to be used only a few times with low dosages such that the risk of induction of tumors by these compounds in a direct way is very small. Additionally, the action of the adjuvant is to stimulate the immune system and not to act as a general immunosuppressant, reducing the risk of the spontaneous formation of lymphoid tumors.

### 17.7.9   Virulence and Neurovirulence Studies

These studies may be needed for live attenuated virus vaccines that have either a theoretical or established potential for reversion of attenuation, also referred to as reversion to virulence [25]. Assessment of potential reversion can be facilitated by genetic or phenotypic markers of attenuation or virulence, and also often involves in vitro or in vivo potency or virulence tests. For new vaccine viruses that have neurotropic activity, neurovirulence testing may be required [3]. It should be noted that for established measles, mumps, rubella, and varicella vaccine strains that have a good safety record, minimal changes in seed lots or in manufacture should not require additional neurovirulence testing [26].

### 17.7.10   Special Considerations for New Synthetic Adjuvants

New synthetic adjuvants are treated as new chemical entities and require a separate toxicological testing program similar to small molecules. This includes single- and repeat-dose toxicity studies in two species, one rodent and one nonrodent, unless a specific scientific rationale can be provided for why a single species might be sufficient [6]. Since adjuvants may stimulate the immune system, tests for the induction of systemic hypersensitivity in appropriate models should be considered. For synthetic chemical-based adjuvants, an assessment of genotoxicity is needed using the standard battery of tests (e.g., potential for gene mutation, chromosome aberrations, and primary DNA damage) [24]. Although pharmacokinetic studies (e.g., determining serum concentrations of antigens) are not required for vaccines [2,3], distribution studies may be of value in understanding the mode of action of the adjuvant. Additionally, the EMEA *Guideline on Adjuvants in Vaccines for Human Use* recommends an in vivo test for pyrogenicity (with the possibility of using alternative in vitro tests for fever-inducing substances if such tests are validated) [6].

## 17.8   QUALITY CONTROL TESTING PROGRAMS

When evaluating the safety of vaccines prior to human clinical trials, it is also important to understand that the nonclinical safety assessment evaluations are only part of the overall safety evaluation of vaccines, which also includes an extensive immunogenicity evaluation in animals and a comprehensive quality control (QC) testing program. Although a detailed discussion on these two components is beyond the scope of this document, a brief description of the QC testing programs requirements are presented below, as many regulatory expectations are also imposed on this type of program.

The development of QC testing programs is usually determined by the key guidelines presented in Table 17.2. In addition to these guidelines, QC tests must be performed in compliance with current Good Manufacturing Practice

**TABLE 17.2  Key Guidelines Pertaining to Vaccine QC Testing Programs**

| Regulatory Agency | Guidance Document | Description | Ref. |
|---|---|---|---|
| FDA (CBER) | Content and Format of Chemistry, Manufacturing and Controls Information and Establishment Description Information for a Vaccine or Related Product (1999) | Provides guidance on the content and format of the chemistry, manufacturing, and controls of a license application for a vaccine | 29 |
| | Points to Consider in the Characterization of Cell Lines Used to Produce Biologicals (1993) | Represents the consensus of CBER staff on characterization of cell lines used to produce biological products (including vaccines) | 30 |
| | Guidance for Industry: Guidance for Human Somatic Cell Therapy and Gene Therapy (1998) | Provides guidance for manufacturers with production, quality control testing, and administration of recombinant vectors for cell-based vaccines and gene-therapy vectors (therapeutic vaccines only, not preventive vaccines) | 9 |
| ICH | Q5A Viral Safety Evaluation of Biotechnology Products Derived from Cell Lines of Human or Animal Origin (1997) | Provides a general framework for virus testing experiments for the evaluation of virus clearance | 31 |
| | Q5C Quality of Biotechnology Products: Stability Testing of Biotechnological/Biological Products (1995) | Outlines stability test procedures | 32 |
| | Q5D Quality of Biotechnology/Biological Products: Derivation and Characterization of Cell Substrates Used for Production of Biotechnological/Biological Products (1997) | Provides guidance on standards for derivation of human and animal cell lines and microbial cells to be used in preparation of biological products (including vaccines) | 33 |
| | ICH Q5E: Comparability of Biotechnological/Biological Products Subject to Changes in Their Manufacturing Process (2004) | This guideline can be used when changes to manufacturing process are made during development and after approval | 34 |
| CFR | 21CFR Part 610 | Provides general biological product guidelines including safety, identity, and purity guidances | 28 |
| | 21 CFR Part 312 | Provides guidelines on information that should be submitted in an investigational new drug application | 35 |
| USP | Viral Safety Evaluation of Biotechnology Products Derived from Cell Lines of Human or Animal Origin. United States Pharmacopeia (USP) | Provides guidance for viral safety evaluation of recombinant vaccines | 36 |
| EP | Biological Tests, European Pharmacopeia (EP) 2.6 | Provides guidelines for biological tests, including tests for sterility, pyrogens, and endotoxins | 37 |
| | Cell Substrates for the Production of Vaccines for Human Use, EP 5.2.3 | Provides guidelines for testing of cell lines used for the production of vaccines for human use | 38 |

415

**TABLE 17.3   Assay Types Typically Included in the QC Testing Programs of Vaccines**

| Assay | Purpose |
|---|---|
| Sterility | Determine bacterial and fungal contamination. |
| Endotoxin | Determine amount of bacterial endotoxin present in vaccine preparations. |
| General safety | Observe animals for lethality and weight loss; performed as part of the final container release and is not always required. |
| Identity | Show that the vaccine contains the intended immunogenic components. |
| Mass | Determine the amount of vaccine antigen per dose. |
| Potency | Estimate the biological activity and immunological strength of a vaccine. |
| Purity | Estimate the amount of process residuals and impurities that are an inevitable consequence of the manufacturing process. |
| Stability | Demonstrate that the vaccine product is stable in the final container. Results can be used to set a shelf life under appropriate storage conditions. |

(cGMP) regulations as detailed in the *Code of Federal Regulations* (CFR), Title 21, Parts 210 and 211 [27]. QC testing programs for vaccines typically contain a panel of release assays that include sterility, endotoxin, general safety, identity, mass, potency, purity, and stability (Table 17.3) [28]. Each of these assays has to be certified/qualified in the formulation in which the vaccine is provided to ensure that the formulation itself does not interfere with the assay.

## 17.9   ADDITIONAL CONSIDERATIONS FOR THE QC TESTING OF VACCINES

In addition to the assays described above, the QC testing program for vaccines manufactured using cell substrates and/or animal-derived components must include viral safety testing assessments, cell substrate characterization, and in some cases, risk assessments for transmission spongiform encephalopathy (TSE) agents.

### 17.9.1   Viral Safety Testing

A major concern with vaccines produced in human or animal cells is the potential for contamination with endogenous viruses that may be present in these cells or with adventitious viruses that may have been inadvertently introduced during development or manufacturing. To ensure freedom from these viruses, the cell banks (master and working), virus seeds (master and working), and the vaccine bulks are subjected to an extensive panel of safety and analytical testing that includes a combination of in vitro, in vivo, and analytical assays [30,31,33,36,38,39]. The extent of the viral safety testing program will depend on the cell line species, tissue source of the cells, history of the cells and original virus isolate, and the culture media ingredients and methods used to manufacture the seeds, among other factors [31].

### 17.9.2  Cell Substrate Characterization

In addition to the viral safety tests, cell substrate banks and bulks should include tests for identity, sterility, and mycoplasmas. Furthermore, new cell substrates should be characterized by karyotype analysis, and tumorigenicity testing in immunodeficient animals [30,33,38].

### 17.9.3  TSE Agents

A more recent concern is potential contamination with TSE agents, also known as prions. Both the FDA and EMEA have issued guidance documents for minimizing the risk of contamination with TSE agents [40–42], and for evaluating the risk of contamination from historic use of animal-derived materials [43–45]. However, while generally there are no regulatory expectations at this point in terms of analytical testing, agencies do rely on the preparation of theoretical risk assessments.

## 17.10  CONCLUSIONS

We have highlighted here the importance of nonclinical safety assessment studies in supporting not only the initiation of human clinical trials, but also their role throughout the clinical development program of vaccines. We have also presented a summary of the regulatory expectations and various aspects that need to be considered during the preclinical development of vaccines to ensure their safety, including, when appropriate, how the nonclinical safety assessment testing strategy is adapted to evaluate the use of novel adjuvants and delivery systems. Although the effectiveness of new adjuvants and delivery systems is still being tested in humans, increasing safety concerns for the use of these novel technologies have led to more stringent nonclinical and clinical regulatory requirements for their use in prophylactic vaccines. It seems almost inevitable, however, that continued advances in immunology and the emerging new technologies will continue to influence vaccine development. Thus, we expect that future licensed vaccines will routinely make use of novel adjuvants and delivery systems. The development of vaccines using these novel technologies, however, will ultimately be dependent on their demonstrated safety profile through the preclinical and clinical stages of development. Finally, it is important to remember that nonclinical safety assessment evaluations are only part of the overall safety evaluation of vaccines, which includes (1) evaluation of vaccines in animal immunogenicity and efficacy studies, (2) toxicology studies in animals, and (3) a QC testing program. These preclinical evaluations, along with the data generated during clinical trials, is what ultimately will support an application for marketing authorization.

### Declaration

Jose A. Lebron and Jayanthi J. Wolf contributed equally to this chapter.

## REFERENCES

1. Gruber, M. Nonclinical safety assesment of vaccines. Presented at the CBER Counter Terrorism Workshop, Bethesda, MD, 2003.
2. EMEA. *Note for Guidance on the Pharmacological and Toxicological Testing of Vaccines*, CPMP/SWP/465/1995, European Medicines Agency, London, 1997.
3. WHO. *WHO Guidelines on Nonclinical Evaluation of Vaccines*, WHO/BS/03.1969, World Health Organization, Geneva, 2003.
4. Preclinical safety evaluation of biotechnology derived pharmaceuticals, S6, *International Conference on Harmonization*, 1997.
5. FDA. *Guidance for Industry: Considerations for Developmental Toxicity Studies for Preventative and Therapeutic Vaccines for Infectious Disease Indications*, Center for Biologies Evaluation and Research, Food and Drug Administration, Washington, DC, 2006.
6. EMEA. *Guideline on Adjuvants in Vaccines for Human Use*, EMEA/CPMP/VEG/134716, European Medicines Agency, London, 2005.
7. EMEA. *Note for Guidance on Phamaceutical and Biological Aspects of Combined Vaccines*, CPMP/BWP/477/1998, European Medicines Agency, London, 1998.
8. FDA. *Guidance for Industry: Considerations for Plasmid DNA Vaccines for Infectious Disease Indications* (draft), Center for Biologics Evaluation and Research, Food and Drug Administration, Washington, DC, 2005.
9. FDA. *Guidance for Industry: Guidance for Human Somatic Cell Therapy and Gene Therapy*, Center for Biologics Evaluation and Research, Food and Drug Administration, Washington, DC, 1998.
10. EMEA. *Note for Guidance on the Quality, Preclinical and Clinical Aspects of Gene Transfer Medicinal Products*, CPMP/BWP/3088/1999, European Medicines Agency, London, 2001.
11. EMEA. *Points to Consider on the Manufacture and Quality Control of Human Somatic Cell Therapy Medicinal Products*, CPMP/BWP/41450/41998, European Medicines Agency, London, 2001.
12. EMEA. *Points to Consider on the Development of Live Attenuated Influenza Vaccines*, CPMP/BWP/2289/2001, European Medicines Agency, London, 2003.
13. *Note for Guidance on the Development of Vaccinia Virus Based Vaccines Against Smallpox*, CPMP/1100/2002, European Medicines Agency, London, 2002.

14. FDA. *Points to Consider in the Production and Testing of New Drugs and Biologicals Produced by Recombinant DNA Technology*, Center for Biologics Evaluation and Research, Food and Drug Administration, Washington, DC, 1985.

15. Good laboratory practice regulations, *Code of Federal Regulations*, Title 21, Part 58 (21 CFR 58).

16. FDA. *Inactive Ingredient Guide*, Division of Drug Information Resources, Food and Drug Administration, Washington, DC, 1996.

17. Lebron, J. Formulation changes during development: strategies for bridging preclinical safety assesment. In *Modern Vaccine/Adjuvant Formulation*, Prague, Czech Republic, 2004.

18. Opalka, D., Lachman, C.E., MacMullen, S.A. et al. Simultaneous quantitation of antibodies to neutralizing epitopes on virus-like particles for human papillomavirus types 6, 11, 16, and 18 by a multiplexed luminex assay. *Clin Diagn Lab Immunol* 2003, **10**(1), 108–115.

19. Casimiro, D.R., Tang, A., Perry, H.C. et al. Vaccine-induced immune responses in rodents and nonhuman primates by use of a humanized human immunodeficiency virus type 1 *pol* gene. *J Virol* 2002, **76**(1), 185–194.

20. Safety pharmacology studies for human pharmaceuticals, S7A, *International Conference on Harmonization*, 2000.

21. Ledwith, B.J., Manam, S., Troilo, P.J. et al. Plasmid DNA vaccines: investigation of integration into host cellular DNA following intramuscular injection in mice. *Intervirology* 2000, **43**(4–6), 258–272.

22. Manam, S., Ledwith, B.J., Barnum, A.B. et al. Plasmid DNA vaccines: tissue distribution and effects of DNA sequence, adjuvants and delivery method on integration into host DNA. *Intervirology* 2000, **43**(4–6), 273–281.

23. Wang, Z., Troilo, P.J., Wang, X. et al. Detection of integration of plasmid DNA into host genomic DNA following intramuscular injection and electroporation. *Gene Ther* 2004, **11**(8), 711–721.

24. Genotoxicity: a standard battery for genotoxicity testing of pharmaceuticals, S2B, *International Conference on Harmonization*, 1997.

25. Ledwith, B. Preclinical development of new vaccines. In *American Pharmaceutical Outsourcing*, 2005, pp. 38–42.

26. WHO. *Final Report: IABS Scientific Workshop on Neurovirulence Tests for Live Virus Vaccines*, World Health Organization, Geneva, 2005.

27. Current good manufacturing practice in manufacturing, processing, packing or holding of drug; General and current good manufacturing practice for finished pharmaceuticals, *Code of Federal Regulations*, Title 21, Parts 210 and 211 (21 CFR 210 and 21 CFR 211).

28. General biological products standards, *Code of Federal Regulations*, Title 21, Part 610 (21 CFR 610).

29. FDA. *Content and Format of Chemistry, Manufacturing and Control of Information and Establishment of Description Information for a Vaccine or Related Product*, Center for Biologics Evaluation and Research, Food and Drug Administration, Washington, DC, 1999.

30. FDA. *Points to Consider in the Characterization of Cell Lines Used to Produce Biologicals*, Center for Biologics Evaluation and Research, Food and Drug Administration, Washington, DC, 1993.

31. Viral safety evaluation of biotechnology products derived from cell lines of human or animal origin, Q5A, *International Conference on Harmonization*, 1997.

32. Quality of biotechnology products: stability testing of biotechnological/biological products, Q5C, *International Conference on Harmonization*, 1995.

33. Quality of biotechnology/biological products: derivation and characterization of cell substrates used for production of biotechnological/biological products, Q5D, *International Conference on Harmonization*, 1997.

34. Comparability of biotechnological/biological products subject to changes in their manufacturing process, Q5E, *International Conference on Harmonization*, 2004.

35. Investigational new drug application, *Code of Federal Regulations*, Title 21, Part 312 (21 CFR 312).

36. Viral safety evaluation of biotechnology products derived from cell lines of human or animal origin, *United States Pharmacopeia* 29th revision, and *National Formulary* 24th ed., 2006, Sec. 1050, pp. 2887–2896.

37. Biological tests, *European Pharmacopoeia*, 5th ed., 2004, Vol. 1, Sec. 2.6, pp. 145–180.

38. Cell substrate for the production of vaccines for human use, *European Pharmacopeia*, 5th ed., 2004, Vol. 1, Sec. 5.2.3, pp. 455–458.

39. WHO. *Requirements for the Use of Animal Cells as in-vitro Substrates for the Protection of Biologicals*, WHO Technical Report Series, No. 878, World Health Organization, Geneva, 1998.

40. EMEA. *Note for Guidance on the Use of Bovine Serum in the Manufacture of Human Biological Medicinal Products*, CPMP/BWP/1793/2002, European Medicines Agency, London, 2003.

41. EMEA. *Note for Guidance on Minimizing the Risk of Transmitting Animal TSE Agents via Human and Veterinary Medicinal Products*, EMEA/410/401/rev 403.draft, European Medicines Agency, London, 2004.

42. EMEA. *Position Paper on Re-establishment of Working Seeds and Working Cell Banks Using TSE Compliant Materials*, EMEA/22314/22002, European Medicines Agency, London, 2002.

43. FDA. *Estimating Risks for vCJD in Vaccines Using Bovine Derived Materials*, Center for Biologics Evaluation and Research, Food and Drug Administration, Washington, DC, 2001.

44. EMEA. *Public Statement on the Evaluation of BSE via the Use of Bovine Origin During the Manufacture of Vaccines*, EMEA/CPMP/476/2001, European Medicines Agency, London, 2001.

45. WHO. *Acceptability of Cell Subtrates for Production of Biologicals*. WHO Technical Report Series, No. 747, World Health Organization, Geneva, 1987.

# 18

# SAFETY EVALUATION OF VACCINE ADJUVANTS

Erik B. Lindblad

## 18.1 INTRODUCTION

Vaccination has been stated to be probably the most cost-effective way to control infectious diseases [1–3]. However, with vaccines administered to healthy individuals, only very few and mild, or preferably no side effects at all, can be tolerated. As a consequence, vaccine formulation scientists have over the last decades aimed at replacing whole-cell vaccines, as well as crude subunit vaccines, with more refined and well-defined formulations. Purified subunit vaccines or vaccine formulations based on synthetically derived peptides, which can be produced in high quantities and with high purity by recombinant technology, for example, are future antigen candidates for new vaccines [4]. However, small peptides, containing a single or a few epitopes, although highly specific and well defined, are frequently poorly immunogenic by themselves when compared to vaccine formulations based on whole microorganisms [5,6].

This is not surprising. Intact microorganisms contain a series of highly conserved structures, known as *pathogen-associated molecular patterns* (PAMPs) [7]. These molecules are now known to act as ligands for Toll-like receptors on antigen-presenting cells, and this receptor–ligand interaction is important in the early phases of establishing adaptive immunity [8].

When the killed, whole-cell, or crude subunit approach is replaced by purified subunit–protein vaccines, the antigen is transformed from a particulate to a solubilized form. This may have a negative impact on antigen targeting [i.e., uptake by antigen-presenting cells (APCs) for processing] [9]. When bacterial

*Vaccine Adjuvants and Delivery Systems*, Edited by Manmohan Singh
Copyright © 2007 John Wiley & Sons, Inc.

DNA is removed from the vaccine preparation by purification steps, the immunostimulating effect of the CpG motifs of the bacterial DNA is lost. When bacterial cell wall components are removed, the immunostimulatory effects of peptidoglycans/muramylpeptides are lost. The same is the case for trehalose dimycolate and lipopolysaccharide (LPS). Finally, the vaccine preparation may, if based on synthetic peptides, contain only a few antigenic epitopes compared to the range of epitopes of the native, infectious agent itself, and a small molecular size of the peptide construct may introduce a limitation in itself. Hence, all such efforts of designing a well-defined and standardized vaccine may lead to a situation where the vaccine has lost important co-stimulatory signaling and a significant part of its protective efficacy.

As a consequence of this new approach, a quest for new potent adjuvants has emerged for such purposes, and the number of new substances with documented adjuvant activity and the literature describing their use have expanded enormously over the last 20 years [10–12]. However, it also introduces a quest for assessment of the safety of traditional as well as future vaccine adjuvants, which is the subject of the present chapter.

## 18.2   GENERAL CONSIDERATIONS

It is a generally accepted principle that toxicity and adjuvant activity must be balanced to obtain maximum immune stimulation with minimal adverse effects [13]. However, the actual acceptance level for adverse reactions depends on whether the adjuvant is intended for use in human or veterinary vaccines. For veterinary applications the acceptance level depends on whether the animal is a companion animal or a livestock animal bred for human consumption. The safety documentation requirements for adjuvants used in human vaccines are, for obvious reasons, higher. When used in preventive medicine the vaccine is administered to healthy persons and in many cases, as part of vaccination programs for children. Here adverse reactions to the adjuvant are not acceptable. With therapeutic vaccines, however, a compromise is not unrealistic. Were therapeutic vaccines against serious human diseases (e.g., HIV/AIDS or cancer) or therapeutic vaccines against viral infections (e.g., HTLV-I or hepatitis C) to be developed that required the help of strong adjuvants to be effective, less strict levels of acceptance for the adjuvant side effects may be acceptable [14]. It would be a question of balancing the profile of vaccination side effects against the general prognosis for the disease if untreated or treated by other therapeutic regimens, many of which themselves are not without side effects.

## 18.3   MECHANISMS BEHIND ADJUVANT SIDE EFFECTS

The majority of adjuvants produce some effects at the injection site, the most frequent being an inflammatory response [15]. For the better tolerated adjuvants, used in practical vaccination, by far the majority of cases lead to transient and negligible symptoms only: mild pain, transient swellings, and so on.

However, among more than 100 different compounds, described as adjuvants in the literature, the vast majority have been shown to be too reactogenic to be used in human as well as veterinary applications. Such adjuvant active substances may nevertheless be valuable tools for studying the immune system as such, including side effects from excessive stimulation of the immune system. The mechanisms behind adjuvant side effects, as described below, comprise both observations from the investigation of such highly reactogenic adjuvants (or cytokines) and observations from significant *overdosing* of classical adjuvants.

Local reactions seen after the use of such adjuvants may range from local pain and erythemas to granulomas, cysts, abscesses, and ulcers, particularly if overdosing the adjuvant beyond the acceptable dose ranges [13,16]. Adverse systemic reactions due to adjuvant- or cytokine-induced stimulation of the immune system, including pyrogenicity [17], flulike symptoms, and autoimmune disorders [18,19], are known from experimental immunology, but are, of course, disqualifying for use of the adjuvant in practical vaccination. A number of observations of side effects seen after vaccination with adjuvanted vaccines must, however, be attributed to the vaccine preservatives (e.g., thiomersal, β-propriolactone, or formaldehyde) or, as mentioned, to bacterial toxins from the antigen preparation [20].

### 18.3.1   Local Reactions: Effect of the Injection Modus

Vaccinations may be given subcutaneously or intramuscularly. Other administration routes, such as the intraperitoneal route known from experimental immunology, are not used in practical parenteral vaccination. Oral vaccination of humans has been practiced against poliovirus since the 1960s [21,22], but this vaccine is not adjuvanted. Quillaja saponin has been used as an adjuvant for oral experimental vaccines [23] and is accepted as a food additive in Europe under code E999 due to low oral toxicity [24–26]. Hence, the potential of using Q. saponin as an adjuvant for oral immunizations is yet to be explored. Nasal immunization may have a future in practical vaccination but is still at the developmental stage [27,28] and is not discussed here.

The injection modus is not without importance in relation to local reactogenicity. When immunizing by the subcutaneous route the vaccine inoculum is introduced into a compartment with numerous sensory neurons (in contrast to the intramuscular compartment). The introduction of a local inflammatory response here may more easily give rise to irritation, itching reactions, and local pain. Also, a transient swelling, as a consequence of the inflammatory focus formed, may be palpable more easily through the skin. After immunizing by the intramuscular route, even a of similar size swelling may be less easily visible and palpable, as it is located in deeper-lying tissue [29]. Some adjuvants (e.g., Q. saponin) which show acceptable safety profiles when administered intramuscularly or subcutaneously in rodents, may cause chemical peritonitis and induce fibrous adherences in the body cavity when injected intraperitoneally.

## 18.3.2 Local Reactions: The Inflammatory Focus

Mineral adjuvants (aluminum- and calcium-based adjuvants) should, along with water-in-oil emulsions, (Freund's-type emulsion adjuvants) be regarded as depot-forming or repository adjuvants. With these adjuvants the formation of a temporary inflammatory focus attracting immunocompetent cells shortly after injection must, more or less, be expected [15]. Upon injection, phagocytic cells and APCs are attracted to the site to phagocytize and clear the inoculum. The local reaction may be negligible if the inoculum is dispersed rapidly from the injection site. However, if the inoculum resides for a prolonged period of time at the injection site (as is the case with repository adjuvants) then in situ accumulation of phagocytic and immunocompetent cells may in some cases manifest itself as an inflammatory focus accompanied by transient swelling, local irritation, and redness. There are observations of aluminum-adsorbed vaccines giving lead to more local reactions than unadsorbed vaccines with plain toxoid [30]. This could in part be explained by the plain toxoid vaccine being dispersed from the injection site before a local reaction was established.

Any visible or palpable reaction at the injection site is in principle non grata, as it hinders the obtaining of a hypothetical and nonreactogenic "ideal adjuvant." However, it is important to realize that the mechanisms described are part of a normally functioning immune system. Hence, it may not be achievable to use repository adjuvants without temporarily also inducing an inflammatory focus around the inoculum [15].

Attempts have been made in recent years to link the presence of a local inflammatory focus in the myofascii [the condition is referred to as *macrophagic myofasciitis* (MMF)] after intramuscular injections of aluminum-adjuvanted vaccines to such conditions as myalgia and muscle fatigue, but also to neurological disorders with no obvious etiological relation to the vaccination [31]. Such correlations are, however, associated with statistical problems. There is a very high vaccination coverage in Western countries. Hence, it is expected statistically that patients suffering from a wide range of etiologically unrelated diseases would all have been vaccinated with aluminum-containing vaccines at some point in their medical history. Another problem is that adequate statistical control groups of nonvaccinated persons may be hard to find in the same population [32].

In a recently published controlled study in primates by Verdier and co-workers in France, it was not possible to detect any histological changes after injection of aluminum-adjuvanted vaccine besides the local inflammatory focus itself, and they found no abnormal clinical signs associated to it [33].

As repository adjuvants give lead to local disturbance of tissue integrity and a transient inflammatory response [15,34], I have previously suggested that this, through elicitation of danger signals and induction of heat shock proteins, in particular hsp70, is after injection of importance for the function of these adjuvants through interaction with Toll-like receptors [35]. Hsp70 has been shown to react with Toll-like receptors, addressing TLR2 and TLR4 [36,37] for pro-inflammatory signal transduction in a CD14-dependent manner. It has also been shown that there was a positive correlation between the extent of disturbance of tissue integrity after the injection of metallic oxides and the magnitude of antibody titers induced [38]. Although the mechanism described is likely to be only one among several mechanisms by which the repository adjuvants act, the theory supports the idea that the induction of an inflammatory focus around the inoculum is of importance for the adjuvant stimulus, and the hypothesis implicates that attempts to eliminate the reactogenicity may have a negative impact on the efficacy of the adjuvant [35].

As pointed out by a specialist in immunotoxicology, Dr. Alexander Batista Duharte, a major question is, however, when to judge that a mild and normal local reaction moves from being a temporary, minor cosmetic problem into a toxic adverse reaction. In some such cases the adverse effects associated with vaccine adjuvants may simply be a consequence of overdosing. With increasing dosage of an adjuvant, in principle, the same immunopharmacological mechanisms may, due to hyperstimulation, cause a subclinical reaction to develop into a clinically significant reaction considered adverse in nature (A. Batista, personal communication). This fact presents a particular problem, as safety screening of adjuvants usually involves deliberate overdosing of the adjuvant.

### 18.3.3 Local Reactions: Local Pain and Tissue Lysis

Local pain is usually a consequence of some degree of tissue damage or disturbance of tissue integrity at the site of injection. For this reason, vaccines should be formulated at physiological pH and osmolarity. The pain can be light (irritation or itching), moderate, or severe, and the duration of the pain may, in principle, reflect the persistence of the stimulus, the time required for tissue repair, and so on. The pain elicited by the adjuvant itself can in some cases be reduced by adding excipients, as has been demonstrated with QS-21 [39]. Morton and Griffiths set up guidelines for the recognition of pain and distress in experimental animals [40].

Interaction with cell membranes leading to lysis has been described for adjuvants that are surface active, such as the saponins from *Quillaja saponaria Molina* [41] and purified fractions thereof (e.g., QS-18). Saponins have a distinct hemolytic activity, traditionally demonstrated on sheep red blood cells, and they bind efficiently to cell membrane–associated cholesterol [41]. Lytic tissue reactions have also been described for mineral oil emulsions prepared from short-chain hydrocarbons [42]. Again, these local reactions are directly dose-dependent, and the doses recommended for the use of these adjuvants reflect taking the local tissue reactogenicity into account. For the saponins the recommended doses [43] reflect a balance between the wish of eliciting a high immune response and a concomitant wish of not introducing so much detergent that local tissue lysis becomes clinically significant.

Pain, irritation, and general discomfort can cause behavioral disturbances as seen in animal models. The symptoms may comprise piloerection, reduced food intake, loss of weight, scratching the site of injection, and other manifestation of distress [40,44,45]. With increasing overdosing of the surface-active components, lytic tissue reactions may range from clinically insignificant over sterile abscess formation to ulceration.

### 18.3.4  Local Reactions: Granulomas and DTH Reactions

Local granulomas are sometimes seen after injection of adjuvant formulations containing non- or slowly degradable particles. Histological examinations of the injection site may show granulomas with activated macrophages, lymphocytes, fibroblasts, and other cells. Granulomas (Figure 18.1) have been reported after the use of Freund's adjuvants (Table 18.1), aluminum salts, and other adjuvants [13,16,46–48]. With aluminum hydroxide the accuracy of the subcutaneous injection itself may play a role [49].

Induction of a local delayed hypersensitivity (DTH) reaction is sometimes seen after the use of adjuvants [50]. It is due to a migration of immunocompetent cells to the inoculation site, involving regulation mediated by CD8[+] T-suppressor cells, CD4[+] Th1 lymphocytes, macrophages, and other cells

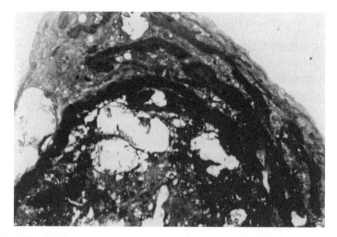

**Figure 18.1**  Histological cross section of a granuloma induced by mineral oil adjuvant in BALB/c mouse. A stratified structure is seen surrounding large vacuoles where the oil droplets resided. (Preparation and photo: E. B. Lindblad and Jens Blom.) (*See insert for color representation of figure.*)

**TABLE 18.1   Observed Side Effects Attributed to the Use of Freund's Adjuvants**

A.  *Freund's complete adjuvant*
Sterile abscesses
Granulomas
Plasma cell neoplasia in BALB/c mice
Ascites in BALB/c mice
Amyloidosis
Adjuvant arthritis in Lewis rats
Experimental allergic encephalomyelitis in guinea pigs

B.  *Freund's incomplete adjuvant*
Sterile abscesses
Granulomas
Plasma cell neoplasia in BALB/c mice
Ascites in BALB/c mice

secreting inflammatory cytokines (e.g., IL-2, IL-3, IL-6, and interferon), accompanied by manifestations such as proliferation of fibroblasts, cellular recruitment, and local edemas [51,52].

### 18.3.5   Systemic Effects

It should be emphasized that systemic adverse effects are not tolerated for adjuvants to be implemented in formulations for practical vaccination. However, systemic adverse reactions have been seen for a number of adjuvant active substances and exogenous cytokines subject to comprehensive laboratory investigation. One such example of a systemic effect is pyrogenicity, which is seen after injection of, for example, muramyl dipeptides (MDPs) or lipopolysaccharides [53,54].

Induction of an antibody response directed toward the adjuvant itself could in fact be seen as a systemic effect. In particular, it could prevent repeated administration of the adjuvant in question to the same vaccinee, due to the risk of inducing Arthus reactions. But the vast majority of adjuvants have not been shown not to elicit antibody responses against themselves: one exception being squalene, a triterpenoid hydrocarbon oil with adjuvant activity [55,56].

Systemic effects may also be a consequence of hyperstimulation of the immune system as a consequence of overdosing the adjuvant or if exogenous cytokines are implemented into the adjuvant formulation. In particular, administration of exogenous IL-2 may be accompanied by a severe systemic reaction, known as the *vascular leak syndrome* (VLS), which is accompanied by hypotension, anasarca, and extravasation of serum protein and fluids, and which in severe cases may lead to organ failure [57].

Another systemic effect following hyperstimulation of the immune system is *flulike symptoms* mediated by TNF, interferons, and IL-1, IL-2, and IL-3.

Flulike symptoms may manifest themselves within hours from having been exposed to an immunostimulating treatment by Bacille Calmette-Guérin (BCG), for example, and they are normally only of temporary duration [58]. The clinical signs comprise, as the name indicates, symptoms resembling those known from flu: fever, malaise, tachycardia, chills, and myalgias.

Although many systemic reactions may be attributed to hyperstimulation of the immune system and excessive release of cytokines, it can be difficult to correlate the readouts from traditional cytokine measurements to clinical safety assessment of vaccine adjuvants. Autoimmune disorders have been seen after the administration of adjuvants and after administrations of certain antigens in combination with potent adjuvants [59–61]. In the latter cases the autoimmune condition is in principle a reaction against an antigen, but the reaction may require adjuvant co-stimulation to manifest itself. For this reason it is imperative that clinical safety assessment is not restricted to the adjuvant alone, but must also comprise antigen–adjuvant combinations in the screening for autoimmune disorders.

The overall conclusion is that no adverse systemic reactions should be accepted after administration of adjuvanted vaccines, and the symptoms mentioned above merely indicate what you should be observant for in preclinical safety assessment of new adjuvants.

## 18.4  SAFETY ASPECTS OF ADJUVANTS USED IN PRACTICAL VACCINATION

In contrast to the high number of substances with documented adjuvant effect, very few have been accepted for use in practical vaccination. For use in human vaccines these comprise aluminum adjuvants, calcium phosphate adjuvants, and recently, MF-59 and liposomes or virosomes.

Billions of doses of aluminum adjuvanted vaccines have been administered over a period of more than 60 years [13,29,62]. Freund's incomplete adjuvant (FIA), a mineral oil-in-water emulsion, was used in human vaccination for a period of approximately 20 years. In Britain alone, approximately 900,000 doses of mineral oil–adjuvanted vaccines have been administered to humans, and more than 500,000 persons worldwide have received FIA-adjuvanted vaccines. But the use of FIA in human vaccines was discontinued in the mid-1960s due to concerns about safety. FIA is now used in veterinary applications only.

In addition, a number of adjuvants have been applied to humans in clinical studies or field trials. Examples of such adjuvants are Maurice Hilleman's formulation, Adjuvant-65 (a mixture of peanut oil, mannide monooleate, and aluminum monostearate) [63], and quaternary ammonium compounds [64,65]. More recently, QS-21 [39], monophosphoryl lipid A (MPL) [66], and Montanide ISA 720 metabolizable oil emulsion adjuvant from SEPPIC have also been under evaluation as candidates for future application in humans.

Emphasis here will be put on the most commonly used adjuvants: aluminum and mineral oil adjuvants, as these are the only adjuvants that have been used in large-scale practical vaccination of both humans and animals.

### 18.4.1  Aluminum Adjuvants

*Aluminum Adjuvants*   Aluminum adjuvants (aluminum hydroxide and aluminum phosphate gel adjuvants) are accepted for use in humans, but there are limitations for the content of aluminum allowed. These limits are 1.25 mg of aluminum per dose in Europe [67], and in the United States, 0.85 mg of aluminum per dose if determined by assay, 1.14 mg if determined by calculation, and 1.25 mg if safety and efficacy data justify it [68]. In veterinary vaccines there is no defined maximum limit for the allowed content of aluminum adjuvants. Here the dose is normally set from a balance between efficacy and local reactogenicity. Valtulini reported that up to 12 mg of Al in vaccines was well tolerated in pigs, whereas vaccine formulations containing 40 mg induced granulomas [47].

*Safety Profile of Aluminum Adjuvants*   Aluminum hydroxide and aluminum phosphate adjuvants have been used for more than half a century and are generally regarded as safe when used according to current immunization schedules [62,69]. In 1993, the U.S. NCVDG Working Group on Safety Evaluation of Vaccine Adjuvants with the participation of U.S. Food and Drug Administration (FDA) representatives concluded that "the extensive experience with this class of adjuvant for vaccine use has indicated that it is safe" [70].

Cases of local reactions after administration of aluminum-adjuvanted vaccines have been reported [13,29,46,49]. They comprise swellings, indurations, erythemas, and cutaneous nodules that can persist for up to eight weeks, sometimes longer [71]. These reports often describe cases of hyposensitization of allergic patients who receive a large number of injections of adsorbed allergenic extracts over a limited period of time, or cases where injections were made incorrectly [49]. In a vaccination program in Sweden, Elisabeth Bergfors and her colleagues found itching local reactions in 0.8% of 76,000 vaccinees [72].

There is no evidence that aluminum adjuvants themselves should be pyrogenic, and there is no evidence of carcinogenicity or teratogenicity attributed to their use. Further, contact hypersensitivity reactions are not commonly seen [69,73]. There are inconsistent observations whether adsorption onto aluminum adjuvants leads to increased or decreased vaccine reactogenicity [74,75]. However, Butler et al. found that adsorption onto aluminum hydroxide (Alhydrogel) significantly reduced the side effects with combined diptheria, tetanus, pertussis (DTP) vaccines [75].

It is conceivable that the acute toxicity is reduced in adsorbed vaccines simply by delaying the release of reactogenic or toxic vaccine constituents, such as pertussis toxin, peptidoglycans from gram-negative cell walls, or LPSs from the injection site. The binding affinity of LPS to aluminum hydroxide is well established and was much higher than to aluminum phosphate [76], probably due to the phosphate content of LPS binding strongly to aluminum hydroxide. Norimatsu found that adsorption of LPS onto aluminum hydroxide prior to injection inhibited or mitigated systemic effects such as the trembling, transient leucopenia, and elevated serum TNFα otherwise observed following intramuscular injection of LPS in saline [77]. Also, the level of IL-6 after administration of LPS was reduced if the LPS was adsorbed to aluminum hydroxide prior to injection [76].

*In Vivo Clearing of Aluminum Adjuvants*   Both humans and animals are being exposed fairly constantly to aluminum compounds from the environment. As a consequence, aluminum is normally found in the blood and serum of humans and animals, whether or not they have been vaccinated using aluminum adjuvants. The major source of this aluminum is apparently oral intake with the food and drinking water, which was reported to be in the magnitude of 5 to 10 mg daily for humans in Britain [78]. This aluminum uptake would be excreted with the urine by persons with normal renal functions.

Previous claims that aluminum adjuvants are not broken down in vivo and subject to excretion have, however, been shown to be incorrect. The clearing in vivo of aluminum adjuvants has been investigated in rabbits using adjuvants prepared from the isotope $^{26}$Al [79]. $^{26}$Al excreted from blood and urine was followed using accelerator mass spectroscopy for a period of 28 days. As early as one hour following intramuscular injection, radioactive-labeled aluminum could be detected in the blood, and it was found that approximately three times more $^{26}$Al was excreted from animals vaccinated with aluminum phosphate than from those vaccinated with aluminum hydroxide. Assumingly, interstitial fluid containing organic acids with an α-hydroxycarboxylic acid, able to chelate aluminum, reacted more readily with aluminum phosphate than with aluminum hydroxide [79]. Following the injection of adjuvant containing 0.85 mg of aluminum, the normal plasma concentration of aluminum in rabbits of 30 ng Al/mL only rose with approximately 2 ng Al/mL during the experiment. According to the calculation of Flarend, a similar aluminum dose injected into humans, provided that similar clearing kinetics existed, would lead to an estimated increase of serum aluminum of only 0.04 ng Al/mL, equaling 0.8% above the normal level of approximately 5 ng Al/mL [79]. Based on these figures, it seems that the amount of aluminum administered via vaccination does not contribute significantly to the general exposure to aluminum in humans and serum levels of aluminum.

### 18.4.2   Calcium Phosphate Adjuvant

Calcium phosphate was used as an adjuvant in vaccines against diphtheria, tetanus, *Bordetella pertussis*, and poliomyelitis [80,81], commercialized by

Institut Pasteur in the IPAD series of vaccines for approximately 25 years, ending in 1988. In addition, calcium phosphate has been used for adsorption of allergens for hyposensitization of allergic patients for decades. Calcium phosphate has so far not been used as an adjuvant in commercial veterinary vaccines.

In Europe the maximum allowed amount of calcium delivered by calcium phosphate–adjuvanted vaccines is 1.3 mg of calcium [67]. There is, however, no obvious toxicological rationale behind limiting the amount of calcium in vaccines to 1.3 mg per dose. Calcium phosphate is a natural constituent of the mammalian organism and is considered safe to use. Further, it has been a component of bone replacement transplants in much higher amounts with no problems [82]. Some of the hesitations to the implementation of calcium phosphate adjuvant in vaccines are due to the relatively low protein adsorption capacity. This could, in some cases, be compensated for by increasing the amount of the adjuvant.

### 18.4.3   Mineral Oil Adjuvants

*Mineral Oil Adjuvants*   Vaccination of humans using mineral oil emulsion adjuvants was practiced for about two decades before discussions of safety aspects stopped this line of work. Following the early work in 1944–1945 [83,84] on influenza vaccines, Jonas Salk introduced a highly refined mineral oil and purified Arlacel A for use in vaccines with very good results. This made it possible to carry out large-scale vaccination against influenza under the auspices of the Commission on Influenza of the U.S. Armed Forces Epidemiological Board [85,86].

*Safety Profile of Mineral Oil Adjuvants*   The early preparations of Arlacel A contained free oleic acid, which was later regarded as a causative agent for sterile abscess formation. Of 6000 patients receiving the polyvalent vaccine by subcutaneous injection in 1951, a total of 17 cases of subcutaneous fluctuant cystic swellings were recorded [87]. In a study by Bell and co-workers using a monovalent influenza vaccine in 2426 individuals, 19 such swellings were recorded [88].

Following the introduction of more refined and purified reagents and changing the route of administration to the intramuscular route, a reduction in the rate of cyst formation was recorded. The Salk mineral oil poliomyelitis vaccine was injected into 15,420 patients by Cuttler and co-workers 8497 of which also received a second injection [89]. Among these 8497 individuals, 109 cases of side effects were recorded.

In Japan, trials using mineral oil–influenza vaccines were initiated in 1960–1961. The study comprised 150 volunteers, and the vaccine showed good efficacy, but in about 10% of cases, muscle indurations were observed [90]. In some cases they persisted for up to one year after injection. Histological

SAFETY EVALUATION OF VACCINE ADJUVANTS

examinations of the indurated loci were undertaken. The main feature was a *foreign body reaction*, oil granulomas with central vacuoles where the oil was assumed to have resided. These vacuoles were surrounded by epitheloid and fibroblast cells with scattered plasma cells. No signs of malignancy were found.

A combined mineral oil–influenza and adenovirus vaccine was given by Meiklejohn to 2303 patients [91]. During a six-month observation period, no local or other reactions were recorded. However, it was found [92] that patients who received mineral oil–adjuvanted influenza vaccines of chick embryo origin in several cases developed complement-fixing antibodies for components of normal chicken allantoic fluid.

Two series of tetanus vaccination field trials in women were carried out in New Guinea by Pittman [93]. A high frequency of abscesses was seen among the 600 women who received the oil vaccine. The peak incidence occurred between six and 10 weeks after injection. After 14 weeks, 199 cases of abscesses (30%) had been reported.

In Great Britain, mineral oil–influenza vaccines were approved for use in 1963 [94]. In the period 1963–1965, about 900,000 doses of FIA-based influenza vaccine were administered. A close follow-up study of that magnitude was not possible, but general practitioners reported at least 40 cases of local reactions, of which nine required surgical attention. Application of mineral oil adjuvants in humans was then discontinued in the UK.

In a joint WHO field trial in 1964 in the Philippines, 143,600 persons were inoculated with mineral oil–cholera vaccine [95]. The oil vaccine proved to have a long-lasting effectiveness but was accompanied in 2934 cases by severe side reactions, ranging from hard indurations at the injection site to sterile abscesses or even ulcers.

Continuous discussions about the safety of using mineral oil adjuvants in humans led the Commission on Influenza of the U.S. Army Medical Research and Development Command to request the National Academy of Sciences to undertake a long-term follow-up study on the adverse effects of the mineral oil influenza vaccines administered nine years earlier. The follow-up study [96] comprised 18,000 army personnel who received mineral oil–influenza vaccine. There was no evidence of increased mortality or increased retirement due to disability among the recipients of the mineral oil vaccine. In 1965 a committee of specialists convened by the FDA restricted the use of mineral oil influenza vaccine in the United States to be administered only to military personnel and groups recognized as high-risk groups of death related to influenza because of a specified preexisting disease status.

Contemporary work by Potter and Robertson Boyce [97] reported that intraperitoneal injection of Freund's incomplete adjuvant was able to induce plasma cell neoplasia in BALB/c mice. This caused additional concern about the safety of mineral oil emulsions in humans, and eventually this line of work was discontinued. At present the traditional mineral oil emulsions are used in veterinary vaccines only.

*In Vivo Clearing of Mineral Oil Adjuvants*    Bollinger studied the clearing of mineral oil as well as mannide monooleate [98,99] in rats and monkeys by use of $^{14}$C-labeled tracers. To study the clearing of mineral oil, he injected $^{14}$C-labeled hexadecane in an emulsion with unlabeled mannide monooleate in the thighs of the animals intramuscularly and subcutaneously. One week after injection, 85 to 98% of the radioactivity remained at the site of injection; after one month, 65 to 75%; after three months, 55 to 65%; and after 10 months, approximately 30% of the labeled oil could still be found at the injection site. Thin-layer chromatography separation was undertaken to tell in which type of lipid the radioactivity was found. It was shown clearly that after 10 months, radioactivity at the injection site was still found in the hydrocarbon fraction. Samples from the major organs showed that there was an increasing level of radioactivity in the liver that peaked after one week to one month. After 10 months the level was back to normal. This was accompanied by a slightly later but analogous rise-and-fall pattern in the radioactivity of the depot fat. No other organs achieved high levels of radioactivity. In the liver, the radioactivity after one month was found predominantly in triglycerides, sterol esters, and free sterols, whereas in the depot fat it was found primarily in triglycerides and free fatty acids. After three months practically all the radioactivity in the depot fat was found in triglycerides, whereas most of the radioactivity at that time in the liver was found as phospholipids [98].

The metabolic fate of mannide monooleate was studied by $^{14}$C labeling of either the mannitol or the oleic acid, and subsequently preparing the mannide monooleate emulsifier with one labeled component prior to incorporating them into a FIA preparation [99]. Clearing from the injection site of the emulsifier took place significantly faster than was the case with mineral oil. After one week approximately only 50% of the radioactivity could be found at the injection site. The [$^{14}$C]mannide monooleate containing inoculum lost about 40% of its in situ radioactivity in about two days. The clearing rate in this case was faster than when [$^{14}$C]oleate mannide was incorporated. When the oleate was labeled, a rise-and-fall pattern of the radioactivity in the liver and depot fat similar to that seen with [$^{14}$C]hexadecane [99] could be found. This was not the case when the mannitol had been labeled. In that case, radioactivity in the inguinal lymph nodes was seen, and a significant amount of radioactivity was excreted with the urine [99].

In conclusion, the general picture was that upon injection of a mineral oil emulsion, the emulsifier tended to leave the inoculum depot faster than the oil itself, giving lead to a coalesced oil depot. Bollinger emphasized, however, that the mannide monooleate had a high free mannitol and oleic acid content, due to incomplete esterification. This could account at least partially for the rapid clearing of mannitol from the injection site. The transport of mannitol appears to take place through the lymphatics. The lipids [i.e., both the oleic acid and the hydrocarbons of the mineral oil (which is very slowly transported away from the inoculum)] can be metabolized in the liver and may end up in the depot fat as triglycerides. The mannitol is excreted with the urine [99].

Attention should be drawn to the fact that the emulsifier used in FIA preparations, mannide monooleate, can be synthesized using oleic acid of animal origin (tallow). This now introduces some concerns in relation to whether or not this material comes from livestock free from prion disease/TSE.

## 18.5 SAFETY REQUIREMENTS FOR FUTURE ADJUVANTS

For the traditional adjuvants with a long track record of use in practical vaccination it is possible to analyze extremely comprehensive cohorts and to analyze for side effects of long-term exposure: in the case of aluminum adjuvants, even over a lifelong time span in humans. Similar statistical comprehensiveness is obviously not available for judging the safety of new adjuvants. New adjuvants should therefore be subject to a comprehensive safety screening in vitro as well as in laboratory animal models before clinical testing could even be considered. One initial parameter to be considered is the species in which the adjuvant is to be applied; another is whether the adjuvant is intended for use in preventive or therapeutic vaccination, as this is of importance for the level of side effects to be accepted [100]. Among other factors to be considered are which antigens the adjuvant is intended to be combined with, as both the adjuvant and the antigen may contribute to the net safety profile of the final formulation, and possible interactions between adjuvant and antigen must be considered. As such, the question of adjuvant safety must be addressed separately for each formulation of the final vaccine. As part of the overall safety profile, teratogenicity and embryo–fetal toxicity screening at the preclinical stage are also relevant. The age of the person to receive the adjuvanted vaccine formulation should be considered. Infants may react differently from adults, and this may have an impact on the dosage recommendations. If new adjuvant formulations are to include products of animal origin, freedom from prions or TSE-affected material must be guaranteed.

Thoughts about using exogenous cytokines as adjuvants introduce a series of additional safety considerations, the nature of which is highly dependent on the particular cytokine in question. These are not discussed here.

## 18.6 ASSAYS FOR TESTING ADJUVANT SAFETY

Over the last decades there have been intense discussions over how to ascertain adjuvant safety. Several distinguished authors have given valuable contributions over the years, not the least Stewart-Tull [101] and Goldenthal et al. [70], yet no firm consensus has been reached and the question is fairly open. In the absence of such consensus, it is, in principle, up to the individual investigators to design their own regimen at the preclinical stage; however, some inspiration is provided below.

### 18.6.1   At the Design Stage

The experimental design should reflect whether the setup should comprise both animal genders and breeding animals (to screen for miscarriages and embryo–fetal toxicity), and whether possible MHC-restriction issues should be considered for inbred strains of mice. Susceptibility to certain side effects may also be strain specific, one example being the susceptibility of BALB/c mice to developing plasma cell neoplasia after administration of mineral oil adjuvants [97]. This means that it may be difficult to extrapolate observations from such strains to other animal models; however, it may also be viewed as meaning that the specific animal model is a very sensitive tool for detecting these side effects in particular. The number of animals in the various experimental groups should be sufficient to allow repeated blood tests during the experiment and to take out animals, for example, to test for DTH reactions.

It is normal – and recommended – to include groups for *high-dose* and *repeated-dose* regimens in the experimental design to provoke the appearance of side effects in small animal groups. Adjuvants intended for use in humans should be tested in nonhuman primates when the safety profile from testing in other animal models does not give rise to concern. Differences between what the regulatory authorities and animal ethics committees in various countries find acceptable may have an impact on the experimental design in terms of the size of the individual animal groups and the duration of the experiments.

### 18.6.2   At the Beginning of the Experiment

Animals in the various experimental groups should be randomized. Animals should be weighed and blood samples should be taken prior to the first injection for control purposes. Pyrogenicity testing in rabbits after intravenous injection followed by monitoring the rectal temperature should be considered as an indicator of the intrinsic pyrogenicity of the adjuvant preparation. But note that particulate adjuvant preparations injected into the ear veins may get trapped in the lung capillaries and lead to embolism. The Ames test has been used to screen new muramyl dipeptide derivative adjuvants for mutagenicity [102].

### 18.6.3   During the Experiment

The *Berlin weight gain test*, ascertaining that young mice injected intraperitoneally with $0.25\,cm^3$ of test solution should not suffer from transient weight loss, could be considered. The difference in weight gain between young mice injected intraperitoneally with an acceptable preparation of test solution and a control group receiving saline should be less that 10.5% [103]. A body weight monitoring of the animals simply by weighing them regularly during the experiment is useful as an indicator of reduced food intake or ascites formation. If ascites production is seen, samples should be taken for further analysis (for gammapathias, Bence-Jones protein, etc.).

Palpation of the injection site should be undertaken once or twice weekly during the entire experiment for ascertaining cystic swellings, indurations, swellings, and granulomas. Depilation of the injection site may be recommended, as it allows observation for erythemas and other skin abnormalities. It is facilitated by the depilation cremes now available. Observations for carcinogenicity should also include macroscopically visible or palpable tumors.

Following intramuscular injections it can be of value to get a readout for muscular tissue damage. Creatine phosphokinase (CPK) release due to leakage from muscular tissue, indicative of tissue damage, can be assayed using a commercially available test kit (kit no. 520 for colorimetric determination of CPK in serum or plasma by the Sigma Chemical Company) [101].

Stewart-Tull suggested a skin test for delayed-type reactivity to be carried out at day 19 after the primary immunization and day 44 after the secondary immunization for standardization purposes [101]. The popliteal lymph node assay has been modified to study chemically induced immune reactions [104,105].

Some serologic parameters are useful to evaluate autoimmune reactions such as formation of immune complexes and autoantibodies in serum (e.g., antinuclear antibodies (ANAs) and antihistone- and anti-single-strand-DNA antibodies). Some researchers advocate screening for elevated levels of cytokines; however, although several ready-to-use test kits are now commercially available, it may be difficult to use the results in actual assessment of adjuvant safety.

### 18.6.4 At the End of the Experiment

All animals should be weighed and bled. Blood tests should include leukocyte counts. Necropsy and histology after euthanizing should be undertaken as a routine at the termination of the trial. Note that due to the short life span of mice, the duration of the observation period may often extend into the later stages of the normal life of the animals. Here an age-declining immune system may manifest itself in various ways (e.g., as an increased incidence of autoimmune disorders), and even in the control groups tumor incidence may be significant in some strains of inbred mice.

### 18.7 CONCLUSIONS

It is common practice that a safety evaluation includes deliberately overdosing the adjuvant. The rationale behind this practice is that with overdosing, sometimes using multiple overdoses, reactions may be provoked in relatively small experimental groups that would otherwise appear only rarely and hence escape detection in small-scale setups. However, to some extent, the normal immunological clearing mechanisms move from clinically insignificant into significant adverse or even toxic reactions with increasing dosage of tissue reactogenic adjuvants. Too often, this fact is overlooked. Similarly, an adjuvant-mediated stimulation of the immune system may turn into hyperstimulation with toxic side effects as a function of increasing the adjuvant dose alone. Nevertheless, the overdosing regimens serve a purpose and seem unavoidable in preclinical safety screening. When properly designed and undertaken and with a cautious interpretation of the results achieved, the *high-dose* and *repeated-dose* regimens provide important information about safety margins in the administration of a new adjuvant and are of value for setting dosage recommendations.

The choice of adjuvants for human use requires much more refined measures and safety evaluations than that of adjuvants intended for experimental or veterinary use. New adjuvant candidates should be evaluated through preclinical assays that address detection of potential induction of local toxicity at the application site, general toxicity, teratogenesis, carcinogenesis, and immunotoxicity (immunosuppression, autoimmunity, or allergy). Use of nonhuman primates for testing should be restricted to adjuvants that have already shown a satisfactory safety profile in other animal species.

It would be of significant value to establish an international database comprising all observations relating to adjuvant safety. Data on the safety of vaccine adjuvants used in licensed veterinary vaccines should also contribute to building up safety records on those adjuvants under consideration for use in other species, including humans.

The development of novel adjuvants and delivery systems should lead to the initiation of preclinical safety studies as early as possible, but it is necessary to keep in mind that these studies do not pick up all adverse events. A particular challenge is to distinguish between mild and normal local reactions as an expected consequence of activation of the immune response, constituting only a temporary, minor cosmetic phenomenon and those reactions that are related to product exposure and are toxicologically relevant. The latter would comprise long-term persistent local reactions, unexpected adverse changes of the immune status, hyperstimulation, and systemic effects [106]. Consequently, the overall profile of adverse side effects should be evaluated carefully, both by validated in vitro methods and in suitable animal models prior to evaluation for clinical use.

## ACKNOWLEDGMENTS

The author wishes to thank Dr. Alexander Batista Duharte, Toxicology and Biomedicine Center, Santiago de Cuba, and Professor Duncan Stewart-Tull, University of Glasgow, for many valuable discussions.

## REFERENCES

1. Foege, W.H., Millar, J.D. & Henderson, D.A. Smallpox eradication in West and Central Africa. *Bull World Health Organ* 1975, **52**(2), 209–222.
2. Rey, M., Guillaumont, P. & d'Intignano, B.M. Benefits of immunization versus risk factors in tetanus. *Dev Biol Stand* 1979, **43**, 15–23.
3. Van Wezel, A.L. Vaccines: quantities, production and costs. *Trop Geogr Med* 1985 Sep, **37**(3), S35–S39.
4. Olive, C., Toth, I. & Jackson, D. Technological advances in antigen delivery and synthetic peptide vaccine developmental strategies. *Mini Rev Med Chem* 2001 Nov, **1**(4), 429–438.
5. Francis, M.J., Fry, C.M., Rowlands, D.J., Bittle, J.L., Houghten, R.A., Lerner, R.A. et al. Immune response to uncoupled peptides of foot-and-mouth disease virus. *Immunology* 1987 May, **61**(1), 1–6.
6. Haro, I. & Gomara, M.J. Different approaches to potentiate the immune response induced by a 12-mer synthetic peptide. *Curr Protein Pept Sci* 2000 Sep, **1**(2), 125–137.
7. Medzhitov, R. & Janeway, C., Jr. Innate immune recognition: mechanisms and pathways. *Immunol Rev* 2000 Feb, **173**, 89–97.
8. Heine, H. & Ulmer, A.J. Recognition of bacterial products by Toll-like receptors. *Chem Immunol Allergy* 2005, **86**, 99–119.
9. Mannhalter, J.W., Neychev, H.O., Zlabinger, G.J., Ahmad, R. & Eibl, M.M. Modulation of the human immune response by the non-toxic and non-pyrogenic adjuvant aluminium hydroxide: effect on antigen uptake and antigen presentation. *Clin Exp Immunol* 1985 Jul, **61**(1), 143–151.
10. Vogel, F.R. & Pruett, M.F. A compendium of vaccine adjuvants and excipients. In *Vaccine Design: The Subunit and Adjuvant Approach*, Powell, M.F. & Newman, M.J., Eds., Plenum Press, New York, 1995, pp. 141–228.
11. Kensil, C.R. & Kammer, R. QS-21: a water-soluble triterpene glycoside adjuvant. *Expert Opin Invest Drugs* 1998 Sep, **7**(9), 1475–1482.
12. Nixon, A., Zaghouani, H., Penney, C.L., Lacroix, M., Dionne, G., Anderson, S.A. et al. Adjuvanticity of stearyl tyrosine on the antibody response to peptide 503–535 from HIV gp160. *Viral Immunol* 1992 Summer, **5**(2), 141–150.
13. Gupta, R.K., Relyveld, E.H., Lindblad, E.B., Bizzini, B., Ben-Efraim, S. & Gupta, C.K. Adjuvants: a balance between toxicity and adjuvanticity. *Vaccine* 1993, **11**(3), 293–304.
14. Moss, R.B., Giermakowska, W., Lanza, P., Turner, J.L., Wallace, M.R., Jensen, F.C., Theofan G., Richierl, S.P., Carlo, D.J. Cross-clade immune responses after immunization with a whole-killed gp120-depleted human immunodeficiency virus type-1 immunogen in incomplete Freund's adjuvant (HIV-1 immunogen, REMUNE) in human immunodeficiency virus type-1 seropositive subjects. *Viral Immunol* 1997, **10**(4), 221–228.
15. WHO. Immunological adjuvants. Report of a WHO scientific group. *Tech Rep Ser* 1976, **595**, 3–40.
16. Lindblad, E.B. Freund's adjuvants. In *Vaccine Adjuvants: Preparation Methods and Research Protocols*, O'Hagan, D.T., Ed., Humana Press, Totowa, NJ, 2000, pp. 49–63.

17. Soszynski, D., Kozak, W. & Szewczenko, M. Course of fever response to repeated administration of sublethal doses of lipopolysaccharides, polyinosinic:polycytidylic acid and muramyl dipeptide to rabbits. *Experientia* 1991 Jan 15, **47**(1), 43–47.

18. Shaw, C.M., Alvord, E.C., Jr., Fahlberg, W.J. & Kies, M.W. Adjuvant–antigen relationships in the production of experimental "allergic" encephalomyelitis in the guinea pig. *J Exp Med* 1962, **115**, 169–179.

19. Pearson, C.M. Development of arthritis, periarthritis and periostitis in rats given adjuvants. *Proc Soc Exp Biol Med* 1956, **91**(1), 95–101.

20. Chaby, R. & Caroff, M. Lipopolysaccharides of *Bordetella* pertussis endotoxin. In *Pathogenesis and Immunity in Pertussis*, Wardlaw, A.C. & Parton, R., Eds., Wiley, New York, 1988, pp. 247–271.

21. Anderson, G.W. & Henderson, D.A. Report on oral polio vaccines. *JAMA* 1965 Jan 18, **191**, 252–253.

22. Franklin, G.C. & Robertson, M.J. A mass vaccination campaign against poliomyelitis using the Sabin oral vaccine. *Public Health* 1965 Jan, **79**, 81–99.

23. Maharaj, I., Froh, K.J. & Campbell, J.B. Immune responses of mice to inactivated rabies vaccine administered orally: potentiation by *Quillaja saponin*. *Can J Microbiol* 1986 May, **32**(5), 414–420.

24. Gaunt, I.F., Grasso, P. & Gangolli, S.D. Short-term toxicity of *Quillaja* extract in rats. *Food Cosmet Toxicol* 1974 Oct, **12**(5–6), 641–650.

25. Phillips, J.C., Butterworth, K.R., Gaunt, I.F., Evans, J.G. & Grasso, P. Long-term toxicity study of *Quillaja* extract in mice. *Food Cosmet Toxicol* 1979 Feb, **17**(1), 23–27.

26. Drake, J.J., Butterworth, K.R., Gaunt, I.F., Hooson, J., Evans, J.G. & Gangolli, S.D. Long-term toxicity study quillaia extract in rats. *Food Chem Toxicol* 1982 Feb, **20**(1), 15–23.

27. Aggerbeck, H., Gizurarson, S., Wantzin, J. & Heron, I. Intranasal booster vaccination against diphtheria and tetanus in man. *Vaccine* 1997 Feb, **15**(3), 307–316.

28. Gizurarson, S., Tamura, S., Kurata, T., Hasiguchi, K. & Ogawa, H. The effect of cholera toxin and cholera toxin B subunit on the nasal mucosal membrane. *Vaccine* 1991 Nov, **9**(11), 825–832.

29. Lindblad, E.B. Aluminium adjuvants: in retrospect and prospect. *Vaccine* 2004 Sep 9, **22**(27–28), 3658–3668.

30. Collier, L.H., Polakoff, S. & Mortimer, J. Reactions and antibody responses to reinforcing doses of adsorbed and plain tetanus vaccines. *Lancet* 1979 Jun 30, **1**(8131), 1364–1368.

31. Authier, F.J., Cherin, P., Creange, A., Bonnotte, B., Ferrer, X., Abdelmoumni, A. et al. Central nervous system disease in patients with macrophagic myofasciitis. *Brain* 2001 May, **124**(Pt 5), 974–983.

32. Lindblad, E.B. Mineral adjuvants. In *Immunopotentiators in Modern Vaccines*, Schijns, V. & O'Hagan, D.T., Eds., Elsevier, New York, 2005, pp. 217–234.

33. Verdier, F., Burnett, R., Michelet-Habchi, C., Moretto, P., Fievet-Groyne, F. & Sauzeat, E. Aluminium assay and evaluation of the local reaction at several time points after intramuscular administration of aluminium containing vaccines in the *Cynomolgus* monkey. *Vaccine* 2005 Feb 3, **23**(11), 1359–1367.

34. Goto, N., Kato, H., Maeyama, J., Shibano, M., Saito, T., Yamaguchi, J. et al. Local tissue irritating effects and adjuvant activities of calcium phosphate and aluminium hydroxide with different physical properties. *Vaccine* 1997 Aug–Sep, **15**(12–13), 1364–1371.

35. Lindblad, E.B. Are mineral adjuvants triggering TLR2/TLR4 on dendritic cells by a secondary cascade reaction in vivo through the action of heat shock proteins and danger signals? Editorial letter. *Vaccine* 2006, **24**, 697–698.

36. Asea, A., Kraeft, S.K., Kurt-Jones, E.A., Stevenson, M.A., Chen, L.B., Finberg, R.W. et al. HSP70 stimulates cytokine production through a CD14-dependent pathway, demonstrating its dual role as a chaperone and cytokine. *Nat Med* 2000 Apr, **6**(4), 435–442.

37. Asea, A., Rehli, M., Kabingu, E., Boch, J.A., Bare, O., Auron, P.E. et al. Novel signal transduction pathway utilized by extracellular HSP70: role of Toll-like receptor (TLR) 2 and TLR4. *J Biol Chem* 2002, **277**(17), 15028–15034.

38. Naim, J.O., van Oss, C.J., Wu, W., Giese, R.F. & Nickerson, P.A. Mechanisms of adjuvancy: I. Metal oxides as adjuvants. *Vaccine* 1997 Aug, **15**(11), 1183–1193.

39. Waite, D.C., Jacobson, E.W., Ennis, F.A., Edelman, R., White, B., Kammer, R. et al. Three double-blind, randomized trials evaluating the safety and tolerance of different formulations of the saponin adjuvant QS-21. *Vaccine* 2001 Jul 16, **19**(28–29), 3957–3967.

40. Morton, D.B. & Griffiths, P.H. Guidelines on the recognition of pain, distress and discomfort in experimental animals and an hypothesis for assessment. *Vet Rec* 1985 Apr 20, **116**(16), 431–436.

41. Bomford, R. Saponin and other haemolysins (vitamin A, aliphatic amines, polyene antibiotics) as adjuvants for SRBC in the mouse: evidence for a role for cholesterol-binding in saponin adjuvanticity. *Int Arch Allergy Appl Immunol* 1980, **63**(2), 170–177.

42. Shaw, C.M., Alvord, E.C., Jr. & Kies, M.W. Straight chain hydrocarbons as substitutes for the oil in Freund's adjuvants in the production of experimental "allergic" encephalomyelitis in the guinea pig. *J Immunol* 1964 Jan, **92**, 24–27.

43. Dalsgaard, K. Assessment of the dose of the immunological adjuvant Quil A in mice and guinea pigs, using sheep red blood cells as model antigen. *Zentralbl Veterinarmed B* 1984 Oct, **31**(9), 718–720.

44. Leenaars, P.P., Hendriksen, C.F., Koedam, M.A., Claassen, I. & Claassen, E. Comparison of adjuvants for immune potentiating properties and side effects in mice. *Vet Immunol Immunopathol* 1995 Sep, **48**(1–2), 123–138.

45. Leenaars, M. Adjuvants in laboratory animals: evaluation of immunostimulatory properties and side effects of Freund's complete adjuvant and alternative adjuvants in immunization procedures. Doctoral dissertation, RIVM, The Netherlands, 1997.

46. White, R.G., Coons, A.H. & Connolly, J.M. Studies on antibody production: III. The alum granuloma. *J Exp Med* 1955 Jul 1, **102**(1), 73–82.

47. Valtulini, S., Macchi, C., Ballanti, P., Cherel, Y., Laval, A., Theaker, J.M. et al. Aluminium hydroxide–induced granulomas in pigs. *Vaccine* 2005 Jun 10, **23**(30), 3999–4004.

48. Lindblad, E.B. & Hau, J. Escaping from the use of Freund's complete adjuvant. In *Progress in the Reduction Refinement and Replacement of Animal Experimentation*, Balls, M., van Zeller, A.-M. & Halder, M.E., Eds., Elsevier Science, Amsterdam, 2000, pp. 1681–1685.

49. Vogelbruch, M., Nuss, B., Korner, M., Kapp, A., Kiehl, P. & Bohm, W. Aluminium-induced granulomas after inaccurate intradermal hyposensitization injections of aluminium-adsorbed depot preparations. *Allergy* 2000 Sep, **55**(9), 883–887.

50. Bomford, R. The comparative selectivity of adjuvants for humoral and cell-mediated immunity. *Clin Exp Immunol* 1980 Feb, **39**(2), 435–441.

51. Kobayashi, K., Kaneda, K. & Kasama, T. Immunopathogenesis of delayed-type hypersensitivity. *Microsc Res Tech* 2001 May 15, **53**(4), 241–245.

52. Marcinkiewicz, J. Cell-mediated immunity: role of IL-3 and IL-6 in the regulation of contact sensitivity reaction. *Folia Histochem Cytobiol* 1990, **28**(3), 107–119.

53. Dinarello, C.A., Elin, A.J., Chedid, L. & Wolff, S.M. The pyrogenicity of the synthetic adjuvant muramyl dipeptide and two structural analogues. *J Infect Dis* 1978, **38**(6), 760–767.

54. Nishio, A. & Kanoh, S. Development changes in the febrile response to endotoxin in rabbit. *Jpn J Physiol* 1980, **30**(4), 645–653.

55. Matyas, G.R., Wassef, N.M., Rao, M. & Alving, C.R. Induction and detection of antibodies to squalene. *J Immunol Methods* 2000 Nov 1, **245**(1–2), 1–14.

56. Asa, P.B., Wilson, R.B. & Garry, R.F. Antibodies to squalene in recipients of anthrax vaccine. *Exp Mol Pathol* 2002 Aug, **73**(1), 19–27.

57. Baluna, R. & Vitetta, E.S. Vascular leak syndrome: a side effect of immunotherapy. *Immunopharmacology* 1997 Oct, **37**(2–3), 117–132.

58. Edelman, R., Palmer, K., Russ, K.G., Secrest, H.P., Becker, J.A., Bodison, S.A. et al. Safety and immunogenicity of recombinant Bacille Calmette–Guerin (rBCG) expressing *Borrelia burgdorferi* outer surface protein A (OspA) lipoprotein in adult volunteers: a candidate Lyme disease vaccine. *Vaccine* 1999 Feb 26, **17**(7–8), 904–914.

59. Pearson, C.M. & Wood, F.D. Studies of polyarthritis and other lesions induced in rats by injection of mycobacterial adjuvant: I. General clinical and pathological characteristics and some modifying factors. *Arthritis Rheum* 1959, **2**, 440–459.

60. Freund, J., Stern, E.R. & Pisani, T.M. Isoallergic encephalomyelitis and radiculitis in guinea pigs after one injection of brain and mycobacteria in water and oil emulsion. *J Immunol* 1947, **57**, 170–194.

61. Whitehouse, M.W., Orr, K.J., Beck, F.W. & Pearson, C.M. Freund's adjuvants: relationship of arthritogenicity and adjuvanticity in rats to vehicle composition. *Immunology* 1974 Aug, **27**(2), 311–330.

62. Clements, C.J. & Griffiths, E. The global impact of vaccines containing aluminium adjuvants. *Vaccine* 2002 May 31, **20**(Suppl 3), S24–S33.

63. Weibel, R.E., McLean, A., Woodhour, A.F., Friedman, A. & Hilleman, M.R. Ten-year follow-up study for safety of adjuvant 65 influenza vaccine in man. *Proc Soc Exp Biol Med* 1973 Sep, **143**(4), 1053–1056.

64. Veronesi, R., Correa, A. & Alterio, D. Single dose immunization against tetanus: promising results in human trials. *Rev Inst Med Trop Sao Paulo* 1970 Jan-Feb, **12**(1), 46–54.

65. Stanfield, J.P., Gall, D. & Bracken, P.M. Single-dose antenatal tetanus immunisation. *Lancet* 1973 Feb 3, **1**(7797), 215–219.

66. Bienzle, U., Gunther, M., Neuhaus, R., Vandepapeliere, P., Vollmar, J., Lun, A. et al. Immunization with an adjuvant hepatitis B vaccine after liver transplantation for hepatitis B-related disease. *Hepatology* 2003 Oct, **38**(4), 811–819.

67. *Pharmacopeia Europa*, 3rd ed. (Suppl), 2000, p. 1304.

68. *Code of Federal Regulations*, Title 21, vol. 7, sec. 610.15 (Constituent materials), rev. Apr 1, 2003.

69. Edelman, R. Vaccine adjuvants. *Rev Infect Dis* 1980 May–Jun, **2**(3), 370–383.

70. Goldenthal, K., Cavagnaro, J.A., Alving, C. & Vogel, F.R. Safety evaluation of vaccine adjuvants: National Cooperative Vaccine Development Meeting Working Group. *AIDS Res Hum Retroviruses* 1993, **9**(Suppl 1), S47–S51.

71. Frost, L., Johansen, P., Pedersen, S., Veien, N., Ostergaard, P.A. & Nielsen, M.H. Persistent subcutaneous nodules in children hyposensitized with aluminium-containing allergen extracts. *Allergy* 1985 Jul, **40**(5), 368–372.

72. Bergfors, E., Trollfors, B. & Inerot, A. Unexpectedly high incidence of persistent itching nodules and delayed hypersensitivity to aluminium in children after the use of adsorbed vaccines from a single manufacturer. *Vaccine* 2003 Dec 8, **22**(1), 64–69.

73. Bohler-Sommeregger, K. & Lindemayr, H. Contact sensitivity to aluminium. *Contact Dermatitis* 1986 Nov, **15**(5), 278–281.

74. Aprile, M.A. & Wardlaw, A.C. Aluminium compounds as adjuvants for vaccines and toxoids in man: a review. *Can J Public Health* 1966 Aug, **57**(8), 343–354.

75. Butler, N.R., Voyce, M.A., Burland, W.L. & Hilton, M.L. Advantages of aluminium hydroxide adsorbed combined diphtheria, tetanus, and pertussis vaccines for the immunization of infants. *Br Med J* 1969 Mar 15, **1**(645), 663–666.

76. Shi, Y., HogenEsch, H., Regnier, F.E. & Hem, S.L. Detoxification of endotoxin by aluminum hydroxide adjuvant. *Vaccine* 2001 Feb 8, **19**(13–14), 1747–1752.

77. Norimatsu, M., Ogikubo, Y., Aoki, A., Takahashi, T., Watanabe, G., Taya, K. et al. Effects of aluminum adjuvant on systemic reactions of lipopolysaccharides in swine. *Vaccine* 1995 Oct, **13**(14), 1325–1329.

78. Martyn, C.N., Barker, D.J., Osmond, C., Harris, E.C., Edwardson, J.A. & Lacey, R.F. Geographical relation between Alzheimer's disease and aluminum in drinking water. *Lancet* 1989 Jan 14, **1**(8629), 59–62.

79. Flarend, R.E., Hem, S.L., White, J.L., Elmore, D., Suckow, M.A., Rudy, A.C. et al. In vivo absorption of aluminium-containing vaccine adjuvants using 26Al. *Vaccine* 1997 Aug–Sep, **15**(12–13), 1314–1318.

80. Coursaget, P., Yvonnet, B., Relyveld, E.H., Barres, J.L., Diop-Mar, I. & Chiron, J.P. Simultaneous administration of diphtheria–tetanus–pertussis–polio and hepatitis B vaccines in a simplified immunization program: immune response to diphtheria toxoid, tetanus toxoid, pertussis, and hepatitis B surface antigen. *Infect Immun* 1986 Mar, **51**(3), 784–787.

81. Relyveld, E.H. Preparation and use of calcium phosphate adsorbed vaccines. *Dev Biol Stand* 1986, **65**, 131–136.

82. Tanzer, M., Gollish, J., Leighton, R., Orrell, K., Giacchino, A., Welsh, P. et al. The effect of adjuvant calcium phosphate coating on a porous-coated femoral stem. *Clin Orthop Relat Res* 2004 Jul(424), 153–160.

83. Friedewald, W.F. Adjuvants in immunization with influenza virus vaccines. *J Exp Med* 1944, **80**, 477–491.

84. Henle, W. & Henle, G. Effect of adjuvants on vaccination of human beings against influenza. *Proc Soc Exp Biol NY* 1945, **59**, 179–181.

85. Salk, J.E., Bailey, M.L. & Laurent, A.M. The use of adjuvants in studies on influenza immunization: II. Increased antibody formation in human subjects inoculated with influenza virus vaccine in a water-in-oil emulsion. *Am J Hyg* 1952, **55**(3), 439–456.

86. Davenport, F.M., Hennessy, A.V. & Bell, J.E. Immunologic advantages of emulsified influenza vaccines. *Mil Med* 1962, **127**, 95–100.

87. Davenport, F.M. Applied immunology of mineral oil adjuvants. *J Allergy* 1961, **32**, 177–189.

88. Bell, J.A., Philip, R.N., Davis, D.J., Beem, M.O., Bergelman, P.M., Engler, J.L. et al. Epidemiological studies on influenza in familial and general population groups: 4. Vaccine reactions. *Am J Hyg* 1961, **73**, 148–163.

89. Cutler, J.C., Lesesne, L. & Vaughn, I. Use of poliomyelitis virus vaccine in light mineral oil adjuvant in a community immunization program and report of reactions encountered. *J Allergy* 1962, **33**, 193–209.

90. Fukumi, H. Effectiveness and untoward reactions of oil adjuvant influenza vaccines. *Symposium Series in Immunobiological Standardization*, Karger, New York, 1967, pp. 237–240.

91. Meiklejohn, G. Adjuvant influenza adenovirus vaccine. *JAMA* 1962, **179**, 594–597.

92. Meiklejohn, G. Personal communication quoted in: Hilleman, MR. Critical appraisal of emulsified oil adjuvants applied to viral vaccines. *Prog Med Virol* 1966, **8**, 157.

93. Pittman, M. Some remarks regarding the reaction provoking properties. *Symposium Series in Immunobiological Standardization*, Karger, New York, 1967, pp. 101–102.

94. Stuart-Harris, C.H. Adjuvant influenza vaccines. *Bull World Health Organ* 1969, **41**, 617–621.

95. Ogonuki, H., Hashizume, S. & Abe, H. Histopathological tests of tissues in the sites of local reactions caused by the injection of oil-adjuvant cholera vaccine. *Symposium Series in Immunobiological Standardization*, Karger, New York, 1967, pp. 125–128.

96. Beebe, G.W., Simon, A.H. & Vivona, S. Follow-up study on army personnel who received adjuvant influenza virus vaccine, 1951–1953. *Am J Med Sci* 1964, **247**, 385–405.

97. Potter, M. & Boyce, C.R. Induction of plasma-cell neoplasms in strain BALB/c mice with mineral oil and mineral oil adjuvants. *Nature* 1962, **193**, 1086–1087.

98. Bollinger, J.N. Metabolic fate of mineral oil adjuvants using $^{14}$C-labeled tracers: I. Mineral oil. *J Pharm Sci* 1970, **59**(8), 1084–1088.

99. Bollinger, J.N. Metabolic fate of mineral oil adjuvants using $^{14}$C-labeled tracers: II. Mannide monooleate. *J Pharm Sci* 1970 Aug, **59**(8), 1088–1092.

100. Batista, A.D. & Lindblad, E.B. Analyzing toxicological aspects of immunological adjuvants for vaccine and experimental use. 2006 (submitted).

101. Stewart-Tull, D.E.S. Recommendations for the assessment of adjuvants (immuno-potentiators). In *Immunological Adjuvants and Vaccines*, Gregoriadis A., Ed., Plenum Publishing Corporation, New York, 1989.

102. Shimada, H., Hattori, C. & Sato, T. Mutagenicity studies of muroctasin. *Arzneimit-telforschung* 1988, **38**(7a), 1031–1033.

103. Berlin, B.S. Tests for biologic safety of Arlacel A. *Ann Allergy* 1962, **20**, 472–479.

104. IPCS. Principles and methods for assessing allergic hypersensitization associated with exposure to chemicals. In *Environmental Health Criteria*, 1999, p. 212.

105. Bloksma, N., Kubicka-Muranyi, M., Schuppe, H.C., Gleichmann, E. & Gleichmann, H. Predictive immunotoxicological test systems: suitability of the popliteal lymph node assay in mice and rats. *Crit Rev Toxicol* 1995, **25**(5), 369–396.

106. Kimber, I. & Dearman, R.J. Immune responses: adverse versus non-adverse effects. *Toxicol Pathol* 2002 Jan–Feb, **30**(1), 54–58.

# INDEX

*Vaccine Adjuvants and Delivery Systems*, Edited by Manmohan Singh
Copyright © 2007 John Wiley & Sons, Inc.